Bacteriology

Bacteriology

Edited by **Henry Evans**

SYRAWOOD
PUBLISHING HOUSE
New York

Published by Syrawood Publishing House,
750 Third Avenue, 9th Floor,
New York, NY 10017, USA
www.syrawoodpublishinghouse.com

Bacteriology
Edited by Henry Evans

International Standard Book Number: 978-1-68286-290-2 (Hardback)

Printed in China.

Contents

Preface

This book provides a comprehensive overview of the vast field of bacteriology, which deals with identification and application of bacteria in various other fields. It provides significant information of this discipline to help develop a good understanding of concepts of different types of pathogenic bacteria, their physiology, cell signalling, environmental prokaryotes, etc. The chapters included present contributions made by scholars from around the globe. Students, researchers, experts and all associated with bacteriology will benefit alike from this book.

Various studies have approached the subject by analyzing it with a single perspective, but the present book provides diverse methodologies and techniques to address this field. This book contains theories and applications needed for understanding the subject from different perspectives. The aim is to keep the readers informed about the progress in the field; therefore, the contributions were carefully examined to compile novel researches by specialists from across the globe.

Indeed, the job of the editor is the most crucial and challenging in compiling all chapters into a single book. In the end, I would extend my sincere thanks to the chapter authors for their profound work. I am also thankful for the support provided by my family and colleagues during the compilation of this book.

Editor

1

Calcitonin gene-related peptide and substance P significantly influence coronary flow rate in gene knockout mice

Hongbao Ma[1, 2]*, Jenny Young[2] and Shen Cherng[3]

[1]Bioengineering Department, Zhengzhou University, Zhengzhou, Henan 450001, China.
[2]Brookdale University and Medical Center, Brooklyn, New York 11212, USA.
[3]Department of Electrical Engineering, Chengshiu University, Niaosong, Taiwan 833, China.

Calcitonin gene-related peptide (CGRP) and substance P (SP) which is potent vasodilator neuropeptides play a counter-regulatory role in several models of experimental hypertension. Langendorff-perfused mouse hearts (n = 328) were used to compare coronary flow rates among wild type (WT), α-CGRP gene knockout (KO) and NK1-KO mice under various pressure loading conditions (20, 30, 40 and 50 mmHg). The aorta of each heart was cannulated and all hearts were perfused with PBS at 37°C. Coronary flow rate was measured by pressure difference of both sides of a capillary. Perfusion was stopped 15 min for ischemia. Deletion of α-CGRP gene resulted in a significant reduction in coronary flow rate for both genders at all pressures. Deletion of NK1 gene resulted in a significant reduction in coronary flow rate for male mice at all pressures, but not for female mice. Coronary flow rate for both WT and α-CGRP-KO mice was consistently lower in female than in male mice, but not for NK1-KO mice. Coronary flow rate in α-CGRP mice was 19.2 and 15.4% lower than that of female and male WT mice, respectively. This effect seems to be gender related with less coronary flow noted in female WT and α-CGRP-KO mice, but not in NK1-KO mice.

Key words: Calcitonin gene-related peptide, flow rate, gene knockout, heart, mouse, neurokinin 1, substance P.

INTRODUCTION

Calcitonin (CT) and calcitonin gene-related peptide (CGRP) are derived from the CT/CGRP gene. CT is a 32 amino acid single chain peptide expressed mainly in the thyroid gland, and CGRP is a 37 amino acid vasoactive neuropeptide that is widely distributed in the central and peripheral nervous systems in mammals (DiPette et al., 1995; Wimalawansa et al., 1996; Ma, 2004; Goadsby, 2008). α-CGRP is produced by the tissue-specific alternative splicing of the primary transcript of the CT/α- CGRP gene and is synthesized almost exclusively in neuronal tissues (Breimer et al., 1988; Rosenfeld, et al., 1983). There is a second CGRP gene, β-CGRP which does not produce CT (Amara et al., 1985).

The two CGRP genes, α and β in rats and I and II in humans, differ in their protein sequences by one and three amino acids respectively, and their biological activities are quite similar in most vascular beds (Breimer et al, 1988).

Substance P (SP) is an 11 amino acids neuropeptide that is abundant in the periphery and the central nervous system, where it is co-localized with other neurotransmitters such as serotonin or dopamine. Also, SP is often co-localized with CGRP in perivascular sensory nerves (DiPette et al., 2000). Three main receptor sub-types for the tachykinins have been identified (NK1, NK2 and NK3).

Though SP acts primarily as a vasodilator, it also can

*Corresponding author. E-mail: mahongbao@zzu.edu.cn or hongbao@gmail.com.

Abbreviations: **BP**, blood pressure; **CGRP**, calcitonin gene-related peptide; **CT**, calcitonin; **ISP**, immunoreactive substance P; **KO**, gene knockout; **NK1**, neurokinin 1; **NO**, nitric oxide; **PBS**, physiologic buffered solution; **SP**, substance P; **WT**, wild type.

contract smooth muscle. Activation of the SP receptor (also known as the neurokinin 1 [NK1] receptor) causes oxidative stress and neurogenic inflammation (Weglicki et al., 1996). As a result, studies typically involve receptor knockouts as opposed to SP gene knockouts to eliminate any possible complication of oxidative stress. As a member of the tachykinin family, SP is derived from tissue-specific alternative splicing of the preprotachykinin I gene and is produced almost exclusively in neuronal tissues (Hokfelt et al., 2000). A prominent site for SP synthesis is the dorsal root ganglia which contain the cell bodies of sensory neurons that terminate centrally in the spinal cord and peripherally on blood vessels (DiPette, et al., 2000). SP and its receptor are widely distributed in the central and peripheral nervous system. Immunoreactive substance P (ISP) containing nerve fibers is widely distributed in the brain as well as peripheral sensory nerve endings, where it is often co-localized and co-released with CGRP (Mau et al., 1999). SP is involved in various physiological activities such as neuromodulation, smooth muscle contraction and the modulation of vascular tone (Stewart-Lee et al., 1989; Whittle et al., 1989; Khurshed et al., 2002). SP acts only in situations with an increased nervous activity, for example, after stress-induced activation of the pituitary-adrenal and sympathetic-adrenal axis. It normalizes stress-induced disorders by maintaining homeostasis in the catecholamine system. The basis of the antistress effect is the modulation of both biosynthesis and release of catecholamines in the adrenals. The homeostasis effect of SP, as well as the antistress effect does not require the complete sequence of the molecule. The N-terminal SP fragment is fully active and has no side effects (Oehme et al., 1987).

SP is a nitric oxide (NO)-dependent vasodilator that influences the relationship of mesenteric arterial arcades flow rate and NO release (Keber et al., 2001). It was reported that CGRP and SP play an important role in the development of pain and hyperalgesia (Greco et al., 2008). Statins act directly on sensory neurons to decrease expression of proinflammatory neuropeptides that trigger neurogenic inflammation, specifically CGRP and SP (Bucelli et al., 2008).

In this study, Langendorff-perfused heart preparations under both non-ischemia and ischemia conditions were used to compare coronary flow rates among wild type (WT) control, α-CGRP gene knockout (α-CGRP-KO) and SP gene knockout (SP-KO) mice with different genders under various pressure loading conditions (20, 30, 40 and 50 mmHg) to determine the effects of α-CGRP and SP on coronary flow rate.

MATERIALS AND METHODS

Animals

Three hundred and twenty-eight (328) mice were used in the study (2 - 3 months, 23.4 ± 2.5 g). Homozygous α-CGRP-KO and NK1-KO mice (C57BL/6, Fayetteville, NC, USA) were obtained from breeding pairs generated by the genetic targeting of the α-CGRP and NK1 genes respectively. These gene knockout mice were provided by Dr. Robert Gagel at the Baylor College of Medicine (Hoff et al., 1998).

Heart weight

To obtain an accurate heart weight is important, but to weigh the heart after adsorption of water usually damages the heart and calculation of the heart weight from body weight can cause substantial error. Instead, we used the watery heart weight to calculate the real heart weight (net weight). The hearts from 60 mice were removed after mice were sacrificed and put in ice cooled physiologic buffered solution (PBS) and the watery hearts from the buffer were weighed immediately. The hearts were adsorbed with paper to obtain the net heart weight.. Ratio of the net heart weight to watery heart weight was obtained from the 60 mice. The watery heart was weighed for each experimental heart and the real heart weight was calculated by the calibration of the 60 mice.

Heart perfusion

Fifty-two mouse hearts were used in the perfusion to measure flow rate (9 female WT, 6 female α-CGRP-KO, 5 female NK1-KO, 10 male WT, 8 male α-CGRP-KO, and 14 male NK1-KO). Mice were injected with heparin (Sigma Chemical Co., St. Louis, MO, USA) at 1000 units/kg (I.P.). Thirty min after heparin injection, mice were anesthetized with pentobarbital at 100 mg/kg mouse (Abbott Laboratories, North Chicago, IL, USA). Hearts were immediately isolated and placed in oxygenated ice cooled PBS. Then the aortas of the isolated hearts were cannulated and hearts placed in an incubator with oxygenized PBS at 37°C. The hearts were perfused with oxygenized PBS by the Langendorff method. The PBS perfusion was performed from aorta to coronary, left atrium and flow out through the left pulmonary veins. The Langendorff-perfused heart preparations were used to compare coronary flow rates among WT, α-CGRP-KO and NK1-KO mice for both genders under various pressure loading conditions (20, 30, 40 and 50 mmHg) (Figure 1).

Flow rate

One hundred and twenty-seven mice were used in the coronary flow rate measurements (19 female WT, 25 female α-CGRP-KO, 17 female NK1-KO, 27 male WT, 20 male α-CGRP-KO and 19 male NK1-KO). Coronary flow rate was measured by the pressure difference of both sides of a capillary in the perfusion line. Flow rate vs. pressure difference was calculated using PBS. The relationship of coronary flow rate and pressure was measured at multiple perfusion pressures and the relationship curve was constructed. The flow rates of WT, α-CGRP-KO and NK1-KO mouse hearts for both genders were compared.

Ischemia

One hundred and forty-one mice were used in the ischemia measurements (18 female WT, 19 female α-CGRP-KO, 17 female NK1-KO, 25 male WT, 38 male α-CGRP-KO, and 24 male NK1-KO). Hearts were exposed to ischemia situation by stopping perfusion for 15 min after the hearts were fixed and perfused one hour with oxygenized PBS by the Langendorff method. As heparin changes the heart behavior in ischemia, heparin was not injected for this test and the blood inside the heart was washed out quickly with PBS after the heart was removed from the mouse. Then the

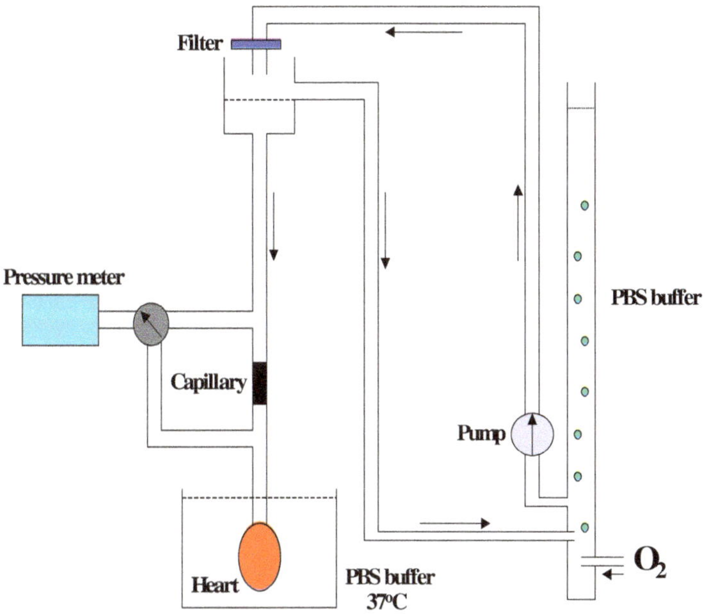

Figure 1. Langendorff Heart Preparation Set Up for Coronary Flow Rate Measurement.

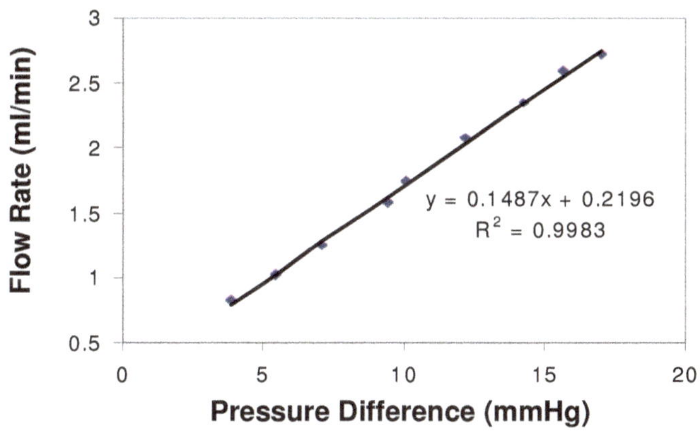

Figure 2. Correlation between Mouse Wet Weight and Net Weight.

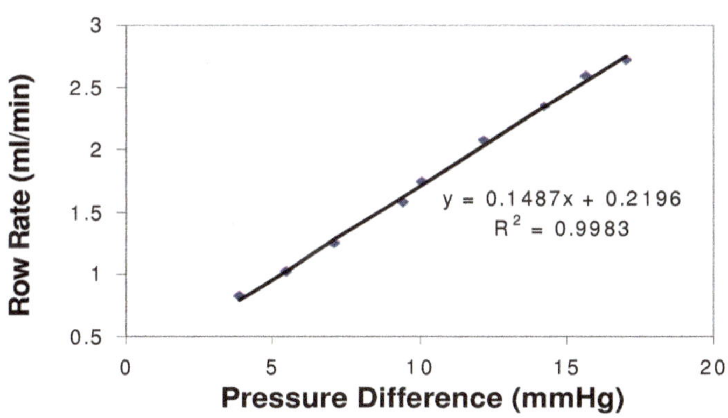

Figure 3. Flow rate vs. Pressure Difference Calculated with PBS.

hearts were perfused with oxygenized PBS. The flow rates of WT and CGRP-KO mice for both genders and SP-KO male mouse hearts were compared.

RESULTS

Heart weight: The average watery and net mouse heart weights obtained from 60 mice were 0.158 ± 0.019 and 0.131 ± 0.016 respectively, and the ratio of the net heart weight to watery heart weight was 0.83. There was a linear relationship of the watery and net heart weight ($y = 0.8316x$; $R^2 = 0.793$) (Figure 2).

Flow rate calibration

There was a linear relationship between the flow rate calibrated by PBS and the pressure difference of both sides of a capillary in the perfusion line ($y = 0.1487x + 0.2196$; $R^2 = 0.9983$) (Figure 3).

Flow rate

As shown in Tables 1 and 2, knockout of the α-CGRP gene resulted in a significant reduction in the coronary flow rate for both genders at all pressures (average of 6.13 ± 1.24 from 7.59 ± 0.95 ml/min/g net heart weight for female, $p < 0.0005$ and 6.80 ± 0.77 from 8.04 ± 1.27 ml/min/g net heart weight for male, $p < 0.01$). Deletion of the NK1 gene resulted in a significant reduction in the coronary flow rate for male mice at all pressures (average of 7.34 ± 1.05 from 8.04 ± 1.27 ml/min/g net heart weight, $p < 0.0008$), but not for female mice. The female mouse coronary flow rate for NK1 increased to 9.18 ± 1.40 ml/min/g net heart weight compared to 7.59 ± 0.95 ml/min/g net heart weight of WT. In addition, the coronary flow rate for both WT and α-CGRP-KO mice was consistently lower in female than in male mice (average of 7.59±0.95 to 8.04 ± 1.27 ml/min/g net heart weight for WT and 6.13 ± 1.24 to 6.80 ± 0.77 ml/min/g net heart weight for α-CGRP-KO, and was significant at higher perfusion pressure), but not for SP-KO mice which had a higher flow rate (average of 9.18 ± 1.40 to 7.34 ± 1.05 ml/min/g net heart weight for NK1, $p < 0.0002$). The data show that the coronary flow rate for α-CGRP mice is 19.2% lower than that of WT mice for females and 15.4% for males. Also, this effect seems to be gender related with less coronary flow noted in the female mice for WT and α-CGRP-KO, but not for SP-KO (Tables 1 and 2).

Ischemia

In this experiment, we measured the ratios of mouse coronary flow rates after and before ischemia to evaluate the heart reaction to ischemia. The ischemia was done by 15 min stopping perfusion. The results showed that

Table 1. Mouse Coronary Flow Rate (ml/min/g net heart weight) vs. Perfusion Pressure.

Perfusion Pressure (mmHg)	WT Female n=25	α-CGRP-KO Female N=19	NK1-KO Female n=17	WT Male n=20	α-CGRP-KO Male n=27	NK1-KO Male n=19
50	11.42±1.42	9.76±1.80	13.75±1.78	12.36±1.75	10.54±0.94	11.51±1.54
40	9.15±1.06	7.73±1.45	11.00±1.57	9.85±1.34	8.23±0.78	8.84±1.22
30	6.31±0.79	4.88±0.92	7.49±1.22	6.52±1.22	5.53±0.75	5.95±0.88
20	3.48±0.53	2.16±0.82	4.48±1.02	3.43±0.78	2.92±0.61	3.08±0.26
Mean	7.59±0.95	6.13±1.24	9.18±1.40	8.04±1.27	6.80±0.77	7.34±1.05

Table 2. P Value between the Different Mouse Coronary Flow Rates.

Perfusion Pressure (mmHg)	F WT /F CGRP-KO	F WT/F NK1-KO	F CGRP-KO/F NK1-KO	M WT /M CGRP-KO	M WT/M NK1-KO	M CGRP-KO /M NK1-KO	F WT/M WT	FCGRP-KO/M CGRP-KO	F NK1-KO/M NK1-KO
50	0.0009	0.00005	0.00000006	0.0001	0.0573	0.0106	0.0312	0.0371	0.0002
40	0.0005	0.0001	0.0000001	0.00003	0.0095	0.0336	0.0335	0.0734	0.00004
30	0.000003	0.0008	0.00000003	0.0016	0.0502	0.0503	0.2503	0.0071	0.00009
20	0.0004	0.0006	0.0000009	0.0098	0.0583	0.1708	0.4173	0.1126	0.00002

Note: F – female; M – male

female α-CGRP-KO mice had a lower ratio than WT mice and that NK1-KO mice had a higher ratio. For males, both α-CGRP-KO and NK1-KO mice had lower ratios than WT mice, and α-CGRP-KO mice showed the lowest (Figure 4, Table 3).

DISCUSSION

CGRP and SP are two vasoactive neuropeptides that play counter regulatory roles in hypertension and coronary flow rate. CGRP causes peripheral arterial dilation, thereby decreasing blood pressure and increasing heart rate in a dose-dependent manner (DiPette, et al, 1995; Wimalawansa, et al, 1996). CGRP is a peptide that influences the blood pressure. CGRP has several important physiologic roles: (1) it is a potent vasodilator and can affect the force and rate of heart beat. (2) it can modulate acetylcholine receptor function at the neuromuscular junction. (3) It has been demonstrated to block tolerance to morphine and (4) it can modulate antigen presentation in Langerhans cells in the skin. Despite these important physiological functions, therapeutic strategies using CGRP have been impeded due to the lack of a cloned CGRP receptor with which ligands could be developed. There is an induction of the blood pressure when animals are lacking CGRP. The CGRP neuropeptide has a marked dose-dependent decrease in blood pressure with increases in heart rate and blood pressure of CGRP gene knockout mice (Gangula et al., 2000). CT gene was first cloned in 1980 (Jacobs et al., 1981) and CGRP was discovered in 1982 by molecular cloning of CT gene (Amara et al., 1982; Rosenfeld et al., 1983). The CT mRNA predominates in the thyroid while the CGRP-specific mRNA appears to predominate in the hypothalamus (brain). The second rat CGRP (β-rCGRP) gene has been discovered in brain and thyroid. This β-rCGRP is different from the α-CGRP by one amino acid (position 1 Ala instead of Ser). The second human CGRP (β-hCGRP) gene has been discovered in medullary thyroid (Tschopp et al., 1985). This β-hCGRP differs from the α-hCGRP by three amino acids located position 3, 22 and 25. Different splicing of primary RNA transcript arouses the translation of CT and CGRP peptides in a tissue-specific manner (Ma, 2004).

According to Okamoto et al. (2003) ketamine and pentobarbital inhibit SP receptor function, and the mechanism of their inhibition on SP receptor function could not be through activation of the protein kinase C pathway and may be due to noncompetitive displacing of SP binding (Okamoto et al., 2003).

α-CGRP plays a major role in coronary blood flow. From the research results here, we showed that flow rates of both α-CGRP-KO and SP-KO mouse hearts are lower than those of WT mouse hearts. In the ischemia condition, the flow rate ratios of both α-CGRP-KO and SP-KO mouse hearts are lower than those of WT mouse hearts for males, but there is no difference between α-CGRP-KO and WT mouse hearts for females. This indicates a difference in females and males when α-CGRP-KO is genetically deleted. The results of this study

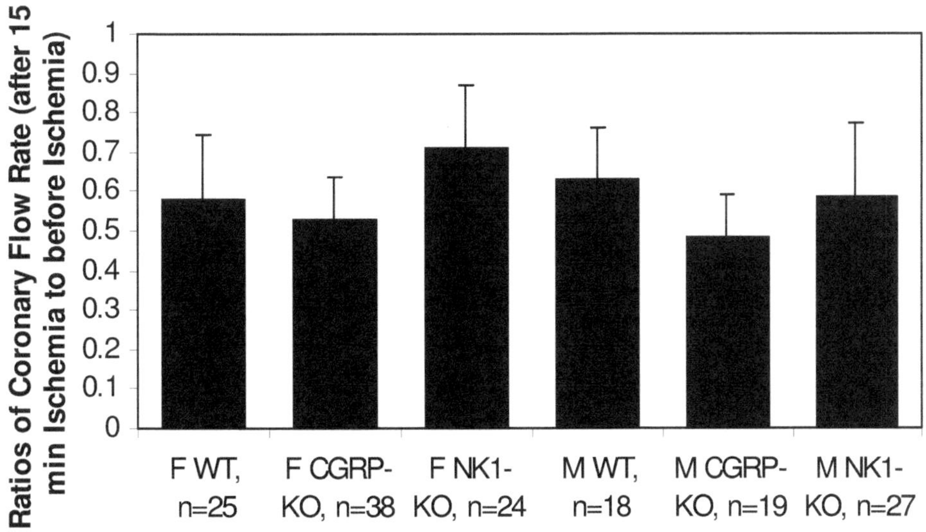

Figure 4. Male Mouse Coronary Flow Rate and Beating after Ischemia.

Table 3. P Value between the Ratio of Mouse Coronary Flow Rates after and before Ischemia.

F WT/F CGRP-KO	F WT/F NK1-KO	F CGRP-KO/F NK1-KO	M WT/M CGRP-KO	M WT/M NK1-KO	M CGRP-KO /M NK1-KO	F WT/M WT	F CGRP-KO/M CGRP-KO	F NK1-KO/M NK1-KO
0.13	0.01	0.0002	0.00005	0.15	0.01	0.01	0.091	0.01

Note: F – female; M – male.

provide information that α-CGRP and SP or their activetors/enhancers may be candidates to improve blood flow rate. These potential pharmaceutical applications could be considered in clinical trials on ischemia or hypertension, etc. The gender correlation of CGRP and CT on flow rate offers information that it may be necessary to treat female and male patients differently. Lastly, it is possible to consider clinical treatment by a combination of CGRP and SP.

REFERENCES

Amara SG, Jonas V, Rosenfeld MG, Ong ES, Evans RM (1982). Alternative RNA processing in calcitonin gene expression generates mRNAs encoding different polypeptide products. Nature. 298 (5871): 240-4.

Anara SG, Arriza JL, Leff SE, Swanson LW, Evans RM, Rosefeld MG (1985). Expression in brain of messenger RNA encoding a novel neuropeptide homologues to calcitonin gene-related peptide. Sci. 229: 1094-9.

Breeze AL, Harvey TS, Bazzo R, Campbell ID (1991). Solution structure of human calcitonin gene-related peptide by 1H NMR and distance geometry with restrained molecular dynamics. Biochem. 30(2): 575-82.

Breimer L, MacIntyr I, Zaidi M (1988). Peptides from the calcitonin genes: molecular genetics, structure and function. Biochem. 255:377-90.

Broad PM, Symes AJ, Thakker RV, Craig RK (1989). Structure and methylation of the human calcitonin/alpha-CGRP gene. Nucleic Acids Res. 17(17): 6999-7011.

Bucelli RC, Gonsiorek EA, Kim WY, Bruun D, Rabin RA, Higgins D, Lein PJ (2008). Statins decrease expression of the proinflammatory neuropeptides calcitonin gene-related peptide and substance P in sensory neurons. J. Pharmacol. Exp. Ther. 324(3): 1172-80.

Changeux JP, Duclert A, Sekine S (1992). Calcitonin gene-related peptides and neuromuscular interactions. Ann. NY Acad. Sci. 657: 361–78.

Copp DH, Cameron EC, Cheney BA, Davidson AGF, Henze KG (1962). Evidence of calcitonin – a new hormone from the parathyroid that lowers blood calcium. Endocrinol. 70: 638-649.

DiPette DJ, Supowit SC (2000). Calcitonin gene-related peptide and hypertension. In: Oparil S, ed. Hypertension. Philadelphia, PA: WB Saunders pp. 182–189.

DiPette DJ, Wimalawansa SJ (1995). Calcium regulating hormones and cardiovascular function. In: Cross JI, Aveoli LV. eds. Cardiovascular Actions of Calciropic Hormones. Baltimore. MD: CRC Press pp. 239-252.

Dong Z, Brennan A, Liu N, Yarden Y, Lefkowitz G, Mirsky R, Jessen KR (1995). Neu differentiation factor is a neuron-glia signal and regulates survival, proliferation, and maturation of rat Schwann cell precursors. Neuron; 15: 585–596.

Eysselein VE, Reeve JR, Sternini C, Cominelli F, Davis WM, Davis MT, Lee TD, Ho FJ, Ridout D, Shively JE (1991). Structural characterization of calcitonin gene-related peptide purified from rabbit intestine. Peptides. 12(2): 289-95.

Falls DL, Rosen KM, Corfas G, Lane WS, Fischbach GD (1993). ARIA, a protein that stimulates acetylcholine receptor synthesis, is a member of the neu ligand family. Cell. 72: 801–815.

Fischbach GD, Rosen KM (1997). ARIA: A neuromuscular junction neuregulin. Annu. Rev. Neurosci. 20:429–58.

Fontaine B, Klarsfeld A, Changeux JP (1987). Calcitonin gene-related peptide and muscle activity regulate acetylcholine receptor α-subunit mRNA levels by distinct intracellular pathways. J Cell Biol. 105:1337–

1342.
Fontaine B, Klarsfeld A, Hokfelt T, Changeux JP (1986). Calcitonin gene-related peptide, a peptide present in spinal cord motoneurons, increases the number of acetylcholine receptors in primary cultures of chick embryo myotubes. Neurosci. Lett. 71:59–65.
Gangula PRR, Zhao H, Supowit SC, Wimalawansa SJ, Dipette DJ, Westlund KN, Gagel RF, Yallampalli C (2000). Increased blood pressure in α-Calcitonin gene-related peptide/calcitonin gene knockout mice. Hypertension 35: 470-5.
Goadsby PJ (2008). Calcitonin gene-related peptide (CGRP) antagonists and migraine: is this a new era? Neurol. 70(16): 1300-1.
Goodearl AD, Yee AG, Sandrock AW Jr, Corfas G, Fischbach GD (1995). ARIA is concentrated in the synaptic basal lamina of the developing chick neuromuscular junction. J. Cell Biol. 130: 1423–1434.
Greco R, Tassorelli C, Sandrini G, Di Bella P, Buscone S, Nappi G (2008). Role of calcitonin gene-related peptide and substance P in different models of pain. Cephalalgia. 28(2): 114-26.
Grinspan JB, Marchionni MA, Reeves M, Coulaloglou M, Scherer SS (1996). Axonal interactions regulate Schwann cell apoptosis in developing peripheral nerve: neuregulin receptors and the role of neuregulins. J. Neurosci. 16: 6107–18
Hoff AO, Thomas PM, Cote GJ, Qiu H, Bain H, Gage RF (1998). Generation of a calcitonin knockout mouse model. Bone. 23:S64.
Hokfelt T, Broberger C, Xu ZQ, Sergeyev V, Ubink R, Diez M (2000). Neuropeptides: an overview. Neuropharmacol. 39: 1337–1356.
Hoppener JW, Steenbergh PH, Zandberg J, Geurts-van-Kessel AH, Baylin SB, Nelkin BD, Jansz HS, Lips CJ (1985). The second human calcitonin/CGRP gene is located on chromosome 11. Hum-Genet. 70(3): 259-63.
Jacobs JW, Goodman RH, Chin WW, Dee PC, Habener JE, Bell NH, Potts JT (1981). Calcitonin messenger RNA encodes multiple polypeptides in a single precursor. Sci. 213: 457-459.
Jones G, Herczeg A, Ruegg MA, Lichsteiner M, Kroger S, Brenner HR (1996). Substrate-bound agrin induces expression of acetylcholine receptor ε-subunit gene in cultured mammalian muscle cells. Proc Natl Acad Sci USA. 93(12): 5985-5990.
Katki KA, Supowit SC, DiPette DJ (2002). Substance P in subtotal nephrectomy-salt hypertension. Hypertension. 39: 389-393.
Keber KM, Mager GM, Miller CE, Nowicki PT (2001). Relationship between flow rate and NO production in postnatal mesenteric arteries. Am. J. Physiol. Gastrointest Liver Physiol. 280:G43-50.
Loeb JA, Fischbach GD (1995). ARIA can be released from extracellular matrix through cleavage of a heparin-binding domain. J. Cell Biol. 130:127–135.
Ma H, Huang R, Abela GS, Shah I, Katki KA, Zhao H, DiPette DJ, Supowit SC (2002). Coronary flow rate is decreased in the α-calcitonin gene-related peptide knockout mouse. Hypertension; 56(Suppl):23.
Ma H (2004). Calcitonin Gene-Related Peptide (CGRP). Nature and Science 2(3): 41-47.
Martinou JC, Falls DL, Fischbach GD, Merlie JP (1991). Acetylcholine receptor-inducing activity stimulates expression of the ε-subunit gene of the muscle acetylcholine receptor. Proc. Natl. Acad. Sci. USA. 88: 7669–7673.
Mau Se (1999). Effects of Substance P on cellular signaling systems in the rat anterior pituitary. Danish Med. Bull. 46: 35–56.
Meier T, Hauser DM, Chiquet M, Landmann L, Ruegg MA, Brenner HR (1997). Neural agrin induces ectopic postsynaptic specializations in innervated muscle fibers. J. Neurosci. 17: 6534–6544.
Meier T, Masciulli F, Moore C, Schoumacher F, Eppenberger U, Denzer AJ, Jones G, Brenner HR (1998). Agrin can mediate acetylcholine receptor gene expression in muscle by aggregation of muscle-derived neuregulins. J. Cell Biol. 141:715–726.
Meyer D, Birchmeier C (1995). Multiple essential functions of neuregulin in development. Nature 378:386–90.
Miyata A, Jiang L, Minamino N, Arimura A (1987). Identification of calcitonin gene related peptide in ovine hypothalamic extract. Neuropeptides 9(1): 75-82.
Moscoso LM, Chu GC, Gautam M, Noakes PG, Merlie JP, Sanes JR (1995). Synapse-associated expression of an acetylcholine receptor-inducing protein, ARIA/heregulin, and its putative receptors, ErbB2 and ErbB3, in developing mammalian muscle. Dev. Biol. 172: 158–169.
New HV, Mudge AW (1986). Calcitonin gene-related peptide regulates muscle acetylcholine receptor synthesis. Nature 323: 809–811.
Oehme P, Hecht K, Faulhaber HD, Nieber K, Roske I, Rathsack R (1987). Relationship of substance P to catecholamines, stress, and hypertension. J. Cardiovasc. Pharmacol. 10(Suppl. 12): S109-11.
Okamoto T, Minami K, Uezono Y, Ogata J, Shiraishi M, Shigematsu A, Ueta Y (2003). The Inhibitory Effects of Ketamine and Pentobarbital on Substance P Receptors Expressed in Xenopus Oocytes. Anesth. Analg. 97(1):104-110.
Rimer M, Mathiesen I, Lomo T, McMahon UJ (1997). γ-AChR/ε-AChR switch at agrin-induced postsynaptic-like apparatus in skeletal muscle. Mol. Cell Neurosci. 9: 254–263.
Rosenfeld MG, Mermod JJ, Amaraa SG, Swanson LW, Swchenko PE, Rivier J, Vale WW, Evans RM (1983). Production of a novel neuropeptide encode by the calcitonin gene via tissue-specitfic RNA processing. Nature 304:129-135.
Sandrock AW Jr, Dryer SE, Rosen KM, Gozani SN, Kramer R, Theill LE, Fischbach GD (1997). Maintenance of acetylcholine receptor number by neuregulins. Science 276(5312): 599–603.
Stewart-Lee A, Burnstock G (1989). Actions of tachykinins in the rabbit mesenteric artery: substance P and [Glp6-L-Pro9] SP6–11 are potent agonists for endothelial neurokinin-1 receptors. Br J Pharmacol. 97:1218–1224.
Trachtenberg JT, Thompson WJ (1996). Schwann cell apoptosis at developing neuromuscular junctions in regulated by glial growth factor. Nature 379:174–177.
Tschopp FA, Henke H, Petermann JB, Tobler PH, Janzer R, Hokfelt T, Lundberg JM, Cuello C, Fischer JA (1985). Calcitonin gene-related peptide and its binding sites in the human central nervous system and pituitary. Proc. Natl. Acad. Sci. USA 82: 248-252.
Uchida S, Yamamoto H, Iio S, Matsumoto N, Wang XB, Yonehara N, Imai Y, Inoki R, Yoshida H (1990). Release of calcitonin gene-related peptide-like immunoreactive substance from neuromuscular junction by nerve excitation and its action on striated muscle. J. Neurochem. 54(3) :1000-1003.
Usdin TB, Fischbach G (1986). Purification and characterization of a polypeptide from chick brain that promotes the accumulation of acetylcholine receptors in chick myotubes. J. Cell Biol. 103: 493–507.
Whittle BJR, Lopez-Belmont J, Press DD (1989). Modulation of the vasodepressor actions of acetylcholine, bradykinin, substance P and endothelin in the rat by a specific inhibitor of nitric oxide formation. Br J. Pharmacol. 98: 646–652.
Wimalawansa SJ (1996). Calcitonin gene-related peptide and its receptors: molecular genetics, physiology, pathophysiology, and therapeutic potentials. Endocr. Rev. 17: 533-585.
Yoshizaki H, Tokamiya M, Okada T (1987). Characterization of picomolar affinity binding sires for CGRP in rat brain and heart. Biochem. Biophys. Res. Commun.146: 443-451.

2

Amidase from plant growth promoting rhizobacterium

Sonia Chacko[1], Pramod W. Ramteke[2*] and Suchit A. John[2]

[1]Department of Biotechnology, Allahabad Agricultural Institute – Deemed University, Allahabad – 211007, India.
[2]Department of Biological Sciences, Allahabad Agricultural Institute – Deemed University, Allahabad – 211007, India.

Ninety three cultures of plant growth promoting rhizobacteria (PGPR) isolated from rhizosphere of *Pisum sativum*, among them one isolate identified as *Pseudomonas putida* was found to be potential amidase producer. The organism exhibited a battery of PGPR traits including enhanced production of plant growth hormone indoleacetic acid (IAA) and siderophore. *P. putida* MTCC 6809 exhibited both intracellular and extra-cellular amidase activity. The organism produced maximum extracellular amidase enzyme at 30°C and pH 7.5 in shaking state. The organism hydrolyzed a wide range of aliphatic amides that included acetamide, propionamide, acrylamide and butyramide. Acrylamide is a known carcinogen, teratogen and neurotoxicant and utilization of acrylamide by *P. putida* MTCC 6809 assume great importance. The organism is also tolerant to number of heavy metals at higher levels. These characteristics make *P. putida* MTCC 6809 an excellent candidate for field application in contaminated soil.

Key words: PGPR, amidase, aliphatic amides, acrylamide, heavy metal tolerance.

INTRODUCTION

Amidases (acylamide amidohyrolases EC 3.5.1.4) catalyze the hydrolysis of the carboxylic amide bonds to liberate carboxylic acid and ammonia (Asano and Lubbehusen, 2000). They are widely distributed in nature and are involved in nitrogen metabolism in both prokaryotic and eukaryotic cells. The ability of amidase to convert the cyanofunctionally into either an amide or an acid, together with regio and stereoselectivity, makes it a good biocatalyst in biotransformation processes. Amidases have been as biocatalysts for the production of ammonium, acrylate and acrylic acid, compound of intense industrial use (Nagasawa and Yamada, 1989).Apart from amide hydrolysis activity, some amidases also exhibit acyl transferase activity (Maestracci et al., 1980). Furthermore, many of these amidases have dual specificities, cleaving amide as well as ester and nitrile bonds (Patricelli and Cravatt, 2000; Pollmann et al., 2003; Cilia et al., 2005). Amidases form an integral part of nitrile converting enzymes and is known to have versatile industrial applications. In addition to several industrial applications, amidase plays an important role in degradation of agricultural and industrial wastes (Shanker et al., 1990).

Enzyme activity is considerably greater in the rhizosphere of plants than in bulk soils which often lends specific advantage to the plant and this activity is due to either a specific flora or the plant root or due to both (Curl and Truelove, 1986; Zahir et al., 2001). Many of the enzymes detected in soil are hydrolases and amidase is one of the important among them. Apart from amide hydrolysis activity, amidases have been shown to be involved in the biosynthesis of indole acetic acid (IAA) an important plant hormone (Pollmann et al., 2003; 2006; Spaepen et al., 2007). During the interaction between IAA-producing bacteria and plants, this phytohormone is used by bacteria as part of their colonization strategy, including phyto-stimulation and circumvention of basal plant defense mechanisms (Spaepen et al., 2007). Hydrolysis of carboryl based pesticides by soil amidases is reported by Kay-Shoemake and co-workers (2000).

Many investigations reported elevated levels of amidase enzyme in soil samples. Owing to the versatile applications, we screened PGPRs isolated from the rhizosphere of *Pisum sativum* that could be a potent source of amidase enzyme. Along with screening, parametric optimization for maximum amidase production is presented here in a promising PGPR strain of *Pseudomonas putida* MTCC 6809. The bacterium could also be the source to

*Corresponding author. E-mail: pwramteke@yahoo.com.

develop desired microbial bioaugumentation for growth of field crops in contaminated soil.

MATERIALS AND METHODS

Microorganisms

Ninety three isolates of PGPR obtained from rhizosphere of *P. sativum* comprising strains of *Pseudomonas* spp. (23), *Bacillus* spp. (19), *Rhizobium* spp. (34) and *Azotobacter* spp. (17) were included in the study.

An amidase producing strain *E. coli* NCIM 2568 was obtained from National Collection of Industrial Microorganisms (NCIM), Pune, India which was included in the study as control.

Screening of amidase producers

A qualitative estimation of amidase producing organism was done using a modified method by Rahim et al., 2003. Screening medium contained the following composition (g/l). K_2HPO_4 , 3.5; KH_2PO_4 ,2.2; $MgSO_4$, 0.05; 0.5ml of trace element solution containing (mg/l) $ZnSO_4.7H_2O$,0.3; $FeSO_4.7H_2O$,0.6; $MnSO_4$,0.3; $CuSO_4.5H_2O$, 0.4; $CaCl_2.2H_2O$, 0.5g/l Phenol red 0.012; Agar,15 and acetamide 20 (as sole carbon and nitrogen source). Acetamide was filter sterilized using Axiva membrane filter assembly and was added to the autoclaved medium. The pH of the medium was adjusted to 6.6 ± 0.2. Different PGP organisms were inoculated into the broth/ agar plates and then incubated for 36 h at 30°C. The broth and plates were regularly observed for colour change that is, from yellow to pink. A potent amidase producing organism was selected for further studies.

Identification of the organism

The promising amidase producer was identified to species level by Microbial Type Culture collection (MTCC), Institute of Microbial Technology (IMTECH), Chandigarh, India and assigned accession number MTCC 6809.

Determination of PGP characters

Ammonia production: The organism was grown in peptone water at 30°C for 4 days and 1 ml of Nessler's reagent was added. Production of ammonia was represented by development of faint yellow colour (Bakker and Schippers, 1987).

Indole acetic acid

The organism was inoculated in LB medium amended with 50 mg ml^{-1} of tryptophan and incubated at 30°C for 24 h. After incubation, 2ml of cell suspension was centrifuged at 1000 rpm for 10 min and 2 - 3 drops of orthophosphoric acid was added to the supernatant along with 4 ml of Solawaski's reagent (50 ml 35% perchloric acid; 1 ml of 0.5 M $FeCl_3$). The tubes were kept at room temperature for 20 min. IAA production was indicated by the development of pink colour. Quantification of IAA was estimated at OD 530 nm using standard IAA graph (Bric et al., 1991).

Catalase production

Bacterial culture was grown in a nutrient agar medium for 18 - 24 h. The cultures were mixed with appropriate amount of H_2O_2 on a glass slide to observe the evolution of oxygen.

Phosphate solubilization

The isolate was tested for phosphate solubilizing ability on Pikovskaya medium (Pikovskaya, 1948) incorporated with tricalcium phosphate (Ca_3 $(PO_4)_5$). Phosphate solubilization was indicated by the formation of a clear halo zone around the bacterial growth after 3 days of incubation.

HCN production

The isolate was streaked on King's B medium amended with 4.4 gl^{-1} of glycine. The plates were covered with sterile filter paper impregnated with 0.5 % picric acid in 2 % sodium carbonate, sealed with parafilm and incubated for 4 days (Bakker and Shippers, 1987). Development of yellow colour on the filter paper indicates the positive result.

Siderophore production

Siderophore production was detected by the universal method of Schwyn and Neilands (1987) using blue agar plates containing the dye Chrom Azurol S (CAS). Orange halos around the colonies on blue were indicative of siderophore production.

Analytical methods

The soluble proteins were estimated by using the method given by Lowry et al., 1951. Cell growth was monitored by determining the optical density of the culture medium at 540 nm using (Genesys6, Thermoelectron) spectrophotometer, the culture was diluted if necessary. Amidase activity was determined by measuring the ammonia released (Nesslerization reaction) spectrophotometrically at 450 nm (Imada et al., 1973). One unit of amidase activity is expressed as the amount of enzyme required to release 1 µmols of ammonia per minute under the assay conditions.

Intracellular amidase production

To check the intracellular amidase production, cells were harvested at different time intervals (8 - 36 h) by centrifuging at 10,000 rpm for 15 min in Sigma centrifuge which was then stored at 4°C for 30 min. 5 gm cells were then dissolved in 20 mM PO_4 buffer pH 7.4 containing 10 mM EDTA. Lysozyme was dissolved at a concentration of 0.5 mg/ml. The mixture is then stirred gently for 30 min at 4°C and 0.1 mg of DNase and 50 µl of 0.2 M $MgCl_2$ were added to it. The mixture was incubated for 15 min and centrifuged at 20,000 rpm for 30 min at 4°C. The clear supernatant was used for enzyme assay.

Parametric optimization for enzyme production

Different physiological (pH of the production medium, temperature of incubation and incubation time) and nutritional parameters (substrate, carbon and nitrogen source, concentration etc.) were optimized for maximum amidase production.

Heavy metal tolerance test

Tolerance to metal ions was determined by minimal inhibitory concentration (MIC) technique by agar dilution method (Luli et al., 1983). Agar plates supplemented with different concentrations of heavy metals (50 - 250 µg/ml) were inoculated aseptically by exponentially growing culture of bacteria. The plates were incubated for 36 h at 30°C, growth of the isolates indicated positive tolerance. The metal used were As, Co, Cd and Cr, Zn, Ni and Cu and Hg. The compounds used were As_2O_3, $CoCl_2$, Cd $(NO_3)_2$, $K_2Cr_2O_7$, $ZnCl_2$, Ni $(NO_3)_2$, $CuCl_3$, and $HgCl_2$.

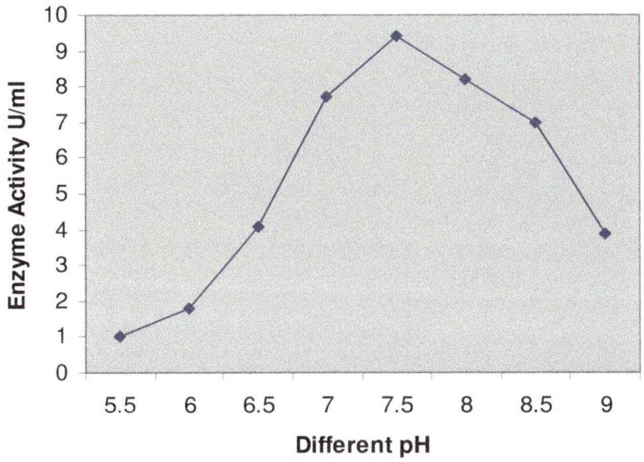

Figure 1. Effect of pH on amidase production.

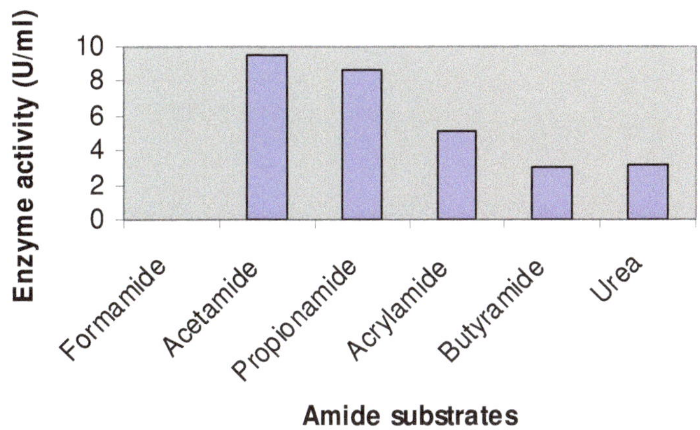

Figure 2. Utilization of substrates in enzyme production.

Plasmid curing

The microorganism was subjected to acridine orange mediated plasmid elimination (Ramteke, 1997). Overnight logarithmically growing culture was inoculated in nutrient broth containing acridine orange 20 μg ml^{-1} incubated at 30°C for 48 h. After incubation the culture was serially diluted and plated in nutrient agar plates. Individual colonies were screened for plasmid-encoded traits.

RESULTS AND DISCUSSION

Among the various PGPR isolated from rhizosphere of *P. sativum*, one isolate identified as *Pseudomonas putida* was found to be potential amidase producer. The pH of the medium was found to increase from 7.0 - 8.3 after 48 h of incubation as ammonia was released due to amide hydrolysis. Since intact cells released ammonia into the medium so the enzymes could be assayed *in situ* (Miller and Gray, 1982). The organism is deposited in Microbial Type Culture Collection (MTCC), India and assigned accession number MTCC 6809.

The organism exhibited a battery of PGPR characterristics. It showed reasonably good production (4.2 μg/ml) of plant growth hormone indoleacetic acid (IAA) and siderophore and was also found positive for the production of ammonia and catalase. The bacterium exhibited stimulation of root and shoot growth in presence of heavy metals (data not shown). Amidases are known in biosynthesis of IAA (Pollmann et al., 2006; Spaepen et al., 2007). Thus, amidase producing bacteria in the plant root rhizosphere that can rapidly convert precursors to biologically active molecules holds promise for increasing crop production.

The enzyme production is greatly affected by pH; hence production medium with different pH was used for amidase production and was assayed for enzyme activity (Figure 1). Amidase production was optimum at pH 7.5. Also the amidase activity due to various pH was found to be statistically significant (SE (m) 0.026; C. D. 0.08).

P. putida MTCC 6809 exhibited both intracellular and extracellular amidase production. However, production of intracellular enzyme was three fold less as compare to extracellular enzyme. The organism produced maximum extracellular enzyme activity after 12 h of incubation at 30°C and pH 7.5 in shaking state.

Utilization of various amide substrates were screened for amidase production (Figure 2). The organism hydrolyzed a wide range of aliphatic amides that included acetamide, propionamide, acrylamide and butyramide. But the organism was not able to hydrolyse, carbon compound formamide. Among the amides tested, acetamide followed by propionamide was found to be the better substrates.

Acetamide and propionamide were further studied to obtain its optimum concentration in enzyme production. Acetamidase activity at 2% w/v concentration was found to be statistically significant (SE (m) 0.026; C. D. 0.082) (Figure 3).

On the other hand the organism produced maximum enzyme at 1 % w/v concentration of propionamide (Figure 4). Also at higher substrate concentrations [3% and 4% (w/v)] it was observed that there was no increase in enzyme production. This might be due to less conversion of amide substrates into products or hydrolysis of the amide compounds to products due to increased amide concentration. Brammer and Clark (1964) reported that amidase production by *P. aeruginosa* was repressed at higher concentration of acetaamide which was used as inducer, as at higher concentration the system was saturated by the inducer.

Amides when used as sole source of carbon and nitrogen, growth of the organism was observed to be less and therefore, the production medium was supplemented with various sources of carbon and nitrogen to enhance both growth and enzyme production. Among the various carbon sources tested glucose followed by fructose emerged the best carbon sources (Figure 5). Lactose was the least suitable carbon source. Enhanced production of enzyme was observed when glucose concentration was at 1.0% (w/v) but there was a significant decrease in enzyme pro-

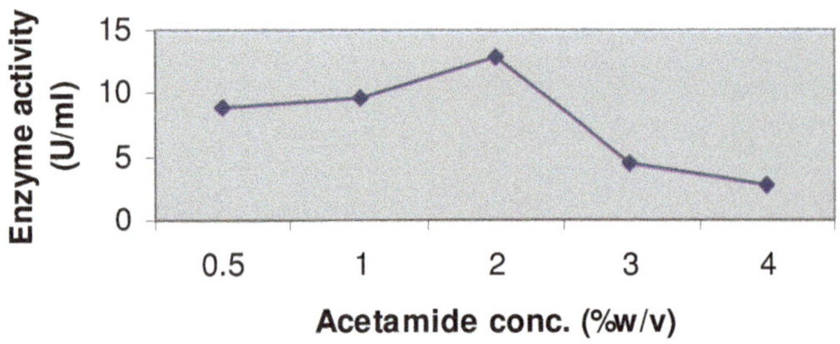

Figure 3. Effect of concentration of acetamide on amide production by *Pseudomonas putida* MTCC6809.

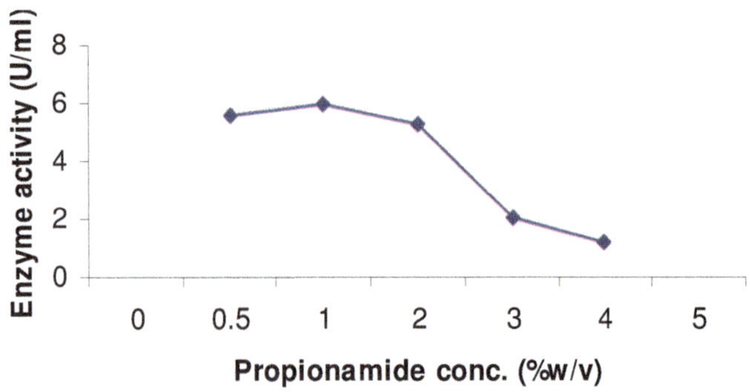

Figure 4. Effect of concentration of propionamide on amidase production *Pseudomonas putida* MTCC6809.

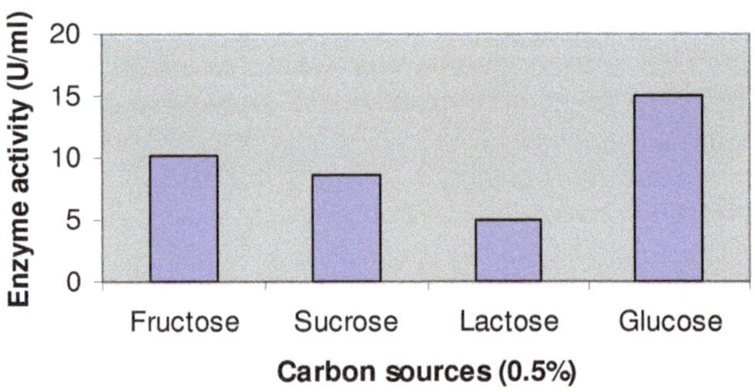

Figure 5. Effect of carbon sources on amidase production.

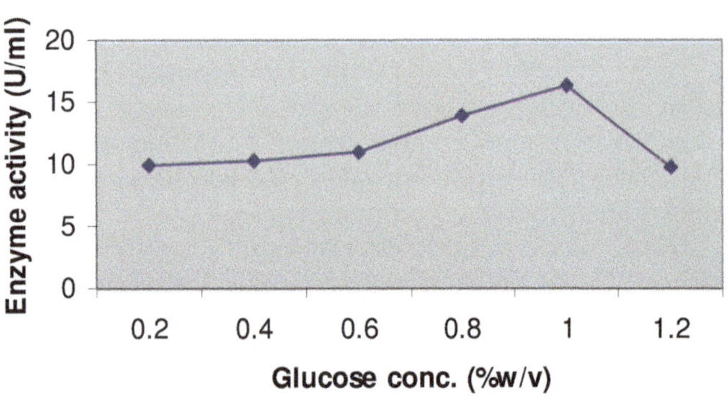

Figure 6. Effect of glucose concentration on amidase production.

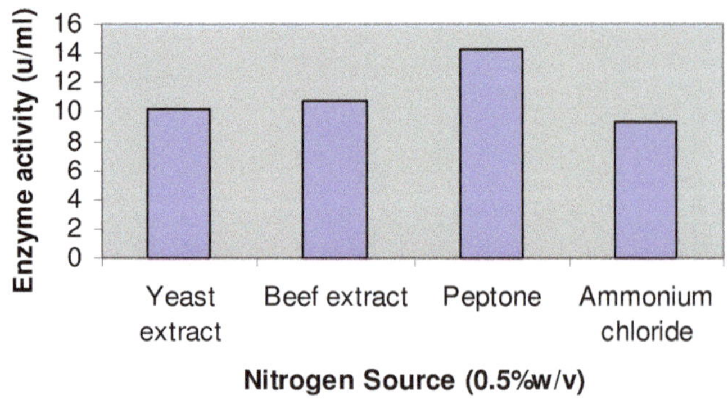

Figure 7. Effect of nitrogen sources on amidase production.

duction when the glucose concentration was raised to 1.2% (w/v) suggesting that the enzyme production was repressed by higher concentration of glucose (Figure 6).

Catabolic repression of amidases by common sugars was also observed in other organisms. Fructose in *Alcaligenes eutrophus* (Friedrich and Mitrenga, 1981) and glucose in *Pseudomonas* spp (Shanker et al., 1990) were found responsible for catabolic repression of amidases. However, contradictory to these reports we observed that production of amidase in *P. putida* MTCC 6809 is not repressed catabolically by fructose and glucose. Also, we checked the various nitrogen sources to enhance amidase production, among them peptone enhanced production of the enzyme (Figure 7).

Bacterial resistance to metal ions seems to be directly related to the presence and adopted response to these elements in the environment (Timoney et al., 1978). Thus, tolerance to heavy metals gives an added advantage of survival in the polluted environment. *P. putida* MTCC 6809 was isolated from the sewage irrigated rhizosphere and was found tolerant to higher concentration multiple heavy metals (Table 1).The organism showed tolerance to high levels (>100 $ugml^{-1}$) of Ni, Pb, Zn and As. Tolerance to Co, Cr, Cu and Cd was at 50 $ugml^{-1}$. However, surprisingly we observed that tolerance to heavy metals were found not curable when acridine orange was used as curing agent suggesting chromosomal traits. Acrylamide is used in numerous industrial processes but its extensive and discriminate usage has lead to the contamination of terrestrial ecosystem. Acrylamide is a known carcinogen, teratogen and neurotoxicant (Dearfield et al., 1988; Klause and Schmahl, 1989). In this context, utilization of acrylamide by *P. putida* MTCC 6809 assumes greater importance. This finding is very important from the point of environmental detoxification of acrylamide.

Polyacrylamide (PAM) is currently used as an irrigation water additive to reduce the amount of soil erosion that occurs during furrow irrigation of crops. Elevated soil amidase activity specific toward the large PAM polymer has been reported in PAM-treated field soils (Kay-Shoemake et al., 2000). Additionally, nitrile compounds are dis-

Table 1. Tolerance to heavy metal.

Heavy metal	Conc. µg/ml	Tolerance
Nickel	200	+
Lead	200	+
Arsenic	200	+
Zinc	100	+
Cobalt	50	+
Copper	50	+
Chromium	50	+
Cadmium	50	+
Mercury	50	+

charged into the environment as industrial waste water, agricultural chemicals, etc. In mineralization of nitriles and corresponding amides, amidases play an important role in addition to nitrile-hydrolyzing enzymes (Asano et al., 1982; Shanker et al; 1990). Hydrolysis of carbaryl by a soil amidase was also demonstrated by Kay- Shoemake and coworkers (2000).

Thus, considering overall characteristics to produce amidase, plant growth promoting traits and to tolerate higher levels of heavy metals makes *Psuedomonas putida* MTCC 6809 an excellent candidate for field application in contaminated soil.

ACKNOWLEDGEMENTS

The authors thank Dr. Bini Toms for her help with statistical analysis. The financial grant from the Department of Biotechnology, Government of India is gratefully acknowledged.

REFERENCES

Asano Y, Lubbehusen TL (2000). Enzymes acting on peptides containing D-amino acid. J. Biosci. Bioeng. 89: 295-306.

Asano Y, Tachibana M, Tani Y, Yamada H (1982). Purification and characterization of amidase which participates in nitrile degradation. Agric. Biol. Chem. 46: 1175-1181.

Bakker AW, Schippers P (1987). Microbial cyanide production in the rhizosphere in relation to potato yield reduction and *Pseudomonas* spp.-mediated plant growth-stimulation. Soil Biol. Biochem. 19: 451-457.

Bric JM, Bostock RM, Silversone SE (1991). Rapid in situ assay for Indole Acetic Acid production by bacteria immobilized on a nitrocellulose membrane Appl. Environ. Microbiol. 57: 535-538.

Brammer WJ, Clark PH (1964). Induction and repression of *Pseudomonas aeruginosa* amidase. J. Gen. Microbiol. 37: 307-309.

Cilia E, Fabbri A, Uriani M, Scialdone CG, Ammendola S (2005). The signature amidase from Sulfolobus solfataricus belongs o the CX3C subgroup of enzymes cleaving both amides and nitriles – Ser195 and Cys145 are predicted to be the active site nucleophiles. FEBS J. 272: 4716-4724.

Curl EA, Truelove B (1986). The rhizozsphere. Springer-Verlag, New York.

Dearfield KL, Abernathy CD, Ottley MS, Brantner JH, Hayes PF (1988). Acrylamide: its metabolism, developmental and reproductive effects, genotoxicity and carcinogenicity. Mutat. Res. 195: 45-77.

Friedrich CG, Mitrenga G (1981). Utilization of aliphatic amides and formation of two different amidases by *Alealigenes eutrophus*. J. Gen. Microbiol. 125: 367-374.

Imada A, Igarsasi S, Nakahoma K, Isono M (1973). Asparginase and glutaminase activities of microorganisms. J. Gen. Microbiol. 76: 85-99.

Kay-Shoemake JL, Watwood ME, Sojka RE, Lentz RD (2000). Soil amidase activity in polyacrylamide-treated soils and potential activity toward common amide-containing pesticides. Biology and Ferility of soils 31: 183-186.

Klaus AN, Schmahl W (1989). Mutagenic and teratogenic effects of acrylamide in the mammalians spot tests. Mutant Res. 226: 157-162.

Kobayashi M, Komeda H, Yanaka N, Nagasawa T, Yamada H (1992). Nitrilase from *Rhodococcus rhodochrous* J1: Sequencing and over expression of the gene and identification of an essential cysteine residue. J. Biol. Chem. 267:20746-20751.

Luli GW, Talanagi JW, Strohl WR, Pfister RM (1983). Hexavalent chromium resistant bacteria isolated from the sediments. Appl. Env. Microbiol. 46: 846-854.

Lowry OM, Rosebrough NJ, Farr HL, Randal RJ (1951). Protein measurement with the Folin phenol reagent. J. Gen. Microbiol. 76: 85-99.

Maestracci M, Thiery A, Arnaud A, Glazy P (1980). A study of the mechanism of the reaction catalysed by the amidase of Brevibacterium sp. R312. Agric. Biol. Chem. 50:2237-2241.

Miller JM, Gray DO (1982). The utilization of nitriles and amides by a *Rhodoccocus* species. J. Gen. Microbiol. 128: 1803-1809.

Nagasawa T, Yamada H (1989). Microbial transformation of nitriles. Trends Biotechnol. 7: 153-158.

Patricelli MP, Cravatt BR (2000). Clarifying the catalytic roles of conserved residues in the amidase signature family. J. Biol. Chem. 275: 19177-19184.

Pikovskaya RI (1948). Mobilization of phosphorus in soil connection with the vital activity of some microbial species. Microbiologiya 17: 362–370.

Pollmann S, Neu D, Weiler EW (2003). Molecular cloning and characterization of an amidase from Arabidopsis thaliana capable of converting indole-3-acetamide into the plant growth harmone, indole-3-acetic acid. Phytochemistry 62: 293-300.

Pollmann S, Muller A, Weiler EW (2006). Many road leads to "auxin": of nitrilases, synthases, and amidases. Plant Biol. 8: 326-333.

Rahim MA, Saxena RK, Gupta R, Sheoran A, Giri B (2003). A novel quick plate assay for acetamidase producers and optimization for its production by *Aspergillus candidus*. Process Biochem. 38: 861-866.

Ramteke PW (1997). Plasmid mediated co-transfer of antibiotic resistance and heavy metal tolerance in coliforms. Ind. J. Microbiol. 37: 177-181.

Schwyn B, Neilands JB (1987). Universal chemical assay for detection and determination of siderophore. Anal. Biochem. 160: 47-56

Shanker R, Ramakrishna C, Seth PK (1990). Microbial degradation of acrylamide monomer. Arch. Microbiol. 154: 192-198.

Skujins JJ (1967) Enzymes in soil, Soil Biochemistry. Vol.1.A.D.Mc Laren and G.H.Peterson (eds.). Marcel Dekker, New York. pp 371-414.

Spaepen S, Vanderleyden J, Remans R (2007). Indole-3-acetic acid in microbial and microorganism-plant signaling. FEMS Microbiol. Rev. 31(4):425-48.

Timoney JF, Port J, Gills J, Spanier J (1978).Heavy-metal and antibiotic resistance in the bacterial flora of sediments of New York Bight. Appl. Environ. Microbiol. 36:465 - 472.

Yamada T, Palm CJ, Brooks B, Kosuge T (1985). Nucleotide sequences of the *Pseudomonas savastanoi* indoleactic acid genes show homology with *Agrobacterium tumefaciens* T-DNA. Proc. Natl. Acad. Sci. USA 82: 6522-6526.

Zahir A, Atteeq M, Arshad M (2001). Soil Enzymes Research: A review Online J. Biol. Sci. 1(5): 299-307.

3

Simple method to study the mechanism of thermal and non thermal bactericidal action of microwave radiations on different bacterial species

Arifa Tahir[1]*, Bushra Mateen[1], Salih Univerdi[2], Ozer KaraGoban[2] and Mehmet Zengin[2]

[1]Microbiology Research Centre, Lahore College for Women University, Lahore Pakistan.
[2]Science department, PakTurk International Colleges, Islamabad, Lahore Pakistan.

The present study was based on the investigations to determine whether the bactericidal effect of microwave radiations on bacteria was either thermal or due to microwaves. It was also investigated which bacterial sp is more sensitive to microwave. Five bacterial species, *Proteus vulgaris PP25D, Staphylococcus aureus bgh010*, Bacillu *subtilis ase98, Escherchia coli TRE04* and *Corynebacterium spN33* were exposed to microwave radiation. These cultures were selected as test organisms due to their importance in food industry as human pathogens. Bacterial suspensions were exposed to microwave radiations (2450 MHz and 800W) for 60, 120 and 180 s to study the effect of heat generated by microwaves. The second experiment was designed to study the effect of microwaves only by maintaining temperature below 40°C. The degree of inactivation at uncontrolled and controlled temperature was compared quantitavely. The viable counts of all cell suspensions were found to reduce greatly with an increase in microwave heating time and temperature. *B. subtilis ase98* showed highest reduction at uncontrolled temperature. No significant reduction of cell density was observed in either cell suspension. The effect of microwave radiations was also studied in terms of morphological changes. No detectable change was observed in cell shape and morphology of colonies except *S. aureus bgh010*. Results indicated that the effect of microwave radiations on bacteria was purely thermal and no detectable change occurred in non-thermal treatment. *B. subtilis ase98* were the most sensitive species for microwave radiations. *B. subtilis ase98* can be used as an indicator bacterium to assess microwaves for sterilization.

Key words: Microwave radiation, thermal and non thermal effect, sensitive bacterial sp.

INTRODUCTION

Innovative approaches are required to inhibit and control food pathogens (Bevilacqua et al., 2008). Numerous studies address the effect of microwave heating on pathogenic microorganisms in foods. Microwave lamps have been reported for the disinfection of bacteria with promising results (Barkhudarov et al., 2007). Bacteria reported to be inactivated by microwave heating include *Bacillus cereus*, *Campylobacter jejuni*, *Clostridium perfringens*, pathogenic *Escherichia coli*, *Enterococcus*, *Listeria monocytogenes*, *Staphylococcus aureus*, and *Salmonella* (Heddleson et al., 1994; Rosenberg and Bogl, 1987; Knutson et al., 1987; Chipley, 1980).

A lot of work has been done on microwaves and their effect on several bacterial species (Vaid and Bishop, 1998). The actual mechanism of bacterial killing is still controversial. It is very difficult to precisely compare the effectiveness of microwave heating to conventional heating based on the literature, because of the different techniques employed or the lack of detail in the methods or materials used, especially in relation to temperature monitoring (Heddleson and Doores, 1994). Yeo et al. (1999) reported that the inactivation of *E. coli* was solely due to thermal effect, while other observed that heat generated during microwave exposure alone is inadequate to fully account for the nature of lethal effect (Barnes and Ho, 1977; Salvatorelli et al., 1996). Thus combined microwave

*Corresponding author. E-mail: arifa.tahir@yahoo.com.

heat may be involved in killing of bacteria. Some researchers used copper cooling coil to maintain temperature. But this treatment was not effective and copper ions were toxic to the bacterial cells and cause inactivation (Cha, 1993; Cooksey, 1990). Other workers also controlled the temperature of the irradiated specimen through various timing, pulsing or cooling techniques (Welt et al., 1994).

Kozempel et al., 2000 designed a new system that was capable of isolating thermal and non-thermal effects of microwave energy. The system was a double tube that allowed input of microwave energy but removed thermal energy with cooling water. Khalil and Villota, 1988 further studied the effect of microwaves (2450 MHz assumed) on injury of *S. aureus* FRI-100. They heated cells at a sub lethal temperature of 50°C and maintained microwave temperature using recirculated cooled kerosene. The aim of present study was to assess the thermal and non-thermal bactericidal action of microwave radiation. In order to investigate the effect of microwave radiations only, an ice bath was used for bacterial suspension to kept temperature below 40°C throughout the exposure to heating in the microwave. After exposure to different temperature cell viability and changes in morphology were studied. The study will help in differentiating the thermal and non thermal effect of microwave radiations.

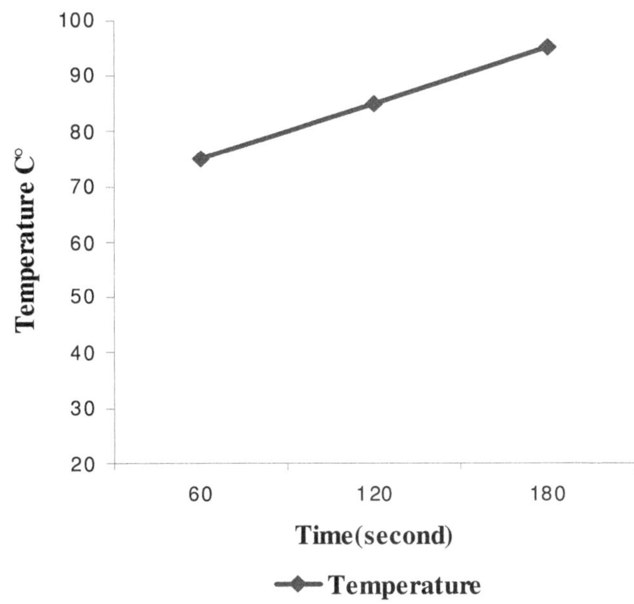

Figure 1. Change in temperature with increase in time of microwave irradiation.

MATERIALS AND METHODS

Collection of bacterial samples

The cultures of *P. vulgaris PP25D, S. aureus bgh010* , *B. subtilis ase98, E. coli TRE04* and *Corynebacterium spN33* were taken with the courtesy of DTL laboratory Lahore, Pakistan. These bacteria were used to study the effects of microwave radiations.

Preparation of inoculum

All the bacterial species were grown in Nutrient Broth in 250 ml cotton plugged conical flask. A loopful of pure bacterial culture was transferred by a sterilized wire loop and suspended into to 50 ml Nutrient Broth in 250 ml conical flask. The flask was then incubated at 35°C and rotated at 150 r.p.m for 20 h in a rotatory shaker (Model No. 3033, GFL, Germany). In order to obtain accurate results all indicator bacteria were cultured and reproduced for several generations, then prepared for the bacterial solution.

An amount of 10 ml of this suspension was used to check absorbance at 600 nm, 0.50 absorbance was used to get the approximately same number of microbes for further treatment.

Microwave source

Microwave treatment process was performed in a microwave oven (DW-131G; Dawlance). The full output power of the equipment was 800 W at 2450 MHz (frequency).

Exposure of bacteria to microwave radiations

Bacterial suspension of 0.50 absorbance was used to make dilutions (1×10^5) with sterilized phosphate buffer of pH 7.0. One of four test tubes (MW-C) represents control sample for viable counting without any treatment while the remaining (MW-60, MW-120, MW-180) were exposed to microwave radiation for 60, 120 and 180 s .

Bacterial suspension were exposed to radiation by placing test tubes in a 100 ml Pyrex beaker within a 500 ml Pyrex beaker containing 100 ml water and placed at the center of rotating plate of microwave oven. Water was used to prevent the evaporation of sample. Then samples were exposed to microwave for 60, 120 and 180 s separately. Temperature was measured after every treatment by measuring the temperature of suspension with the help of sterilized thermometer. The correlation between the microwave radiation time and temperature changes in bacterial suspension is shown in Figure 1.

In order to see the effect of microwave radiation only the bacterial suspension were taken in a test tube and placed in a 500 ml Pyrex beaker containing ice to keep the temperature below 40°C and placed at the center of rotating plate of microwave oven. Then samples were exposed to microwave for 60,120 and 180 s separately. In this series of experiment, there will be no effect of heat to these bacterial cultures and thus independent effect of radiation could be noted.

Measurement of viable cell count

After exposure the 0.1 ml sample from each of three test tubes containing treated bacterial suspensions were transferred into sterilized petriplates (100 mm) separately and about 15 ml of nutrient agar (maintained at about 40°C) was poured into each of petriplate. In order to mix the sample with medium, the petridishes were rotated uniformly clockwise and the medium was allowed to solidify and then incubated for about 20 h at 35°C.

Measurement of cell density

Cell density was measured at 600 nm using a spectrophotometer (UTECH PRODUCTS IN C.N.Y.12203.U.S.A).

Morphology changes

The slides were made from all the samples and stained (Gram Staining) to check morphological changes after treatment under simple microscope.

Data Analysis

The statistical analysis of the data was carried out using the Minitab v 13.0 software. Paired t- test was used to compare the data. The level of significance was taken at P = 0.05. The results are average of triplicates.

RESULTS

Effect of microwave radiations to *Escherchia coli TRE04*

E. coli is a rod shaped gram negative bacteria. They form pink colonies on MacConkey agar. The viable count of *E. coli TRE04* determined by pour plate method was 50 ± 4.36 × 10^5 CFU/ml before irradiation. The average reduction in viable count of *E. coli TRE04* after 60, 120 and 180 s of microwave exposure at uncontrolled temperature was 21.3 ± 2.08 × 10^5, 4 ± 1 × 10^5 and 2 ± 1× 10^5 CFU/ml respectively. Highly significant (P = 0.024) reduction was found in all experimental sample as compared to control. It was found that there was slight reduction in viable count of samples subjected to microwave irradiation at controlled temperature (<40°C). The observed values of samples after microwave exposure for 60, 120 and 180 s at controlled temperature was 47 ± 3.5 × 10^5, 45 ± 4.58 × 10^5 and 42 ± 2 × 10^5 while the control was 50 ± 4.36 × 10^5 CFU/ml. Non-significant (P = 0.067) reduction was found in all experimental samples as compared to control (Figure 2). No detectable change in colony morphology and cell shape was observed.

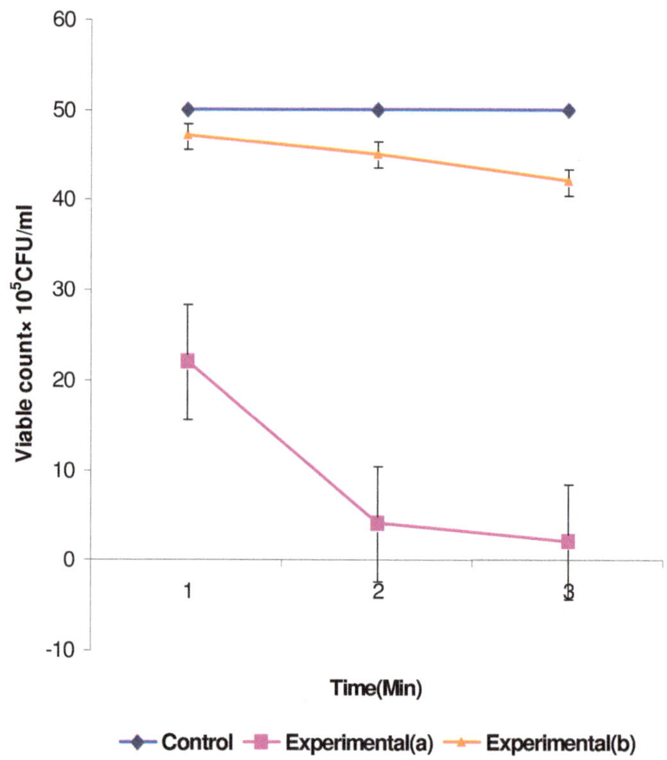

Figure 2. Change in viable count of *Escherchia coli* after microwave irradiation for 60, 120 and 180 s (a) at uncontrolled temperature (b) at controlled temperature.

Effect of microwave irradiation on *Bacillus subtilis ase98*

B. subtilis is a rod shaped Gram positive spore former bacteria. They form white convex shaped colonies on Nutrient agar. *B. subtilis ase98* was found highly sensitive to thermal treatment. The viable count of *B. subtilis ase98* was determined by pour plate method. The average reduction in viable count of *B. subtilis (*69 ± 4.36 × 10^5 CFU/ml before irradiation) after 60, 120 and 180 s microwave exposure at uncontrolled temperature was found to be 30 ± 3 × 10^5, 15 ± 2.65 × 10^5 and 5 ± 1.73 × 10^5 CFU/ml respectively. The reduction found in all experimental samples was highly significant (P = 0.01).

It was observed that there was slight reduction in viable count of samples subjected to microwave irradiation at controlled temperature (< 40°C). The observed values of viable count after microwave exposure for 60, 120 and 180 s at controlled temperature was 65 ± 1.73 × 10^5, 64 ± 4 × 10^5 and 58 ± 2 × 10^5 CFU/ml while in control, viable count was 69 ± 2 × 10^5 CFU/ml (Figure 3). Non-significant (P = 0.09) reduction was found in all experimental samples. No detectable change in colony morphology and cell shape was observed

Effect of microwave irradiation on *Staphylococcus aureus bgh010*

S. aureus was round shaped Gram positive cocci. They form yellowish slimy and convex shaped colonies on Nutrient agar. A gradual reduction was observed in viable count of *S. aureus bgh010* (67 ± 3 × 10^5 CFU/ml before irradiation) with increase in exposure time to microwave radiation (Figure 4). The observed values of reduction in viable count of *S. aureus bgh010* after 60,120 and 180 s microwave exposure at uncontrolled temperature was found to be 38 ± 2 × 10^5, 12 ± 3 × 10^5 and 2 ± 1 × 10^5 CFU/ml respectively. Significant (P = 0.044) reduction was found in all suspensions.

According to observed values there was small reduction in viable count of specimens subjected to microwave irradiation at controlled temperature (< 40°C). The observed values of viable count after microwave exposure for 60, 120 and 180 s at controlled temperature was 63 ± 2 × 10^5, 62 ± 2 × $10^5$and 58 ± 2 × 10^5 while in control, viable count was 67 ± 3 × 10^5 CFU/ml. Non-

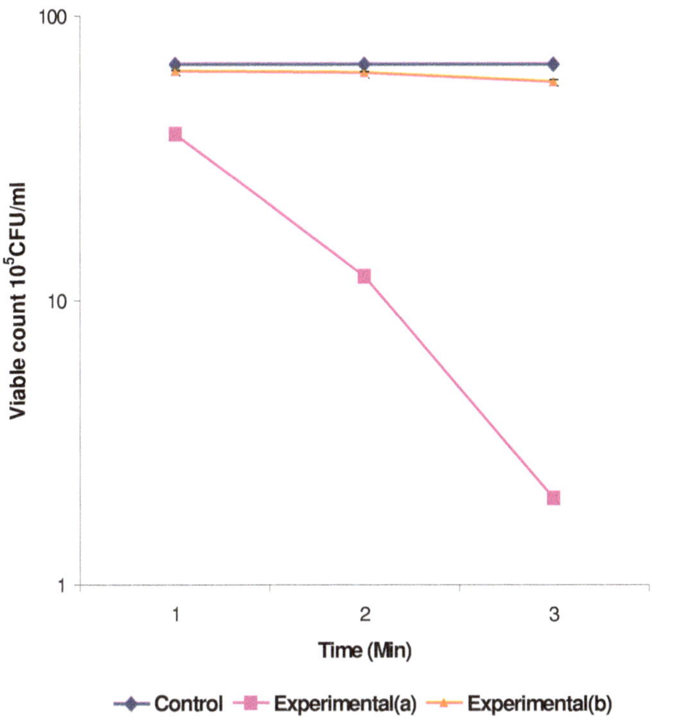

Figure 3. Change in viable count of *Bacillus subtilis* after microwave irradiation for 60, 120 and 180 s (a) at uncontrolled temperature (b) at controlled temperature.

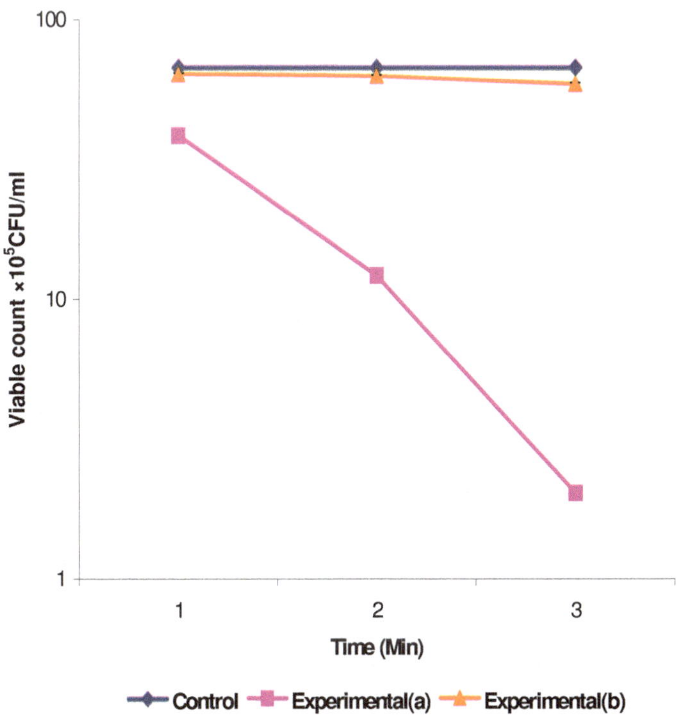

Figure 4. Change in viable count of *Staphylococcus aureus* after microwave irradiation for 60,120 and 180 s (a) at uncontrolled temperature (b) at controlled temperature.

significant (P = 0.059) reduction was found in all experimental samples (Figure 4). The cells grown on petriplate after exposure form white while the untreated sample formed yellowish colonies. The change was observed in both irradiated samples at uncontrolled and controlled temperature. No visible change in cell shape was observed.

Effect of microwave irradiation on *Proteus vulgaris PP25D*

P. vulgaris PP25D was a rod shaped Gram negative bacteria. They form white irregular shaped colonies on Nutrient agar. The viable counts of all cell suspensions (57 ± 3 × 10^5 CFU/ml before irradiation) were found to reduce relatively with an increase in microwave heating time and temperature (Figure 5). The average reduction in viable count of *P. vulgaris PP25D* after 60, 120 and 180 s after microwave exposure at uncontrolled temperature was 27 ± 4.58 × 10^5, 20 ± 3.79 × 10^5 and 12 ± 2 × 10^5 CFU/ml respectively. Highly significant (P = 0.01) reduction was found in all experimental sample. The observed values of viable count after microwave exposure for 60, 120 and 180 s at controlled temperature was 54 ± 4 × 10^5, 50 ± 4 × 10^5 and 46 ± 4 × 10^5 while in control, viable count was 57 ± 3 × 10^5 CFU/ml . Non-significant (P = 0.09) reduction was found in all experimental samples. No detectable change in colony morphology and cell shape was observed.

Effect of microwave irradiation on *Corynebacterium sp N33*

Corynebacterium sp N33 was a rod shaped Gram positive bacteria. They form white irregular shaped and rough colonies on Nutrient agar. The average reduction in viable count of *Corynebacterium sp N33* (65 ± 3 × 10^5 CFU/ml before irradiation) after 60, 120 and 180 s after microwave exposure at uncontrolled temperature was 46 ± 4 × 10^5, 35 ± 2 × 10^5 and 25 ± 2.65 × 10^5 CFU/ml respectively. Highly significant (P = 0.03) reduction was found in all experimental sample. The observed values of viable count after microwave exposure for 60, 120 and 180 s at controlled temperature was 62 ± 3.61 × 10^5, 60 ± 2 × 10^5 and 57 ± 2.65 × 10^5 while in control, viable count was 65 ± 3 × 10^5 CFU/ml. Non-significant (P = 0.067) reduction was found in all experimental samples as compare with control (Figure 6). No detectable change in colony morphology and cell shape was observed.

Effect of microwave radiation on cell density

Change in cell density of bacterial suspensions was observed by taking its absorbance value at 600 nm. The average reduction in cell density of all cell suspensions after microwave irradiation was not detectable (data not

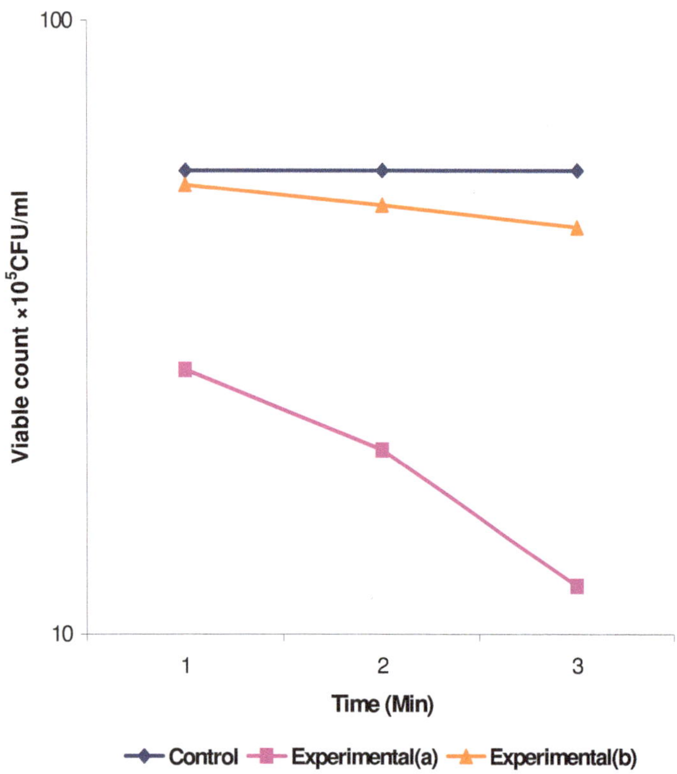

Figure 5. Change in viable count of *Proteus vulgaris* after microwave irradiation for 60, 120 and 180 s (a) at uncontrolled temperature (b) at controlled temperature.

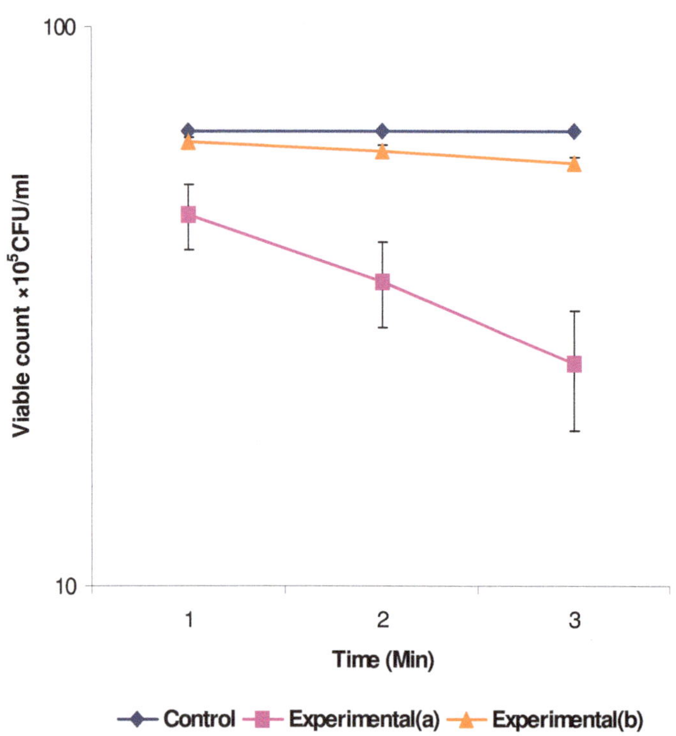

Figure 6. Change in viable count of *Corynebacterium* after microwave irradiation for 60, 120 and 180 s (a) at uncontrolled temperature (b) at controlled temperature.

shown). The rate of reduction was lower in both irradiations at uncontrolled and controlled temperature.

DISCUSSION

It was observed that the viable cell count reduced relatively with an increase in time of microwave irradiation at different temperatures. It might be due to microwave energy absorption by biological materials with increase in temperature. It was observed that the exposure up to 90°C resulted in a highest reduction of the viable count in all bacterial strains. The highest reduction ratio in the viable counts was observed when the temperature was increased from 70 to 90°C. Our findings are in agreement with other researchers (Woo et al., 2000; Jeng et al., 1987; Yeo et al., 1999), who reported that the reduction effect on exposing species was due to higher increase in temperature. The temperature was increased because substances with high dielectric constant such as water absorb microwaves and convert the energy to heat. As the temperature of the material increases, its ability to absorb microwaves also goes up, increasing the temperature. This can lead to uncontrolled heating (Zhang et al., 2000). Microbial lethality of microwave radiation may be due to the penetration of electromagnetic waves into a biological wet material, heating up the intra and extra cellular fluids by the transfer of energy from polar water molecules and dissolved ions. This results in the generation of heat within the material itself due to molecular activity (Mudgett and Schwartzberg, 1982).

It was observed that exposure of bacterial suspensions to microwaves for different time periods did little or no damage in terms of the viability. The non-thermal effects of the microwave energy are of no significance. Our findings are in agreement with other workers (Ramaswamy et al., 2000). It was found that bacteria respond differently to inactivation by microwaves. Microwave treatment has been reported to cause protein denaturation and aggregation in cytoplasm as well as to induce heat shock proteins, which may cause microbial inactivation.

It was very interesting to observe that Gram positive bacteria were found more sensitive to microwave radiations than Gram negative. It might be due to the difference in chemical composition of cell walls. It has been reported that the lipid content of gram negative bacterial cell wall make them more resistant to microwave. *B. subtilis ase98* was found highly sensitive to microwave radiation because it showed highest reduction in viable count. *B. subtilis ase98* should be considered as an indicator bacterium for inactivation by microwave energy (Deng et al., 1990).

The sensitivity of cells to lyses by microwave irradiation was also investigated at 600 nm. It was observed that cell density in all microwave heated cell suspensions slightly decreased for increasing and controlled temperature. The reduction in cell density in all cell suspensions was not significant. Our findings are in agreement with other workers (Woo et al., 2000). This insignificant change in cell density may be due to the fact that the microwave-treated cells were not completely lysed even when they were inactivated by microwave radiation, and thus the cell density did not decrease.

During the current research work a change in bacterial colony was observed, white colonies of *S. aureus bgh010* were observed after microwave irradiation while yellowish colonies were observed before irradiation. Similar types of changes in character were reported by other workers (Dreyfuss and Chipley, 1980; Atmaca et al., 1996). These changes might be due to microwave radiations effect on the genetic make up of the bacterial cells as destruction of DNA has been reported by Kakita et al., 1995. Other bacterial species did not show any detectable change in colonial morphology and cell shape.

Conclusion

It was evident from the present study that microwave radiations might affect the microorganisms up to significant level. Thus, the microwave can be used after maintaining the temperature to control the required microbes. Microwave radiations caused high reduction in viable count of bacteria with increase in exposure time. But no significant change was found by irradiation at controlled temperature, which shows that the effect of microwaves was highly thermal. *B. subtilis ase98* can be considered as an optimum indicator bacterium for microwave sterilization.

REFERENCES

Atmaca S, Akdag Z, Dasdag S, Celik, S (1996). Effect of microwaves on survival of some bacterial strains. Acta. Microbiol. Immunol. Hung. 43: 371–378.

Barkhudarov EM, Christofi N, Kossyi IA, Misakyan MA, John SJ, Taktakishvili IM (2007). Killing bacteria present on surfaces in films or in droplets using microwave UV lamps. World J. Microbiol. Biotechnol. 24: 761-769

Barnes FS, Ho CL (1977). Model for some nonthermal effects of radio and microwave E-fields on biological membranes. IEEE Trans. Micro Theor. Tech. 25:742–746.

Bevilacqua A, Sinigalia M, Corbo MR (2008) Alicyclobacillus acidoterrestris: New methods for inhibiting spore germination. 125: 103-110

Cha JS, Cooksey DA (1993). Copper hypersensitivity and uptake in Pseudomonas syringae containing cloned components of the copper resistance operon. Appl. Environ. Microbiol. 59: 1671-1674.

Chipley JR (1980). Effects of microwave irradiation on microorganisms. Academic Press Inc. Adv. Appl. Microbiol. 26: 129-145.

Deng SF, Wu Q, Qiu LH (1990). Research on sterilizing several common bacteria by microwave cabinet (in Chinese). J. Harbin. Inst. Tech. 1: 60–63.

Dreyfuss MS, Chipley JR (1980). Comparison of effects of sublethal microwave radiation and conventional heating on the metabolic activity of *Staphylococcus aureus*. Appl. Environ. Microbiol. 39:13-16.

Heddleson RA, Doores S, Anantheswaran RC, Kohn G (1996). Viability of *Salmonella spp.,* Staphylococcus aureus and Listeria monocytogenes. J. Food Prot. 58: 813–818.

Jeng DKH, Kurt A, Kaczmarek AGW, Balasky G (1987). Appl. Environ. Microbiol. 53: 2133-2137.

Kozempel MF, Cook RD, Scullen OJ, Annous BA (2000). Development of a process for detecting non-thermal effects of microwave energy on microorganisms at low temperature. J. Food Process. 24(4): 287-301.

Knutson KM, Marth EH, Wagner MK (1987). Microwave Heating of Food. Lebensm Wiss Technol. 20:101-110

Mudgett RE, Schwartzberg HG (1982). Microwave food processing: pasteurization and sterilization - a review. AICHE Symp. 78(218): 1-11.

Ramaswamy H, Koutchma T, Tajchakavit S (2000). Enhanced thermal effects under microwave heating conditions. International Conference of Engineering and Food [ICEF-8]. Puebla, MX.

Rosenberg U, Bögl W (1987). Microwave pasteurization, sterilization, blanching, and pest control in the food industry. Food Technol. 41(6): 92-9

Salvatorelli G, Marchetti MG, Betti V, Rosaspina S, Finzi G (1996). Comparison of the effects of microwave radiation and conventional heating on Bacillus subtilis spores. Microbios. 87: 169–174.

Vaid A, Bishop AH (1998). The destruction by microwave radiation of bacterial endospores and amplification of the released DNA. J. Appl. Micro. 85: 115–122.

Welt BA, Street JA, Tong CH, Rossen JL, Lund DB (1993). Utilization of microwaves in the study of reaction kinetics in liquid and semi-solid media. Biotechnol. Prog. 9: 481-487

Woo IS, Rhee IK, Park HD (2000). Differential damage in bacterial cells by microwave radiation on the basis of cell wall structure. Appl. Environ. Microbial. 66: 2243–2247.

Yeo CBA, Watson IA, Stewart-Tull DES, Koh VHH (1999). Heat transfer analysis of *Staphylococcus aureus* on stainless steel with microwave radiation. J. Appl. Microbiol. 87: 396-401

Zhang H, Datta AK, Taub I, Doona C (2000). Experimental and numerical investigation of microwave sterilization of solid foods. J. AIChE. 56: 99-103.

4

A microbiological technique for the separation of *Hibiscus sabdariffa* L. fibers

Pilanee Vaithanomsat[1]*, Poom Phusanakom[2], Waraporn Apiwatanapiwat[1] and Molnapat Songpim[1]

[1]Enzyme Technology and Waste Management Research Unit, Kasetsart Agricultural and Agro-Industrial Product Improvement Institute (KAPI), Kasetsart University, 50, Chatuchak, Bangkok 10900, Thailand.
[2]Toyota Motor Asia Pacific Engineering and Manufacturing Company Limited, 99 Moo 5, Ban Ragad, Amphur Bang Bo, Samutprakan 10560, Thailand.

This study is related to a process for the separation of *Hibiscus sabdariffa* L. fibers using living microorganisms. The selected fungal strains, *Datronia* sp. and *Oligoporous* sp., were isolated from natural sources in Thailand and were previously shown to be capable of degrading lignin in pulp and paper industrial wastewater. The process described here is comprising the fungal-treated fibrous of *H. sabdariffa* L. barks in certain conditions. The fungal inoculum size, retting time and retting temperature were important in the experiments and the retting efficiency was based on the visual characteristics of resulting fibers. In addition, the pectinolytic enzymes released from the fungi in retting conditions were shown to correspond to the separation of *Hibiscus sabdariffa* L. fibers.

Key words: *Hibiscus sabdariffa* L., fibers, separation, microbiological process.

INTRODUCTION

Hibiscus sabdariffa L., an ecological plant became a topic of the world due to its characteristics with growth speed quicker than normal plants, powerful absorption of carbon dioxide more than normal plants and strong water purification power. The advantage of kenaf has been known for a long time butits production has gradually decreased because of the development of less expensive items made by industrialized synthetic fibers. Due to its coarseness and stiffness, the traditional products of kenaf are package bag, cordage and ropes, which are low-grade products. The combination with plastic fibers has drawn attention as an alternative in a variety of industries. However, retting or a treatment degrading the pectin-rich middle lamella connecting adjacent fiber cells to release bast fibers is the predominant problem in preparation (Yu and Yu, 2007). Soaking kenaf stem or bark in water is a common way of kenaf fiber extraction. During retting, microbial activity causes a partial degradation of the components that bind tissues together, thereby separating the cellulosic fibers from non-fiber tissues (Akin et al., 2007). This technique is extremely dependent on the weather and water source, and therefore soaking time varies and is very difficult to control. As a result, a better natural fiber retting method has been sought for some time (Sharma and Faughey, 1999; Akin et al., 2001). Normally, stored premature wood is rapidly attacked by microorganisms, mainly fungi (Lopez et al., 2007). White-rot fungi are fungi belonging to the class Basidiomycetes capable of degrading these wood polymers (Lopez et al., 2007). The use of fungi not only aims at making production more effective but also at an environmentally friendly production (Li et al., 2009). As a result, by a suitably controlled fungal treatment it is possible to apply such a technique for plant fiber separation. WO/2003/080812 on novel white rot fungus and use thereof in wood pretreatment disclosed a novel white rot fungus, *Physisporinus rivulosus* strainT241i, and the biopulping process associated with paper making, utilizing said fungus. The fungus strongly reduced the amount of wood extractives. The first article about application of the white-rot fungi, *Datronia* sp. and *Oligoporous* sp., for specific purposes was in 2006 when Apiwattanapiwat and coworkers (Apiwattanapiwat et al., 2006) isolated *Datronia* sp. and *Oligoporous* sp. from natural

*Corresponding author. E-mail: aappln@ku.ac.th.

Oligoporous sp. (a) *Oligoporous* sp. (b) *Datronia* sp. (a) *Datronia* sp. (b)

Figure 1. The characteristics of *Hibiscus sabdariffa* L. fibers after being retted by *Oligoporous* sp. and *Datronia* sp. in liquid (a) and solid (b) treatments for 4 weeks.

sources in Thailand. Its ability in decolorization of waste water from pulp and paper mill by decomposing or degrading extractives such as lignin compound was investigated. Li et al. (2008) applied the low cost method, namely bag retting with white-rot fungi, to hemp fiber. The resulting fibers showed an increase in tensile strength. The aim of this study is to evaluate the feasibility of white-rot fungi, *Datronia* sp. and *Oligoporous* sp., in environmentally-friendly, efficient retting processing of *H. sabdariffa* L.

Experimental

Raw material and microbial culture

H. sabdariffa L., which was collected from the Khonkhan province in Thailand, was used in this study. After harvesting, the non-retted green outer barks were separated by hand and dried under sunlight. Before subjecting to the experiments, the barks were sterilized (121 °C, 15 min) to get rid of all the contaminants. The cultures of white-rot fungi, *Dratronia* sp. and *Oligoporous* sp., were kindly supplied by Apiwattanapiwat et al. (2006), to be used in this study. The fungi were subcultured and maintained in Potato Dextrose Agar (PDA) slants and stored at 5 ± 1 °C until use. To prepare the inoculum, the fungi were transferred onto a fresh PDA plate and incubated at 30 °C for 7 days. After that, they were inoculated in Erlenmeyer flasks containing sterilized basal media (g/L: glucose 10.0; K_2HPO_4 1.0; $MgSO_4.7H_2O$ 0.5; KCl 0.5; $FeSO_4.7H_2O$ 0.01; NH_4NO_3 1.75) and incubated at 30 °C for 7 days and kept for further experiments.

Fiber separation by liquid-fungal treatment

Sterilized dried non-retted barks were inoculated with the fungal suspension of either *Dratronia* sp. or *Oligoporous* sp., and incubated at 30 °C in plastic bags to promote the growth for 4 weeks. The amount of fungal inoculum was varied between 5 - 25% on a dry weight basis of barks, different retting period i.e. 1, 2, 3 and 4 weeks under different temperature that is, 30, 40 and 50 °C were studied. Soaking wastewater was sampled during the testing period to determine the activity of pectinolytic enzyme. Barks were subjected to sterilization to get rid of all remaining fungi. The fungal-retted *H. sabdariffa* L. fibers were washed with clean water and dried. Evaluation of the retting was based on the visual separation characteristics of the obtained fibers.

Fiber separation by solid-fungal treatment

White-rot fungal inoculums, as mentioned above, were filtered through Whatman filter paper No.1 to obtain fungal cells. Sterilized dried non-retted barks were inoculated with either *Dratronia* sp. or *Oligoporous* sp. cell pellets and incubated at 30 °C in plastic bags to promote the growth for 4 weeks. The amount of fungal inoculum varied between 5 - 25% on a dry-weight basis of bark. The retting period (1, 2, 3 and 4 weeks) and temperature (30, 40 and 50 °C) were also studied. The appropriate buffer was added into the reaction in order to extract substances produced by the fungi. The retted *H. sabdariffa* L. barks were subjected to sterilization to get rid of all remaining fungi. Activity of pectinolytic enzyme which was released by the fungal strains in the soaking wastewater was quantitatively determined. The resulting *H. sabdariffa* L. fibers were washed with clean water and dried. The evaluation, of the retting degree, based on the visual fiber separation characteristics was determined.

Measurement of pectinolytic enzyme and single fiber tensile testing

Pectinase enzyme was determined as pectate lyase (PAL) activity using the method of Silva *et al.* (1993).The single *H. sabdariffa* L. fibers resulting from each treatment were tensile strength tested in accordance with the ASTM Standard Test Method and Young's Modulus for High-Modulus Single Filament Materials (ASTM D3379-75, 1989) and smell tested according to the instruction of Toyota Motor Asia Pacific Engineering and Manufacturing Company Limited.

RESULTS AND DISCUSSION

The effect of retting parameters on fiber separation by fungal treatment

Figure 1 demonstrated the *Hibiscus sabdariffa* L. fibers obtained from the 4-week incubation with selected fungal strains. It showed that the fibers resulted from the solid treatment were softer and better separated when compared with those from the liquid treatment. This might be due to the solid treatment providing a more preferable growing environment for the fungi (Lopez et al., 2007). It could also be explained that fungal secretes essential for fiber separation were released in sufficient quantity (Yu and Yu, 2007). Figure 2 illustrated the effects of retting time period (1, 2, 3 and 4 weeks) by the selected fungi in the solid treatment on the characteristics of fibers. The results indicated that more fiber separation was obtained when the retting period was longer as also suggested by Yu and Yu (2007). Figure 3 showed the effects of retting

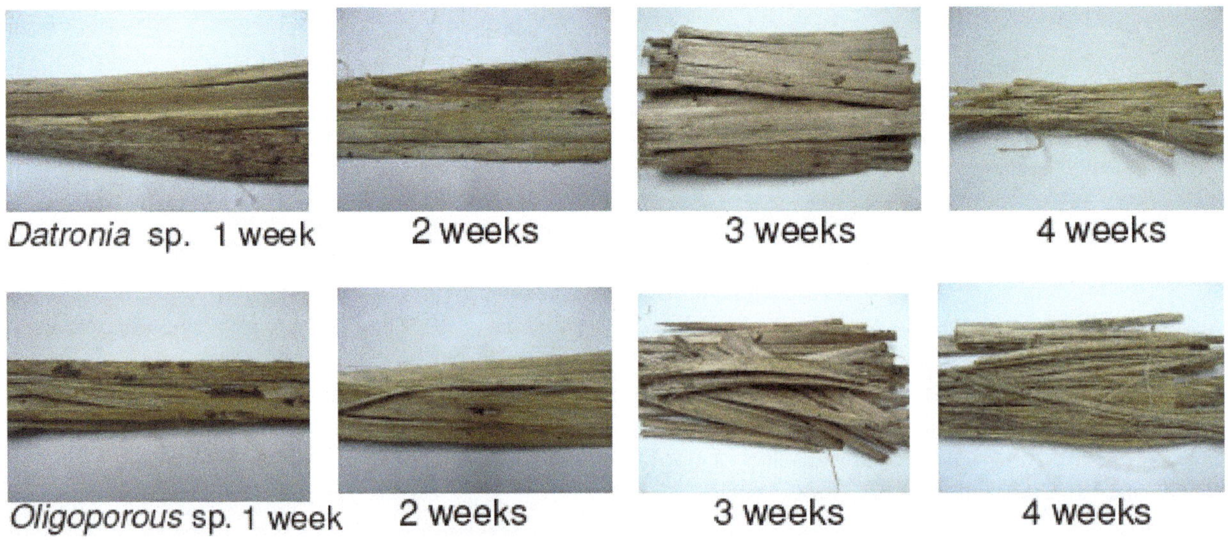

Figure 2. The effects of retting time period (1, 2, 3 and 4 weeks) by *Datronia* sp. and *Oligoporous* sp. at 30°C in the solid treatment on the characteristics of *Hibiscus sabdariffa* L. fibers.

Figure 3. The effects of retting temperature (30, 40 and 50°C) by *Datronia* sp. and *Oligoporous* sp. in the solid treatment for 4 weeks on the characteristics of *Hibiscus sabdariffa* L. fibers.

temperature (30, 40 and 50℃) by the selected fungi in the solid treatment for 4 weeks on the characteristics of fibers. Unsurprisingly, the best fiber separation by both fungi was obtained under the temperature at 30℃. This was also supported by the study of Yu and Yu (2007) in that 30 - 32℃ was an appropriate temperature for the retting of kenaf fibers by an unknown isolated fungus. Temperature was one of the most important factors affecting microbe growth. When temperature was too low, the metabolism speed was low or microbes were dormant (Yu and Yu, 2007). While the temperature raised, microbes grew faster. However, if temperature was too high, the growth of microbes would be limited (Yu and Yu, 2007).

Figure 4 showed the effects of fungal inoculum size (5, 10, 15, 20 and 25%) by the selected fungi at 30℃ in the solid treatment for 4 weeks on the characteristics of fibers. The results demonstrated that better fiber separation was achieved when the inoculum size increased. This also corresponded to the study by Yu and Yu (2007). However, when too many microbes existed in the retting condition, nutrients would be limited (Yu and Yu, 2007). Apart from the visual characteristics of resulting fibers, the smell of the fibers was also considered for retting efficiency. It demonstrated that the best condition to obtain least smell fibers was achieved by *Datronia* sp. under the solid treatment. In conclusion, the most appro-

Figure 4. The effects of fungal inoculum size (5, 10, 15, 20 and 25%) by *Datronia* sp. and *Oligoporous* sp. at 30°C in the solid treatment for 4 weeks on the characteristics of *Hibiscus sabdariffa* L. fibers.

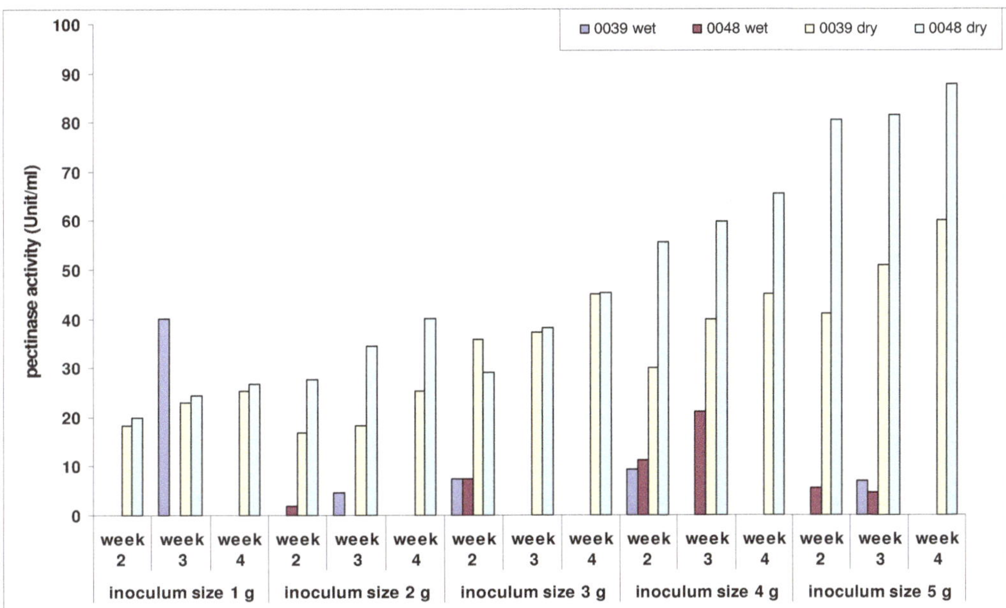

Figure 5. The effects of inoculum size and retting time on released pectinolytic enzymes of *Oligoporous* sp. and *Datronia* sp. after solid and liquid retting for 4 weeks.

priate condition for *H. sabdariffa* L. fiber separation was incubation at 20% *Datronia* sp. at 30°C under the solid treatment for 4 weeks.

Furthermore, the activity of pectinolytic enzymes sampled from the soaking wastewater of each fungal treatment also corresponded to the incubation period and fungal inoculum size (Figure 5). It showed that the activity of pectinolytic enzymes was higher as the fungal inoculum size increased. Likewise, the increase of the incubation period of H. sabdariffa L. barks with selected fungi positively resulted in higher pectinolytic enzymes activity. As a result, the increase in pectinolytic enzyme activity was related to the better efficiency of fiber separation. Thus, the released pectinolytic enzymes from the fungi might play an important role in the separation of *H. sabdariffa* L. fibers. However, the mechanism of these white-rot fungi for fiber separation seemed to be appropriate in only solid treatment which was the natural environment for the fungal growth.

Fiber smell testing

The results from smell test of *H. sabdariffa* L. fibers obtained from variation of the fungal inoculum sizes demon-

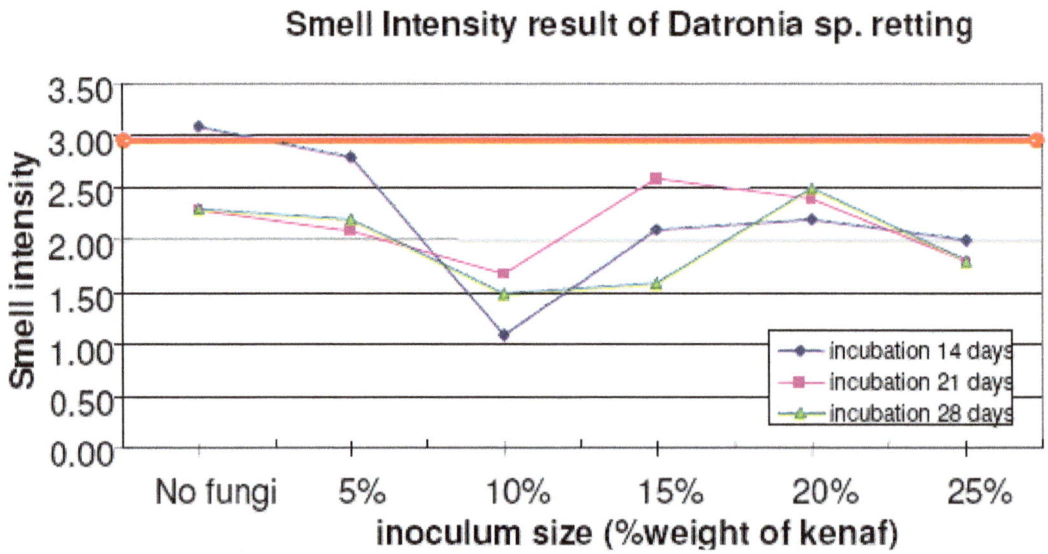

Figure 6. The effect of fungal inoculum size on the smell intensity of retted fibers by *Datronia* sp. at 30°C in solid treatment for 4 weeks.

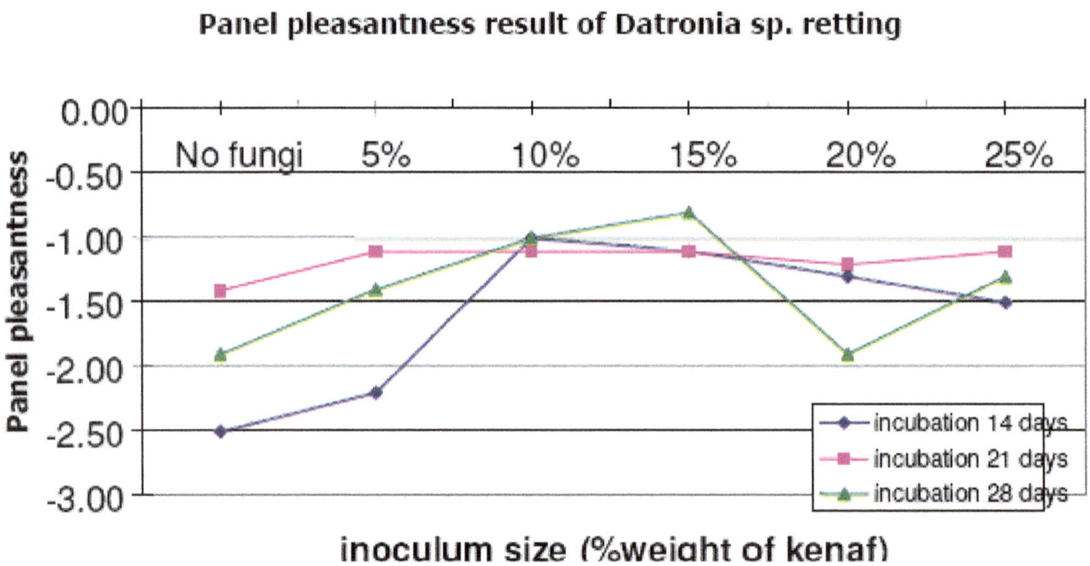

Figure 7. The effect of fungal inoculum size on pleasantness of the panels on retted fibers by *Datronia* sp. at 30°C in solid treatment for 4 weeks.

strated that the solid treatment by *Datronia* sp. produced fibers with most satisfactory of the panels. The comparison between liquid and solid treatments indicated that more pleasantness of the panels was obtained from the solid treatment. The results also illustrated that the fungal inoculum size did not directly affect the smell of retted fibers. Separation of the fibers, at fungal inoculum size 20% gave most fiber separation. The result from solid treatment was summarized and illustrated in Figures 6 - 9. The results showed that the smell intensity of treated fibers decreased as the fungal inoculum size decreased. The pleasantness of treated fiber smell also varied with the fungal inoculum size. Larger inoculum size the more pleasantness of the panels. It could be concluded that the most optimum condition should be at 20% *Datronia* sp. under the solid treatment.

Conclusion

This work indicates the possibility of using white-rot fungi in *H. sabdariffa* L. fiber separation. The fungal treatment of *H. sabdariffa* L. bark with strain *Datronia* sp. and *Oligo-*

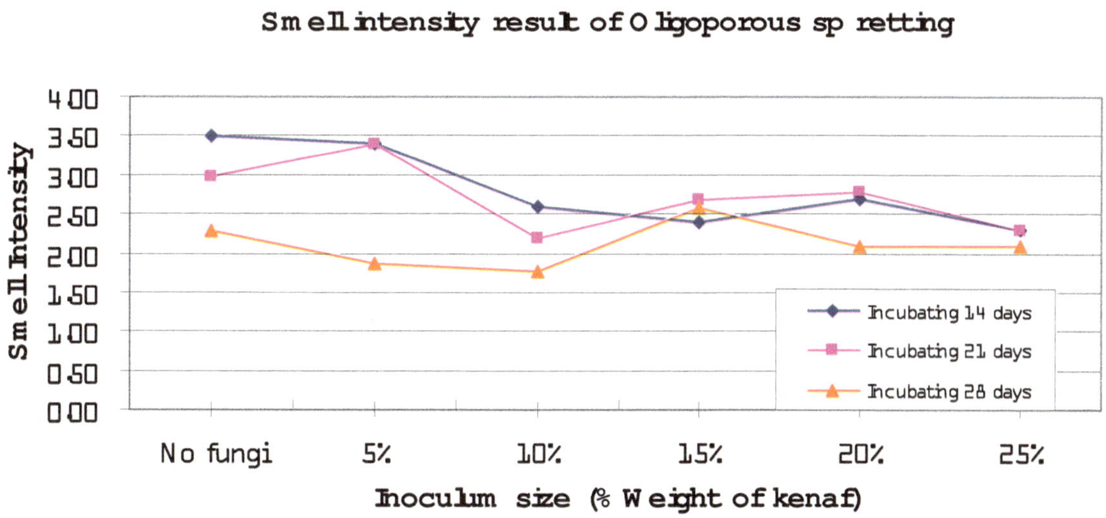

Figure 8. The effect of fungal inoculum size on the smell intensity of retted fibers by *Oligoporous* sp. at 30°C in solid treatment for 4 weeks.

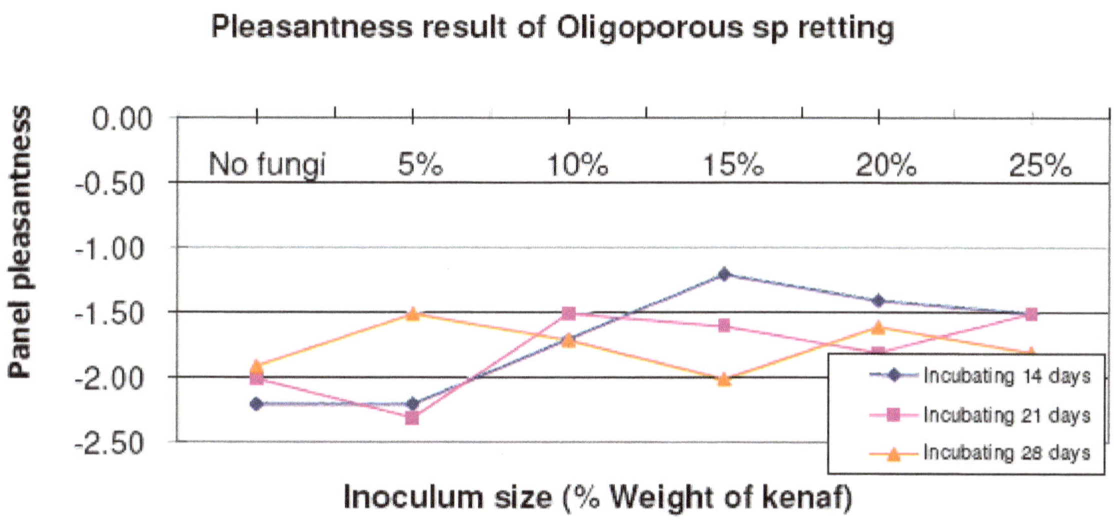

Figure 9. The effect of fungal inoculum size on pleasantness of the panels on retted fibers by *Oligoporous* sp. at 30°C in solid treatment for 4 weeks.

porous sp. caused an increase in the solubility of wood bark extractives such as pectin already after a treatment for 4 weeks as observed in the fiber separation. The increase in the solubility of pectin is likely to indicate that, as a consequence of the fungal treatment, changes occurred in the structure of pectin, promoting the separation of fibers (Akin et al., 2007). These changes are positively correlated with the production of pectinolytic enzyme as shown.

ACKNOWLEDGEMENT

The authors would like to thank Toyota Motor Asia Pacific Engineering & Manufacturing Company Limited for financial support.

REFERENCES

Akin DE, Foulk JA, Dodd RB, McAlisterIII DD (2001). Enzyme retting of flaxand characterization of processed fibers. J. Biotechnol. 89: 193-203.

Akin DE, Condon B, Sohn M, Foulk JA, Dodd RB, Rigsby LL (2007). Optimization of enzyme-retting of flax with pectate lyase. Ind. Crops Prod. 25: 136-146.

Apiwattanapiwat W, Siriacha P, Vaithanomsat P (2006). Screening of fungi for decolorization of wastewater from pulp and paper industry. Kasetsart J. Natr. Sci. 40(5): 215-221.

Evans JD, Akin DE, Foulk JA (2002). Flax-retting by polygalacturonase

containing enzyme mixtures and effects on fiber properties. J. Biotechnol. 97: 223-231.13
Li Y, Pickering KL, Farrell RL (2008). Analysis of green hemp fibre reinforced composites using bag retting and white rot fungal treatments. Ind. Crops Prod. 29: 420-426.
Lopez MJ, Carmen MD, Garcia V, Suarez-Estrella F, Nichols NN, Dien BS, Moreno J (2007). Lignocelulose-degrading enzymes produced by the ascomycete *Coniochaeta ligniaria* and related species: Application for alignocellulosic substrate treatment. J. Enzyme Microbial. Technol. 40: 794-800.
Sharma HSS, Faughey GJ (1999). Comparison of subjective and objective methods to assess flax straw cultivars and fibre quality after dewretting. Ann. Appl. Biol. 135: 495-501.
Silva DO, Attwood MM, Tempest DW (1993). Partial purification and properties of pectin lyase from *Penicillium expansum*. World J. Microbiol. Biotechnol. 9: 574-578.
Yu H, Yu C (2007). Study on microbe retting of kenaf fiber. Enzyme and Microbial Technology 40: 1806-1809.
Zhang J, Henriksson G, Johansson G (2000). Polygalacturonase is the key component in enzymatic retting of flax. J. Biotechnol. 81: 85-89.

Comparison of *in vitro* activity of imipenem productions on bacterial isolates from Hashemi Nezhad Tehran hospitalized patients

Amirmorteza Ebrahimzadeh Namvar[1], Babak Asghari[1], Shoshtari Zadeh[2] and Abdolaziz Rastegar Lari[1,3]*

[1]Antimicrobial Research Center, Faculty of Medicine, Iran University of Medical Science, Iran.
[2]Hashemi Nezhad Hospital of Tehran, Iran University of Medical Science, Iran.
[3]Medical Microbiology Department, Faculty of Medicine, Iran University of Medical Science, Iran.

Carbapenams are one of the ß-lactamase antibiotic resistances to penicillin that were introduced in 1980 to the medical world. Imipenem- a semi-synthetic derivative of thienamycin- is one of the most important broad spectrum ß-lactamase in carbapenems. The irregular usage of drugs, especially in these types of gram-negative bacteria will lead to a serious problem in the treatment of pathogenic bacteria. Comparison of *in vitro* activity of imipenem productions is the aim of this study. 135 strains of various gram positive and gram negative bacteria collected from Hashemi Nejad Hospital in Tehran were studied. Initially strains were identified by phenotypic methods; then Disk Diffusion and MIC methods based on instructions of Forum France Microbiology were used. From 135 samples, urine samples (36%) were the most and trachea (4%) was the lowest. Among gram-negative bacteria isolated, both *E. coli* and *Pseudomonas* (37%) had most common and *Enterobacter* (1%) was the lowest. By disk diffusion method, Supranem, Taynam and Mast pharmaceutical products, showed similar result, but the results related to internal company disks were different. 20% of the total bacteria studied had MIC over 8 mg (resistant). Due to results of this study using standard dicks and antibiotic powder (Imipenem/Cilastatin sodium) (Supranem) or (Tienam) is recommended.

Key words: ß-lactamase, antibiogram, imipenem.

INTRODUCTION

Carbapenems were introduced in 1980 and now are commonly applied as the last choice in treating crucial infections caused by multidrug-resistant strains of gram negative bacilli. These antibiotics are stable to ß-lactamase generated by gram-negative bacilli. The carbapenems are a class of ß-lactamase antibiotics different from the penicillins by the replacement of a carbon to sulfur atom and by the addition of a double bond to the five member rings of the penicillin nucleus. Carbapenems block the elongation and cross linking function of the peptidoglycan cell wall by binding to bacterial peptidases (Pastel, 1986; Bonfiglio et al., 2002; Barza, 1985; Barry et al., 1985). Imipenem were one of the most important broad spectrum ß-lactamase in Carbapenems. Imipenem, also referred to as N-formimidoyl thienamycin monohydrate, is a semi-synthetic derivative of thienamycin, the parent compound produced by the filamentous bacterium *Streptomyces cattleya*. Cilastatin is a competitive, reversible and specialized inhibitor of dehydropeptidase-I enzyme, the renal enzyme which metabolizes and inactivates imipenem (Bonfiglio et al., 2002; Barza, 1985; Barry et al., 1985). It has no intrinsic antibacterial activity and does not affect the antibacterial activity of imipenem. Combination of imipenem/cilastatin against an infrequently wide spectrum of pathogens makes it mainly

*Corresponding author. E-mail: lari@iums.ac.ir.

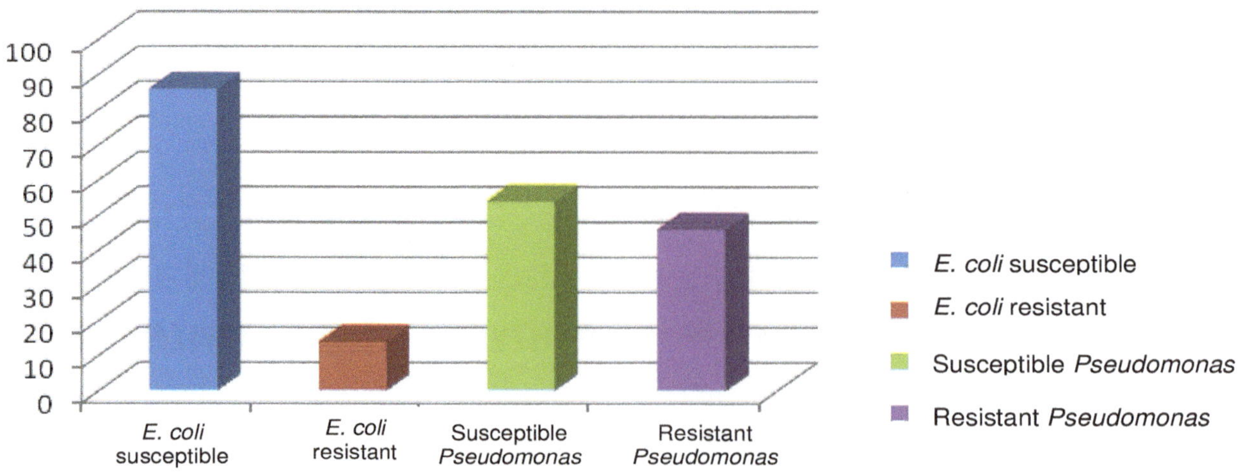

Figure 1. *E-coli* and *Pseudomonas* antimicrobial susceptibility test by supranem powder.

beneficial in the treatment of polymicrobic mixed aerobic/anaerobic infections (Queenan and Bush, 2007; Zhanel et al., 2007; Oteo et al., 2008; Kohler et al., 1999). Resistance to carbapenems emerged from 1990 and has been reported in gram negative bacilli. Multi-drug resistance (including carbapenem) in gram-negative bacteria established a crucial problem as a result of the lack of therapeutic choices and the capability transfer of antibiotic resistance to other virulent pathogens. Offered carbapenems in Islamic Republic of Iran are meropenem and imipenem. Combination of imipenem/cilastatin is used in the treatment of the following infections due to susceptible organisms: Intra-abdominal infections, lower respiratory tract infections, gynecological infections, septicemia, genitourinary tract infections, bone and joint infections, skin and soft tissue infections and endocarditis. Also, the mentioned compound is used in the treatment of mixed infections caused by susceptible strains of aerobic and anaerobic bacteria (Queenan and Bush, 2007; Zhanel et al., 2007; Oteo et al., 2008; Kohler et al., 1999; Jones et al., 2005).

MATERIALS AND METHODS

135 clinical isolates of various strains of gram positive and gram negative bacteria (*E. coli, Pseudomonas, Streptococci, Staphylococci, Entrococci, Proteus, Klebsiella pneumonia, Enterobacter*) from hospitalized patients in Hashemi Nezhad Hospital of Tehran from April 2008 to November 2009 were studied. The identification of the bacteria was determined by using routine bacteriology methods. Therefore antimicrobial susceptibility patterns of clinical isolates were determined by standard NCCLS methods.

Antibiogram method

The antimicrobial susceptibility test was performed by disk diffusion method as recommended by France Microbiology Institute protocol, 2007. The antibiotics used were trimethoprim sulfamethoxazole (SXT), ciprofloxacin (CIP), amikacin (AN), ofloxacin (OFX), nalidixic acid (NA), cefotaxime (CTX), vancomycin (VA), amoxicillin (AMX), nitrofurantoin (FD), ampicillin (AM), ceftizoxime (ZOX) , ceftriaxon (CRO) provided from Pad tan Teb Co, Iran. Then all strains were examined for *in vitro* imipenem susceptibility test by using the following disks which were made from supranem, tineam and Pad tan Teb Co powders; each of these disks were compared with the others.

Determination of MIC

The MICs for all isolates were determined by Agar dilution method according to standard NCCLS by using supranem powder from Exir Co and tienam powder from Merck Co (van Ogtrop et al., 1998; Castanheira et al., 2009; NCCLS, 2003, 2001; Clinical and Standards Institute (CLSI), 2007).

RESULTS

In total, during the study period 135 gram positive and gram-negative bacterial infections were isolated from 242 patients. Gram positive and gram-negative bacteria were most frequently recovered from urine (36%), followed by surgical-site cultures (32%), blood (28%) and (4%) trachea. The most percentage of clinical isolates, which includes *E.coli* and *Pseudomonas* were predominant among clinical strain, while *Enterobacter* has minimum percentage. Among gram-negative bacteria isolated, both *E. coli* and *Pseudomonas* had most common (37%) and *Enterobacter* was the lowest (1%) by disk diffusion method. Supranem, Taynam and Mast pharmaceutical products, showed the same result (Figures 1 and 2); however, the results related to internal company disks

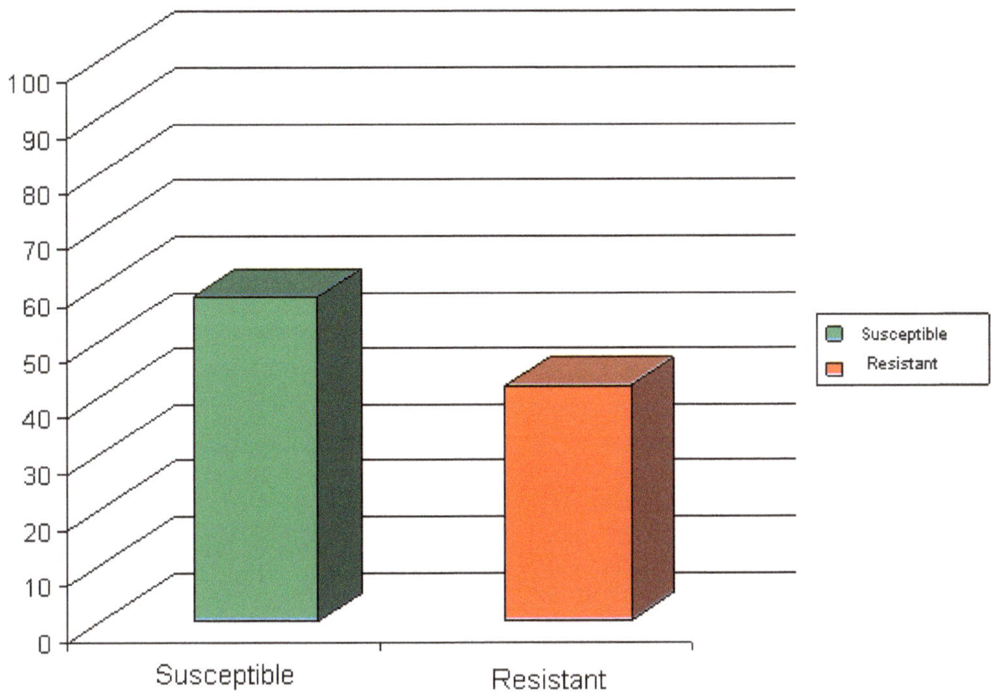

Figure 2. *Pseudomonas* antimicrobial suscesptibility test by tienam powder.

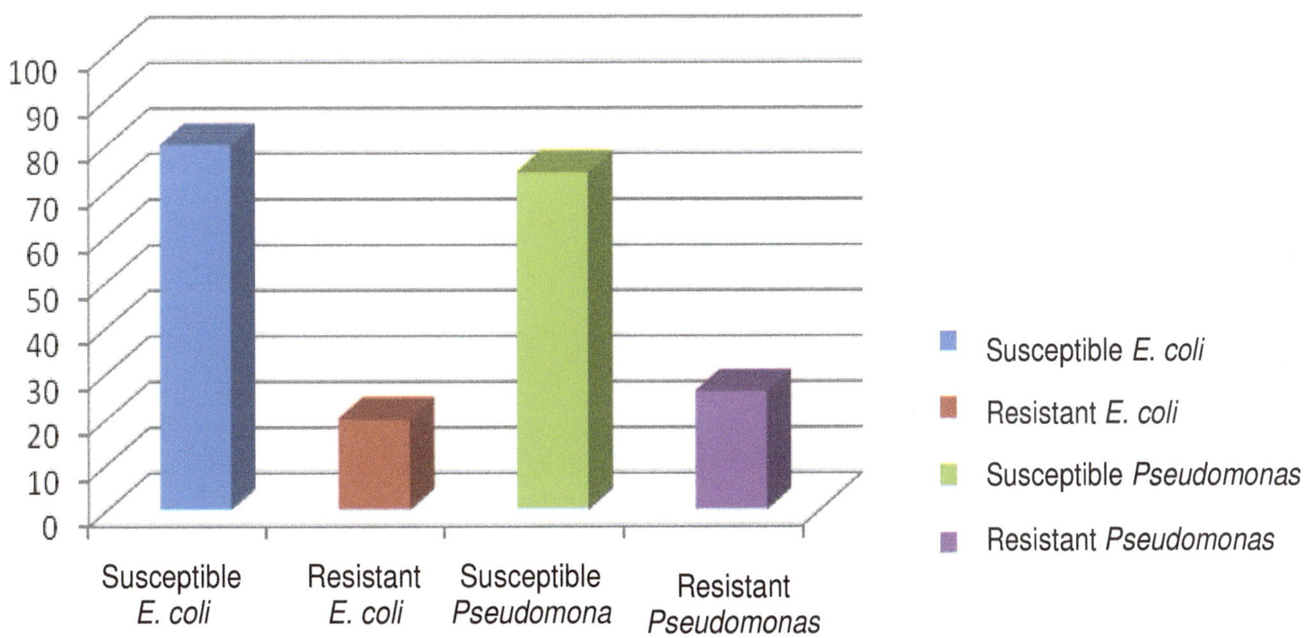

Figure 3. MIC results of *E. coli* and of *Pseudomonas* by using supranem powder (S<8 mg/l) (R>8 mg/l).

were different. 20% of the total bacteria studied had MIC over 8 mg (resistant) (Figure 3).

The antimicrobial susceptibility test result from disk diffusion.

DISCUSSION

Resistance to antibiotic drug therapy is an increasing public health problem in all populations. In recent years, through the abuse and misuse of antibiotics, many bacteria have developed resistance to the variety of antibiotics. This pattern of resistance can be different in various populations and therefore, each of them needs to be specially programmed for reduction of resistance to antibiotics especially those most commonly used for treatment. Most studies show that imipenem is the most effective agents against gram-negative bacteria especially enterobacteriaceae and pseudomonas (Kohler et al., 1999). Comparison of the results of our study and other similar studies in different regions in Iran and even other countries shows that the susceptibility of some bacteria such as *E. coli*, *Salmonella,* and *Proteus* to imipenem is satisfactory. The results of our study demonstrated that more than 80% of *E.coli* was sensitive to antimicrobial susceptibility test by disk diffusion made from supranem and tienam. The result of antimicrobial susceptibility test for gram-positive bacteria for both supranem and tienam powder were the same. Therefore antimicrobial susceptibility for all isolates was performed by Mast Co disk provided, and similar results were obtained. Our studies show supranem and tienam have the same effective spectrums; however, both native disks showed completely different results. On the other hand, determination of imipenem antimicrobial activity with both powder by MIC method shows that only 20% of strains have MIC>8. As a matter of fact supranem and tienam had a similar effect on studied strains, hence avoiding the use of non standard disks is recommended. These antibiotics are frequently used to treat infections caused by multi drugs-resistant strains of gram negative bacteria especially *P. aeruginosa* and *E. coli.*

In conclusion, it seems that the infection control measures to limit the emergence of imipenem resistance are important issues in all population especially in developing countries. Because of the major health problems caused by antibiotic resistance in the last few years, attempts have been made to organize a national surveillance program in these countries.

REFERENCES

Pastel DA (1986). Imipenem-cilastatin sodium, a broad-spectrum carbapenem antibiotic combination. Clin. Pharm., 5(9):719-36.

Bonfiglio G, Russo G, Nicoletti G (2002). Recent developments in carbapenems. Expert Opin. Investig. Drugs Rev., 11(4):529-44.

Barza M (1985). Imipenem: First of a new class of beta-lactam antibiotics. Ann. Int. Med., 103(4): 552-60.

Barry AL, Jones RN, Thornsberry C, Ayers LW, Kundargi R (1985). Imipenem (N-formimidoyl thienamycin): *In vitro* antimicrobial activity and beta-lactamase stability. Diagn. Microbiol. Infect. Dis., Mar, 3(2):93-104.

Queenan AM, Bush KC (2007). The versatile β-lactamases, Clin. Microbiol. Rev., 20: 440-458.

Zhanel GG, Wiebe R, Dilay L, Thomson K, Rubinstein E, Hoban DJ, Noreddin AM, Karlowsky JA (2007). Comparative review of the carbapenems. Drugs, 67(7): 1027-1052.

Oteo J, Delgado-Iribarren A, Vega D, Bautista V, Rodríguez MC, Velasco M, Saavedra JM, Pérez-Vázquez M, García-Cobos S, Martínez-Martínez L, Campos J (2008). Emergence of imipenem resistance in_clinical *Escherichia coli* during therapy. Int. J. Antimicrob. Agents, 32(6):534-7. Epub 2008.

Kohler J, Dorso KL, Young K, Hammond GG, Rosen H, Kropp H, Silver LL (1999). *In vitro* activities of the potent, broad-spectrum carbapenem MK-0826 (L-749,345) against broad-spectrum β-lactamase- and extended-spectrumβ-lactamase-producin *Klebsiella pneumoniae* and *Escherichia coli* clinical isolates. Antimicrob. Agents Chemother., 43: 1170–1176.

Jones RN, Sader HS, Fritsche TR (2005). Comparative activity of doripenem and three other carbapenems tested against Gram-negative bacilli with various β-lactamase resistance mechanisms, Diagn. Microbiol. Infect. Dis., 52: 71–74.

Van Ogtrop ML, Andes D, Craig WA (1998). *In vivo* antimicrobial activity of MK-0826, a new carbapenem, against various penicillin-resistant pneumococci. In: Program and abstracts of the 38th Interscience Conference on Antimicrobial Agents and Chemotherapy. Am. Soc. Microbiol., Washington D.C. Abstr., F 48.

Castanheira M, Jones RN, Livermore DM (2009). Antimicrobial activities of doripenem and other carbapenems against *Pseudomonas aeruginosa*, other non fermentative bacilli, and *Aeromonas* spp. Diagn. Microbiol. Infect. Dis., Apr., 63(4): 426-33. Epub.

NCCLS (2003). Performance standard for antimicrobial susceptibility testing: Disk diffusion. In: Document M100-S13 (M2), NCCLS, Wayne, PA.

NCCLS (2001). Methods for dilution antimicrobial susceptibility tests for bacteria that grow aerobically. In: (5th ed.), NCCLS document M7-A5, NCCLS, Wayne, PA.

Clinical and Standards Institute (CLSI) (2007). Methods for antimicrobial susceptibility testing of anaerobic bacteria approved standard (Seventh edition, M11-A7.), CLSI, Wayne, PA.

6

Utility of IS*1245*-IS*1311* based PCR typing system for *Mycobacterium avium* isolates obtained from clinical and environmental sources

G. P. S. Jadaun[1], Prashant Upadhyay[1], Zafar Ahmed[1], Ram Das[1], Deepti Parashar[2], D. S. Chauhan[1], V. D. Sharma[1] and V. M. Katoch[3]*

[1]Department of Microbiology and Molecular Biology, National JALMA Institute for Leprosy and Other Mycobacterial Diseases(Indian Council of Medical Research), P. O. Box 1101, Dr. M. Miyazaki Marg, Tajganj, Agra 282001, India.
[2]National Institute of Virology (ICMR) Pune, India.
[3]Department of Health Research, Ministry of Health and Family Welfare, Government of India and Director General, Indian Council of Medical Research Post Box No. 4911 Ansari Nagar, New Delhi-110029.

In the present study, twenty four *Mycobacterium avium* isolates obtained from clinical (sputum) and environmental (water and soil) samples were analyzed by a PCR typing method. Gel electrophoresis patterns showed that a single PCR reaction generated 21 profiles. Based on the relatedness of the PCR band patterns six isolates could be grouped in three clusters. Results have the implication that the PCR based system could be used as a simple and rapid method for molecular typing of *M. avium* isolates with high discriminatory index.

Key words: IS*1245*, IS*1311*, *Mycobacterium avium*, PCR typing.

INTRODUCTION

Non-tuberculous mycobacteria (NTM) are ubiquitous environmental organisms. The important species of NTM associated with human disease are: *Mycobacterium avium*, *M. intracellulare*, *M. kansasii*, *M. marinum*, *M. ulcerans*, *M. scrofulaceum*, *M. smegmatis* etc. Water and soil have been reported as common reservoirs of NTM. These mycobacteria are naturally resistant to most of the anti-tuberculosis drugs but may be sensitive to some aminoglycosides, fluoroquinolones or new macrolides. Among the NTM, members of *M. avium* complex (which included two species *M. avium* and *M. intracellulare*) are present in all natural sources and can be isolated from water, soil, air, plants, house dust, food samples, and even from animals such as chickens and pigs (Inderlied et al., 1993). These organisms, particularly *M. avium* isolates, are the most important among the NTM that have assumed clinical prominence in the post-AIDS era (Katoch, 2004).Till date, no human-to-human transmission of *M. avium* has been established, and environmental sources of *M. avium* infections have been described (Inderlied et al., 1993; Katoch, 2004).

Isolation and identification of *M. avium* relies on methods used to detect bacilli as well as blood cultures. Strain typing methods are often employed for epidemiological investigations of bacterial infections. Various methods described for typing of *M. avium* strains include serotyping (Thoen et al., 1975), phage typing (Crawford and Bates, 1985), multilocus enzyme electrophoresis (Yarkus et al., 1992), pulsed-field gel electrophoresis (Pestel-Caron et al., 1999), restriction fragment length polymorphism and hybridization with specific probes (Guerrero et al., 1995), random amplified polymorphic DNA (RAPD) (Matsiota-Bernard et al., 1997) and PCR based typing (Picardeau and Vincent, 1996). Among the described methods for *M. avium* typing, pulsed-field gel electrophoresis is considered as the 'Gold Standard' (Yoder et al., 1999); however, the procedure is technically cumbersome and hence would not be useful for a laboratory with limited resources.

A PCR based typing method has been described by Picardeau and Vincent (1996) that targets two insertion elements present in the *M. avium* genome, namely IS*1245* and IS*1311*. This system is based on amplifica-

*Corresponding author. E-mail : vishwamohan_katoch@ yahoo mail.com.

tion of the DNA between closely spaced copies of these repeat elements. Thus, agarose gel electrophoresis of the PCR products gives a band pattern corresponding to the DNA fingerprint of the isolate. This method was proposed as a rapid technique for investigating small numbers of *M. avium* isolates and the results were comparable to PFGE (Pestel-Caron et al., 1999). In the present study we have investigated the utility of this PCR based system for molecular typing of *M. avium* isolates obtained from clinical and environmental sources.

MATERIALS AND METHODS

M. avium isolates

A total of 24 *M. avium* isolates and one reference *M. avium* strain (TMC-724) were taken from the collection of Mycobacterial Repository Centre of our Institute. Among the *M. avium* isolates included in this study eight were obtained from sputum samples (HIV status not known) collected in Delhi (n = 4), Chennai (n = 2), Jaipur (n = 1), and Lucknow (n = 1). Sixteen isolates were obtained from environmental sources (Parashar et al. (2004)). Environmental isolates included those obtained from water (n = 12) and soil (n = 4) samples collected in Agra. Isolates were maintained on LJ medium and freshly subcultured before being used for further investigations.

DNA isolation

Colonies were scraped from the LJ slant and suspended in 400 µl TE (10 mM Tris, 1 mM EDTA pH 8.0). Cells were killed by boiling in a water bath at 95°C for 20 min. DNA was purified as per a previously described procedure (van Embden et al., 1993). DNA was dissolved in TE and kept at -20°C until use. For parallel sets of PCR reactions, DNA was prepared from the isolates exactly as described by Picardeau and Vincent (1996). Briefly, colonies were taken in 50 µl of TE containing 1% Triton X-100 and boiled in a water bath for 30 min. Cells were pelleted and the supernatant was used as such in PCR without further purification.

IS*1245*-IS*1311* based PCR typing

Molecular typing based on PCR targeting IS*1245* and IS*1311* was done as described earlier (Picardeau and Vincent, 1996). Briefly, amplification reactions were performed with 50 µl volume containing 1X Taq polymerase buffer, 1.5 mM $MgCl_2$, 200 µM (each) dNTPs, 1 µM (each) primers (PA, 5′-CAGAGCCTCAGGCGA-3′, and PB, 5′-CAGAGCCTCACGCGGA-3′) and 2U of *Taq* polymerase. Two microlitre of template DNA was added to the reaction mix. The 35 PCR amplification cycles carried out in a GeneAmp® 9700 thermal cycler (Applied Biosystems, Foster City, CA, USA) were denaturation at 95°C for 1 min, annealing at 55°C for 1 min, extension at 72°C for 2 min with a final extension at 72°C for 10 min. Reactions were kept at 4°C until use PCR products were electrophoresed on 1.5% agarose gel (Bangalore Genei), stained with ethidium bromide and scanned by Gel Documentation System (Bio-Rad Laboratories, USA). The gel electrophoresis band patterns were analyzed and band sizes were calculated using the accompanying analysis software Quantity One. DNA of reference *M. avium* strain (TMC-724) was used with each set of PCR reactions and its band patterns were used as control to compare different agarose gels.

RESULTS AND DISCUSSION

In the present study, band patterns of *M. avium* isolates observed after PCR typing were polymorphic (Figure 1). Band patterns of *M. avium* isolates were analyzed based on the number of PCR bands and their size range (Table 1). This analysis shows that there were a minimum of two bands to a maximum of nine bands. The range of numbers of DNA bands in clinical and environmental isolates was 3 - 9 and 2 - 8, respectively. In contrast to another study from India, reporting zero numbers of bands in eight clinical *M. avium* isolates (out of 65 isolates analyzed) (Kumar et al., 2006), none of the isolates in our study had zero bands. Molecular sizes of the DNA bands were in the range of 100 bp to 2174 bp indicating that results of PCR typing system can easily be analyzed after agarose gel electrophoresis. Both intense and faint bands were observed after agarose gel electrophoresis, although band intensity was not considered as a discriminatory factor (Picardeau and Vincent, 1996). Among twenty four isolates analyzed in this study, a total of twenty one band patterns were observed after gel electrophoresis. Eighteen isolates had unique band patterns. Based on the relatedness of band patterns, six *M. avium* isolates obtained from water were grouped in three clusters (two isolates in each cluster) (Table 2). Isolates of clusters A and B had identical PCR band patterns specific to their clusters. Two isolates of cluster C showed four shared bands while three bands were unique to either one or another isolate suggesting that these isolates were closely related rather than identical. On the basis of these results a discriminatory index of the PCR based typing system was calculated to be 0.989 (Hunter and Gaston, 1988). The results of the present study are based on the isolates at our Mycobacterial Repository Centre and do not represent the true picture of the isolates from this region; hence, the results need to be interpreted cautiously from epidemiological point of view.

Earlier study carried out by Yoder et al. (1999) had established links between *M. avium* isolates obtained from patient and food samples using the same PCR based typing method. In our study we were not able to obtain patient isolates from the Agra region, and therefore the same comparison was not possible. Still the study is significant as it reconfirms the utility of PCR typing method for DNA fingerprinting of *M. avium* isolates obtained from both patient and environmental isolates in different settings (Yoder et al., 1999). Moreover, to the best of our knowledge, this is the first study that describes the application of this PCR typing system for *M. avium* isolates obtained from soil and water samples.

RAPD analysis of *M. avium* is another PCR based system that has been used by investigators for comparing strains (Matsiota-Bernard et al., 1997). However, major limitation of RAPD is its poor reproducibility (Matsiota-Bernard et al., 1997; Penner et al., 1993). Previously Pestel-Caron et al. (1999) have reported that the PCR

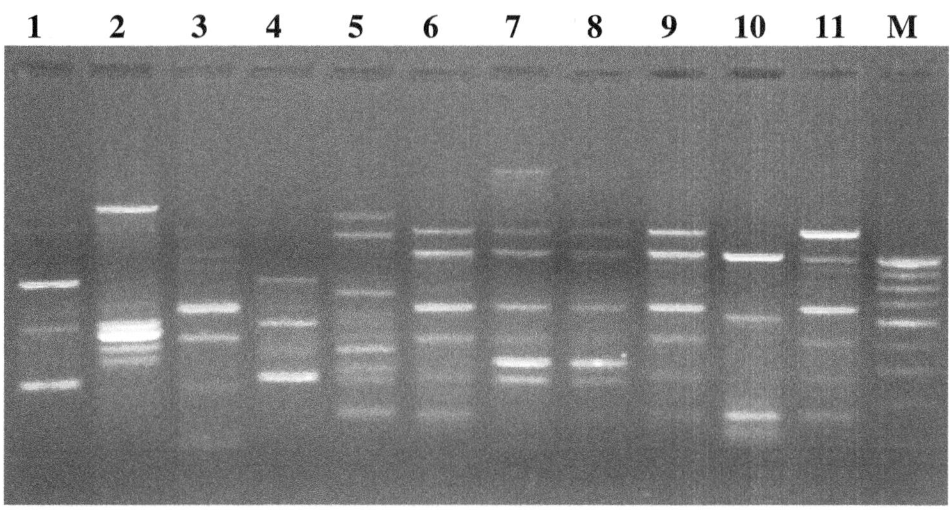

Figure 1. PCR typing patterns of *M. avium* isolates after agarose gel electrophoresis and ethidium bromide staining. Lane 1, TMC-724; lanes: 2 to 11, isolates of *M. avium*; M, 100 bp DNA ladder

Table 1. Analysis of PCR typing results.

M. avium source	No. of PCR bands	Molecular sizes (bp) of PCR bands
Sputum	3 - 9	117 - 1866
Water	3 - 7	100 - 2174
Soil	2 - 8	181 - 1278

Table 2. Environmental *M. avium* isolates showing relatedness based on PCR band patterns.

Cluster	Isolate code	Source	Molecular sizes (bp) of PCR bands						
A	ICC-1330	Water	1324	1091	786	688	525	384	272
	ICC-1622	Water	1301	1091	786	676	525	384	272
B	ICC-1339	Water	2174	1309	1099	682	513	433	368
	ICC-1345	Water	2136	1309	1080	671	513	433	368
C	ICC-1309	Water	1078	741		572	354	282	204
	ICC-1332	Water	1056	761	648		351		212

ICC Indian culture collection.

typing system can generate DNA fingerprints of *M. avium* isolates with 90% reproducibility. In view of these observations it has been thought that PCR typing system may serve as an alternative DNA fingerprinting method for the characterization of *M. avium* strains. Moreover, we performed PCR in parallel reactions using DNA samples from the same *M. avium* isolates isolated either according to the procedure of Picardeau and Vincent (1996) or the procedure of van Embden et al. (1993). The results of these parallel PCR reactions showed similar band pattern after agarose gel electrophoresis. Hence, we propose that tedious DNA purification steps need not be performed before processing for the PCR typing method, which further simplifies the whole procedure of this typing system.

Based on these results we conclude that PCR typing based on IS*1245* and IS*1311* offers a rapid and technically less demanding approach for characterization of *M. avium* strains obtained from both clinical as well as environmental sources. This approach of strain characterization may particularly be useful for resource-poor settings in developing countries that might have majority of infections due to *M. avium*. Further validation of this method requires confirmation by testing of more clinical strains by different laboratories.

ACKNOWLEDGEMENTS

We acknowledge the financial support provided by Department of Biotechnology, Govt. of India. LEPRA, UK is gratefully acknowledged for providing reagents for the experimental work in our laboratory. GPSJ was supported by a Senior Research Fellowship grant from the Council of Scientific and Industrial Research, New Delhi, India.

REFERENCES

Crawford JT, Bates JH (1985). Phage typing of the *Mycobacterium avium-intracellulare-scrofulaceum* complex: a study of strains of diverse geographic and host origin. Am. Rev. Respir. Dis. 132: 386-389.

Guerrero C, Bernasconi C, Burki D, Bodmer T, Telenti A (1995). A novel insertion element from *Mycobacterium avium*, IS*1245*, is a specific target for analysis of strain relatedness. J. Clin. Microbiol. 33: 304-307.

Hunter PR, Gaston MA (1988). Numerical index of the discriminatory ability of typing systems: an application of Simpson's index of diversity. J. Clin. Microbiol. 26: 2465-2466.

Inderlied CB, Kemper CA, Bermudez LEM (1993). The *Mycobacterium avium* complex. Clin. Microbiol. Rev. 6: 266-310.

Katoch VM (2004). Infections due to non-tuberculous mycobacteria (NTM). Indian J. Med. Res. 120: 290-304.

Kumar S, Bose M, Isa M (2006). Genotype analysis of human *Mycobacterium avium* isolates from India. Indian J. Med. Res. 123: 139-144.

Matsiota-Bernard P, Waser S, Tassios PT, Kyriakopoulos A, Legakis NJ (1997). Rapid discrimination of *Mycobacterium avium* strains from AIDS patients by random amplified polymorphic DNA analysis. J. Clin. Microbiol. 35: 1585-1588.

Parashar D, Chauhan DS, Sharma VD, Chauhan A, Chauhan SVS, Katoch VM (2004). Optimization of procedures for isolation of mycobacteria from soil and water samples obtained in northern India. Appl. Environ. Microbiol. 70: 3751-3753.

Penner GA, Bush A, Wise R, Kim W, Domier L, Kasha K, Laroche A, Scoles G, Molnar SJ, Fedak G (1993). Reproducibility of random amplified polymorphic DNA (RAPD) analysis among laboratories. PCR Methods Appl. 2: 341-345.

Pestel-Caron M, Graff G, Berthelot G, Pons J, Lemeland J (1999). Molecular analysis of *Mycobacterium avium* isolates by using pulsed-field gel elctrophoresis and PCR. J. Clin. Microbiol. 37: 2450-2455.

Picardeau M, Vincent V (1996). Typing of *Mycobacterium avium* isolates by PCR. J. Clin. Microbiol. 34: 389-392.

Thoen CO, Jahnagin JL, Champion ML (1975). Micromethod for serotyping strains of *Mycobacterium avium*. J. Clin. Microbiol. 1: 469-471.

van Embden JDA, Cave MD, Crawford JT, Dale JW, Eisenach KD, Gicquel B, Hermans P, Martin C, McAdam R, Shinnick TM, Small PM (1993). Strain identification of *Mycobacterium tuberculosis* by DNA fingerprinting: recommendations for a standardized methodology. J. Clin. Microbiol. 31: 406-409.

Yarkus MA, Reeves MW, Hunter SB (1992). Characterization of isolates of *Mycobacterium avium* serotypes 4 and 8 from patients with AIDS by multilocus enzyme electrophoresis. J. Clin. Microbiol. 30: 1474-1478.

Yoder S, Argueta C, Holtzman A, Aronson T, Berlin OGW, Tomasek P, Glover N, Froman S, Stelma G Jr. (1999). PCR comparison of *Mycobacterium avium* isolates obtained from patients and foods. Appl. Environ. Microbiol. 65: 2650-2653.

7

Isolation of *Enterobacteriaceae* and *Pseudomonas spp.* from raw fish sold in fish market in Khartoum state

Sanaa O. Yagoub

Department of Microbiology and Molecular Biology, Faculty of Science and Technology, El Neelain University, P. O. Box 12702, Khartoum, Sudan. E-mail: soyagoub08@ hotmail.com.

***Enterobacteriaceae* were isolated from gills, skin, muscles and the intestine of 83 out of 150 (55%) randomly collected fishes, the most dominants isolates were *E. coli*, *Citrobacter spp*, *Enteriobacter spp* and *Klebsiella spp*. This together with the highly pathogenic *Enterobacteriaceae* including *Salmonella spp* and *Shigella spp*. *Proteus spp*, and *Alklegens* spp. Potential pathogenic organisms were also among the isolates. On the other hand *Pseudomonas spp* was isolated from 62% of randomly collected fishes. The number and percentages of the isolated bacteria were compared according to seasons. The total bacterial count, coliform count and *E. coli* count were estimated from all parts of collected samples. The negative impacts of the presence of *Enterobacteriaceae* and *Pseudomonas spp* in fishes were discussed based on their potential pathogenic effect toward public health and their role to enhance rapid spoilage of fishes.**

Key words: *Enterobacteriaceae*, fisheries, public health, *Pseudomonas*.

INTRODUCTION

Seafood derived from wild fish as well as farmed fish has always been an important source of protein in the human diet. On a global scale, fish and fish products are the most important source of protein and it is estimated that more than 30% of fish for human consumption comes from aquaculture (Håstein et al., 2006). According to the Center for Food Safety and Applied Nutrition in Washington (2001), most fish related food borne illness are traced to *Salmonella*, *Staphylococcus spp.*, *Escherichia spp.*, *Vibrio parahemolyticus, Clostridium perfringens, Clostridium botulinum E,* and *Enteroviruses.* Thampuran et al. (2005) reported that the microbial quality of the tilapia indicated that all tissue samples except muscle tissues were contaminated with fecal coliform were *Escherichia coli* is the most common contaminant and is often encountered in high numbers. The presence of *E. coli* as well as verotoxigenic *E. coli* O157:H7 in fish meal was investigated by (Ayulo et al., 1994; Hwang et al., 2004; Thampuran et al., 2005; Ristori et al., 2007). Ristori et al. (2007) isolated *Aeromonas spp., Plesiomonas shigelloides, Vibrio cholerae* 01, *Vibrio parahaemolyticus, and Vibrio vulnificus* from different organs of fishes. It was found that the hygienic quality and freshness of fish and shellfish decreased in summer, especially for clam and mussel (Hwang et al., 2004).

Bacterial microbiota associated with fresh raw shrimp was *Aeromonas, Pseudomonas, Vibrio, Flavobacterium* and *Serratia* (Jeyasekaran et al., 2006).

Arannilewa et al. (2006) found that the total coliform count range in fish was between 3.0×10^3 - 7.5×10^6 with increasing values, as the duration of storage increases. Hood et al. (1983) found that fecal coliform levels were above the recommended wholesale level suggested by the National Shellfish Sanitation Program (less than or equal to 230/100 g). El Hadi et al. (2004) detected the presence of eight potentially pathogenic *Vibrio* species, with overall incidence in the samples as 4.6% for *V. cholerae*, 4.7% for *V. parahaemolyticus*, 6.0% for *V. vulnificus*, 11% for *Vibrio alginolyticus*, 9.9% for *Vibrio metschnikovii*, 1.3% for *Vibrio mimicus*, 13% for *Vibrio damsela*, 7.6% for *Vibrio fluvialis*, and 52% for a combined population of all of the above.

Heinitz et al. (2000) found that 10% of imported and 2.8% of domestic raw seafood was positive for *Salmonella. Enterococcus sp* and *Aeromonas sp*, fecal and total coliform, the presence of *Listeria sp* and *Salmonella spp* from the external surface of tilapias were shown by Morales et al. (2004).

Håstein et al. (2006) outlined and discussed the hazards and challenges associated with handling fish during

Table 1. Number and percentage of isolation of *Enterobacteriacaea and Pseudomonas spp* from collected 150 samples of raw fish.

Isolated organism	Positive growth		Negative growth	
	Number	Percentage	Number	Percentage
Enterobacteriacaea	83	55.3%	67	44.7%
Pseudomonas	93	62%	57	38%

Table 2. Number and percentage of *Enterobacteriacaea and Pseudomonas spp* isolated from samples collected at different seasons.

Type of organism	Winter	Summer	Autumn
Enterobacteriacaea	20 (40%)	30 (60%)	33 (66%)
Pseudomonas	32 (64%)	21 (42%)	40 (80%)
Mixed (*Enterobacteriacaea* and *Pseudomonas*)	19 (38%)	14 (28%)	17 (34%)

farming and capture and the environmental contaminants in seafood that may pose a risk to human health.

MATERIALS AND METHODS

Area of the study

This study was aimed to evaluate the hygienic quality and freshness of fresh fish (*Tilapia nilotica* Linn) and to investigate the occurrence of *Enterobacteriaceae* and *Pseudomonas spp* as indicator for fish quality. Fishes were collected randomly at 2007 from Khartoum State fish markets during the winter (January - February) and summer (March and April) and autumn (August-September) seasons. Skin surface, gills, intestinal tract and raw fish flesh were examined for each raw fish. One hundred and fifty raw fish samples were collected as 50 fish samples at each season.

Sampling

Skin: Sample from different locations of 150 raw fish of the skin was taken by rubbing the sterilized cotton swab over the skin and then inoculating into the nutrient broth.

Gills: The sterilized cotton swab was wiped against the gill filaments by lifting the operculum with the help of a pair forceps. The sample was inoculated in the nutrient broth as well as swabbed on nutrient agar. A part of the gill filament removed aseptically was also placed in a separate (nutrient broth, MacConkey broth and Selinite F broth tubes) in order to isolate all the bacteria present on the gill filaments which might have escaped contact with the swab. The examined gills were taken from 150 raw fish.

Intestine and muscles: This was done by cutting a part of intestine and muscle after sterilizing with red hot scalped and inoculation in the media (nutrient broth, MacConkey broth and Selinite F broth tubes). The samples included 150 intestines and 150 muscles from 150 raw fish.

Bacteriological examination: Preparation of the media, Isolation and identification of the bacteria were done according to Cheesbourgh (1984). Sterilization of the media was done by autoclaving at 121°C for 15 min. All samples were incubated at 37°C for 24 - 48 h.

Counting was done according to Plate count method, media used for Total bacterial count and *Pseudomonas* count (nutrient agar and Muller and Hinton), coliform count (MacConkey agar) *E. coli* count (EMB agar).

RESULTS

Isolation of organisms

Out of 150 collected samples 83 (55.3%) and 93 (62%) showed positive isolation of *Enteriobacteriaceae* and *Pseudomonas spp.* respectively. Mix isolation (isolation of both *Enterobacteriaceae* and *Pseudomonas spp* from the same sample) were recorded in 50 (40.3%) of the positive samples Table 1.

Table 2 showed that the highest isolation of both *Enteriobacteriaceae* and *Pseudomonas spp.* was recorded in autumn season as 66 and 80% respectively; this was followed by summer for *Enteriobacteriaceae* (60%) and winter for Pseudomonas *spp* (64%). The highest mixed isolation was recorded in samples collected at winter season (38%) while the highest percentage for negative isolation of both *Enteriobacteriaceae* and *Pseudomonas spp.* were shown in samples that collected at summer season (30%).

Identification of different species of isolated bacteria

The number and percentage of bacterial species which were isolated from different parts of collected fishes were shown in Table 3. Four hundred and fifty-two isolates of *Enterobacteriacaea* were recovered from all parts of collected fishes, the highest number of isolates was recovered from gills as 234 isolates (41.4%) and the lowest isolation was obtained from the muscles as 46 (8.1%).

Table 3. Number and percentage of *Enterobacteriacaea and Pseudomonas spp* isolated from 150 samples of each parts of collected fishes.

Organism	Gills	intestine	Skin	Muscles	Total No. of the isolates &%
E. coli	50 (29.2%)	33 (23.4%)	16(17%)	6 (13.0%)	105 (23.2%)
Enterobacter spp.	7 (4.1%)	4 (2.8%)	0 (0%)	0 (0%)	11 (2.4%)
Klebsiella spp.	22 (12.9%)	13 (9.2%)	9 (9.6%)	5 (10.9%)	49 (10.8%)
Citobacter spp.	8(4.7%)	12 (8.5%)	4(4.3%)	4 (8.7%)	28 (6.2%)
Salmonella spp.	2 (1.2%)	18 (12.8%)	0 (0%)	8 (17.4%)	28 (6.2%)
Shigella spp	0 (0%)	8 (2.7%)	2 (2.1%)	0 (0%)	10 (2.2%)
Alkaligenes spp	37 (21.6%)	12 (8.5%)	25 (26.6%)	14 (30.4%)	88 (19.5%)
Serritia spp	0 (0%)	0 (0%)	11 (11.7%)	0 (0%)	11 (2.4%)
Proteus spp	16 (9.4%)	11 (7.8%)	12 (12.8%)	7 (15.2%)	46 (10.2%)
Providencia spp	12 (7.8%)	23 (16.3%)	8 (8.5%)	2 (4.3%)	45 (10%)
Pseudomonas spp.	63 (55.8%)	31 (27.4)	19 (16.8%)	0 (0%)	113
Favibacterium spp	11 (6.4%)	4 (2.8%)	7 (7.4%)	0 (0%)	22 (4.9%)
Morexilla spp.	6 (3.5%)	3 (2.1%)	0 (0%)	0 (0%)	9 (2%)
Total	234 (41.4%)	172 (30.3%)	113 (20%)	46 (8.1%)	565

Table 4. Mean count of the bacteria present at different parts of examined fishes.

Collected parts	Total Bacterial Count (TBC) Cfu/ml	Coliform count Cfu/ml	*E. coli* count Cfu/ml	*Pseudomonas* count Cfu/ml
Gills	$3 \times 10^6 - 7 \times 10^9$	$3 \times 10^3 - 5 \times 10^5$	$7 \times 10^2 - 9 \times 10^4$	$5 \times 10^7 - 2 \times 10^9$
Intestine	$1.5 \times 10^5 - 1.6 \times 10^8$	$3 \times 10^3 - 7.5 \times 10^6$	$3 \times 10^2 - 4 \times 10^5$	$7 \times 10^5 - 3 \times 10^7$
Skin	$3 \times 10^7 - 4 \times 10^9$	$3 \times 10^2 - 2 \times 10^2$	$1 \times 10^2 - 7 \times 10^3$	$4 \times 10^5 - 3 \times 10^7$
Muscles	$7 \times 10^4 - 2 \times 10^6$	$1 \times 10^2 - 4 \times 10^4$	$3 \times 10^2 - 7 \times 10^3$	0

Table 3 also showed that *E. coli* was the dominant isolates as 105 isolates (23.2%) and the lowest isolates were *Shigella spp* and *Morexiella spp* with percentages a 2.2 and 2% respectively.

Pseudomonas spp, was obtained as 113 these were 63 (55.8%) isolated from gills, 31 (27.4%) isolates from the intestine and 19 (16.8%) isolates from skin. muscles samples revealed negative growth for this organisms (Table 3).

Counting of the bacteria

Table 4 showed the mean of total bacterial counts, coliform count, *E. coli* count and *Pseudomonas spp* count at different parts of collected fishes, the gills showed the highest number of bacteria; this was followed by skin then muscles.

DISCUSSION

Fishery products are important not only from a nutritional point of view, but also as an item of international trade and foreign exchange earner for a number of countries in the world. Fish and shellfish are highly perishable, and prone to vast variations in quality due to differences in species, environmental habitats, feeding habits. In addition, they can also function as carriers of several microbial and other health hazards. Therefore maintenance of quality is of utmost importance in production and trade of fishery products. Most of current quality control techniques are time consuming and cumbersome.

Although only a few infectious agents in fish are able to infect humans, some exceptions exist that may result in fatalities. However, the greatest risk to human health is due to the consumption of raw or insufficiently processed fish and fish products. In this study isolation of Enterobacteriacaea as 55% from collected samples indicated public health hazards and concern, particularly isolation of some highly pathogenic agents such as Salmonella *spp., Shigella spp.*, and the pathogenic *E. coli.* and potential pathogenic organisms such as *Klebsiella spp., Citrobacter spp., Proteus spp.* The isolation of these groups of organisms indicated faecal and environmental pollution and these supported the findings of Yagoub et al. (2004) who isolated pathogenic and potential pathogenic organisms from tap water that originated from Nile river. This also confirms the findings of Koutsoumanis and Nychas (2000); Gonzalez-Podriguez et al. (2001) and

Herrera et al. (2006) who isolated similar organisms from fish and fish products. The isolation of *Alkaligenes* which are widely distributed in soil, water and intestinal tract of vertebrates and play an important role in the spoilage of fish and fish meat support the findings of Collins et al. (1989).

The isolation of *Pseudomonas spp* from 63% of collected fish samples is of highly importance because this bacterium plays a considerable role as potential pathogenic bacteria for human and as an indicator of food quality as spoilage organism. This is in accord with previously mentioned by Jeyasekaran et al. (2006) and Koutsoumanis and Nychas (2000) who identified *pseudomonads* as a good spoilage index.

The highest isolation of microorganisms during the autumn season might be due to rain drainage that carries polluted soil to river water when the presence of high organic substances promoted multiplication of the organisms. The high percentage of the presence of Enterobacteriacaea in summer and *Pseudomonas spp* in winter season might be partially attributed to the suitability of the temperature for the survival and multiplication of these bacteria.

This study revealed that raw fish sold in fish market in Khartoum state could be a source of food- borne bacterial pathogens. Improvements in handling and processing are needed to minimize the prevalence of the pathogenic bacteria. The total viable *Enterobacteriaceae* counts strongly suggest the urgent need to improve the quality control systems. The present results may be considered as additional knowledge to enhance proper controlling of the storage life of fish, and fish product quality.

REFERENCES

Arannilewa ST, Salawu SO, Sorungbe AA, Ola-Salawu BB (2006). Effect of frozen period on the chemical, microbiological and sensory quality of frozen Tilapia fish (Sarotherodun galiaenus). Nutr. Health 18(2):185-92.

Ayulo AM, Machado RA, Scussel VM (1994). Enterotoxigenic Escherichia coli and *Staphylococcus aureus* in fish and seafood from the southern region of Brazil. Int J. Food Microbiol. 24(1-2): 171-8.

Center for Food Safety and Applied Nutrition, (2001). Food and Drug Administration. 3rd edition, Washington pp.145-166,

Cheesbrough M (1984). Medical Laboratory for Tropical Countries, First ed., Green Britain of the University Press Cambridge. UK

El Hadi N, Radu S, Chen CH, Nishibuchi M (2004). Prevalence of potentially pathogenic Vibrio species in the seafood marketed in Malaysia. J. Food Prot. 67(7): 1469 - 1475.

Gonzalez-Rodriguez MN , Sanz JJ, Santos JA, Otero A, Garcia-lopez ML (2001). Bacteriological Quality of aquaculture freshwater fish portions in prepackaged trays stored at 3 degrees C. J. Food Prot. 64 (9): 1399 - 1404.

Håstein T, Hjeltnes B, Lillehaug A, Utne Skåre J, Berntssen M, Lundebye AK (2006). Food safety hazards that occur during the production stage: challenges for fish farming and the fishing industry. Rev. Sci. Technol. 25(2): 607-625.

Heinitz ML, Ruble RD, Wagner DE, Tatini SR (2000). Incidence of Salmonella in fish and seafood. J Food Prot. 63(5): 579-592.

Herrera FC, Santos JA, Otero A, Garcia-Lopez ML (2006). Occurrence of Foodborne pathogenic bacteria in retails prepackage portions of marine fish in Spain. J. Appl. Microbiol. 100(3): 527-36.

Hood MA, Ness GE, Blake NJ (1983). Relationship among fecal coliforms, *Escherichia coli* and *Salmonella* spp. in shellfish. Appl. Environ. Microbiol. 45(1): 122-6.

Hwang DF, Huang YR, Lin KP, Chen TY, Lin SJ, Chen LH, Hsieh HS (2004). Investigation of hygienic quality and freshness of marketed fresh seafood in Northern Taiwan. Shokuhin Eiseigaku Zasshi. 45(5): 225-30.

Jeyasekaran G, Ganesan P, Anandaraj R, Jeya Shakila R, Sukumar D (2006). Quantitative and qualitative studies on the bacteriological quality of Indian white shrimp (*Penaeus indicus*) stored in dry ice. J. Food Microbiol. 23(6): 526-533.

Koutsoumanis K, Nychas GJ (2000). Application of systemic experimental procedure to develop a microbial model for rapid fish shelf life predictions. Int. J. Food Microbiol. 60(2-3): 171-184.

Morales G, Blanco L, Arias ML, Chaves C. (2004). Bacteriological evaluation of fresh tilapia (*Oreochromis niloticus*) coming from the northern region of Costa Rica. Arch Latinoam Nutr. 54(4): 433-437.

Ristori CA, Iaria ST, Gelli DS, Rivera IN (2007). Pathogenic bacteria associated with oysters (*Crassostrea brasiliana*) and estuarine water along the south coast of Brazil. Int. J. Environ. Health Res. 17(4): 259-269.

Thampuran N, Surendraraj A, Surendran PK (2005). Prevalence and characterization of typical and atypical Escherichia coli from fish sold at retail in Cochin, India. J. Food Prot. 68(10): 2208-2211.

Yagoub SO, Ahmed TM (2004). Pathogenic Microorganisms in fresh water samples collected from Khartoum central market. Sudan J. Vet. Sci. Anim. Husbandry 43(1-2): 32-37.

Conjugational transfer and survival of plasmid encoding silver and antibiotic resistance genes of *Acinetobacter baumannii* BL54, *E. coli* K12 J53.2 transconjugants and pseudomonas transformants in different soil microcosms

Mohammad Reza Shakibaie[1]*, Pershant Dhakephalker[2], Balu Parnab Kapadnis[3] and Balu Ananda Chopade[4]

[1]Department of Microbiology and Immunology and Infectious Diseases and Tropical Medicine Research Comittee, Kerman University of Medical Sciences, Kerman, Iran.
[2]Department of Microbiology, Agarkar Agriculture Institute, Pune, India.
[3]Department of Microbiology, University of Pune, Ganeshkhind Road, Pune - 411007, India.
[4]Institute of Bioinformatics and Biotechnology, University of Pune, Ganeshkhind Road, Pune - 411007 Maharashtra State, India.

In this investigation we tried to transfer plasmid encoded silver and antibiotic resistance genes from *Acinetobacter baumannii* BL54 to *E. coli* k12 J53.2 and Pseudomonas (a soil microflora) by conjugation and transformation in different soil microcosms and study the survival of the isolated bacterium in each soil. Clay loam, fine clay, sandy and clay soils were collected from different area of mahrashtra in India. Microcosm was developed for each type of soil in the glass tube (150 x 25 mm) with 2 g soil moistened with 1 ml 0.5% sterile saline. The conjugation frequency was lowest in sterile clay soil with frequency of 0.2 x 10^{-6}, while, it was maximum in clay loam soil with frequency of 0.6 x 10^{-6}. Similarly, in non-sterile soil microcosms, the rate of conjugation was highest in clay loam soil with frequency of 0.09 x 10^{-6} while was lowest in sandy soil (0.03 x 10^{-6}). Rapid death of the organisms was observed within 9 days of incubation in presence of selection in sandy soil, while in fine clay, survival of the organisms was extended beyond 11 days. The Pseudomonas transformant survived for more than 40 days in presence of selection. From above results it can be concluded that plasmid mediated silver and antibiotic resistant genes were transferred in different soil by conjugation process. However, the rate of conjugation was affected by soil type. Soil transformant considerably survived in fine clay containing high amount of organic carbon and neutral pH as compared to sandy soil.

Key words: Conjugation, soil type, survival, plasmid, *Acinetobacter baumannii.*

INTRODUCTION

Soil is a macrohabitat containing minute amounts of nutrients, and is subjected to temporal and spatial variation (Barkey et al., 1985). Plasmids have been found in many genera of soil bacteria (Chopade et al., 1985; Devanas et al., 1986) and play an important role in sur-vival of soil organisms in presence of metals or antibiotics in the soil environment. Conjugation in the soil showed that plasmid encoding gene for degradation of 3-chlorobezoate was transferred to indigenous soil flora (*Pseudomonas*) with

*Corresponding author. E-mail: mr_shakibaei@kmu.ac.ir.

Table 1. Chemical analysis of soil samples used in this study.

Soil type	Source	pH	Electric conductivity Ohm $^{-1}$ /Cm $^{-1}$	Organic carbon (%)	P (gm %)	K (gm %)	Organic matter (%)	Total soluble salt (%)	CaCO3	Silt (%)
Clay loam	Pune	8.1	0.2	0.65	0.0012	0.007	0.72	0.8	0.9	30
Fine clay (Black)	Pimperi	7.5	0.2	0.79	0.0014	0.001	1.3	0.8	0.96	18.75
Clay	Mahabale shwar	6.3	0.12	0.549	0.008	0.001	0.94	0.6	0.88	16
Sandy	mumbi beach	8.6	0.15	0.04	0.001	0.0006	0.04	-	-	50

P= Phosphorous
K= Potassium
Soils were collected from 15cm depth above locations and subjected to analysis as described in text.
The above results are average of two independent experiments

high frequency (Hirkala and Germida, 2004). Talbot et al. (1980) reported transfer of antibiotic resistance plasmids among strains of *Klebsiella* in sandy soil. Population dynamics of Hg^{2+} resistant bacteria showed that out of 76 isolates, 6 were capable to transfer this heavy metal resistance to the recipient cells in the soil (Naik et al., 1994; Pertsova et al., 1984). Appli-cation of Cd^{2+} in laden sewage and grass land soil showed that cd^{2+} resistant bacteria could not be able to compete with normal flora of the soil upon the long term exposure to cadmium (Poluektova et al., 2008). Devanas et al. (1986) suggested that conjugative plasmids may be the most important means of gene transfer in the soil as it was appeared a considerable decrease in number of some *E. coli* strains when recovered from sterile and non sterile soil microcosms as they lost the plasmids. Van Elsas et al. (1989) reported the survival of *P. fluorescence* containing plasmid RP4 and pRK501 and their stability in two soil microcosms. However, soil type profoundly effected the host survival. Indeed the indigenous microbiota of the soil may directly affect the conjugation process in the soil (Shakibaie et al., 1999, 1998). Field and soil microcosm studies on the survival and conjugation of a *P. putida* strain bearing a recombinant plasmid, pADPTel (Pertsova et al., 19847).

Majority of studies on transfer of the plasmids in soil microcosms have been conducted in Enterobacteriaceae (Krasovsky and Stotzky, 1987) and there is a paucity of information exist about conjugation and survival of plasmid mediated antibiotic and metal resistance bacteria in the soil. In previous study (Shakibaie et al., 1999, 1998, 2003) we isolated a species of *Acinetobacter baumannii* encoding conjugative plasmid responsible for silver and antibiotic resistance genes. In present investigation we tried to transfer this plasmid by conjugation to *E. coli* k12 J53.2 and by transformation to *Pseudomonas* (a soil microflora) in different sterile and non sterile soil microcosms and study survival of these plasmids encoding bacteria and Pseudomonas transformant in different soil microcosms.

MATERIALS AND METHODS

Collection of soil samples and developing soil microcosm

Clay loam, fine clay (forest, black), sandy and clay (red, latertic) soils were collected around different area of mahrashtra provenience in India. Thirty gram of each soil was removed by spade from 15 cm depth and collected in polythene bag. The soils were dried over night at 30°C and subjected to grinding with pestle followed by sieving through 0.242 mm pore size sieve. The processed soils were subjected to physicochemical analysis as shown in Table 1. Organic carbon was determined by Walkey Black potassium dichromate oxidation method and available phosphorus was determined by Olson method (Redford et al., 1981). The soils were then divided in two parts and one was sterilized by autoclaving at 121°C for one hour. Microcosm was developed for each type of soil in the glass tube (150 x 25 mm) with 2 g soil moistened with 1 ml 0.5% sterile normal saline and used for gene transfer and survival studies.

Conjugation in soil

Conjugation was carried out in each sterile and nonsterile soil as described by Naik et al. (1994). Briefly, 1 ml O/N growth of *A. baumannii* (10^7/ ml) in Luria -Bertani medium as donor and 1 ml of *E. coli* K12 J53.2 (Rif ^{r}) as recipient were added to each soil microcosm and incubated for 24 h at 30°C. With help of a sterile glass rod approximately 20 mg of soil was withdrawn and suspended in sterile saline and serially diluted. 0.2 ml aliquot of each dilution was spread in to Lactose electrolyte deficient medium (CLED) agar selective for transconjugant (Ag 200 µg/ml + Te 100 µg/ml + ampicillin 200 µg/ml and Rif 100 µg/ml). The frequency of conjugation was then calculated as number of colonies grown in this medium divided by number of recipient cells.

Transformation and plasmid stability

Transformation and stability of plasmid were determined by shakibaie et al. (1998) as previous described. Briefly, Plasmid DNA was prepared with the Qiagen plasmid purification kit (Qiagen, Hilden, Germany). Overnight culture of *A. baumannii* BL54 were grown in LB broth with 100 µg AgNO3, washed twice in 0.9% NaCl and adjusted to a titer of 1×10^9 ml^{-1}. To 20 µl of this suspension 10 µl of plasmid DNA in 10 mM Tris-HCl, pH 8.0, 1 mM EDTA and 50 mM ice cold $CaCl_2$ was added and the mixture was spotted onto LB

Table 2a. Conjugation between *A. baumannii* and *E. coli* K12 J53.2 in different sterile soil microcosms.

Soil type	*A. baumannii* (CFU/g soil)	*E. coli* K12 J53.2 (CFU/g soil)	Frequency of conjugation	Marker co-transferred
Fine clay	1.6×10^{7}	2.2×10^{7}	0.3×10^{-6}	Te, Ag, Amp
Sandy	2×10^{7}	1.2×10^{7}	0.3×10^{-6}	Ag, Te, Amp
Clay loam	3×10^{7}	2.2×10^{7}	0.6×10^{-6}	Amp, Te, Ag
Clay	1.2×10^{7}	3×10^{7}	0.32×10^{-6}	Te, Ag, Amp

Table 2b. Conjugation between *A. baumannii* and *E. coli* K12 J53.2 in different non sterile soil microcosms.

Soil type	*A. baumannii* CFU/g soil	*E. coli* K12 J53.2 (CFU/g soil)	Frequency of conjugation	Marker co-transferred
Fine clay	1.2×10^{7}	3.1×10^{7}	0.07×10^{-6}	Te, Ag, Amp
Sandy	2.1×10^{7}	2×10^{7}	0.03×10^{-6}	Ag, Te, Amp
Clay loam	6.2×10^{7}	1.6×10^{7}	0.09×10^{-6}	Amp, Te, Ag
Clay	2×10^{7}	3.8×10^{7}	0.07×10^{-6}	Te, Ag, Amp

A control of conjugation was carried out to check presence of spontaneous mutation. It was below limits of detection.

agar. Incubation was for 20 h at 37°C. Appropriate dilutions of the resuspended cells were plated on LB agar containing 100 µg $AgNO_3$ + 100 µg Rif (titer of transformants) and Rifampicin (titer for recipients). Colonies were counted following incubation for 48 h at 37°C.

Survival of *A. baumannii* BL54, *E. coli* K12 transconjugant and Pseudomonas transformant in different soil microcosms

2 ml of over night growth of each isolate (10^{8} CFU/ml) was inoculated separately in to 4 g of sterile and nonsterile (Fine clay, clay loam, clay and sandy) soil microcosms. All tubes were incubated up to 44 days. At two days interval 5 mg soil was taken from each soil microcosm and suspended in 10 ml 0.85% sterile saline and serially diluted. 0.1ml of each dilution was spread into CLED agar medium containing 25 µg/ml Tetracycline and 200 µg/ml $AgNO_3$ and into CLED agar. After 24 h of incubation at 30°C, total viable count of bacteria in each soil then was calculated.

RESULTS

Chemical composition of different soil microcosms used in this study is shown in Table 1. Table 1 indicates, the forest soil contain considerable amount of organic matter as well as phosphorous and neutral pH while, sandy soil contain minimum amount of nutrient as well as alkaline pH (8.6). The results of conjugation in sterile and nonsterile soil microcosms are shown in Table 2a, b. The conjugation frequency was lowest in sterile Clay soil with frequency of 0.2×10^{-6}, while, it was maximum in clay loam soil with frequency of 0.6×10^{-6}. Similarly, in non sterile soil microcosms (Table 2b), the rate of conjugation was highest in clay loam soil with frequency of 0.09×10^{-6} while was lowest in sandy soil (0.03×10^{-6}). This indicate that transfer of metal and antibiotic resistance genes indeed occur in different soil microcosms and soil type influenced the frequency of conjugation. Concurrent testing of transconjugant in presence of ampicillin 50 µg/ml + Rif 100 µg/ml, Te 50 µg/ml + Rif and Ag 200 µg/ml + Rif showed simultaneous transfer of these genes to the recipient cells. Transfer of plasmid from *A. baumannii* to *Pseudomonas* a soil microflora had occurred by transformation process (data not shown) with frequency of 0.4×10^{-6}. Co-transfer study of transformant indicated simultaneous transfer of above antibiotic resistant markers to *Pseudomonas*.

The survival of *A. baumannii, E. coli* K12 J53.2 transconjugant and Pseudomonas transformant was studied in different types of soil as shown in Figures 1 and 2. Rapid death of the organisms was observed within 9 days of incubation in presence of selection in sandy soil, while in fine clay soil survival of the organisms was extended beyond 11 days. This decrease in plasmid bearing cells was apparently the result of partitioning of the plasmid without duplication in daughter cells. The above conclusion was supported from comparatively high CFU of the organisms at the start of experiment as shown in Figure 1. This indicated that expression of resistant genes on plasmid was inhibited after long incubation under starvation condition. The above results were further supported by study of the viable count of pseudomonas transformant in presence and absence of selection as shown in Figure 2. The organism survived for more than 44 days in the absence of selection and for40 days in presence of selection. This indicates that expression of these resistant genes was also depending on physiology of the organism as well. Since *Pseudomonas* is as soil

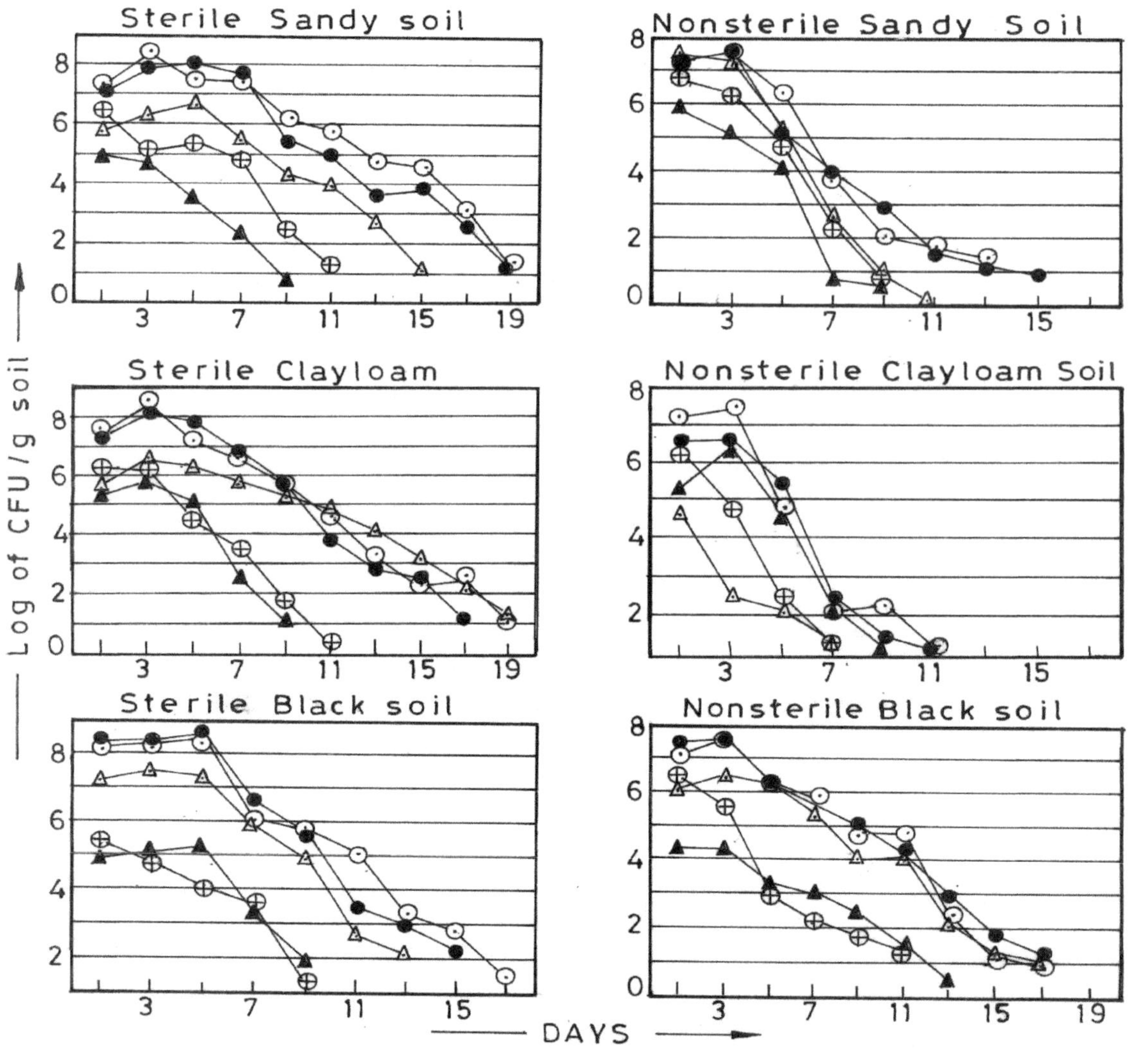

Figure 1. Survival of *A. baumannii* and *E. coli transconjugant* in sterile and nonsterile sandy, clay, clay loam and fine clay (forest) soil microcosms.
⊙ ; *A. baumannii* plated on CLED agar, •; *E. coli* plated on CLED agar, □; *A. baumannii* plated on CLED agar + 200µg/ml $AgNO_3$,▲; *E. coli* transconjugant plated on CLED agar + 200µg/ml $AgNO_3$, ▲; *E. coli* transconjugant plated on CLED agar + 50µg/ml Te.

isolate it better adapted in starvation condition, while, *A. baumannii* and *E. coli* transconjugant being zymogenous were not able to survive in starvation under pressure of antibiotic and metal containing medium.

DISCCUSSION

The major concern about introduction of plasmid containing organisms in soil is not only their potential adverse ecological impacts on the homeostasis of soil but also on the bioremediation of pollutants like heavy metals from different type of soil. Krasovky and Stotzky, (1987) demonstrated that indigenous microflora of soil directly effects the survival of plasmid bearing bacteria in nonsterile soil.

In our study plasmid encoding Acinetobacter was quite stable in all soil tested. The stability was not depending on type of soil but it was depend on type of host. Chemical composition of different soil microcosms used in this study indicates that, the forest soil contain considerable amount of organic matter as well as phosphorous and neutral pH (7.5) while, sandy soil contain minimum amount of nutrient as well as alkaline pH (8.6). Therefore, it is appropriate to suggest that genetic markers could not be expressed up on long term in starvation condition in sandy soil. This was supported by study of population dynamics of Acinetobacter, *E coli* transconjugant and Pseudomonas transformant in different types of soil microcosms. This also indicated that plasmid have no role in long term survival of the organisms in soil. Infact, viable count declined rapidly when cells were plated in

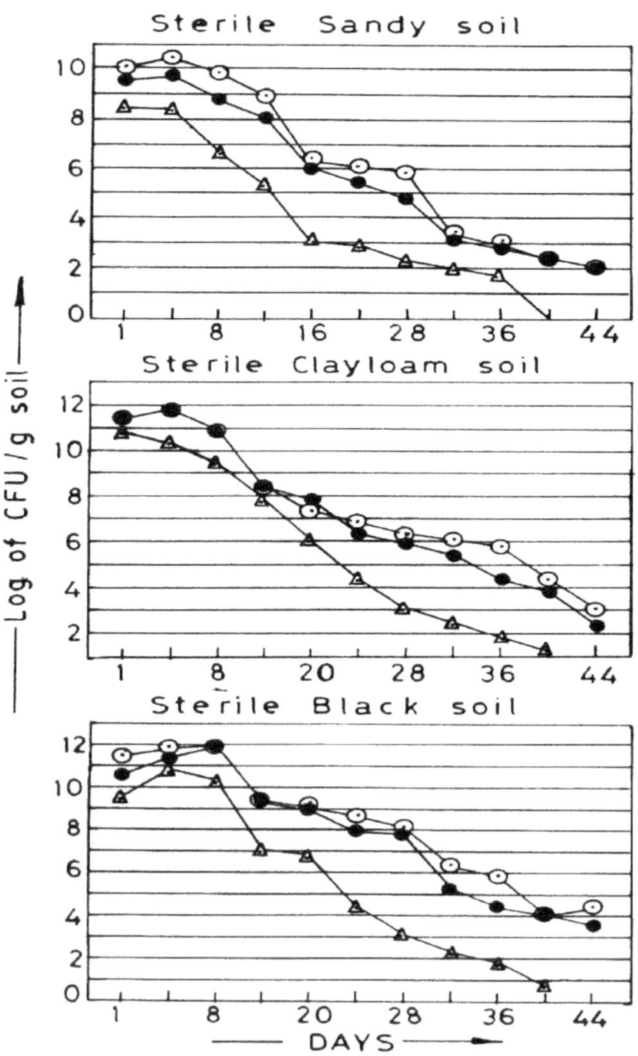

Figure 2. Survival of Pseudomonas transformant (an indigenous soil microflora) in sandy, clay loam and fine clay (forest) soil microcosm
⊙; Organism plated on CLED agar, •; Organism plated on CLED agar + 50 µg/ml Te, ▲; Organism plated on CLED agar + 200µg/ml $AgNO_3$

metal and antibiotic containing medium when plasmid was present in the cells. The viable count decreased rapidly in sandy soil, while in forest soil it was prolonged by few days.

Endogenous energy source is an important factor in microbial survival and it may be that autochthonous bacteria have a mechanism for reducing metabolic rate in order to enhance their survival under starvation condition (Trevors, and Oddie, 1986). Shakibaie et al. (1999) reported that accumulation of silver is an energy dependent process. It is quite possible nutrients were not available in soil and affect long term survival of the organisms. At present, research is carrying out by our group in this regard. Recently, horizontal transfer of antibiotic resistant genes was studied in sewage and lake water by Shakibaie et al. (2009). It was observed that the rate of conjugation was two fold high in sewage than in lake water. The physicochemical parameters of water also were contributed to gene transfer by conjugation.

Conclusion

From above results it can be concluded that metal and antibiotic resistant genes were transferred in different soil types by conjugation process. However, the rate of conjugation was depending on soil type as well as type of host. The plasmid containing soil microbiota transformant considerably survived under selective and non selective conditions and soil type also play role in the survival of isolated organisms as well.

AKNOWLEDGMENTS

The authors would like to thank Department of Microbiology Staff University of Pune, India and Kerman University of Medical Sciences, Iran for providing necessary facilities for this research.

REFERENCES

Barkey T, Tripp SC, Olson BT (1985). Effect of metal rich sewage sludge application on bacterial communities of grass land soil. Appl. Environ. Microbiol. 49: 333-337.

Chopade BA, Wise PJ, Towner KJ (1985). Plasmid transfer and behavior in *A. calcoaceticus* EBF 65/65. J. Gen. Microbiol. 131: 2805-2811.

Devanas MA, Rafaeli-Fshkol D, Stotzky G (1986). Survival of plasmid containing strains of *E.coli* in soil. Effect of plasmid size and nutrient on survival of hosts and maintenance of plasmids. Curr. Microbiol. 13: 269-277.

Hirkala DL, Germida JJ (2004). Field and soil microcosm studies on the survival and conjugation of a *Pseudomonas putida* strain bearing a recombinant plasmid, pADPTel. Can. J. Microbiol. 50(8):595-604.

Krasovsky VN, Stotzky GC (1987). Conjugation and genetic recombinetion in *Escherichia coli* in sterile and no sterile soil. Soil. Biol. Biochem.19: 631-638.

Naik AG, Bhat LN, Chopade BA, Lynch JM (1994).Transfer of broad host range antibiotic resistance plasmids in soil microcosms. Curr. Microbiol. 28: 209-215.

Pertsova RN, Kunc F, Golovleva LA (1984).Degradation of 3-chlorobenzoate in soil by *Pseudomonas* carrying a biodegradative plasmid. Folia Microbiol. 29: 242-247.

Poluektova EU, Gagarina E Yu, Shilovskii IP, Fedorina A, Nezametdinova V Z. Rodionova S A, Prozorov AA (2008). Molecular analysis of some genes from plasmid p19 of the *Bacillus subtilis* 19 soil strain involved in conjugation. Russian. J. Genet. 44 (5): 539-545.

Redford AJ, Oliver J, Kelly WJ, Reanney DC (1981). Transformable resistance to mercuric and phenyl mercuric ions in soil bacteria. J. Bacteriol. 147: 1110-1112.

Shakibaie MR, Dhakephalker PK, Kapadnis BP, Chopade BA (1999). Removal of silver from waste water effluents by *Acinetobacter baumannii* BL54. Can. J. Microbiol. 45: 995-1000.

Shakibaie MR, Dhakephalker PK, Kapadnis BP, Chopade BA (1998). Plasmid mediated silver and antibiotic resistance in *Acinetobacter baumannii* BL54. Irn. J. Med. Sci. 23: 30-36.

Shakibaie MR, Dhakephalker PK, Kapadnis BP, Chopade BA (2003). Silver resistance in *Acinetobacter baumannii* BL54 occurs through

binding to a Ag- Binding protein. Iran. J. Biotech. 1(1): 41-46.
Shakibaie MR, Jalilzadeh K, Yamakanamardi SM (2009). Horizontal transfer of antibiotic resistance genes among gram negative bacteria in sewage and lake water and influence of some physico-chemical parameters of water on conjugation process. J. Environ. Biol. 30(1): 45-49.
Talbot HW, Yamamato DK, Smith MW, Seudler RJ (1980). Antibiotic resistance and its transfer among clinical and non clinical Klebsiella strains in botanical environments. Appl. Environ. Microbiol. 39: 97-104.
Trevors TJ, Oddie KM (1986). R-plasmid in soil and water. Can. J. Microbiol. 33:191-198.
Van Elsas JD, Vanoverbeek LS, Stardoub ME (1989). Survival of *Pseudomonas fluorescence* containing plasmid RP4 or pRK2501 and plasmid stability after introduction in two soils with different texture. Can. J. Microbiol. 35: 951-959.

9

Correlation of plasmid with drug resistance of clinical isolates of *Escherichia coli*

Mohammad Shahriar* and Nishat Zareen Khair

Department of Pharmacy, University of Asia Pacific, House 73, Road 5A, Dhanmondi R/A, Dhaka, Bangladesh.

According to the latest data from the Centers for Disease Control and Prevention (CDC), the six ESKAPE bacteria are responsible for two thirds of all health care-associated infections (HAIs) including *Escherichia coli. Klebsiella* species and their gram-negative cousin *E. coli* together accounted for 18 percent of all HAIs in 2006 to 2007, and a growing proportion of these two bad bugs carry resistance to a remarkable spectrum of antibiotics. Of the very few drugs in late-stage development, none works by a novel mechanism. The aim of our study is to correlate the plasmids with drug resistance of clinical isolates of *E. coli.* Twenty two clinical isolates of *E. coli* were collected from different diagnostic centers of Dhaka and their antimicrobial susceptibility pattern was tested. Seven multidrug resistant isolates of *E. coli* were selected and their antibiotic susceptibility pattern was tested before and after curing. Our study revealed that 100% of the isolates (22 isolates) were resistant to cephalexin, cephradine, oxacillin, penicillin and vancomycin. 95.45% of the isolates were resistant to ciprofloxacin, cloxacillin and imipenem. 90.91% isolates (21 isolates) were resistant to erythromycin. 81% of the isolates (18 isolates) were resistant to amoxicillin. 72.72% (16 isolates) were resistant to co-trimoxazole, 63.64% (14 isolates) were resistant to tetracycline. 31.81% (7 isolates) were resistant to ceftriaxone and neomycin. 22.73% (5 isolates) were resistant to gentamicin and only 13.64% of the isolates (3 isolates) were resistant to chloramphenicol. There was no significant difference in the antibiotic susceptibility pattern before and after curing indicating no correlation between plasmid and drug resistance in the 7 isolates of *E. coli.* However, isolate number 5 which was resistant to gentamicin, neomycin and imipenem became sensitive after curing. Similarly sample number 1 became sensitive to rifampicin and imipenem and sample number 7 became sensitive to imipenem after curing.

Key words: Multi-drug resistant, *Escherichia coli*, antimicrobial susceptibility, plasmid, curing.

INTRODUCTION

According to the Infectious Diseases Society of America in the January 2009 highlighted the impact of the ESKAPE pathogens (*Enterococcus faecium, Staphylococcus aureus, Klebsiella pneumoniae, Acinetobacter baumannii, Pseudomonas aeruginosa,* and *Enterobacter* species) as a group of particularly troublesome bacteria having the ability to "escape" the effects of current antimicrobial agents (Boucher et al., 2009). Drug resistance is an alarming problem worldwide and it is spreading rapidly due to overuse, self medication, and non-therapeutic use of antimicrobials (Slama et al., 2005). Antimicrobials themselves act as a selective pressure which allows the growth of resistant bacteria within a population and inhibits susceptible bacteria (Levy, 1994). Drug resistance property in bacteria is usually borne in R-plasmids, which can be disseminated, to diverse population and regions causing world wide problems. R-plasmids from resistance strains of an organism may transfer to a sensitive counterpart, which can show the same drug resistance in the donor strain. *E. coli* are commonly found in the lower intestine

*Corresponding author. E-mail: shahriar_12@yahoo.com. Ext. 146

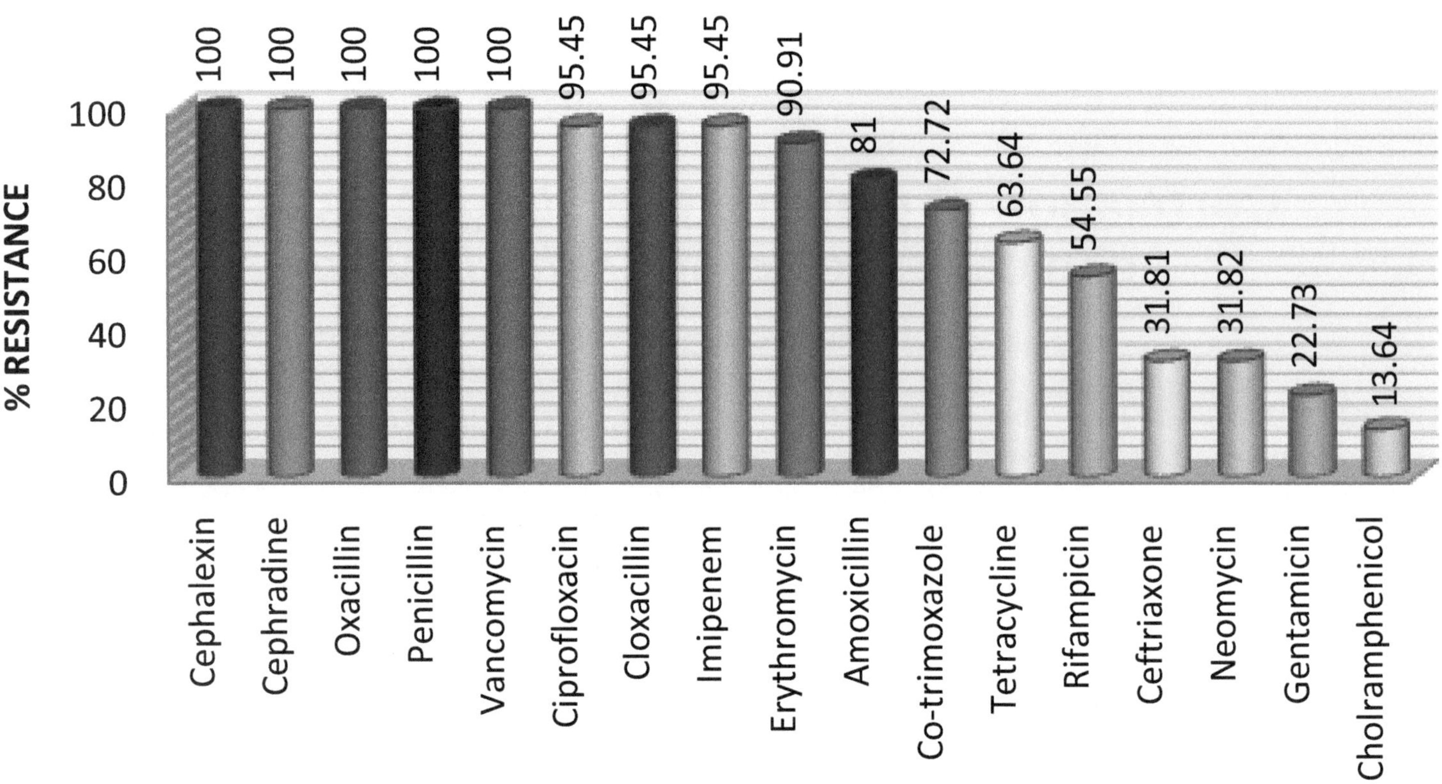

Figure 1. Percentage resistance of *E. coli* to different antimicrobials.

of warm-blooded organisms (endotherms) as normal flora of the gut, and can benefit their hosts in many ways but there are over 700 antigenic serotypes of *E. coli* some of which can food poisoning in humans, and serious illness like gastroenteritis, urinary tract infections, and neonatal meningitis (Todar, 2008). New strains of *E. coli* evolve through the natural biological process of mutation, *E. coli* and related bacteria possess the ability to transfer DNA via bacterial conjugation, transduction or transformation, which allows genetic material to spread horizontally through an existing population (Brüssow e. al., 2004). In this study, plasmid profiles and drug resistance pattern of multi drug resistant clinical isolates of *E. coli* were studied before and after curing to find out whether the resistant gene which makes *E. coli* resistant to multiple drugs (antimicrobials) is present in the plasmid DNA or in the chromosomal DNA. *E. coli* is a very common organism and their resistance to multidrug is alarming for mankind.

MATERIALS AND METHODS

Collection of isolates

Twenty two isolates of *E. coli* were collected from different diagnostic centers of Dhaka. The samples were collected from pathological specimen of stool, pus and urine. The isolates were identified on the basis of their colony morphology, microscopic morphology and biochemical tests for reassurance and reconfirmation of the strain.

Antimicrobial susceptibility test

The antimicrobial susceptibility pattern of *E. coli* to 17 antibiotics namely amoxicillin, Am (30 µg/disc); ceftriaxone, Ci (30 µg/disc); cephalexin Cp (30 µg/disc); cephradine, CV (25 µg/disc); chloramphenicol C (30 µg/disc); ciprofloxacin, Cf (5 µg/disc); cloxacillin, CX (1 µg/disc); co-trimoxazole, Co (25 µg/disc); erythromycin, E (15 µg/disc); gentamicin, G (10 µg/disc); neomycin, N (30 µg/disc); oxacillin, OX (1 µg/disc); penicillin, P (10 µg/disc); rifampicin, R (5 µg/disc); tetracycline, T (30 µg/disc); vancomycin, VA (30 µg/disc) and imipenem, I (10 µg/disc) was determined by disc diffusion Kirby Bauer (A. Bauer et al., 1966) method as per recommendation of National Committee for Clinical Laboratory Standards (NCCLS, 1997) and the result was recorded in Figure 1 for comparing.

Antimicrobial susceptibility test of *E. coli* in presence of 1% SDS solution

From the twenty two isolates 7 isolates of *E. coli* were selected and cultured in Luria broth in presence of 1% SDS solution and in absence of SDS solution. SDS solution is used as the curing agent. And antibiotic susceptibility test for these 7 samples was carried out in absence of SDS and in presence of 1% SDS by disc diffusion

Table 1. Antimicrobial susceptibility pattern of seven selected isolates of *E. coli* without SDS solution.

E. coli isolates	Zone of inhibition (mm)																
	Am^{30}	Ci^{30}	Cp^{30}	CV25	C30	Cf^{5}	CX1	Co^{25}	E^{15}	G^{10}	N^{30}	OX^{1}	P^{10}	R^{5}	T^{30}	VA^{30}	I^{10}
(1) E- 516	R	S	S	R	S	R	R	R	R	R	S	R	R	R*	S	R	R*
(2) *E. coli* -2	R	S	S	R	S	S	R	S	R	S	S	R	R	S	R	R	R
(3) E-371	R	R	R	R	S	R	R	R	R	S	S	R	R	R	R	R	R
(4) E-210	R	R	R	R	S	R	R	R	R	S	S	R	R	R	R	R	S
(5) E-294	R	S	R	R	R	R	R	R	R	R*	R*	R	R	R	R	R	R*
(6) E-653	R	R	R	R	S	R	R	R	R	R	S	R	R	R	R	R	S
(7) E-480	R	S	R	R	R	R	R	R	R	S	S	R	R	R	R	R	R*

Table 2. Antimicrobial susceptibility pattern of seven selected isolates of *E. coli* cultured in Luria broth in the presence of 1% SDS.

E. coli isolates	Zone of inhibition (mm)																
	Am^{30}	Ci^{30}	Cp^{30}	CV25	C30	Cf^{5}	CX1	Co^{25}	E^{15}	G^{10}	N^{30}	OX^{1}	P^{10}	R^{5}	T^{30}	VA^{30}	I^{10}
(1a) E- 516	R	S	S	R	S	R	R	R	R	R	S	R	R	S*	S	R	S*
(2b) *E. coli*-2	R	S	S	R	S	S	R	S	R	S	S	R	R	S	R	R	R
(3c) E-371	R	R	R	R	S	R	R	R	R	S	S	R	R	R	R	R	R
(4d) E-210	R	R	R	R	S	R	R	R	R	S	S	R	R	R	R	R	S
(5e) E-294	R	S	R	R	R	R	R	R	R	S*	S*	R	R	R	R	R	S*
(6f) E-653	R	R	R	R	S	R	R	R	R	R	S	R	R	R	R	R	S
(7g) E-480	R	R	R	R	R	R	R	R	R	S	S	R	R	R	R	R	S*

Kirby Bauer (Bauer et al., 1966) method using the same 17 antimicrobials. The result was recorded in Tables 1 and 2. Then plasmid profile analysis of the above samples was done using gel electrophoresis plasmid isolation.

To isolate plasmid DNA from bacteria a scaled-up miniprep followed by additional purification was done. This results in relatively large amounts (several micrograms) of very pure plasmid DNA.

Separation of plasmid DNA by agarose Gel electrophoresis

Electrophoresis experiments were carried out in horizontal 1.5% agarose gel. The gel was viewed on an ultraviolet (UV) light box (short wave, ultraviolet products, Inc., San Gabriel, California, USA, 254 nm). The photographs were taken under UV illumination (using Sony cyber shot 5.1 megapixels, USA) as shown in Figure 2.

RESULTS

The results of antimicrobial susceptibility of *E. coli* to 17 antibiotics are represented in Figure 1, where percentage resistance is taken in the y-axis and different antimicrobials agents are taken in the x-axis. Then seven multi-drug resistant isolates of *E. coli* were selected, they are: E- 516, *E.coli*-2, E-371, E-210, E-294, E-653 and E-480. Then antibiotic susceptibility pattern test was carried out with these seven samples in presence of 1% SDS and in absence of SDS. The results are shown in Figure 1 and 2.

Gel electrophoresis

Plasmid profile analysis indicated that bands for plasmid DNA were absent, when multi-drug resistant isolates of *E. coli* were cultured in Luria both in presence of 1% SDS confirming that when the isolates were cultured in presence of 1% SDS, all its plasmids come out of the cells due to the formation of small pores. So no bands for plasmids were obtained in the electrophoretic pattern. But, when the same isolates were not treated with 1% SDS, the electrophoretic pattern showed the presence of plasmid DNA in Figure 2.

Discussion

Resistance to antimicrobials is highly prevalent in bacterial isolates worldwide, particularly in developing countries. Normal intestinal flora is a reservoir for resistance genes; the prevalence of resistance in commensal *E. coli* is a useful indicator of antibiotic resistance in bacteria in the community. Oral administration of antibiotics can also influence the normal intestinal micro flora and can lead to an overgrowth of resistant the correlation between antibiotic resistance and plasmid profile may indicate that the genetic information is plasmid borne (Myaing et al., 2005).

In our six months period of study, it was observed that

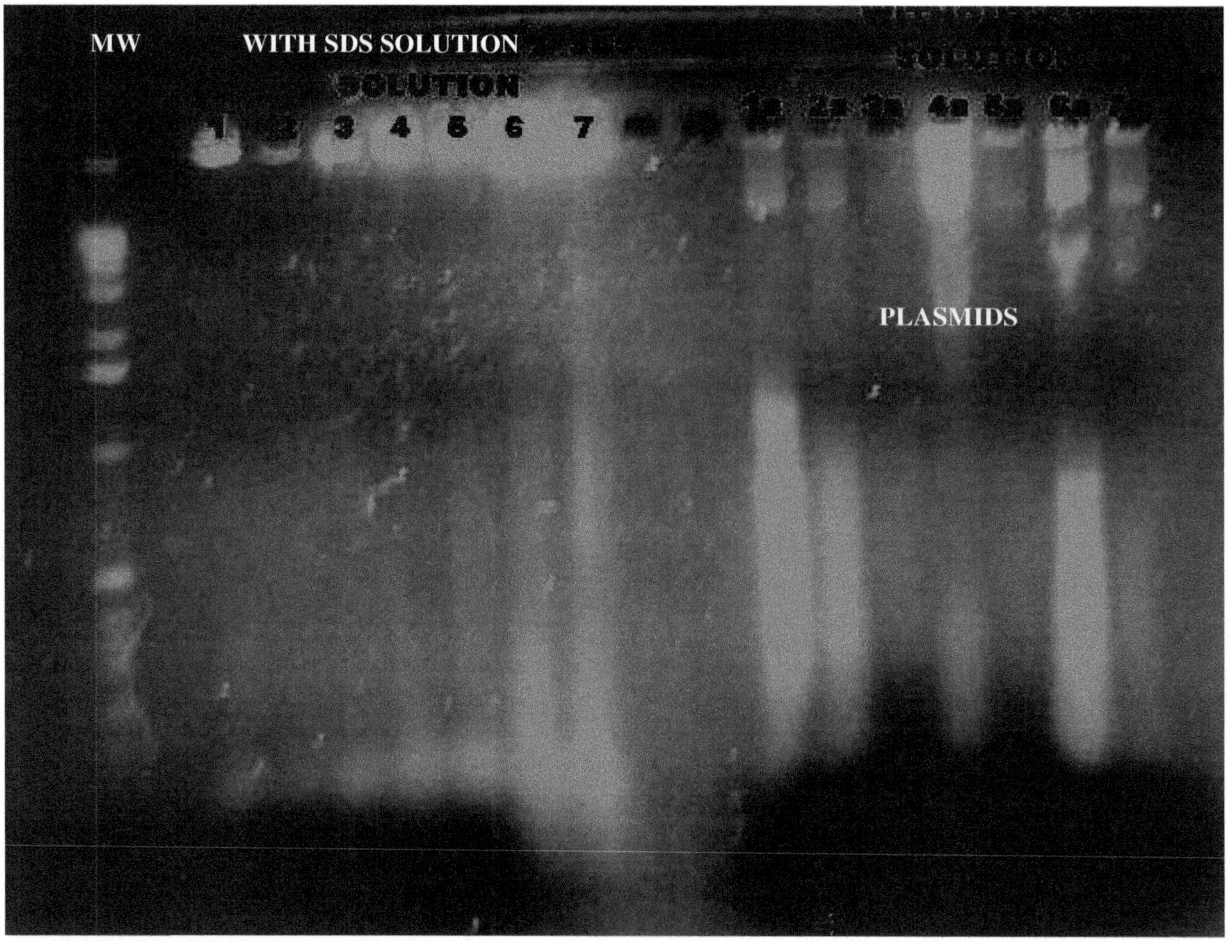

Figure 2. Agarose (1.5%) gel electrophoresis of plasmid DNA from *E. coli* isolates grown in Luria broth in absence and in presence of 1% SDS.

the isolates of *E. coli* collected from different pathological specimen showed different degree of sensitivity to different antimicrobials. Seventeen different antimicrobial agents were used to test the susceptibility. Most of the *E. coli* isolates of this study are resistant to the first line antibiotics that are commonly prescribed by the physicians. All the isolates (100%) were found to be resistant to cephalexin, cephradine, oxacilline, penicillin and vancomycin. 95.45% of the isolates were resistant to ciprofloxacin, cloxacillin and imipenem. 90.91% isolates were resistant to erythromycin. 81% of the isolates were resistant to amoxicillin. 72.72% resistant to co-trimoxazole, 63.64% to tetracycline. 31.81% of the isolates were resistant to ceftriaxone and neomycin. 22.73% is resistant to gentamicin and only 13.64% of the isolates were resistant to chloramphenicol.

However, in the present study isolate number 5 was resistant to Gentamicin, Neomycin and Imipenem when cultured in absence of 1% SDS solution but became sensitive when cultured in presence of 1% SDS solution. Since isolate number 5 has lost its plasmid (Figure 2) after being cultured in presence of 1% SDS (that is, the SDS treated sample lacks plasmid inside the cells) it appears that gentamicin, neomycin and imipenem resistant genes may be present in the plasmid. Similar phenomenon was observed in case of isolate numbers 1 and 7. Isolate number 1 becomes sensitive to rifampicin and imipenem and isolate number 7 becomes sensitive to imipenem after being cultured in presence of 1% SDS, though both the isolates were resistant to respective antibiotics when cultured in absence of SDS. So it appears that in sample number 1, 5 and 7 the resistant gene for the respective antibiotics may be present in the plasmids.

Conclusion

In the study, the result obtained indicates that there appears to be no correlation between plasmid and antibiotic resistant of *E. coli* though there might be exceptions. So, further studies are needed to be carried out in future to confirm this issue. If the gene responsible for multidrug resistance can be located, genetic engineering and further research can be done to prevent *E .coli* from becoming resistance.

REFERENCES

Bauer A, Kirby WMJ, Sherris C, Truck M (1966). Antibiotic susceptibility testing by a standard single disc method. Am. J. Clin. Pathol., 44: 493.

Boucher HW, Talbot GH, Bradley JS (2009). Bad bugs, no drugs: no ESKAPE! An update from Infectious Diseases Society of America. Clin. Infect. Dis., 48: 1–12.

Brüssow HH, Canchaya C, Hardt WD (2004). Phages and the evolution of bacterial pathogens: from genomic rearrangements to lysogenic conversion. Microbiol. Mol. Biol. Rev., 68(3): 560–602.

Levy SB (1994). Balancing the drug-resistance equation. Trends Microbiol., 2(10): 341-2.

Myaing TT, Saleha AA, Arifah AK, Raha AR (2005). Antibiotic resistance and plasmid carriage among *Escherichia coli* isolates from chicken meat in Malaysia applications of gene-based technologies for improving animal production and health in developing countries, pp. 521-527.

National Committee for Clinical Laboratory Standard (1997). Methods for dilution antimicrobial tests for bacteria that grow aerobically, 3rd end. approved standard. NCCLS, Pennsylvania. Document, M7-A3.

Slama TG, Amin A, Brunton SA (2005). A clinician's guide to the appropriate and accurate use of antibiotics: The Council for Appropriate and Rational Antibiotic Therapy (CARAT) criteria. Am. J. Med., 118 (Suppl 7A): 1S–6S.

Todar K (2008). Pathogenic *E. coli*. Online Textbook of Bacteriology. Department of Bacteriology, University of Wisconsin–Madison, USA.

10

Factors influencing multi-drug resistant and plasmid DNA harbouring *Aeromonas hydrophila* isolated from Lake Manzala, Egypt

Mahmoud M. M. Zaky[1]*, Fathy A. Mansour[2] and Kenneth M.Persson[3]

[1]Botany Department, Faculty of Science, Port-Said University, Egypt.
[2]Botany Department, Faculty of Science, Mansoura University, Egypt.
[3]Water Resources Engineering, Lund University, Sweden.

Lake Manzala is considered one of the most important coastal Lakes in Egypt. This study revealed the highly polluted condition of water and fish samples taken from three of the more important sites (Kapoty, Bashtier and Mataryia areas), representative to human activity and different ecosystems in the Lake water environment. Total suspended solids (TSS), ammonia and nitrates, of the samples taken showed figures exceeding the national and international standards. The samples from the study yielded high levels of bacterial content, which were found through plate counts and by using different growth media such as agar and endo agar. Total viable bacteria (TVB) reached more than 10^4 cfu/ml in water samples and 10^5 cfu/g in fish samples, particularly in Kapoty and Mataryia areas. Faecal coliform counts reached 10^2 cfu/ml in water samples and 10^3 cfu/g in fish samples. Statistical analysis revealed a positive correlation between *Aeromonas* spp. counts and some minerals, such as chlorides, calcium and magnesium (r = 0.9, 0.8, and 0.6 respectively; $p < 0.001$), which revealed the interactions and adaptability of such organisms to environmental factors. A total of 88 isolates from *Aeromonas* differential agar were identified using the API 20E system; 17% of these isolates were *Aeromnas hydrophila*, and most of these strains were resistant to three or more antibiotics including pencillinG, ampicillin and chloramphenicol. Isolation of plasmid DNA from these strains of *A. hydrophila*, showed that most harboured plasmids ranging from less than 2.2 - 4 Kbp. Such findings could explain the adaptability and multi-drug resistance of such bacterial pathogens, which may have an impact on the course of infection and active causal agents of diarrhoeal diseases in this important coastal area of Egypt on the Mediterranean Sea.

Key words: *Aeromonas hydrophila*, multi-drug resistance, plasmid DNA, Lake Manzala.

INTRODUCTION

Lake Manzala is considered one of the most important coastal Lakes in Egypt. It is situated in the northeastern part of the country, on the Delta of the Nile River bordering of the Mediterranean Sea. It receives a high load of different polluted waste water effluents including sewage water, industrial and agricultural waste from many drains, such as Bahr El-Bakar, Ramsis, El-Sirw and Hadous. The effluents from these drainages negatively affect the Lake environment to the eutrophic level as well as human health among communities around the Lake.

Aeromonas spp. particularly *Aeromonas hydrophila* and *Aeromonas sobria* are widely distributed in fresh, brackish and saline water environments, as well as in lakes and sewage water; they have also been isolated from drinking water and food, such as meat, fish, shellfish, raw milk and vegetables (Araujo et al., 1990; Gonzalez et al., 2002; Hanninen et al., 1997; Kirov et al., 1990; Montesw

*Corresponding author. E-mail: zakymahmoud@yahoo.co.uk.

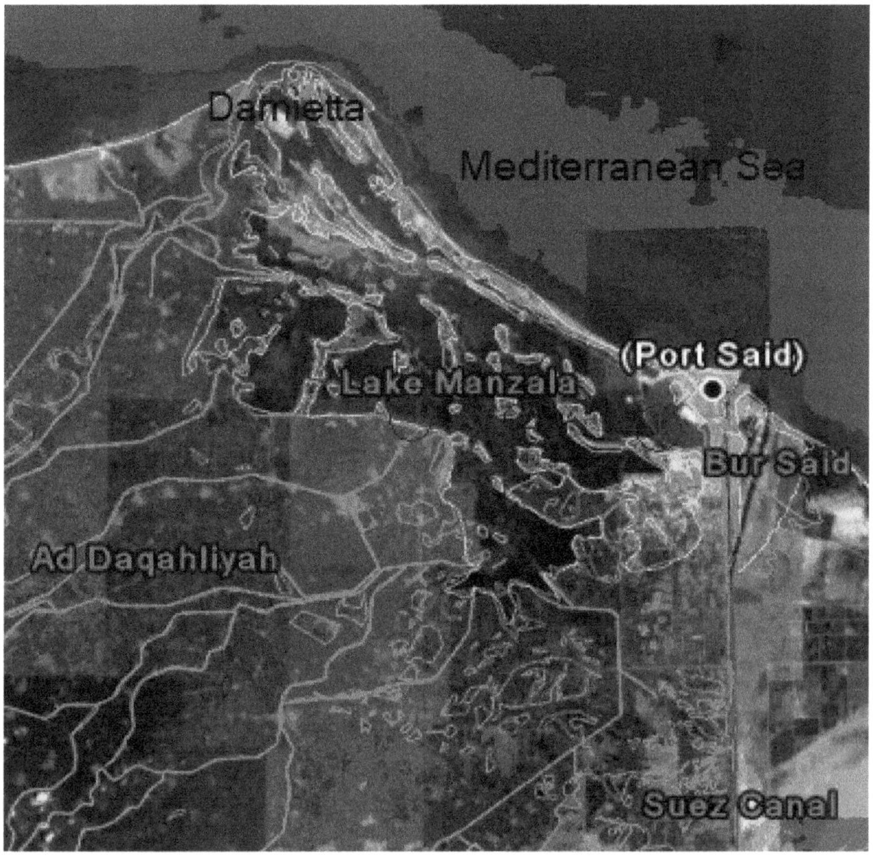

Figure 1. Map of Lake Manzala, showing different sampling sites and sources of pollution.

1990; Montesw et al.,1999; Ortega et al.,1996; Ringo and Vadstein, 1998).

MATERIALS AND METHODS

Sampling location

Lake Manzala is a coastal shallow lake, which consists of 30 basins with depths ranging from 0.7 - 1.5 m and the navigation canal is about 3 m deep (Maclaren et al., 1980). It is situated east of the Nile's Delta between the Damitta branch of the Nile River and the Suez Canal. The Mediterranean Sea is immediately north of the narrow coast, which separates the two water bodies. Lake Manzala did not exist before the 6th century - before then it was an area of cultivated land with fertile soil until a tectonic plate movement caused an earthquake that created the lake (Maclaren et al., 1980). Lake Manzala is quite different today, the construction of canals and drains, such as Bahr El-Bakar, Sirw, Ramsis and Hadous drains have created eutrophic conditions and low salinities in portions of the lake. The area closest to the drain outlets is characterized by brackish water in the south and west; and by saline water in the northeast area near the sea outlets (El-Gamil outlet). The sampling localities were selected close to the most important communities around the Lake and to the areas where the discharge of different types of pollutants occur, which negatively affect the Lake's condition and human health in the area (Figure 1).

Samples were taken from the end of the junction canal, which connects the Suez Canal with Lake Manzala, and the source of pollution in this area is most likely from the Port-Said city. Effluents such as sewage water and different types of industrial waste from many factories are disposed of in the Lake. This site is close to EL-Kapoty village which is a fishermen village in Port-Said and where raw sewage is disposed directly into the lake water, and has been designated as the Kapoty site. The Bashtier sampling site is considered a mid-point between the El-Kapoty and the Mataryia areas; it receives water currents from different directions resulting in high water levels. The depth reaches three meters and is part of the navigation canal. The area has many islets which are inhabited by people who work in fishing and raise animals. The Mataryia area is considered as the fresh-water part of the Manzala Lake; however, it receives high amounts of different types of pollutants. Untreated sewage from the El Dakahlyia governorate disposed of six million m^3/ day of industrial and agricultural waste from the El-Siwr, Hadous, Ramsis and Bahr El-Bakar drains. The drains empty into the El Genka reservoir, a part of the lake that is characterized by vast vegetation composed of reeds and other aquatic plants like the water hyacinth. This area is particularly important for fishing, especially the fishing of *Tilapia* spp.

Sampling methods

One litre water samples was taken from all three sites (Kapoty, Bashtier and Mataryia) on Lake Manzala in clean sterile glass bottles and transported on ice from the lake to the laboratory within six hours for direct examination (Tucker et al., 1990). Fish samples were collected from the Bashtier and Mataryia sites utilizing gill nets and samples were immediately packed in sterile polyethylene cases and preserved on ice. All collected samples were transported and prepared for immediate bacteriological analysis.

Preparation of samples

Water samples were directly diluted with 0.8% saline distilled water for bacterial counting using the dilution and pour plate methods. The fish were dissected and 1 g of the intestinal contents was aseptically stripped out with sterile forceps. Samples of gill lamellae and intestinal contents (1 g) were aseptically resuspended in 100 ml of 0.1% w/v chilled peptone water (pH 7.0), homogenized and poured with different dilutions into Petri dishes (Zaky, 1995).

Physicochemical characterization of water samples

The temperatures of water samples were determined by using an ordinary mercuric thermometer. The pH values were determined using a combination of pH electrode (Eil series 1180) and pH meter (Eill, Model 7030) calibrated against pH 4, 7, and 9 buffers. The salinity of the water was determined using the refractometer model Leica No. 10419. The dissolved oxygen of water was determined using the Oxygen meter Model YSI 58. Chemical analysis of the water for determination of Total Suspended Solids (TSS), ammonia, nitrates, calcium, magnesium and chlorides was done using the recommended standard methods of water analysis (APHA, 1995).

Microbiological analysis

Counts of total viable bacteria were evaluated using nutrient agar, which contains 5 gm of peptone; 5 gm of sodium chloride; 3 gm of beef extract and 15 gm of agar per liter of distilled water used. Incubation was at 37°C for 24 h. Counts of faecal coliforms were conducted with Endo agar medium composed of 10 gm of lactose, 10 gm of peptone, 3.5 gm of dipotassium hydrogen phosphate, 2.5 gm of sodium sulphite, 0.5 gm basic fuchsin, and 15 gm of agar per 1 L of distilled water used. Incubation was at 44°C for 24 h. Counts of *Aeromonas* spp. were determined utilizing *Aeromonas* differential agar media, which contains 15 gm of peptone from casein, 1.6 gm sodium sulphite, 0.25 gm of basic fuchsin, 7.75 gm of disodium hydrogen phosphate, and 15 gm of agar, per 1 L of distilled water. The pH was 7.5, and it was incubated at 37°C for 24 h. Typical colonies of *Aeromonas* spp. were taken from the selective media and identified according to the recommended API 20 E system (bioMerieux) for identification of the family Enterobacteriaceae.

Resistance to antimicrobial agents

The resistance to antimicrobial agents was determined by the MIC method. The bacterial strains of *A. hydrophila* were grown overnight in nutrient broth, then 1 ml of the broth was injected into a Petri dish and nutrient agar media (50°C) was poured on top. Different antibiotic discs (Oxoid) were inserted into the plates before complete solidification of the media and incubated at 37°C for 24 h. Results were obtained by observing and measuring the clear zones around each antibiotic disc. The absence of clear zones was an indication of the resistance of the bacteria to the antibiotics. The antibiotics used were chloramphenicol 30 mcg, ampicillin 10 mcg, penicillin G 10 mcg, streptomycin 10 mcg and gentamycin 10 mcg (Zaky, 1995).

Isolation of plasmid DNA

The alkaline lysis method was used in mini preparations as follows: 5 ml of medium was inoculated with a serial colony and incubated at 37°C overnight with vigorous shaking; 1.5 ml of the culture was poured into an Eppendorf tube and centrifuged. The remainder of the overnight culture was stored at 4°C. The solution was poured off and the last drop was removed with a Pasteur pipette. The pellet was resuspended by vortexing in 100 µl of an ice-cold solution consisting of 50 mM glucose, 10 mM EDTA, 25 mM tris-HCL (pH.8.0), 4 ml / ml lysozyme, and stored for 5 min at 37°C in a water bath. A 200 µl of freshly prepared solution of 0.2 M NaOH, 1% SDS was added and the contents were mixed by inverting the tube gently, two or three times and then stored on ice for 5 min. A volume of 150 µl of an ice cold solution of potassium acetate (pH 4.8) was added and prepared as follows: 11.5 ml of glacial acetic acid and 28.4 ml of water were added to 60 ml of 5 M potassium acetate, the resulting solution was mixed 10 times and stored on ice for 5 min, and the supernatant was transferred to a fresh tube. Two volumes of ethanol were added at room temperature and mixed gently by hand and left at room temperature for 2 min; the solution was then centrifuged for 30 min in an Eppendorf centrifuge at room temperature. The supernatant was removed and the tube was left in an inverted position on a paper towel to allow all the fluid to drain; 10 - 50 ml of TE (pH 8.0) containing DNase free pancreatic RNase (20 µg/ml) was added and vortexed briefly, then 20 µl of the solution was removed to a new Eppendorf tube, and 3 µg of loading buffer was added and then injected into the gel. The remainder of the preparation was stored at -20°C (Maniatis et al., 1984).

Statistical analyses

The correlation matrix is the relationship between a numbers of data sets that are scaled to be independent of the unit of measurement. It is used to determine whether data sets move together; that is whether large values of one set are associated with large values of the other (positive correlation); whether small values of one set are associated with large values of the other (negative correlation); or whether the values of the sets are unrelated (Fowler and Cohen, 1996).

RESULTS

Physical and chemical properties were estimated for all water samples collected from the Kapoty, Bashtier and Mataryia sites seasonally from autumn to summer. Most samples taken from these sites exceed the limits of pollution standards, such as TSS, ammonia and nitrate, respective to their different levels of salinity (Table 1).

Bacteriological counts were higher in the summer months; particularly in the water samples of most sites, with seasonal variation in counts. Total viable bacterial counts (Figure 2) reached a mean value of 20 and 112 cfu.ml^{-1}. Faecal coliform counts (Figure 3) reached a mean value of 776 cfu.ml^{-1} and *Aeromonas* spp. counts (Figure 4) reached a mean value of 206 cfu.ml^{-1}. There were obvious variations of total viable bacteria counts (Figure 5); faecal coliforms (Figure 6) and *Aeromonas* spp. (Figure 7) in different anatomical parts of fish from different sampling sites. The highest mean value counts (102756.3 cfu.g^{-1}) of total viable bacteria were recorded in the intestine of fish taken from the Mataryia area (Table 2).

The correlation matrix of the environmental data and *Aeromonas* spp. counts has shown variation of correlations

Table 1. Mean values of seasonal variations of different physico-chemical parameter in different sites (Mean ± standard deviation and range).

Parameter	Kapoty	Mataryia	Bashtier
Temp (°C)	21.7 ± 1 (21-23)	21.5 ± 1 (21 -23)	21 ± 1 (20 – 22)
pH	9.2 ± 1 (8.2 – 10.2)	8.2 ± 1.3 (6.9 -9.9)	8.2 ± 1 (7.2 -9.2)
Salinity	16 ± 1 (15-17)	4.3 ± 1 (3.3 – 5.3)	8.3 ± 1 (7.3 – 9.3)
D.O (mg/l)	3 ± 1 (2 – 4)	4.3 ± 1 (1.2 – 2.2)	3.4 ± 1 (2.4 -4.4)
TSS (mg/l)	127 ± 1 (126 - 128)	50 ± 10 (40 - 60)	67 ± 1 (66 - 68)
Ammonia (mg/l)	1.2 ± 0.1 (1.1-1.3)	2.0 ± 1 (1 – 3)	1.9 ± 0.1(1.8 –2)
Nitrate (mg/l)	1.6 ± 0.1 (1.5 – 1.7)	1.3 ± 0.8 (0.5 – 2)	0.1 ± 0.0 (0.11 – 0.13)
Chlorides (mg/l)	5500 ± 1000 (4500 -6500)	1300 ± 0.0 (1300 -1300)	2300 ± 533.9 (1400 – 2700)
Calcium (mg/l)	1088 ± 11 (1077 - 1099)	210 ± 10 (200 – 220)	260 ± 10 (250 – 270)
Magnesium (mg/l)	433 ± 11 (422 - 444)	73 ± 1 (72 – 74)	230 ± 10 (220 – 240)

Table 2. Mean values of seasonal variations of different bacterial species in different sites (Mean ± standard deviation and range).

	Kapoty	Mataryia	Bashtier
T.V.B	4700 ± 100(4600 -4800)	20150 ± 50 (20100 -20200)	8900 ± 4273.2 (1500 – 11500)
F.C	776 ± 1(775 - 777)	151 ± 1 (150 – 152)	59 ± 1 (58 – 60)
Aeromonas spp.	183 ± 1(182 - 184)	206 ± 1 (205 – 207)	152 ± 1 (151 – 153)

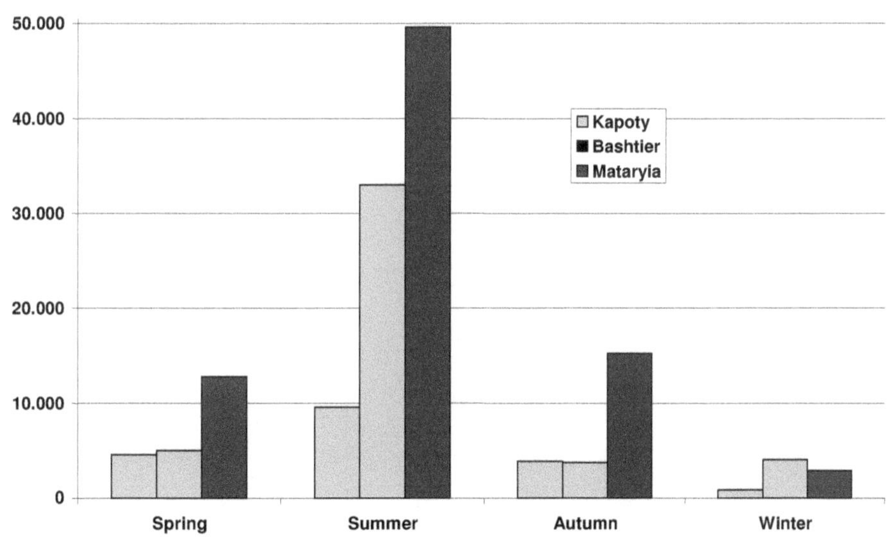

Figure 2. Total viable bacterial counts of water samples taken from different sites of Lake Manzala (cfu ml^{-1}).

correlations in different sites. The data from the Kapoty site demonstrated that there were significant strong correlations between *Aeromonas* spp. counts with ammonia, nitrate and magnesium (r = 0.96036, 0.883244 and 0.866242, respectively, p <0.001), and significant strong correlations with calcium, chlorides and salinity (r = 0.780722, 0.635003 and 0.548988, respectively, p < 0.001). The data from the Bashtier demonstrated that there were significant strong correlations between *Aeromonas* spp. counts with calcium, magnesium and salinity (r = 0.996895, 0.849515 and 0.812784, respectively, p < 0.001) and significant strong correlation

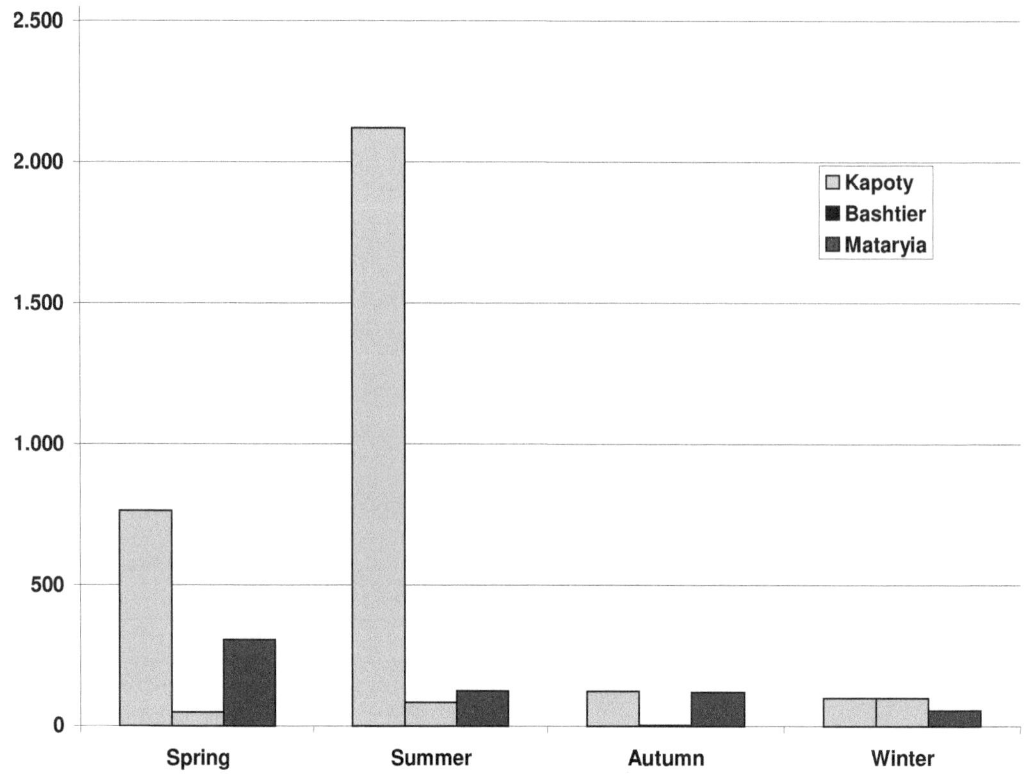

Figure 3. Counts of faecal coliforms of water samples taken from different sites of Lake Manzala (cfu ml^{-1}).

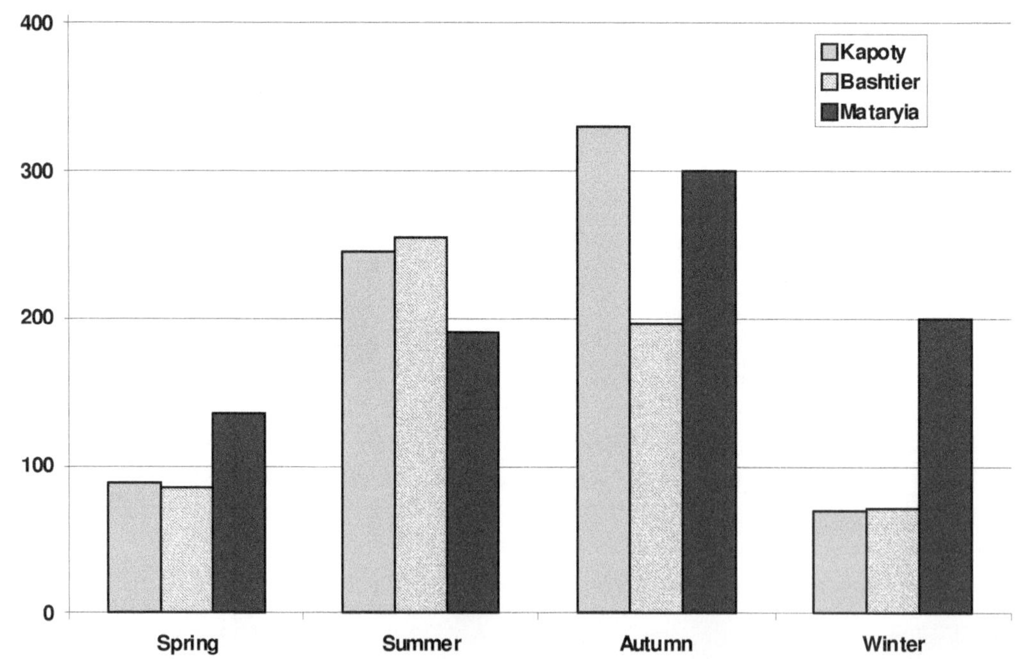

Figure 4. Counts of *Aeromonas* spp. of water samples taken from different sites of Lake Manzala (cfu ml^{-1}).

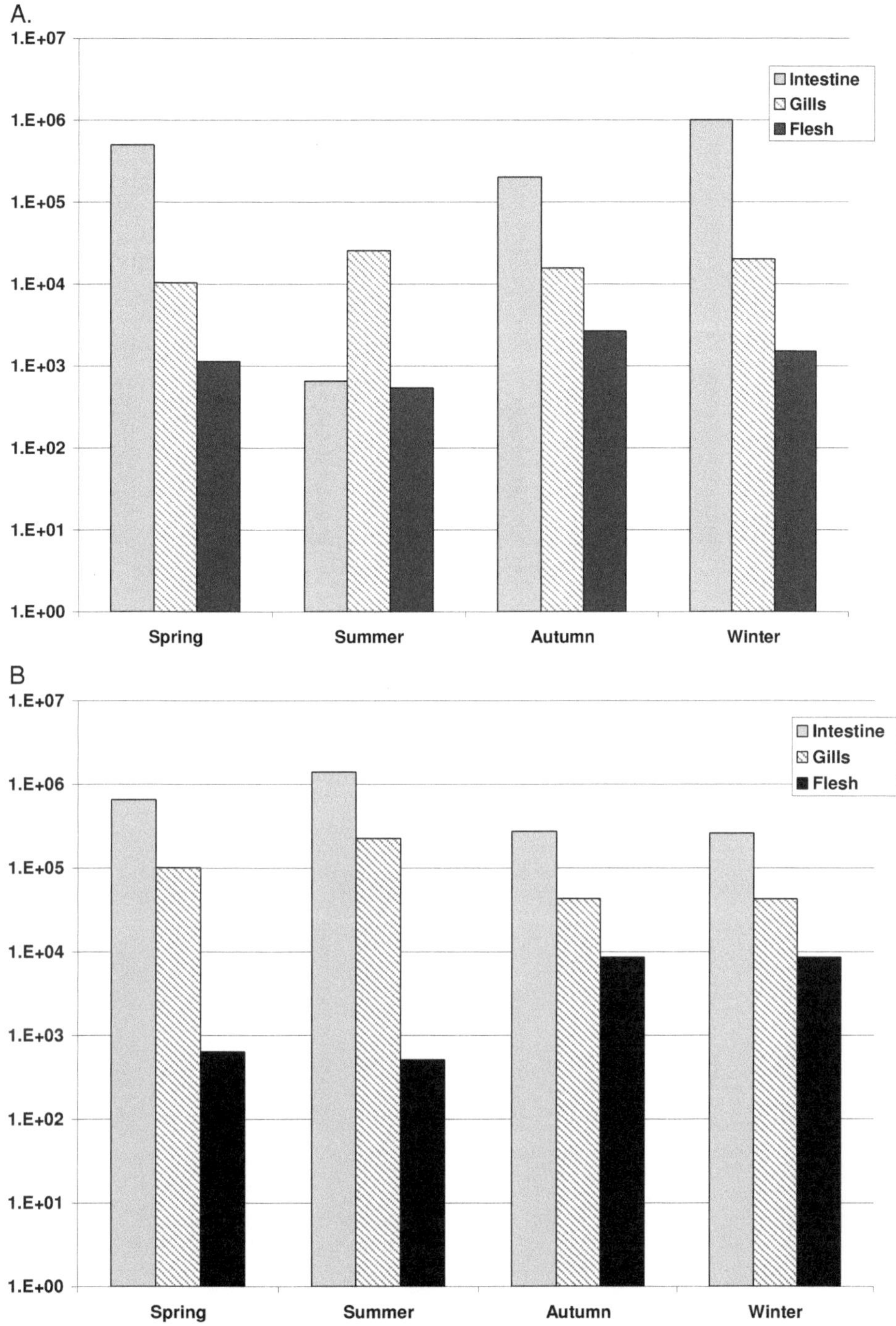

Figure 5. Counts of total viable bacteria (cfu g^{-1}) of different anatomical parts (intestine, gills and flesh) of fish taken from A) Bashtier area and B) Mataryia area of Lake Manzala (cfu g-1).

with chlorides ($r = 0.795329$; $p < 0.001$). The data from the Mataryia demonstrated that there were significant strong correlations between *Aeromonas* spp. counts with nitrate and magnesium ($r = 0.912488$ and 0.828159, respectively, $p < 0.001$) and there was a significant strong correlation with salinity ($r = 0.521413$, $p < 0.001$).

Antibiotic resistance of *A. hydrophila*

Antibiogram figures showed variation of the susceptibility and resistance of different strains isolated from water and fish. Some strains were only resistant to ampicillin and penicillin G, others were resistant to three or more antibiotics

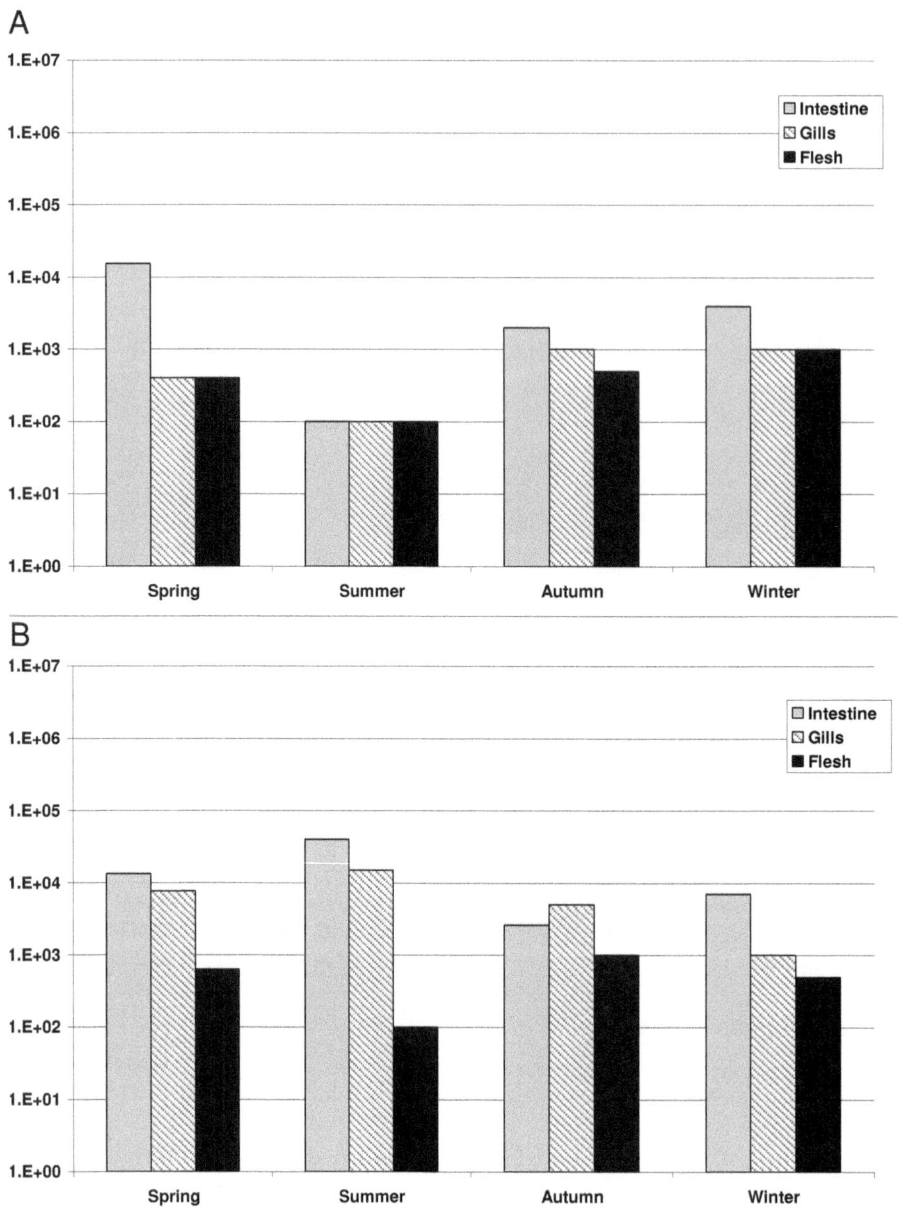

Figure 6. Counts of faecal coliforms of different anatomical parts (intestine, gills and flesh) of fish taken from A-(Bashtier area) and B- (Mataryia area) in Lake Manzala.

antibiotics and two strains isolated from fish were completely resistant to the five antibiotics used in this study (Table 3).

Plasmid DNA profiling of *A. hydrophila*

The Plasmid DNA profiling using agarose gel showed variation of the different strains isolated from water and fish in number and size with regard to λDNA digested with Hind III, where some strains that harbored plasmids less than 2 Kb_P, others had plasmids more than 2.3 Kb_p and others were equal to or more than 4.3 Kb_P (Table 4 and Figure 8).

DISCUSSION

There are many studies that have established the polluted condition of Lake Manzala, as the worsening of physicochemical and bacteriological conditions have adversely affected the lake's ecosystem. For example, during the 1921 - 1926 the lake water was marine brackish but now is brackish fresh (Awadi and Khalil, 1989)

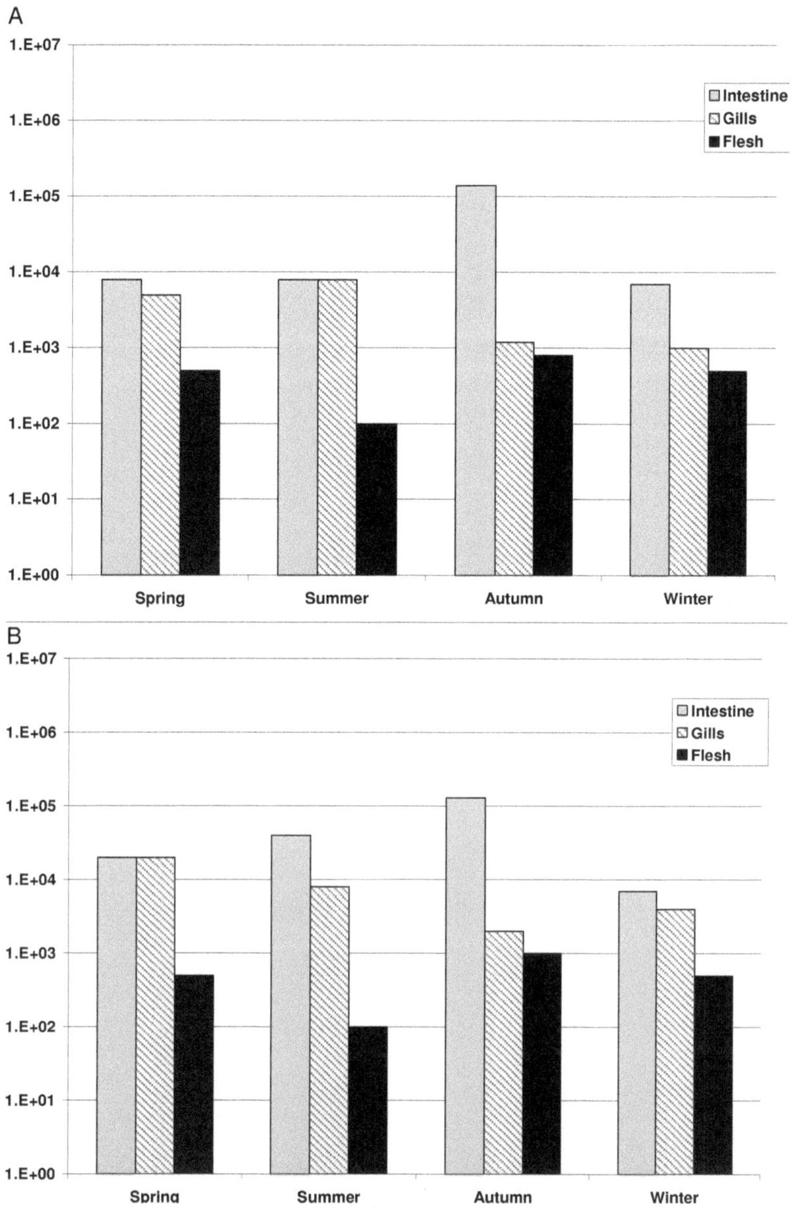

Figure 7. Counts of *Aeromonas* spp. of different anatomical parts (intestine, gills and flesh) of fish taken from A-(Bashtier area) and B- (Mataryia area) in Lake Manzala.

Table 3. Mean values of different bacterial counts (cfu.g^{-1}) in different anatomical parts of fish in tow sampling sites of Lake Manzala.

Sampling site/ anatomical parts	Total viable bacteria	Faecal coliforms	*Aeromonas* spp.
Bashtier			
Intestine	425.675	5,345	40.750
Gill	17.933	625	3.800
Flesh	1.458	500	475
Mataryia			
Intestine	646.648	15,750	49.250
Gill	102.756	7,162.5	8.500
Flesh	4.553	558.75	525

Table 4. Antibiogram of the multi-drug resistant *A. hydrophila* isolated from water and fish taken from Lake Manzala.

Strain No.	Source	Ampicillin	Penicillin G	Gentamycin	Streptomycin	Chloramphenicol
8W	Water	R	R	S	S	S
18W	Water	R	R	S	S	S
22W	Water	R	R	S	S	S
29W	Water	R	R	S	S	S
30W	Water	R	R	R	S	S
32W	Water	R	R	S	R	S
42W	Water	R	R	S	S	S
45W	Water	R	R	S	S	S
50W	Water	R	R	S	R	R
52W	Water	R	R	S	R	S
62W	Water	R	R	S	R	R
70W	Water	R	R	S	S	S
72W	Water	R	R	S	R	S
74W	Water	R	R	S	R	R
4F	Fish	R	R	S	S	S
12F	Fish	R	R	S	S	S
30F	Fish	R	R	R	R	S
43F	Fish	R	R	S	R	S
51F	Fish	R	R	R	R	R
63F	Fish	R	R	S	R	S
71F	Fish	R	R	R	R	R

R-Resistant; S- Susceptible.

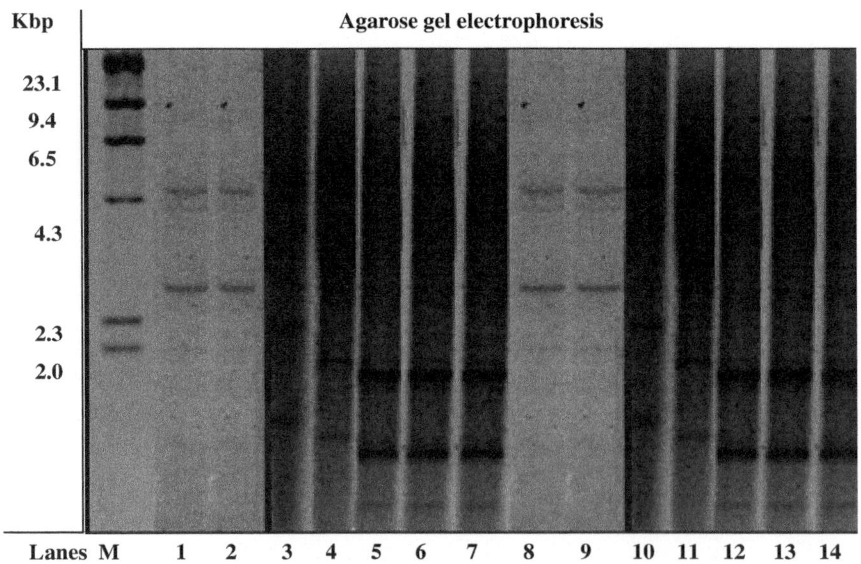

Figure 8. Plasmid profiling of multi-drug resistant *A. hydrophila* isolated from water and fish of Lake Manzala (M: λ DNA digested with Hind III, Lanes from 1 - 14 represented AH strains No. 8, 42, 62 and 70 W, 71, 63 and 30 F, 52, 72, 50 and 74 W, 51, 43 and 12 F, respectively, where W: from water and F: from fish).

1989). The effluents of fresh water from drains like Bahr El-Bakar which oppose the current from the Mediterranean Sea resulting in slow water movement to the point where the Lake is almost stagnant or at least is without currents. The heavy loads of suspended solids and organic matter have raised the bed of the lake and have created shallow water depths. This study revealed the high densities of microbial populations particularly those of bacteria.

This study revealed high records of pollutants, like ammonia, the presence of these can be explained by the anoxic conditions created by the depletion of oxygen. In addition, high suspended particles whether of organic or mineral origin adsorb bacteria to their surfaces. The adsorption of debris particles not only provides microbes with a more favourable nutritional environment than those found in free water but also neutralizes inhibitory and toxic substances. Thus, the suspended particles have a favourable effect on bacterial growth. Faecal coliform counts, well known water pollution indicators, were performed on the different anatomical parts of the fish samples from the Kapoty and Mataryia areas. These counts were used in the study to assess and understand the level of pollution of Lake Manzala. The deterioration of the environmental conditions in Lake Manzala can be attributed to the people inhabiting the region and their role as an element of the ecosystem. The complaints and ever increasing recorded cases of diarrhea, gastroenteritis, kidney failure and other diseases may be partially explained by the aforementioned lake condition. Not only the Lake Manzala water samples but also the fish samples were found to have very high pathogenic bacteria contents (Uerbia et al., 2003; Zaky, 1995).

Aeromonas spp. is always found in faecal polluted water such as waste water ponds, fresh and brackish eutrophic estuaries and lagoons in other parts of the world, as well as in the water, sediment, soil and fish of Lake Manzala's water environment (Chao et al., 2003; Marcel et al., 2002; Sumpono et al., 2003). The statistical analysis of previous studies revealed a strong correlation between counts of *Aeromonas* spp. and faecal coliforms in Lake Manzala, a brackish water environment. Thus *Aeromonas* spp. can be a powerful tool for the assessment of the microbial pollution in brackish water environments (Zaky, 2005; Zaky et al., 2005). In other studies *Aeromonas* spp. dominated other species, both in water and fish samples taken from Lake Manzala, where they represented 44% of the total; this result is in conformity with other studies which described the Mediterranean region as an ideal environment for the proliferation of *Aeromonas* spp., and with the established correlation between the distribution patterns of *Aeronomas* spp. and level of faecal pollution in waters, particularly of the lakes in this area (Araujo et al.,1990; Ben-Dan and Stone,1991).

In this study a strong positive correlation between counts of *Aeromonas* spp. and minerals such as calcium and magnesium was found (r-0.996895 and 0.849515, respectively) as well as with salinity and chlorides (r-0.812784 and 0.795329, respectively). Such positive correlations indicate the importance of the brackish water environment in physiological and genetic terms as well as in the behaviour of the aforementioned microorganism. The manipulation of the growth media (SS Agar, TCBS and Endo Agar) and its effect on the growth of *Aeromonas* spp. attests to the ability of the species to adapt to different environments. Aeromonas spp. shares similar characteristics with the members of the Enterobacteriaceae family more specifically with the *Salmonella* and *Shigella* spp. (Delaat, 1984). The haemolysins produced by *A. hydrophila* and *A, sobria* (Hirono et al., 1992; Tucker et al., 1990) are virulence-linked factors. *A. hydrophila* is a well-known fish pathogen that causes a serious disease in fish and ultimately leads to death; the ability of these species to live on fish maximizes the health risks they pose to humans (Popoff, 1984; NitetoandEllis, 1986). *A. hydrophila* is probably partially responsible for the decrease in the number of fish caught in Lake Manzala because of its toxic activity and as a common species in the Lake Manzala environment (Gurwith et al., 1978; Zaky, 1995; Mansour et al., 2005). *A. hydrophila* and *A. sobria* are opportunistic pathogens that have cytotoxic effects. These pathogenic bacteria thrive in polluted water environments like those found in many developing countries. Their roles as human pathogenic agents causing diarrhoeal and other gastrointestinal diseases have been repeatedly established in studies (Ashbolt, 2004; Dumontet et al., 2003; Pianietti et al., 2004).

In this study, *A. hydrophila* showed a resistance pattern to different antibiotics, particularly ampicillin and penicillin. The resistance is an indication of the presence of β-lactamases, common in bacterial pathogens in polluted water environments in this area, and common in the Egyptian environment. These environments are rich in nutrients like calcium, magnesium and chlorides, and there is a strong correlation between the existence of these nutrients and *Aeromonas* spp. counts in brackish water environments (Zaky, 2006, unpublished). The presence of these minerals stretches bacterial cell walls facilitating the transfer of genetic material, particularly the transfer of antibiotic resistance plasmids.

The different strains of *A. hydrophila* found in the Lake; particularly strains No. 50 W which harbours four plasmids, two of them more than 2 and 4.3 Kbp and is resistant to four antibiotics; and strain No. 52F which is resistant to five antibiotics, and harbours three plasmids less than 2 Kbp, could explain the effect that the water environment has on genetic transfer among bacteria, which could be congruent with environmental changes and factors that enable such processes. The antibiotic resistance found in the fish tissue samples may be due to conditions controlled by the physiological processes and the immune systems of the fish. These conditions may cause the loss of some plasmids or enhance the incorporation of these with chromosomal DNA. The stabilization of drug-resistance is essential for such bacteria to express virulent characteristics (Heesemann et al., 1993; Rakin et al., 1999). The morbidity of diarrhoeal and gastrointestinal diseases has increased in the populations residing along the coastal areas of Egypt where the staple is seafood. The increase of such illnesses is presumably due to the prevalence of toxic *Aeromonas* spp. The importance of studying the mode of infection of pathogenic *A. hydrophila,* the epidemiology of diarrhoeal and gastrointestinal diseases along with the accurate clinical diagnosis of diarrhoeal cases in this area is evident.

REFERENCES

APHA. (1995). Standard methods for examination of water and waste-water, 19th ed. Washington, American Public Health Association.

Araujo RM, Pares R, Lucena F (1990). The effect of terrestrial effluents on the incidence of the *Aeromonas* spp. in costal waters. J. Appl. Bacteriol. 69: 439-44.

Ashbolt NJ (2004). Microbial contamination of drinking water disease outcomes in developing regions. Toxicology. May 20; 198 (1-3): 229-38.

Awadi MHM, Khalil MT (1989). Upgrading of physico-characteristics of Lake Manzala, Egypt. National Res. Centre, Giza (Egypt), water pollution control Dept.

Ben TB, Stone L (1991). The distribution of faecal pollution indicator bacteria in Lake Kinneret. Wat.Res.Vol.25.No. 3. pp. 363-270

Chao KK, Chao CC, Chao WL (2003). Suitability of the traditional microbial indicators and their enumerating methods in the assessment of fecal pollution of subtropical freshwater environments. J Microbiol Immunol Infect., 36(4): 288-93

Delaat ANC (1984). Microbiology for allied health professions. Lea and febiger. Pp 199-201.

Dumontet S, Pasquale V, Mancino M, Normanno G, Krovacek K (2003). Incidence and characterisation of *Aeromonas* spp. in environmental and human samples in southern Italy. New Microbiol., 26(2): 215-2.

Fowler J, Cohen L (1996). Practical statistics for field Biology. John Willy and Sons., England.

Gonzalez-Serrano CJ, Santos JA, garcia-Lopez ML, Otero A (2002). Virulence markers in *Aeromonas hydrophila* and *Aeromonas veronii* biovar sobria isolates from freshwater fish and from diarrhoea case. J. Appl Microbiol., 93(3): 414-9.

Gurwith M, Cumberhatch, n., Sack, B. (1978).Exotoxin produced by *Aeromonas hydrophila*. Abst, Clin. Res, 26, 369A.

Hanninen ML, Oivanen, p. and Hirvela-Koski, V. (1997). *Aeromonas* species in fish, fish-eggs, shrimp and freshwater. *Int J Food Microbiol*; 34 (1): 17-26.

Heesemann J, Hatke, K., Vocke, T, Saken E, Rakin A, Stojiljkovic I, Berner R (1993). Virulence of *Yersinia enterocolitica* is closely associated with sidrophore production, expression of an iron-responsible outer membrane protein of 65000 Da and pesticin sensitivity. Mol. Microbiol.8: 397-408.

Hirono I, Aoki T, Asaoowd T, Kozaki S (1992). Nucleotide Sequences and characterization of haemolysin genes from *Aeromonas hydrophila* and *Aeromonas sobria*, Microbial pathogenesis. 13: 433-446.

Kirov SM, Anderson MJ, McMeekin TA (1990). A note on *Aeromonas* spp. from chickens as possible food-born pathogens. *J. App. Bacteria*. 68: 4, 327-337.

Krovacek K, Faris A, Baloda SB, Lindberg T, Peterz M, Mansson I (1992). Isolation and virulence profiles of *Aeromonas* spp. From different municipal drinking water supplies in Sweden. *Food Microbiology*. 9: 215-22.

Maclaren JF, Stevenson M, Kellogg R (1980). Lake Manzala study (Aquaculture site selection report). Ministry of development and new communities and United Nations development programme, office of projects execution.

Maniatis T, Fritsch EF, Sambrook J (1982). Molecular cloning, a laboratory manual. Cold spring Harbor laboratory, pp.149-171.

Mansour FA., Zaky MM, Salem MA (2005). Environmental and health significance of toxigenic *Aeromonas sobria* and *Aeromonas hydrophila* in polluted water of Lake Manzala,Egypt. 26 th African health sciences congress. November 28 th-December 1 th, Ain Soukhna, Egypt.

Marcel KA, Antoinette AA, Mireille D (2002). Isolation and characterization of *Aeromonas* species from an eutrophic tropical estuary. Mar Pollut Bull.Dec; 44(12): 1341-4.

Montesw M, Perez MJ, Niteto TP (1999). Numerical taxonomy of gram-negative, facultative anaerobic bacteria isolated from skin of turbot (*Scophthalmus maximus*) and surrounding water. Szst Appl. Microbiol., 22(4): 604-18.

Niteto TP, Ellis AE. (1986). Characterization of extra cellular mettalo and serine-proteases of *Aeromonas hydrophila* strains B 51. J.G.M. 132.

Ortega C, Muzquiz JL, Fernandez A, Ruiz I, De Blas I, Simon ME, Alonso JL (1996). Water quality parameters associated with *Aeromonas* spp.- affected hatcheries. Vet Res., 27(6): 553-60.

Pianietti A, Sabatini L, Bruscolini F, Chiaverini F, Cecchetti G (2004). Faecal contamination indicators, *Salmonella*, *Vibrio* and *Aeromonas* in water used for the irrigation of agricultural products. Epidemiol Infect. Apr; 132(2): 231-8.

Popoff M (1984). *Aeromonas*. In Krieg, N.R., Holt, J.G (eds). Bergy s manual of systematic bacteriology. Williams and Wilkins, Vol 1, pp. 545-548.

Rakin A, Noelting C, Schubert S, Heesemann J (1999). Common and specific characteristics of the high-pathogenecity island of *Yersinia enterocolitica*. Infect. Immun., 67 (10): 5265-74.

Ringo E, Vadstein O (1998). Colonization of *Vibrio pelagius* and *Aeromonas caviae* early developing turbot (*Scophthalmus maximus* L.) larvae. J. Appl Microbiol; 84 (2): 227-33.

Sumpono PP, Belan A., Forestier C, Lavedrine B, Bohatier J (2003). Effect of Diuron on aquatic bacteria in laboratory-scale wastewater treatment ponds with special reference to *Aeromonas* species studied by colony hybridization. Chemosphere, Jan; 50(3): 445-55.

Tucker AD, Parker MW, Tsernoglou D, Buckley JT (1990). Crystallization of perform of Aerolysin, a Hole-forming toxin from *Aeromonas hydrophila*, *J. Mol. Biol.* 212:561-562.

Zaky MM (1995). Microbiological and Toxicological study of the environmental pollution of Lake Manzala. M.Sc. Thesis. Suez Canal University, Egypt.

Zaky MM (2004). Study on the microbial pollution and molecular biology of some pathogenic *Yersinia* species in Lake Manzala

Comparison of Norwegian poultry, waterborne and clinical isolates of *Campylobacter jejuni* by ribotyping

Olav Rosef[1]*, Algimantas Paulauskas[2], Nils Grude[3], Camilla Haslekås[1] and Andrew Jenkins[3]

[1]Telemark University College, Hallvard Eikas Plass 1, 3800 Bø i Telemark, Norway.
[2]Vytautas Magnus University, Kaunas, Lithuania.
[3]Unilabs Telelab AS, Skien, Norway.

Campylobacteriosis is the most frequently occurring food and waterborne gastroenteritis in Norway and 90% of cases are caused by *Campylobacter jejuni*. Campylobacteriosis is epidemiologically associated with use of non-disinfected water and consumption of contaminated poultry. In order to provide molecular epidemiological support for this association isolates from local poultry and water were compared with human isolates from domestic and import infections using ribotyping. Automatic riboprints were performed with the *PstI* restriction enzyme and RiboPrinter® and compared with a library of recognised patterns (DUP-IDs). Patterns were further compared with each other using GelCompar software. The isolates from human clinical cases showed high heterogeneity. DUP-IDs found among human isolates were also found in 2 isolates from poultry, DUP-PSTI-1146 and DUP-PSTI-2061. Two human isolates were like isolates from water, DUP-PSTI-2073 and DUP-PSTI-1122, but these were from patients infected abroad. This study provides limited support for the importance of poultry as a source of infection and illustrates the need for studies of much larger scope to encompass the huge diversity of strains and sources of *C. jejuni*.

Key words: *Campylobacter,* similarity, clustering, genotyping, riboprint.

INTRODUCTION

The incidence of human cases of *Campylobacter* infection has increased and recently exceeded that of *Salmonella* in many European countries (http://www.efsa.europa.eu/en.html) and in the United States (Altekruse et al., 1999) and is the most frequently reported cause of bacterial gastrointestinal illness. Interest has been focused on the identification of risk factors and effective control measures. The pathogenic thermophilic campylobacters, represented by *Campylobacter jejuni, C. coli, C. lari,* and *C. upsaliensis,* are zoonotic with many animals serving as reservoirs for human diseases including rabbits, rodents, wild birds, sheep, horses, cows, pigs, poultry and domestic pets (Rosef et al.,1983; Kapperud and Rosef, 1983; Brown et al., 2004; Johnsen et al., 2006a). Although much work has been done to identify the environmental and animal reservoirs of *Campylobacter* knowledge of pathogenesis and epidemiology is still incomplete. Case-control studies have identified a range of different risk-factors for infection and consumption of chicken is the most frequently reported risk-factor (Stafford et al., 2008). In industrialized countries, cases of human campylobacteriosis are about 90% due to *C. jejuni* and 10% due to *C. coli* (Kapperud, 1994).

Campylobacter are apparently fragile organisms that in the environment are unable to grow in the presence of air or multiply outside warm-blooded hosts and are highly susceptible to a number of environmental conditions (Park, 2002). Murphy et al. (2006) summarize the mechanisms by which *Campylobacter* spp. adapt to stress conditions and thereby increase their ability to survive on food and in the environment.

In addition to epidemiological studies, reliable and powerful typing methods for comparing campylobacters are necessary in order to gain more insight into infection routes. Traditionally, phenotyping methods such as serotyping and biotyping have been used. The drawbacks of these methods are their restricted resolution, the lack of

*Corresponding author. E-mail: olav.rosef@hit.no.

specific reagents for serotyping and a large proportion of untypable strains. Using genotyping to determine genetic relatedness is necessary to provide data for a better understanding of the epidemiological aspects. Broman et al. (2004) found a high degree of diversity among *C. jejuni* isolates from migrating birds with pulse-field gel electrophoresis (PFGE) using the restriction enzymes *Sma*I and *Kpn*I. A high degree of diversity was found in outdoor environmental isolates (Johnsen et al., 2006a) and among isolates from cattle (Johnsen et al., 2006b) by using *Bgl*II and *Mfe*I and amplified fragment length polymorphisms (AFLPs). Ge et al. (2006) used both PFGE with the restriction enzyme *Sma*I and riboprinting with *Pst*I for genotyping and found genetic diversity among isolates from retail meats. They found similarity among human and environmental strains. Multilocus sequence typing (MLST) has been used by Lévesque et al. (2008) to investigate the genetic diversity among *C. jejuni*. Nielsen et al. (2000) compared Riboprinting to other typing methods and showed it to be intermediate in discrimination with good correspondence with other methods. It has the advantage that the suppliers provide a continually updated database of patterns (DUP-ID) for the restriction enzyme *Pst*I, allowing the patterns identified to be equated to those found by other users of the RiboPrinter®.

In Norway and other developed countries contaminated water, poultry meat and contact with domestic animals have been identified as risk factors for campylobacteriosis (Kapperud et al., 1992; Bryan and Doyle, 1995, Blaser, 1997; Rautelin and Hänninen, 2000; Kapperud et al., 2003). In this study we use automated ribotyping (Riboprinting) to compare *C. jejuni* isolates from poultry, river water and humans infected in Norway and abroad in order to investigate whether the epidemiological data can be corroborated by molecular typing.

MATERIALS AND METHODS

Bacterial strains

Clinical *C. jejuni* isolates from patients with diarrhoea were from stool samples sent to Unilabs Telelab for routine analysis. Patients were from the county of Telemark or from adjoining regions of the neighbouring county of Vestfold. Samples were spread on Preston agar plates (Mast, Merseyside, UK) and incubated in a microaerophilic atmosphere for 48 h at 40°C. *C. jejuni* were identified by colony morphology, motility, oxidase test and hippurate degradation. Isolates were randomly selected among 108 *C. jejuni* isolated in the period March to September 2008 according to the following criteria. Samples lacking clinical information on symptoms or where the patient had been infected were excluded. Only one sample per patient was included. Equal (±2) numbers of isolates from patients infected in Norway and abroad were selected. Among the latter group, equal numbers of infections from Europe, America, Africa and Asia were chosen. Twenty two isolates were from domestic infections while 20 were infected abroad. 34 *C. jejuni* were from 2 poultry flocks in Telemark isolated over a period of 2 months in 2006 and 17 isolates were collected throughout 2006 from Bø river in Telemark County. The isolates were kept frozen at -70ºC in Microbank™ (Pro-Lab Diagnostics, Canada) and re-cultivated on CCDA agar (Oxoid Ltd, Basingstoke, Hampshire, United Kingdom) for further examination.

PCR-assay

Species identification by multiplex PCR

The colony multiplex-PCR test described by Wang et al. (2002) was used for species confirmation. This test generates PCR products of characteristic length from the *hipO* gene of *C. jejuni*, the *glyA* gene of *C. coli, C. lari and C. upsaliensis* and the 23s RNA gene of all these species, which serves as an internal control. A loopful of material from a single colony was suspended in 500 µl of water, heat treated at 100ºC for 10 min and centrifuged. 5 µl of the supernatant was used as a template. PCR Reactions were run on the iCycler (BioRad, Hercules, CA) in a 50 µl volume using QIAGEN Multiplex PCR mastermix (QIAGEN, Hilden, Germany). The PCR program was: 15 min sample denaturation at 95°C, followed by 30 cycles of {94°C, 30s; 56°C, 90s; 72°C, 60s}, then 72°C, 7 min after which the samples were cooled to 4°C until analysis. This is a reoptimalised protocol, differing from that described by Wang et al. (2002) in having a longer sample denaturation step, lower denaturation temperature, a longer annealing step at a lower temperature and longer extension time. PCR products were separated by agarose gel electrophoresis in 0.5xTBE at 135V for 1 hour on a 1.5% agarose gel, stained with ethidium bromide (Maniatis et al., 1982) and imaged under UV illumination using Gene Genius gel documentation equipment (Syngene, Cambridge, UK). Previously characterised *C. coli, C. jejuni, C. lari* and *C. upsaliensis* strains were used as positive controls and reference standards.

Riboprinting

Ribotyping was performed using the DuPont (Wilmington, DE) Qualicon RiboPrinter®. Single colonies from a 24 h culture on blood agar plates were suspended in sample buffer and heated at 80ºC for 15 min. After the addition of lytic enzymes, samples were transferred to the RiboPrinter®. Further analysis for the restriction of DNA, including use of the *Pst*I enzyme (20.000 U/ML, New England Biolabs Inc., Ipswich, MA), was carried out automatically. The riboprint profiles were aligned according to the position of a molecular size standard and compared with patterns stored in the DuPont Identification library of the RiboPrinter®. The identification (DUP-ID) of an isolate was called when the corresponding patterns matched one of the library patterns with a similarity of > 0.85. The *Pst*I ribotype patterns were automatically assigned a DuPont identification number (e.g. DUP-PST1-1182) by the RiboPrinter®, which was confirmed by visual inspection. The ribotype profiles were further analyzed with the GelComparII® software (Bio Numerics, Applied Maths Inc., Gent, Belgium) using the Pearson correlation and default settings for optimalization (2.0%) and position tolerance (1.00%) for genetic similarity and dendrograms were generated by the unweighted pair group method with arithmetic averages (UPGMA). These isolate-to-isolate comparisons were used to define "ribogroups". The similarity threshold for an isolate joining a ribogroup is an adaptive value between 0.90 and 0.96, depending on the size of the ribogroup.

RESULTS

Multiplex PCR confirmed all isolates as *C. jejuni*. Isolates from human cases infected in (I) or outside Norway (U) showed high heterogeneity (Figure 1). 25 of the isolates (60%) were assigned 18 different DUP-IDs from the

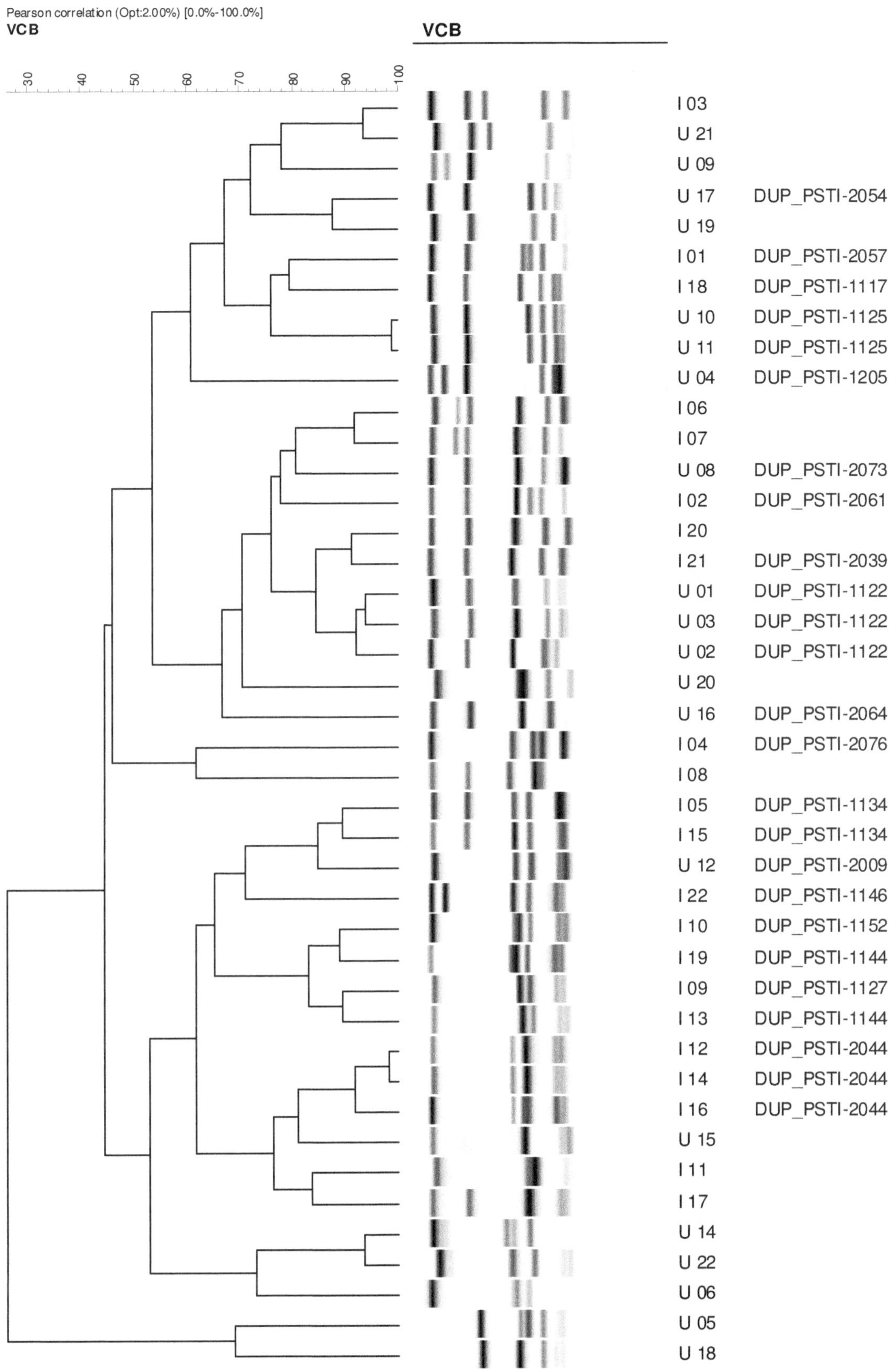

Figure 1. Dendrogram of 42 *Campylobacter jejuni* isolates from patients with diarrhoea infected in Norway (I) and abroad (U).

library. 3 isolates were DUP-PSTI-1122 (U) and DUP-PSTI-2044 (I) 2 isolates were DUP-PSTI-1125 (U), DUP-PSTI-1134 (I) and DUP-PSTI-1144(I) (Figure 1). Single occurrences of DUP-PSTI-1117, 1127, 1146, 1152, 2039, 2057, 2061 and 2076 and of DUP-PSTI-1205, 2009, 2054, 2064 and 2073 were found among the domestic and foreign isolates respectively. One patient had acquired campylobacteriosis caused by DUP-PST1-1125 in Denmark and another in Thailand. Of the 17 isolates from water 4 could be assigned a DUP-ID (DUP-PSTI-1130, 1122, 2000 and 2073). Strains DUP-PSTI-2073 and DUP- PSTI-1122 were found among the both human water isolates as shown in Figure 2 (subcluster c). Of 34 isolates from poultry 10 (29%) were assigned a DUP-ID (DUP-PSTI-1122, 1125, 1205, 2009, 2054, 2064 and 2073), 2 of which, DUP-PSTI-1146 and DUP-PSTI-2061 were also found among the human strains as shown in Figure 2 (subclusters a and c). 14 of the closely related strains in cluster c were isolated from poultry. A similarity of human and water isolates was found in all subcluster (a - e), while similarity between human and poultry was found in subcluster a and c (Figure 2).

DISCUSSION

The majority of the developed countries have experienced an increase in human campylobacteriosis during the last 20 years and it is currently the most frequently reported bacterial enteric infection in many developed countries, including Norway with 2875 reported cases in 2008 (MSIS report 2008). The incidence of campylobacteriosise in Telemark County was 41.2 per 100.000 while the national incidence was 59.8. Most cases of campylobacteriosis are sporadic. In Norway, as in other developed countries, studies of outbreaks as well as case-control studies have identified contact with animals, outdoor barbecuing, consumption of contaminated water, poultry meat and contaminated ready-to-eat foods as significant risk factors for acquiring campylobacteriosis (Kapperud et al., 1992; Bryan and Doyle, 1995; Blaser, 1997; Rautelin and Hänninen, 2000; Kapperud et al., 2003). The isolates from the human sources in this study showed a high heterogeneity. It was not possible to trace the clustered isolates from Norway, DUP-PST1-1144 (I) and DUP-PST1-2044 (I) to outbreaks. The high diversity of strains giving campylobacteriosis makes it difficult to identify virulent strains.

Campylobacters are frequently isolated from water and water supplies and have been the cause of infection in reported outbreaks in many countries including Norway (Mentzing, 1981; Vogt et al., 1982; Rosef and Mork, 1985; Dahl and Melby, 1987; Melby et al., 1991; Koenraad et al., 1997). Waterborne outbreaks associated with contaminated drinking water are common in the Nordic countries (Hänninen et al., 2003). In an earlier study, Rosef et al. (2001) isolated *C.* spp. in 53.3% of water samples from the Bø River. High degree of contamination has also been reported in Finland (Hörman et al., 2004) and New Zealand (Savill, 2001). Waterfowl are regular visitors to most surface water. Because of a higher body temperature, which is favourable for campylobacters, birds have different faecal flora than mammals and normally a lower content of faecal coliforms. 2 waterborne outbreaks of campylobacteriosis caused by geese (*Anser brachyrhynchus*) in Norway are described by Varslot et al. (1996). It is likely that contamination from birds plays a major role in contamination and transmission routes into water. Case-control studies have revealed non-disinfected drinking water and barbecuing as health hazards (Kapperud et al., 1992).

Domestic animals and wild birds may effectively spread *Campylobacter* by contaminating surface water sources. In many parts of the developed and developing word, undisinfected surface water is used for human consumption. The isolates in this study represent surface river water. A prospective case – control study suggest that the waterborne route of infection may be the common underlying pathway linking infection in humans, poultry, other domestic animals and wild birds (Kapperud et al., 2003). We found similarity of isolates from water and humans infected abroad with ribotypes, DUP-PSTI-1122 and DUP-PSTI-2073. Isolates from both human and water were found in subclusters a – e as shown in the dendrogram (Figure 2). Automatic riboprinting to identify isolates from a waterborne outbreak has been used (Nielsen et al., 2000) and shown to be a good tool for epidemiological studies. A study of ribotypes of *C. jejuni*, *C. coli* and *C. lari* from water formed the strains in a main cluster. These isolates are normally distributed among birds (Rosef et al., 2009).

Broiler fowl are commonly colonized by *Campylobacter*, being symptomless carriers of the organism (Newell and Fearnley, 2003). The reported prevalence varies from 5% (Hofshagen and Kruse, 2005) in Norway to 90% in Germany (Atanossava and Ring, 1999). Infected broilers have high levels of *Campylobacter* in the intestines (Newell and Fearnley, 2003) and large diversity has been found between and within broiler flocks (Petersen et al., 2001). One infected flock in this study showed strain stability, Figure 2, subcluster c. Poultry carcasses contaminated through the slaughtering process represent a risk of human infection. We found 2 poultry strains, DUP-PST-1146 and DUP-PST1-2061 similar to human isolates acquired locally. The similar isolates were found in subcluster a and c (Figure 2).

The finding that 2 locally acquired human infections were caused by strains with ribotype patterns identical to poultry isolates provides tentative support for the epidemiological evidence linking contaminated poultry to campylobacteriosis. On the other hand, 3 of the 4 human infections which were similar to water-borne isolates were acquired abroad and most of the human isolates were not similar to any of the poultry or water isolates. This may be a consequence of the limited scale of this study. The

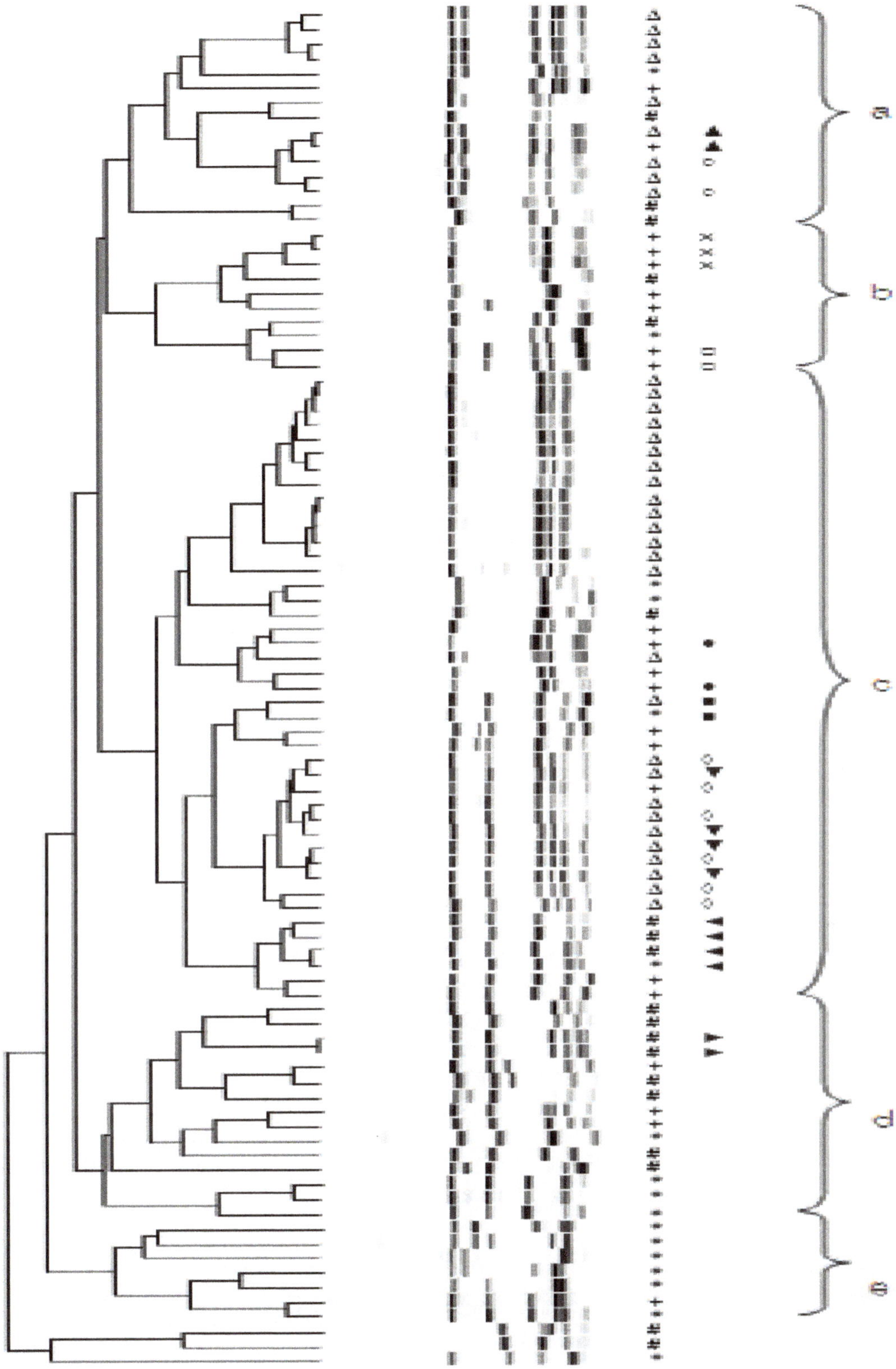

Figure 2. GelCompar dendrogram of riboprint patterns for poultry (Δ), water (*) and human *C. jejuni* isolates from domestic (+) and foreign infections (#). Recognised riboprint patterns (DUP-ID) occurring more than once are DUP-PSTI-1146 (▲), DUP-PSTI-2050 (○),DUP-PSTI-2044 (×), DUP-PSTI-1134 (□), DUP-PSTI-1144 (•), DUP-PSTI-2073 (■), DUP-PSTI-2061 (◊), DUP-PSTI-1131 (▼), DUP-PSTI-1122 (◄) and DUP-PSTI-1125 (►).

poultry and water isolates encompassed 2 poultry farms and 1 river while the catchment area for clinical samples was the entire county of Telemark and the proximal part of the neighbouring county of Vestfold. The great diversity of isolates and potential sources means that a study on a much larger scale would be required to definitively corroborate the epidemiological findings.

ACKNOWLEDGEMENTS

We thank Kristin Fagerland, Aud Stølan and Elisabeth Moen Bråthen for technical assistance with ribotyping of the isolates.

REFERENCES

Altekruse SF, Stern NJ, Fields PI, Swerdlow DL (1999). *Campylobacter jejuni* – An emerging foodborne pathogen. Em. Infect. Dis. 5: 28-35

Atanassova V, Ring C (1999). Prevalence of *Campylobacter* spp. In poultry and poultry meat in Germany. Int. J. Food Microbiol. 51: 187-190.

Blaser MJ (1997). Epidemiologic and clinical features of *Campylobacter jejuni* infections J. Infect. Dis. 176: S2 103-105.

Broman T, Waldenström J, Dahlgren I, Carlsson I, Olsen B (2004). Diversities and similarities in PFGE profiles of *Campylobacter jejuni* isolated from migrating birds and humans. Appl. Environ. Microbiol. 70: 6501-6511.

Brown PE, Christensen OF, Clough HE, Diggle PJ, Hart CA, Hazel S, Kemp R, Leatherbarrow AJH, Moore A, Sutherst J, Turner J, Williams NJ, Wright EJ, French NP (2004). Frequency and spatial distribution of environmental *Campylobacter* spp. Appl. Environ. Microbiol. 70: 6501-6511.

Bryan FL, Doyle MP (1995). Health risks and consequences of *Salmonella* and *Campylobacter jejuni* in raw poultry. J. Food Prot. 58: 326-344.

Dahl OP, Melby K (1987). En vannbåren epidemi forårsaket av *Campylobacter jejuni.* (A waterborne outbreak caused by *Campylobacter jejuni*). Tidsskr. Nor Lægeforen.107: 349-351.

Ge B, Girard W, Zhao S, Friedman S, Gaines SA, Meng J (2006). Genotyping of *Campylobacter* spp. from retail meats by pulsed-field gel electrophoresis and ribotyping. J. Appl. Microbiol. 100: 175-184.

Hänninen ML, Haajanen H, Pummi T, Wermundsen K, Katila ML, Sarkkinen H, Miettinen I, Rautelin H (2003). Detection and typing of *Campylobacter jejuni* and *Campylobacter coli* and analysis of indicator organisms in three waterborne outbreaks in Finland. Appl. Environ. Microbiol. 69: 1391-1396.

Hofshagen M, Kruse H (2005). Reduction in flock prevalence of *Campylobacter* spp. in broilers in Norway after implementation of an action plan. J. Food Prot. 68: 2220-2223.

Hörman A, Rimhanen-Finne R, Maunula L, von Bonsdorff CH, Torvela N, Heikinheimo A, Hänninen ML (2004). *Campylobacter* spp., *Giardia* spp., *Cryptosporidium* spp., noroviruses and indicator organisms in Surface Water in Southwestern Finland, 2000-2001. Appl. Environ. Microbiol. 70: 87-95.

Johnsen G, Kruse H, Hofshagen M (2006a). Genetic diversity and description of transmission routes for *Campylobacter* on broiler farms by amplified-fragment length polymorphism. J. Appl. Microbiol. 101: 1130-1139.

Johnsen G, Zimmerman K, Lindstedt B-A, Vardund T, Kapperud G (2006b). Intestinal carriage of *Campylobacter jejuni* and *Campylobacter coli* among cattle from South-western Norway and comparative genotyping of bovine and human isolates by amplified-fragment length polymorphism. Acta Vet. Scand. 1: 4.

Kapperud G, Skjerve E, Bean NH, Ostroff SM, Lassen J (1992). Risk factors for sporadic *Camylobacter* infections: results of a case - control study in southeastern Norway. J. Clin. Microbiol. 30: 3117-3121.

Kapperud G (1994) *Campylobacter* infection. Epidemiology, risk factors and preventive measures. Tidsskr. Nor Laegeforen (In Norwegian) 114: 795-799.

Kapperud G, Rosef O (1983). Avian Wildlife reservoar of *Campylobacter fetus* subsp. *jejuni*, *Yersinia* spp., and *Salmonella* spp. in Norway. Appl. Environ. Microbiol. 45: 375-380.

Kapperud G, Espeland G, Wahl E, Walde A, Herikstad H, Gustavsen S, Tveit I, Natås O, Bevanger L, Digranes A (2003). Factors associated with increased and decreased risk of *Campylobacter* infection: A prospective case-control study in Norway. Am. J. Epidemiol. 158: 234-242.

Koenraad PMFJ, Rombouts FM, Notermans SHW (1997). Epidemiological aspects of thermophilic *Campylobacter* in water-related environments: A review. Water Environ. Res. 69: 52-63.

Lévesque S, Frost E, Arbeit RD, Michaud S (2008). Multilocus sequence typing of *Campylobacter jejuni* isolates from humans, chickens, raw milk, and environmental water in Quebecc, Canada. J. Clin. Microbiol. 46: 3404-3411.

Maniatis T, Fritsch EF, Sambrook J (1982). Molecular Cloning: A Laboratory Manual. Cold Spring Harbour Laboratory, Cold Spring Harbour, NY.

Melby K, Gondrosen B, Gregusson S, Ribe H, Dahl OP (1991). Waterborne Campylobacteriosis in Northern Norway. Int. J. Food Microbiol. 12: 151-156.

Mentzing LO (1981). Water borne outbreak of *Campylobacter* enteritisdis in Central Sweden. Lancet 2: 352-354.

MSIS –årsrapport (2008). Meldesystem for smittsomme sykdommer. (MSIS- year report, Norwegian Surveillance System for Communicable Diseases. The Norwegian Institute of Public Health, www.fhi.no) Oslo, Norway.

Murphy C, Carroll C, Jordan KN (2006). Environmental mechanisms of the foodborne pathogen *Campylobacter jejuni.* J. Appl. Microbiol. 100: 623-632.

Newell DG, Fearnley C (2003). Sources of *Campylobacter* colonization in Broiler chickens. Appl. Environ. Microbiol. 69: 4343-4351.

Nielsen EM, Engberg J, Fussing V, Petersen L, Brogren CH, On SLW (2000). Evaluation of phenotypic and genotypic methods for subtyping *Campylobacter jejuni* isolates from humans, poultry and cattle. J. Clin. Microbiol. 38: 3800-3810.

Park SF (2002). The physiology of *Campylobacter* species and its relevance to the role as foodborne pathogens. Int. J. Food Microbiol. 74: 177-188.

Petersen L, Nielsen EM, On SLW (2001). Serotype and genotype diversity and hatchery transmission of *Campylobacter jejuni* in commercial poultry flocks. Vet. Microbiol. 82:141-154.

Rautelin H, Hänninen ML (2000). *Campylobacters*: The most common bacterial enteropathogen in the Nordic countrie. Ann. Med. 32: 440-445.

Rosef O, Gondrosen B, Kapperud G, Underdal B (1983). Isolation and characterization of *Campylobacter jejuni* and *Campylobacter coli* from domestic and wild mammals in Norway. Appl. Environ. Microbiol. 46: 855-859.

Rosef O, Mork AV (1985). (Contaminated well water as cause of *Campylobacter*- diarrhoea) Kontaminert brønnvann som årsak til *Campylobacter*-diaré. (in Norwegian) Nor.Vet. Tidsskr. 97: 831-832.

Rosef O, Rettedal G, Lågeide L (2001). Thermophilic campylobacters in surface water: a potential risk of campylobacteriosis. Int. J. Environ. Health Res. 11: 321-327.

Rosef O, Paulauskas A, Stølan A, Bråthen EM (2009). Diversity of thermophilic Campylobacter isolated from the Bø river, Southeast Norway. Vet. Med. Zoot. 45: 45-53.

Savill MG, Hudson JA, Ball JD, Klena JD, Scholes P, Whyte RG, McCormick RE, Jancovic D (2001). Enumeration of *Campylobacter* in New Zealand recreational and drinking waters. J. Appl. Microbiol. 91:38-46.

Stafford RJ, Schluter PJ, Wilson AJ, Kirk MD, Hall G, Unicomb L, OzFoodNet Working Group (2008). Population-attributable risk estimates for risk factors associated with *Campylobacter* infection, Australia. Em. Inf. Dis. 14: 895-901.

Vogt RL, Sours HE, Barrett T, Feldmann RA, Dickinson RJ, Witherell L (1982). *Campylobacter* enteritis associated with contaminated water. Ann. Intern. Med. 96: 292-296.

Varslot M, Resell J, Fostad IG (1996). Vannbåren campylobacterinfeksjon-trolig forårsaket av kortnebbgjess. (Water-borne outbreaks of *Campylobacter* gastroenteritis likely due to pink-footed geese in Norway in 1994 and 1995) Tidsskr. Nor. Laegeforen. 116: 3366-3369.

Wang G, Clark CG, Taylor TM, Pucknell C, Barton C, Price L, Woodward DL, Rodgers FG (2002). Colony multiplex PCR assay for identification and differentiation of *Campylobacter jejuni, C. coli, C. lari, C. upsaliensis* and *C. fetus*. subsp. *fetus* J. Clin. Microbiol. 40: 4744-4747

Colonization and antibiotic susceptibility pattern of methicillin resistance *Staphylococcus aureus* (MRSA) among farm animals in Saudi Arabia

Mohammad A. Alzohairy

Department of Medical Laboratories, College of Applied Medical Sciences, Qassim University, Qassim, Saudi Arabia. E-mail: dr.alzohairy@gmail.com.

Methicillin- resistant *Staphylococcus aureus* (MRSA) is a major human pathogen that causes severe morbidity and mortality worldwide. Additionally MRSA is widely spread in different animals. There are a growing number of MRSA cases have been reported in dogs, cats, horses, sheep, and other animals indicating the animal health treat too. To assess the frequency of MRSA among animals in Qassim region, a total of 400 samples were collected from camels, sheep, cows, and goats from 334 *Staphylococci* recovered, 158 (47.3%) were coagulase positive *Staphylococcus*, among them 90 (57%) were MRSA and 68 (43%) were methicillin-sensitive *Staphylococcus aureus* (MSSA). The reaming strains 176 (52.7%) were coagulase negative *Staphylococcus*, including 32 (18.2%) were methicillin-resistant coagulase negative *Staphylococcus* and 144 (81.8%) were methicillin-sensitive coagulase negative *Staphylococcus*. High rate of MRSA and MRCoNS were isolated from camel and sheep while lower rates were observed in goat and cow. Multi drug resistance (MDR) rate among MRSA and MRCoNS isolates was high. MRSA strains are highly prevalent among animals in Qassim region and they may play a potential role of disseminating pathogens between animal and human as well as to the community. Detection of MRSA will be essential for early prevention and control of community acquired infections.

Key words: Prevalence, methicillin resistant *staphylococcus aureus*, multidrug resistance, antibiotic susceptibility.

INTRODUCTION

Pathogenic microorganisms resistant to commonly used antibiotics is a worldwide concern. In this regard, the bacterial pathogen *Staphylococcus aureus* is one of the most important bacteria, particularly its methicillin-resistant strains. After the introduction of methicillin in the 1960s, methicillin resistance was found in strains of *S. aureus* (Grundmann et al., 2006). Soon after, methicillin-resistant *Staphylococcus aureus* (MRSA) in human populations has been reported worldwide becoming a significant cause for both hospital and community associated infections. Currently, MRSA is causing a significant morbidity and motility worldwide (Van Loo et al., 2007).

Occurrence of MRSA in animals has been shown by various previous studies (Walther et al., 2008; Rich and Roberts, 2006; Weese et al., 2006a; O'Mahony et al., 2005; Van Duijkeren et al., 2004). MRSA prevalence have been increasingly documented in farm or domestic animals including sheep, goats, horses, and cattle as well as in different pet or companion animals such as dogs and cats (Walther et al., 2008; Saleha et al., 2010). Indicating that MRSA has emerged as veterinarian and potentially zoonotic pathogen. The new findings of MRSA among animals have promoted an extensive studies to address the issue of MRSA colonization and transmission to human particularly those with close contact to animals (Khanna et al., 2010).

There are number of studies indicating domestic animals as a potential source of human infections (Grundmann et al., 2006; Weese et al., 2006b; O'Mahony et al., 2005; Seguin et al., 1999; Weese et al., 2006). Similarly, there have some reports indicating the domestic animals were the source of human infections particularly

Table 1. Total number of *staphylococcus* isolated from animals.

Bacterial isolate	No. of isolate	Percentage (%)
Staphylococcus aureus	158	(47.3)
Coagulase negative staphylococc (CNS)	176	(52.7)
Total	334	(100)

human in close contact with animals either through the nature of their occupations or keeping animal as pet. Thus, it is crucial to understand the presence of MRSA among certain population animals for better preventive measures.

This study was carried out to assess the prevalence level of *Staphylococcus* strains in different life stock animals including; camels, sheep, coats, and cows in Qassim region, Saudi Arabia and to analyze the *in vitro* antimicrobial susceptibility pattern of isolated strains. Findings of this study address the potential reservoir of MRSA infection in human, thus signifying a potential public health importance and global interest on MRSA.

MATERIALS AND METHODS

Study population

From January to April 2010, nasal swabs were collected from 400 different farm animals of various breeds in Qassim regions, Saudi Arabia. The study population included 100 camels, 100 sheep, 100 goats, and 100 cows. All samples (n = 400) were taken from healthy animals.

Collection and culturing of nasal swabs

Sterile, individually warped cotton-tipped swabs were used for nasal swabbing of the different animal. Swabs were inserted into the anterior nares and rubbed very well by rotating 5 times over the inner well of nasal septum by trained person. Swabs were then stored at stored at 4°C until processing, which occurred within 24 h of collection. Swabs collected were cultured on Mannitol Salt agar (selective medium for *S. aureus*) by streaking as per the conventional technique. The culture plates were incubated at 37°C for 24 to 48 h.

Identification of *S. aureus*

The suspected *Staphylococcus* colonies-yellow colonies showing Mannitol fermentation and non-yellow colonies (mannitol negative) were selected and subject to Gram staining and sub-cultured into nutrient agar slopes. The isolates showing Gram-positive cocci in clusters were subjected to coagulase test (Staphase Oxoid) by slide and test tube technique.

Identification of MRSA by a chrom agar plate method

For the identification of the MRSA among the isolates of *S. aureus*, ChromID MRSA (selective media) (OXOID) was used. All cultures showing bright blue colored growth were taken as MRSA positive strains, while all others are recorded as MSSA strains. As controls, all strains were also inoculated on MH Agar and incubated simultaneously. Further, methicillin-resistance was confirmed through penicillin binding protein 2a (PBP2a) latex agglutination test (PBP20 Test Kit, Oxoid, Hants, UK).

Antibiotics susceptibility testing

All nasal isolates of *S. aureus* were subjected to *in-vitro* antimicrobial testing method on Muller-Hinton agar, using fresh nutrient broth culture and antibiotic discs (OXOID) as previously described (Khadri and Alzohairy, 2010). Briefly, the zone of inhibition around the discs were measured and interpreted as sensitive, moderately sensitive, and resistant using the interpretation chart supplied by the antibiotic disc manufacturers (OXOID). The disk-diffusion method (Oxoid, Basingstoke, UK) was used to determine the susceptibility of all isolates to oxacillin, methicillin, gentamicin, vancomycin, rifampicin, ciprofloxacin, co-trimoxazole, fusidic acid, and tetracycline, according to Clinical and Laboratory Standards Institute (CLSI-2005) guidelines *S. aureus* ATCC 26923 was used as a standard control strain.

RESULTS

Based on the identification methods used, a total of 334 Staphylococcal strains were isolated from different animal. Out of them, 158 (47.3%) strains were coagulase positive staphylococci (CPS) and 176 (52.7%) isolates were coagulase negative staphylococci (CNS) as illustrated in Table 1. Among the (CPS) strains, 90 (57%) were MRSA and 68 (43%) were Methicillin-Sensitive *Staphylococcus aureus* (MSSA). The reaming isolates (CNS) were divided into 70 (39.7%) as Methicillin-Resistant Coagulase Negative *Staphylococcus* (MRCoNS) and 106 (60.3%) as Methicillin-Sensitive Coagulase Negative *staphylococcus* (MSCoNS).

MRSA in camel

A total number of 96 *staphylococcus* isolates were recovered from camels. Out of them, 54 (56.2%) were coagulase positive *staphylococcus* and 42 (43.7%) were coagulase negative *staphylococcus*. Among the total number of bacterial strains obtained from animals, 32 (35.5%) isolates were MRSA and 22 (32.3%) isolates were methicillin-sensitive *Staphylococcus aureus* (MSSA). The coagulase negative *Staphylococcus* strains sub divided into methicillin-resistant coagulase negative

Table 2. Total distribution of *Staphylococcus* in animals.

Animal	Total no.	*S. aureus*	Coagulase-negative *Staphylococcus* (CNS)
Camels	96	54	42
Sheep	88	38	50
Cows	72	36	36
Goats	78	30	48
Total	334	158	176

Table 3. Frequency of MRSA and MSSA in animals.

Specimen	Frequency of MRSA and MSSA			
	MRSA (n=90)	MRSA (%)	MSSA (n=68)	MSSA (%)
Camel	32	35.5	22	32.3
Sheep	26	28.9	12	17.6
Cow	14	15.5	22	32.3
Goat	18	20	12	17.6
Total	90	100	68	100

Table 4. Frequency of MRCoNS and MSCoNS in animals.

Specimen	Frequency of MRCoNS and MSCoNS			
	MRCoNS (n=32)	MRCoNS (%)	MSCoNS (n=144)	MSCoNS (%)
Camel	8	25	34	23.6
Sheep	14	43.75	36	25
Cow	4	12.5	32	22.2
Goat	6	18.75	42	29.2
Total	32	100	144	100

staphylococcus 8 (19%) (MRCoNS), and Methicillin-sensitive coagulase negative *staphylococcus* 34 (81%) were (MSCoNS) as illustrated in Tables 3 and 4. *Staphylococci* strains isolated from nasal cavity of camels showed a high rate of multi-drug resistance (MDR). As shown in Tables 7, The majority of MDR strains were methicillin-resistant staphylococcus isolates when compare to methicillin sensitive staphylococcus isolates both in coagulase positive staphylococcus and coagulase negative staphylococcus, the total 18 number of MDR strains were obtained from camel among these 14 were MRSA and 4 were MRCoNS isolates display a multidrug resistant pattern.

MRSA in sheep

Only few studies have reported the prevalence of MRSA in sheep. In this study, 88 *Staphylococcus* isolates were recovered from sheep. Out of them, 38 (43.2) isolates were coagulase positive *Staphylococcus* and 50 (56.8) isolates were coagulase negative *Staphylococcus* (Table 2). Among the coagulase positive *Staphylococcus*, 26 (28.9%) were MRSA and 12 (17.6%) were MSSA (Table 3). Out of 32 coagulase negative *Staphylococcus*, 14 (43.75%) were methicillin-resistant coagulase negative *Staphylococcus* (MRCoNS) and 36 (25%) were methicillin-sensitive coagulase negative *Staphylococcus* (MSCoNS) (Table 4). Among sheep isolates, only 8 (23.5%) strains were multi-drug resistance MRSA strains and 6 (42.8%) strains were multi-drug resistance MRCoNS (Table 7).

MRSA in cows

72 *Staphylococcus* isolates were recovered from cows. Out of them, 36 (50%) were coagulase positive *Staphylococcus* and 36 (50%) were coagulase negative *Staphylococcus*. Among the coagulase positive *Staphylococcus* 14 (15.5%) were MRSA and 22 (32.3%) were MSSA (Table 3). Out of 36 coagulase negative *Staphylococcus* 4 (12.5%) were methicillin-resistant coagulase negative *Staphylococcus* (MRCoNS) and 32

Table 5. Antibiotic resistance pattern of MRSA and MSSA.

Antibiotic	Percent of isolates' resistance to antibiotics			
	MRSA (n=90)	MRSA (%)	MSSA (n=68)	MSSA (%)
Penicillin	90	(100)	50	(73.5)
Oxacillin	90	(100)	00	(00)
Erythromycin	76	(84.4)	30	(44.1)
Cephalothin	70	(77.8)	36	(52.9)
Ciprofloxacin	50	(55.5)	18	(26.5)
Tetracycline	54	(60)	28	(41.2)
Gentamycin	72	(80)	24	(35.3)
Co-Trimoxazole	74	(82.2)	40	(58.8)
Vancomycin	0	(00)	0	(00)

(22.2%) were methicillin-sensitive coagulase negative *Staphylococcus* (MSCoNS) (Table 4). Multi-drug resistance rate among cows isolates was relatively low compare to the other animals with only 6 (17.6) MRSA isolates and 2 (14.2) MRCoNS isolates (Table 7).

MRSA in goats

Thirty isolates recovered from goats were determined as *S. aureus,* and another 48 isolates as coagulase-negative *Staphylococci.* Through the oxacillin screening agar, 26 strains were determined as methicillin-resistant isolates including 18 isolates (20%) as MRSA and 6 (18.8%) isolates as coagulase-negative (MRCoNS) (Table 3 and 4). On the other hand, 54 goat isolates were methicillin sensitive including 12 (17.6) isolates determined as coagulase-positive *Staphylococcus* and 42 (29.2%) isolates as coagulase-negative *Staphylococcus* (Tables 3 and 4). Among MRSA isolates, only 6 (17.6%) strains were multidrug resistant and 2 (14.2%) strains were multi-drug resistance MRCoNS strains (Table 7).

DISCUSSION

The presence of MRSA in animals and humans is of public health concern. Where the multi-drug resistant MRSA strains disseminate from animal to human and vice versa in Community (DeNeeling et al., 2007; Smith et al., 2008; Persoons et al., 2009). Several reports shows the presence of MRSA in a variety of domestic species including dogs (Pak et al., 1999; van Duijkeren et al., 2004), cats (Bender et al., 2005) horses (Anzai et al., 1996; Hartmann et al., 1997), sheep (Goni et al., 2004) pigs (Voss et al., 2005) and chickens (Lee, 2003) leading to an upsurge of reports and interest in MRSA colonization and infection in animals. Screening the prevalence of MRSA will be of much use in early prevention and control of community acquired infections. Hence, this study was carried out to address the prevalence of MRSA among domestic animals in Qassim region, Saudi Arabia and to describe the antibiotics susceptibility pattern of these isolates collected from January to April, 2010.

Of the isolates tested in this study, coagulase positive *Staphylococcus* accounted for 47.3% where as coagulase negative *Staphylococcus* accounted for 52.7%. In camels, this study is the first to report an unexpected high rate of MRSA colonization in Saudi Arabia reaching up to 35.5%. However, colonization rate was lower in sheep (28.8%), goats (20%), and cows (15.5%). Different pattern was observed with the occurrence rate of MRCoNS colonization with sheep being the highest (43.7%) followed by camels (25%). The similar report on the emergence of MRSA in animals was also reported in different parts of world from Pakistan (Farzana et al., 2004), Korea (Moon et al., 2007), and Hungary (Juhasz-Kaszanyitzky et al., 2007), except the colonization of MRSA in camel. In another study, several isolates from cows, goats, and sheep were isolated and compared by (Ben Zakour et al., 2008b). Our findings are also supported by those of above reports.

Antibiotic susceptibility profile of the isolates tested in this study demonstrated high resistant of MRSA strains to penicillin and oxacillin followed by erythromycin, Co-trimoxazole, gentamycin, and cephalothin. Resistant rate of MRSA strains was higher than those sensitive to methicillin (Table 5). This phenomena was reported else where (Tahnkiwale et al., 2002). Similarly, the resistant rate of MRCoNS strains was higher than those sensitive to methiicillin MSCoNS (Table 6). On the other hand, all tested strain in this study were sensitive to vancomycin.

The highest rate of multidrug resistant MRSA strain was observed in isolated recovered from camel (41.1%). Strains isolated from sheep, cows, and goats had a closely similar rate (23.5%, 17.6%, and 17.6% respectively). The pattern of multidrug resistant is different among MRCoNS strains with sheep isolates being the highest (42.8%) followed by camels, cows, and goats (28.5, 14.2 and 14.2% respectively). These findings go in agreement with previously reported studies (Goni et al., 2004; Quddoumi et al., 2006; Stastkova et al., 2009). While another study described evaluation of MRSA

Table 6. Antibiotic resistance pattern of MRCoNS and MSCoNS.

Antibiotic	Percent of isolates' resistance to antibiotics			
	MRCoNS (n=32)	MRCoNS (%)	MSCoNS (n=144)	MSCoNS (%)
Penicillin	32	100	96	66.7
Oxacillin	32	100	00	00
Erythromycin	24	75	50	34.7
Cephalothin	22	68.7	58	40.3
Ciprofloxacin	12	37.5	30	20.8
Tetracycline	20	62.5	38	26.4
Gentamycin	10	31.2	18	12.5
Co-Trimoxazole	26	81.2	62	43
Vancomycin	0	00	0	00

Table 7. Distribution of Multi-drug resistant MRSA and MRCoNS in animals.

Specimen	Multi-drug resistant MRSA and MRCoNS			
	MRSA (n=34)	MRSA (%)	MRCoNS (n=14)	MRCoNS (%)
Camel	14	41.1	4	28.5
Sheep	8	23.5	6	42.8
Cow	6	17.6	2	14.2
Goat	6	17.6	2	14.2
Total	34	100	14	100

infection and colonization in household pets, and transmission of MRSA between animals and humans (Baptiste et al., 2005).

In conclusion, this study confirms that staphylococci are one of the most common pathogens associated with animals. Over 26% of tested strains were confirmed to be MRSA with high multidrug resistant rate in isolates recovered from camels. It must be emphasized that there is a need to monitor the presence of MRSA in animals similar to that being done in humans so as to prevent further spread of MRSA in both sets of populations.

ACKNOWLEDGEMENTS

For sample collection, the author would like to thank Dr Abdullah Almusnid. Also, he would like to express his deep appreciation for the significant input and fertile discussion contributed by Dr. Habeeb Khadri and Ala a Eldin Bakr. This work was supported by a grant from the Scientific Research Centre at Qassim University, Saudi Arabia 2009.

REFERENCES

Anzai T, Kamada M, Kanemaru T, Sugita S, Shimizu A, Higuchi T (1996). Isolation of methicillin-resistant Staphylococcus aureus (MRSA) from mares with metritis and its zooepidemiology. J. Equine Sci., 7:7–11.

Baptiste KE, Williams K., Williams NJ, Wattret A, Clegg PD, Dawson S, Corkill JE, O'Neill T, Hart CA (2005) . Methicillin-resistant *Staphylococci* in companion animals. Emerg. Infect. Dis., 11: 1942–1944.

Bender JB, Torres SMF, Gilbert SM, Olsen KE, LeDell KH (2005). Isolation of methicillin-resistant *Staphylococcus aureus* from a non-healing abscess in a cat. Vet. Record, 157: 388–38.

Ben Zakour NL, Sturdevant DE, ,Even S, Guinane CM, Barbey C, Alves PD, Cochet MF, Gautier M, Otto M, Fitzgerald JR, LeLoirY, (2008b) .Genome- wide analysisofruminant Staphylococcus aureus reveals diversificationofthe core genome.J.Bacteriol.,190: 6302–6317

Clinical and Laboratory Standards Institute (2005). Guidelines by. CLSI/NCCLS - CLSI informational supplement. Approved standard M100-S15 Wayne, PA; 2000

De Neeling A, van den Broek M, Spalburg E, van Santen- Verheuvel M, Dam-Deisz W, Boshuizen H, van de Giessen H, van Duijkeren, Huijsdens X (2007). High prevalence of methicillin resistant Staphylococcus aureus in pigs. Vet. Microbiol., 122(3-4):366-72.

Farzana K, Shah SN, Jabeen F(2004) .Antibiotic resistance pattern against various isolates of *Staphylococcus aureus* from raw milk samples. J. Res. Sci., (Bahauddin Zakariya University, Multan, Pakistan) 15:145–151.

Goni P, Vergara Y, Ruiz J, Albizu I, Vila J. and Gómez-Lus R (2004). Antibiotic resistance and epidemiological typing of Staphylococcus aureus strains from ovine and rabbit mastitis. Int J Antimicrob Agents., 23(3): 268-72.

Grundmann H, Aires-de-Sousa M, Boyce J, Tiemersma E (2006). Emergence and resurgence of methicillin-resistant *Staphylococcus aureus* as a public health threat. Lancet, 368: 874–885.

Hartmann F, Trostle SS, Klohnen AAO (1997) . Isolation of a methicillin-resistant *Staphylococcus aureus* from a postoperative wound infection in a horse. J. American Vet. Med. Assoc., 211: 590–592.

Juhasz-Kaszanyitzky E, Janosi S, Somogyi P, Dan A, van der Graaf-van Bloois L, van Duijkeren E, Wagenaar JA (2007). MRSA transmission between cows and humans. Emerg. Infect. Dis. 13: 630-32.

Khadri H, Alzohairy M (2010). Prevalence and antibiotic susceptibility pattern of methicillin-resistant and coagulase-negative staphylococci in a tertiary care hospital in India. Int. J Med. And Medi. Sci.; 2(4) pp 116-120.

Khanna T, Friendship R, Dewey C, Weese J (2010). Methicillin-resistant *Staphylococcus aureus* colonization in pigs and pigs farmers.; Vet. Microbiol., 128: 298-303.

Lee J H, (2003). Methicillin (oxacillin)-resistant Staphylococcus aureus strains isolated from major food animals and their potential transmission to humans. App. Environ. Microbiol., 69, 6489–6494.

Moon JS, Lee AR, Kang HM, Lee ES, Kim MN, Paik YH, Park, YH, Joo YS, Koo HC (2007). Phenotypic and genetic antibiogram of methicillin-resistant staphylococci isolated from bovine mastitis in Korea. J. Dairy Sci., 90, 1176– 1185.

O'Mahony R, Abbott Y, Leonard F, Markey B, Quinn P, Pollock P, Fanning S, Rossney A (2005). Methicillin-resistant *Staphylococcus aureus* (MRSA) isolated from animals and veterinary personnel in Ireland. Vet Microbiol. 109:285–96.

Pak SI, Han HR, Shimizu A (1999). Characterization of methicillin resistant Staphylococcus aureus isolated from dogs in Korea. Korean J. Vet. Med. Sci. 61, 1013–1018.

Persoons D, Van Hoorebeke S, Hermans K, Butaye P, De Kruif A, Haesebrouck F, Dewulf J (2009). Methicillin-resistant Staphylococcus aureus in poultry. Emerg. Infect. Dis., 15(3):452-3.

Quddoumi S, Bdour S, Mahasneh A (2006). Isolation and characterization of methicillin-resistant *Staphylococcus aureus* from livestock and poultry meat. Ann. Microbiol., 56 (2) 155-161.

Rich M, Roberts L (2006). Methicillin-resistant *Staphylococcus aureus* MRSA in companion animals. Vet. Rec., 159:535–6.

Saleha A Zunita Z (2010). Methicillin Resistant Staphylococcus aureus (MRSA): An Emerging Veterinary and Zoonotic Pathogen of Public Health Concern and Some Studies in Malaysia. J. Anim. Vet. Adv. 9(7): 1094-1098.

Seguin J, Walker R, Caron J, Caron J, Kloos W, George C, Hollis R, Jones R P, faller M (1999) . Methicillin-resistant *Staphylococcus aureus* outbreak in a veterinary teaching hospital: Potential human-to-animal transmission. J Clin. Microbiol., 37:1459–1463.

Smith T, Male M, Harper A, Kroeger J, Tinkler G, Moritz E, Capuano A, Herwaldt L and Diekema, D (2008). Methicillin-resistant Staphylococcus aureus strain ST398 is present in Midwestern U.S. swine and swine workers. PLoS One; 4(1):e4258.

Stastkova Z, Karpiskova S, Karpiskova R (2009). Occurrence of methicillin-resistant strains of *Staphylococcus aureus* at a goat breeding farm. Veterinarni Medicina. 54, 2009 (9): 419–426.

Tahnkiwale S, Roy S, Jalgaonkar S (2002). Methicillin resistance among isolates of *Staphylococcus aureus*. Antibiotic sensitivity pattern phage typing, 56 (7): 330-334.

Van Duijkeren E, Box A, Heck M, Wannet W Fluit A (2004). Methicillin-resistant staphylococci isolated from animals. Vet. Microbiol., 103:91–7. DOI: 10.1016/j.vetmic.2004.07.014.

Van Loo I, Huijsdens X, Tiemersma E, de Neeling, A (2007). Emergence of methicillin-resistant *Staphylococcus aureus* of animal origin in humans. Emerg. Infect. Dis., 13: 1834-1839.

Voss A, Loeffen F, Bakker J, Klaassen C, Wulf M (2005) . Methicillin-resistant Staphylococcus aureus in pig farming. Emerg. Infect. Dis., 11: 1965–1966.

Walther B, Wieler L, Friedrich A, Hanssen A, Kohn B, Brunnberg L, Lübke-Becker A (2008). Methicillin-resistant *Staphylococcus aureus* (MRSA) isolated from small and exotic animals at a university hospital during routine microbiological examinations. Vet. Microbiol., 127:171–8.

Weese J, Caldwell F, Willey B, Kreiswirth B, McGeer A, Rousseau J, Low D (2006 a) . An outbreak of methicillin-resistant *Staphylococcus aureus* skin infections resulting from horse to human transmission in a veterinary hospital. Vet. Microbiol., 114:160–164.

Weese J, Dick H, Willey B, McGeer A, Kreiswirth B, Innis B, Low D (2006b). Suspected transmission of methicillin-resistant *Staphylococcus aureus* between domestic pets and humans in veterinary clinics and in the household. Vet. Microbiol., 115:148–55. DOI: 10.1016/j. vetmic.2006.01.004.

Biodegradation model on effect of some physicochemical parameters on aromatic compounds in fresh water medium

C. P. Ukpaka

Department of Chemical/Petrochemical Engineering, Rivers State University of Science and Technology, Nkpolu, P.M.B. 5080, Port Harcourt, Nigeria. E-mail: chukwuemeka24@yahoo.com.

In this paper, mathematical models were developed to simulate the inhibition effect of some physicochemical parameters on the biodegradation rate of aromatic compounds in the presence of low and high concentration of salinity. The inhibitive effects of salinity, in this case, were investigated. The aromatic compounds were obtained from an oil-servicing firm in Port Harcourt, and the microbial analysis was conducted on the water samples for the purpose of identification, isolation and characterization of the *Pseudomonas putida*. A head-to head comparison of the degradation rates of toluene, phenol and benzene in fresh medium was done based on the data obtained from experiments conducted. The result showed that salinity did not inhibit the degradation rate of toluene and benzene. However, phenol was significantly affected by salinity. The aromatic compounds removal from water solution varied depending on the conditions, that is, the type of compound, the composition of the water solution and the conditions of their exposure. The results obtained from this investigation was compared with Suietlik et al. (2002) work which revealed 22 to 28% reduction in aromatic compounds degradation while the present research shows 45-75% reduction in aromatic compounds under investigation for a period of 2 weeks (21 days) exposure. The parameters form the bedrock for further improvement of the kinetic models and also, serve as an outline for possible pilot-scale bioremediation by engineers.

Key words: Kinetic, inhibition, biodegradation, model, salinity.

INTRODUCTION

This paper gives a theoretical and an experimental perspective of biodegradation of petroleum hydrocarbon pollutants in aquatic environments. However, the investigation is target to derive a kinetic model that can predict the rate of product inhibition due to the concentration of salinity, Dissolved Oxygen (DO), Biochemical Oxygen Demand (BOD) and other physicochemical parameters as presented in the paper.

Kinetic modeling is the study of physical, chemical and biological systems that change with time and is a simultaneous system of differential equations associated algebraic equation that defines the state variables and rate laws for a particular physical, chemical or biological system (which in this case is biodegradation and its inhibition of petroleum hydrocarbon (aromatics) (Hutchinson et al., 1979; Geyer et al., 1981; Dauta 1982; Sims and Overcash, 1983; Vysotskaya and Bortun, 1984; Neter et al., 1985; Nested and Giesy, 1987; Zepp, et al., 1987; Miller et al., 1988; O'Brien, 1991; Lemaire et al., 1992; Narbonne et al., 1992; IARC, 1993; Mekenyan et al., 1994; Mekenyan et al., 1995; Djomo et al., 1995; Djomo et al., 1996; Suietlik et al., 2002; Ukpaka 2004, 2005; Marrot et al., 2006). The model is a quantitative statement of theory as to how the real-life system operates. Models embody the principle of cause and effect, and are based on descriptions of the physicochemical processes that comprise the biological system under study (Eisler 1987; DeVoogt, et al., 1991; Levenspiel,

1999; Kreyzig, 1999; Okoh, 2006; Park and Marchaldn, 2006; Knightes and Peter, 2006; Egberongbe et al., 2006; Ukpaka, 2010).

Investigation conducted on bioremediation of contaminants in petroleum hydrocarbon revealed that aromatic hydrocarbon is a big problem today (Belcher et al., 1970; Levenspiel, 1999; Richardson and Peacock, 1991; Zhu et al., 2001; Prasad, 2000; Reardon et al., 2002; Ukpaka, 2004; Djomo et al., 2004; Newsted and Giesy, 1987; Ukpaka, 2004). This is because; the degradation of one component can be inhibited by other compounds in the mixture, as presented in an experiment that were conducted by various research groups (Chryplewicz-Kalwa, 1999; Thomas and Li, 2000; Mill et al., 1981; Swallow et al., 1995; Kochany and Maguire, 1991; Lee et al., 1978; Kowalczyk et al., 2000; Gernjak et al., 2003; Morrison and Boyd, 1974; Buitron et al., 1998; Lante et al., 2000; Alexievaa et al., 2004; Pai et al., 1995; Hidalgo et al., 2002; Prieto et al., 2002). Bioremediation is by far, the most effective means of treating the petroleum hydro-carbon contaminants from the environment. Although petroleum hydrocarbons decompose faster in chemical and physical methods of treatment, their by-products would go a long way to damage its environment, as well as destroying the aquatic life. Even since the success of bioremediation in the clean up of the oil tanker Exxon valdez oil spill of 1989, in Prince William Sound, interest has grown significantly in biodegradation (Halsall et al., 1997; Zerbe et al., 1994; Adamczewska et al., 2000; Okoh, 2006).

The increase of salinity leads to a sharp drop in the growth rate of microorganisms, compared to other influence of the physicochemical parameters as studied in this research. This is only temporary as the culture recovers and adapts to its harsh environment. This effect, however, is difficult to quantity as it is transient, and applies to mixed culture (Mordocco et al., 1999; Monteiro et al., 2000; Gonzalez et al., 2001; Sa and Boaventura, 2001; Annadurai et al., 2002; Chung et al., 2003; Pazarlioglu and Telefoncu, 2005; Margesin et al., 2005; Geyer, 1981; Hao et al., 2002). Inhibition can be illustrated as when an enzyme is introduced in a reactant (substrate) and the process causes the slow down, pauses or completely tops further enzyme-substrate reaction (Watanabe et al., 1996; Banerjee, 1997; Kumaran and Paruchuri, 1997; Wang and Loh, 1999; Kibret et al., 2000; Bandhyopadhyay et al., 2001; Hao et al., 2002; Godjevargova et al., 2003; Timur et al., 2003; Kumar and Kumar, 2004; Aksu and Gonen 2004; Nuhoglu and Yalcin, 2005; Park and Marchland, 2006; Newsted and Giesy, 1987). This paper covers both theoretical and experimental approach, in the biodegradation of aromatic compounds and their mixtures in fresh water medium. The parameters that influence the microbial activities in the remediation of these samples were highlighted. Kinetic modeling, rate of reaction, petroleum hydrocarbon degradation, decays rate and product inhibition kinetics. It is believed that these terms are key to understanding this paper.

THE MODEL

Microbial growth kinetics

Microbial culture consumes substrate as energy source as well as it incorporates the substrates into its own cellular material, and also in the synthesis of a product. However, it would be necessary to mention the law of conservation of matter. This is because the yield is used to relate the quantities of materials consumed and produced in a particular reaction.

The yield coefficient is the ratio of the mass of product synthesized, to that of the reactant consumed. This is usually a constant quantity. However, in relation to this paper, the yield coefficient has to be more closely defined. Yield coefficient may appear not to be a constant quantity and it may rather be a function of time as well as the physicochemical parameters. This is as a result of the change in composition of the microbial cell, in which such a situation can be derived from a very sensitive experiment.

Mathematically, yield coefficient is written in terms of concentration as:

$$Y = \frac{\Delta x}{\Delta S} \qquad (1)$$

The Equation (1) was first developed (Ukpaka, 2010). The negative sign depicts a decrease in substrate or nutrient concentration. This is as a result of an increase in biomass concentration. It should be noted that in biodegradation of aromatic compounds via microbes, the products alcohol, (e.g. catechol) organic acids etc would clearly mean that there would be two yield coefficients for the biomass, which is related to the feed, and the yield coefficient which would relate to the quantity of alcohol, and other subsequent products of biodegradation of aromatic compounds. The yield coefficient for biomass is given as:

$$Y_{x/S} = \frac{\Delta x}{\Delta S} \qquad (2)$$

The yield coefficient for products of biodegradation of petroleum hydrocarbons is given as:

$$Y_{p/S} = \frac{\Delta x}{\Delta S} \qquad (3)$$

In the case of multi-substrate biodegradation, for products 1, 2 ..., i, ... N resulting from microbial action, each has a yield coefficient associated with it in Equation (4)

$$Y_{p/S} = \frac{\Delta P_i}{\Delta S} \quad (4)$$

From the law of conservation of matter,

$$\sum_{i=1}^{n} Y_i / S = 1$$

It should be noted though that when the yield is based on the mass of product and substrate consumed, then the balance may appear to be contravened. In this, case yield coefficients above 1 are obtainable. This is because the reaction may well incorporate other materials into the product, as well as yield coefficient of substrate, this would increase the relative mass of product. In the subsequent as discussed on the paper, yield coefficient relating to biomass production will be measured experimentally, in order to remove it from the overall yield coefficient, in order to revert to the normal unity (1.0) value of yield coefficient.

$$Y_G = \frac{\Delta X}{\Delta S_G} \quad (5)$$

Therefore, the material balance for the consumption of substrate can be written as;

$$\begin{bmatrix}\text{Total consumption}\\ \text{of substrate}\end{bmatrix} = \begin{bmatrix}\text{Substrate used}\\ \text{for growth}\end{bmatrix} + \begin{bmatrix}\text{Substrate used}\\ \text{for maintenance}\end{bmatrix}$$

Expressing the above material balance in terms of mathematical application gives

$$\frac{dS}{dt} = \frac{dX/dt}{Y_G} + mX \quad (6)$$

Dividing both sides of the Equation (6) by X we have:

$$\frac{\frac{dS}{dt}}{x} = \frac{\frac{dX/dt}{Y_G}}{x} + \frac{mX}{x} \quad (7)$$

$$\frac{1}{x}\frac{dS}{dt} = \frac{1}{X}\frac{dX}{dt}\frac{1}{Y_G} + m \quad (8)$$

where $\frac{1}{x}\frac{dX}{dt}$ = specific growth rate of the biomass.

This is usually denoted byμ. Dividing through the Equation (8) with $\frac{1}{x}\frac{dX}{dt}$, we have the expression in Equation (9)

$$\frac{\frac{dS}{dt}}{\frac{dx}{dt}} = \frac{1}{Y_G} + \frac{m}{\mu} \quad (9)$$

Decay rate model

The mathematical model of the decay rate of a microbial culture, which will be used for the experiments in the next chapter, makes use of the concept of doubling time, this is an exponential growth pattern. It can be said that growth rate is directly proportional to the existing population. In most cases, the cell number N, is usually replaced by cell mass X. The exponential growth of microbes is given by this equation:

$$\frac{dX}{dt} = \mu X \quad (10)$$

This is the Malthus' law which can be resolved by rearranging Equation (10) as well as applying the mathematics tool of integration we have

$$\int_{x_0}^{x} \frac{dX}{X} = \int_0^t \mu dt \quad (11)$$

$$\ln\left(\frac{X}{X_0}\right) = \mu t \quad (12)$$

$$\frac{X}{X_0} = e^{\mu t} \quad (13)$$

$$X = X_0 e^{\mu t} \quad (14)$$

Exponential growth occurs only for a limited time during the course of the development of microbial culture, with a fixed supply of nutrients.

Effect of salinity on microbial growth rate

Increase in salinity leads to a sharp, drop in growth rate of microbes. This is only temporary, as the culture recovers after adaptation. The effect is however, difficult to quantity. It is transient, and applies to mixed culture. The model below was developed by Park and Marchland (2005), to explain the effect of salinity on the maximum specific growth rate of the biomass:

$$\mu = \frac{(\mu_m - l_s)s}{K_s + S + (S^2 / K_l)} \tag{15}$$

The mathematical expression for the salinity inhibition constant can be expressed as:

$$I_S = \frac{I_S^*(\%\ NaCl)}{0.01 + \%NaCl} \tag{16}$$

The Park and Marchland's model was modified, with simpler parameters introduced to show the relationship between the maximum specific growth rate and the salinity. A basis of 1% salinity level was used, a specific growth rate of 0.01 (h^{-1}), and the estimated density of fresh water of 1.027 g/cm^3 were also used as a basis. However, this is a hypothesis that would be tested in the experiment:

Salinity is directly proportional to the density of sea water, but inversely proportional to the maximum specific growth rate, as shown below in Equation (17).

$$S_a \quad \alpha \; \rho \; \alpha \; \frac{1}{\mu_m} \tag{17}$$

$$S_a = D_\alpha \left(\rho / \mu_m \right) \tag{18}$$

Fixing the values:

$$0.01 = \frac{D_\alpha (1.027)}{0.01} \tag{19}$$

$\therefore D_\alpha$ = 0.00009737(h^{-1})

Substituting the constant above into the Equation (18) into Equation (15) yield;

$$\mu = \frac{(\mu_m - D_\alpha)S}{K_s + S + (S^2 / K_1)} \tag{20}$$

Henry's equation

Henry's equation will be used to calculate the masses and volumes of volatile hydrocarbons, which are present in both gas and liquid phases in the bioreactor. Since the microbial growth rate depends on the liquid-phase substrate concentration only, and biomass yield depends on the change in total mass of substrate. The equations can be seen in Equations (21) and (22)

$$M_{TOT} = M_L + M_G \tag{21}$$

$$M_{TOT} = M_L \left[1 + \left(\frac{H}{RT} \right) \left(\frac{V_G}{V_L} \right) \right] = M_L \alpha \tag{22}$$

However, in a case where the temperature and volume of liquid hydrocarbon is constant (which could be possible in the experiment), then the rate of substrate consumption shall be depicted by Equation (23) below:

$$\alpha \frac{dS_L}{dt} = -\frac{\mu(S_L)x}{Y_x / s} \tag{23}$$

Product kinetic model

Microbial products are formed as a result of a variety of processes, which occur within the cells of a microbial culture. The cell itself may be the desired product in some cases, or it may be that the product is for due to its growth.

However, many important microbial products are not growth-associated. There may exist mechanisms within the cells, which operate to produce a particular material. As a result, the yield coefficient $Y_{P/s}$ is of little value in predicting the course of product accumulation, except when the overall productivity of a complete batch operation is considered.

$$\frac{dP}{dt} = \delta \frac{dX}{dt} + \varepsilon X \tag{24}$$

where, quantities δ and ε are pH-dependent,

$\delta \frac{dX}{dt}$ is growth-associated

εX is cellular activity due to maintenance functions

Dividing the equation though by X (biomass concentration)

gives the relationship in terms of specific rates, as seen in Equation (25);

$$\frac{1}{X}\frac{dP}{dt} = \delta\frac{1}{X}\frac{dX}{dt} + \varepsilon \quad (25)$$

Substituting μ for specific growth rate $\frac{1}{X}\frac{dX}{dt}$:

$$\frac{1}{X}\frac{dP}{dt} = \delta\mu + \varepsilon \quad (26)$$

The model above shall be known as Product Formation model. Its authenticity tested in the experiments of the subsequent chapter. Values of δ and ε may be estimated by calculating the slope, and the intercept of the plot of the specific rate of product formation $\left(\frac{1}{X}\right)\left(\frac{dP}{dt}\right)$ against μ. The second model for product formation kinetics that shall be tested is was developed by various research groups (Dauta, 1982; Neter et al., 1985; Hutchinson et al., 1979; Geyer et al., 1981; Zepp et al., 1987).

$$\frac{dP}{dt} = \sum_i A_i e^{-k_i t} \quad (27)$$

Concentration of product accumulated in the broth over the lifetime of the cell per unit mass of the cell is given as;

$$P_t = \int_0^t \sum_i A_i e^{-k_i t}\, d\tau \quad (28)$$

In the case of a range of cells of all possible ages, $[\xi(t)]$ and culture time = t.

Then overall cell concentration,

$$X_t = \int_0^t \xi(t)\,d\tau \quad (29)$$

Overall product concentration in the fermenter is given as

$$P = \int_0^t \xi\tau\left(\int_0^t \sum_i A_i e^{-k_i t} d\tau\right) \quad (30)$$

Batch reactor equations

In batch reactor process microorganisms are added in small quantity with the required amount of nutrient as an activator in the system. Therefore, the material balance equation for batch growth of microorganisms can be written as:

$$\begin{bmatrix}\text{Formatiorby}\\ \text{biochemich}\\ \text{reaction}\end{bmatrix} = \begin{bmatrix}\text{Accumulatin}\\ \text{within the}\\ \text{system}\end{bmatrix} \quad (31)$$

Therefore, in the case of the biomass, the mathematical expression is given as:

$$R_x V = \mu V X = \frac{dx}{dt}V \quad (32)$$

For the substrate:

$$R_s V = \frac{dS}{dt}V \quad (33)$$

The yield is used to relate to the equations above, thus:

$$Y_{x/s} = \frac{\Delta X}{\Delta S}$$

Therefore,

$$Y = \frac{\frac{dX}{dt}}{\frac{dS}{dt}}$$

From the yield equation, which is:

$$\frac{\frac{dX}{dt}}{\frac{dS}{dt}} = \frac{1}{Y_{x/s}} + \frac{m}{\mu},$$

$$\frac{dS}{dt} = \frac{dx}{dt}\frac{1}{Y}$$

$$\therefore \quad Y = \frac{dS}{dt} = \frac{dX}{dt} \quad (34)$$

Citing the original yield Equation (2), then this new

equation becomes:

$$Y\frac{dS}{dt} = -\frac{dX}{dt} \quad (35)$$

From Malthus' theory in Equation (10)

$\frac{dX}{dt} = \mu X$, substituting it into Equation (35), becomes

$$YR_s = -\mu X \quad (36)$$

Making R the subject of the formula we have

$$R_s = -\frac{\mu X}{Y} \quad (37)$$

$$\frac{dS}{dt} = -\frac{\mu X}{Y} \quad (38)$$

Since;

$$Y = -\frac{\Delta X}{\Delta S}$$

Then

$$\therefore \quad Y = \frac{X - X_0}{S_0 - S} \quad (39)$$

$$S_0 - S = \frac{X - X_0}{Y}$$

Rearranging the equation,

$$S = S_0 - \frac{X - X_0}{Y} \quad (40)$$

If the growth rate follows the Monod model, then in the equation $\frac{dX}{dt} = \mu X$, μ will be substituted for the Mond equation. Therefore:

$$\frac{dX}{dt} = \left(\frac{\mu_m S}{K_S + S}\right)X \quad (41)$$

Substituting the Equation (40) into (41) will give:

$$\frac{dX}{dt} = \left(\frac{\mu_m\left(S_0 - \frac{X - X_0}{Y}\right)}{K_S + \left(S_0 - \frac{X - X_0}{Y}\right)}\right)X \quad (42)$$

Integrating this equation using the separable differential principle:

$$\int_{x_0}^{x} \frac{K_S + \left(S_0 - \frac{X - X_0}{Y}\right)}{\mu_m\left(S_0 - \frac{X - X_0}{Y}\right)}\frac{dX}{X} = \int_0^t dt \quad (43)$$

The resulting equation will be:

$$\frac{(K_S Y + S_0 Y + X_0)}{\mu_m (S_0 Y + X_0)} \text{In}\left(\frac{X}{X_0}\right) + \frac{K_S Y}{\mu_m (S_0 + X_0 - X)} = t \quad (44)$$

Note that Equation (44) above represents the biomass.

For the case of the substrate concentration:

$$\frac{(K_S Y + S_0 Y + X_0)}{\mu_m (S_0 Y + X_0)} \text{In}\left(1 + \frac{(S_0 - S)Y}{X_0}\right) + \frac{K_S Y}{\mu_m (S_0 Y + X_0)} \text{In}\left(\frac{S}{S_0}\right) = t \quad (45)$$

MATERIALS AND METHODS

Description of sampling site/station (Fresh water source)

The location of sampling site is the new Calabar River, which passes through Choba Community. It flows from Aluu through to Ibefaway at Emohua Local Government area and then to Kalabari area and lined up with the Niger River. The river is used for bathing, drinking and washing. Samples were obtained from the following station/site:

Site 1: Aluu community which the up stream with little contamination from the inhabitants through activities like washing of cloths and plates and dumping of used water from domestic works, bathing and dredging activities.
Site II: Sport close to the Wilbros Company just directly under the bridge. This station is polluted with a lot of diesel oil from the barges. Also some meters away are the local dredges and standard dredgers, carrying out their dredging activity.
Site III: Is the Choba Community, which is located a few kilometers

from the Wilbros Company. Here the diesel-oil also pollutes the water and a lot of human waste is seen here because the inhabitants defecate, bath, wash and dumping of refuse on this site. Other activities like fishing and swimming are also carried out.

Collection of samples

Sub-surface water sample were collected at three (3) different locations or station from the New Calabar Rivers, in sterile sample bottles. This was taken from the sub surface water and the samples collected were not up to the brine of the sample bottles. The sample were immediately kept in an ice bag and taken to the laboratory within two hours of collection. Samples were collected and analysis conducted in the following parameters: Dissolved Oxygen (DO), Biochemical Oxygen Demand (BOD), salinity, pH and conductivity.

Temperature

Mercury in glass thermometer was dipped into the water for 3 min and then the result gotten was read and recorded. The unit for temperature is Celsius/Centigrade.

Dissolved oxygen (DO)

The dissolved oxygen (DO) was determined by the modified Azide or Winkler's method (APHA 1985). To A 70 ml BOD bottle filled with sample, 0.5 ml Winkler II reagent was added, stopper placed and mixed for complete dissolution of precipitate. A 50 ml portion was placed in an Erlenmeyer flask, 5 drops of freshly prepared starch solution added and filtrated with 0.025 $Na_2S_2O_3$ (sodium thiosulfate) solution. The titration was continued to the first disappearance of the blue colour. DO mg/L was calculated using:

$$\frac{V\ x\ N\ x\ 8000}{\text{Mass of sample}}$$

Biochemical oxygen demand (BOD)

Water samples collected in the same way as the DO were incubated at 20°C for five days. At the end of the incubation period, the samples were treated in the same manner as the DO samples stated previously to determine the dissolved oxygen. To ensure presence of oxygen the BOD samples were diluted before incubation and the DO of the dilution water determined. DO at day 5 was determined as in dissolved oxygen above and the BOD_5 calculated using the following (A-B) x DF.

Conductivity, pH and salinity

Measurement for pH, salinity, and conductivity were done using Horiba water Clucker (model U-10) after calibrating the instrument with the standard Horiba solution. The units of measurement, for salinity and conductivity are in percentage (%) and µs/cm respectively.

Experiment to determine total herrotrophic bacteria

Media

The media used for isolation of the organisms are: nutrient agar, manifold salt agar, and petroleum hydrocarbon agar. The solution for the nutrient agar contained 28 g of powdered nutrient agar and 1 L of distilled water Manitol contained NaCl (0.55 mg) and 1 L of distilled water.

Petroleum hydrocarbon media contained 15 g of powdery agar, dissolved in 1 L of distilled water, 0.5 g of NH_4Cl, 5 g of Na_2HPO_4, 0.5 g of $KHPO_4$, 5 ml of toluene. The media were filtered with a filter paper of hole-size 20 µm.

Chemicals

NaCl powdered nutrient agar (Biology Laboratory, RSUST, Port Harcourt), Toluene (Oil Test Group of Companies, Port Harcourt), distilled water (Biology Laboratory, RSUST, Port Harcourt), glyccrol (Biology laboratory, RSUST, Port Harcourt). All chemicals used for media preparation were reagent grade.

Apparatus

(i) Hot oven; (ii) conical flask, (iii) test tubes; (iv) cotton wool; (v) aluminum tool, (vi) thermometer; (vii) Petri dishes; (viii) filter paper (20 µm hole size); (x) pipette (sterile); (x) glass spreader (sterile); (xi) Bijor storage bottle; (xii) Microscope (xiii) wire loop; (xiv) wooden spatula.

Experimental procedure

All glass waves were sterilized in a hot air oven for 1 h at 160°C and left to cool. Culture was extracted and re-grown in another Petri dish. Plate count method was used to count the number of colonies of bacteria formed. The diluent was then prepared using normal saline.

Analytical methods

One ml of the original was transferred in triplicate into test tubes containing 9 ml of normal saline. The sample was then inoculated. The pure culture was isolated, using Gram's stain method to prepare the microorganism for observation under the microscope. The following tests were conducted: (1) Biochemcial test, (2) oxidase test, (3) Voges–Pros Kaner test, (3) Catalase test, (4) Methyl red test, sugar fermentation test and (5) Coagulase test. The tests were conducted on the colony of the test organism. This is in order to identify the total heterotrophic bacterium that is present in the colony.

Experiment to determine the degradation rate of aromatic hydrocarbon content

Aim

The aim of this experiment is to determine the rate of biodegradation of aromatic hydrocarbons, using aromatic hydrocarbon–utilizing bacteria. The bacterium isolated is *Pseudomonas putida* (from *Pseudomonas* sp).

Micro-organism used

Pseudomonas putida was gotten from the isolation of total petroleum hydrocarbon utilizing bacteria from samples of fresh water. The isolation and enumeration of the microorganisms was carried out at the Biology Laboratory of the Rivers State University

Table 1. Enumeration of total aerobic heterotrophic bacteria of water sample for the period (cfu/ml).

Weeks	Site I	Site 2	Site 3	Total
Week 1	2.1×10^4	6.5×10^4	3.0×10^4	11.6×10^4
Week 2	1.1×10^4	1.2×10^4	1.1×10^4	3.4×10^4
Week 3	6.0×10^4	2.0×10^4	6.8×10^4	14.8×10^4
Total	9.2×10^4	9.7×10^4	10.9×10^4	29.8×10^4

Table 2. Enumeration of petroleum hydrocarbon utilizing bacteria (cfu/ml).

Weeks	Site I	Site 2	Site 3	Total
Week 1	2.4×10^2	6.1×10^2	5.2×10^2	13.7×10^2
Week 2	3.3×10^2	3.9×10^2	6.51×10^2	13.3×10^2
Week 3	4.0×10^2	6.6×10^2	4.6×10^2	15.2×10^2
Total	9.7×10^2	16.6×10^2	15.9×10^2	42.2×10^2

The table shows the heterotrophy and petroleum hydrocarbon utilizing bacteria from Sites 1 to 3.

of Science and Technology, Port Harcourt.

Media

The media used was an aqueous nutrient agar. It contained 1litre of distilled water, 15 g of powdered nutrient agar, dissolved in 1litre of distilled water, in order to form a solution. It also contained about 0.5 g of NH_4Cl, 5 g of Na_2HPO_4, 5.0 g/l of KH_2PO_4. The benzene, toluene and phenol were added.

Chemicals

Benzene (Oil Test Group of Companies, Port Harcourt, HPLC Grade), Toluene (Oil Test Group of Companies, Port Harcourt, HPLC Grade), Phenol (Oil Test Group of Companies, Port Harcourt, H_2SO_4 acid (Biology Laboratory RSUST P.H), (CH_3CN), (Biology Laboratory, Port Harcourt, over 99.5% purity), chlorofoam and p-xylene (Baxter GC Grade, Institute of Pollution Studies, RSUST, Port Harcourt), De-ionized water (RSUST Biology Laboratory).

Apparatus used

The following apparatus were used during the investigation: GC spectrophotometer, correlated to biomass concentration; Filter paper (0.33 µm); Gas chromatograph; Thermometer; 2 ml screw cap vials; 25 µl gas-tight syringe and 10 ml Test tubes.

Method used

Cell concentration was measured as optical density of 600 mM (OD_{600}) using GC spectrophotometer. It was correlated to biomass concentration. 0.22 µm filtered samples were used as optical density blanks. De-ionized water was used as the OD blank for toluene (1.00 OD_{600} = 1000 mg/L). Benzene, Toluene and Phenol concentrations were mentioned by gas chromatography. Aqueous samples were extracted (0.75 ml of sample to 0.75 ml of chlorofoam continuing 25 mg/l of p-xylene as an internal standard).

The chloroform layer was removed and analyzed using an HP 5890 II gas chromatography equipped with a mass selective detector (HP 5971A). Samples were stored at 4°C in 2 ml screw cap vials with Teflon-lived rubber septa until analysis. Chloroform was used to extract the Benzene, Toluene and Phenol standard solutions.

The gas phase concentration for the volatile hydrocarbons (Benzene and Toluene) was determined through gas chromatography.

Samples were injected into the gas chromatography equipment using 25 µl gas tight syringe. Aqueous intermediated were formed in the biodegradation experiments.

RESULTS AND DISCUSSION

Enumeration of total heterotrophic bacteria

Table 1 shows the aerobic heterotrophic bacteria count of the water sample from Site 1 to Site 3 and from week one (1) to week three (3), also the mean count of the bacteria in cfu/ml was determined.

The values of the physiochemical constituent of the various stations during the 3 weeks of study are as shown in Table 1. The values for temperature ranging form 19.5 to 20.02°C with a mean of 19.6 ± 0.3, pH values ranging from 6.7 to 7.3 with mean of 7.1 ± 0.5. Dissolved oxygen values ranging from 5.1 mg/L to 6.1 mg/L with mean of 5.7 ± 0.6, Biological Oxygen Demand with values ranging from 12.5 mg/L to 15.3mg/L with mean value of 14.4 ± 1.7, salinity has no values, whereas electrical conductivity has values ranging from 14 to 21.0 µs/cm.

Enumeration of petroleum hydrocarbon utilizing bacteria

Table 2 and Table 3 enumerates the count of petroleum hydrocarbon utilizing bacteria in the different sites.

Table 3. Test results from sample sites (Fresh water medium)

Locations	T°C	pH	DO(mg/L	BOD(mg/L)	Salinity $^0/_0$	Electrical conductivity
	Fresh	Fresh	Fresh	Fresh	Fresh	Fresh
Week I						
Location 1	19.0	6.1	4.5	18.3	0.0	17
Location 2	19.0	6.3	4.1	18.3	0.0	15
Location 3	19.0	6.2	4.1	18.3	0.0	15
Week II						
Location 1	20.0	8.2	6.9	12.2	0.0	21
Location 2	20.0	7.9	7.3	9.1	0.0	16
Location 3	20.0	8.9	5.3	12.2	0.0	15
Week III						
Location 1	19.5	7.7	6.8	15.4	0.0	18
Location 2	19.5	7.5	6.z5	10.1	0.0	14
Location 3	20.0	6.0	5.9	15.3	0.0	16
Total	176.2	63.7	51.4	129.2	0.0	147
	19.6	21.2	5.7	14.4	0.0	16.3
	0.3	0.5	0.6	1.7	0.0	0.6

Heterotrophic and petroleum hydrocarbon utilizing bacteria

Table 4 shows the result gotten from the different constituents for temperature; week 2 has the highest value.

Biodegradation of toluene in fresh water media

The plot in Figure 1 illustrates the rate of toluene degradation with increase in time. It is seen that increase in time yield decrease in toluene concentration. The variation in toluene concentration can be attributed to variation in time as well as the physicochemical parameter. The micro-organisms within 3 h degraded significant amount of substrate. However, the rate of degradation soon started declining from the 5th h. This lasted for about five hours. The degradation soon stopped after the 11th h.

In the microbial population plot depicted in Figure 2, the microbes experienced lag for about 4 h. This was followed by an accelerated growth rate, which lasted for 2.5 h. Their growth became static from the 8th hour, which lasted for two hours. They began to die from the 10th h.

Line weaver-Burk plot has been constructed below in Figure 3 in order to derive the various parameters for the Michaelis-Menten equation. The equation y = mx + c can be derived thus:

$y = 157.21\,x - 6.7076$,
Slope = 157.21,

$1/\,V_{max}$ or Intercept c = -6.7076, Therefore, V_{max} = 0.1491 mg/l per h.
$K_m/V_{max.}$ = 157.21, Therefore, K_m = 23.44 mg/l per h.

Raw data for biodegradation of phenol

It can be seen from Figure 4, the substrate-time plot of phenol that it experienced a period of steady growth for over 20 h. This was soon followed by the steadily decreasing consumption of the substrate for about 18 h. In the microbial growth plot depicted in Figure 5, the microbes underwent a lag phase for about 16 h. This was followed by an exponential phase of 20 h. The exponential growth soon declined from the 25th h.

Lineweaver-Burk plot has been constructed below, in Figure 6 in order to derive the various parameters for the Michaelis-Menten equation. The equation y = mx + c can be derived thus:

$y = 8026.7x - 191.94$,
Slope = 8026.7,

$1/V_{max}$ or Intercept c = -191.94, Therefore, V_{max} = 0.0053mg/l per h.
$K_m/V_{max.}$ = 8026.7, Therefore, K_m = 41.82mg/l per h.

Table 4. Isolates of heterotrophic and Petroleum/aromatic hydrocarbon utilizing bacteria.

Isolate designation		ISO 13	ISO 12	ISO 11	ISO 10	ISO 9	ISO 8	ISO 7	ISO 6	ISO 5	ISO 4	ISO 3	ISO 2	ISO 1
Morphology		Cocci	Rods	Rods	Rods	Cocci	Cocci	Rods	Rods	Cocci	Rods	Rods	Rods	Rods
Gram reaction		+	-	-	+	-	+	-	-	-	-	+	-	-
Catalase		+	+	+	+	+	-	+	+	+	+	+	+	+
Oxidase		-	-	+	+	+	+	+	-	+	-	+	+	+
Conagulase		-	-	-	-	-	-	-	-	-	-	-	-	-
Mority		-	+	+	+	+	-	+	+	+	+	+	+	-
Methyl red		-	-	-	-	+	-	+	+	+	+	-	-	-
Vooges Prokase	**Total Heterotrophic**	-	-	-	+	-	-	-	-	-	-	+	-	-
Glucose		-	+	+	+	+	+	+	+	+	+	+	+	+
Lactose		-	+	+	-	+	+	+	+	+	+	-	+	+
Sucrose		-	+	+	+	+	+	+	+	+	+	+	+	+
Possible organism		*Micrococcus*	*Escherichia coli*	*Chromo bacterium*	*Bacillus sp*	*Pseudomonas sp*	*Streptococcus sp*	*Pseudomonas sp*	*Escherichia coli*	*Pseudomonas sp*	*Escherichia coli*	*Bacillus sp*	*Chromo bacterium*	*Flavo bacterium*

Raw data for benzene

The plot in Figure 7 depicts the degradation rate of benzene in fresh water; it can be seen from the substrate-time plot of benzene that in about 8.5 h, 23 mg/l of substrate had been consumed. The degradation rate decreased till it stopped completely after 12 h. The degradation plot of benzene is slightly similar to that of toluene.

In microbial growth plot depicted in Figure 8, the Microbes underwent lag for about 3 h. This was followed by an exponential phase of 13 h. They began to die out by the 10th h.

Raw data for benzene

The plot in Figure 7 depicts the degradation rate of Benzene in fresh water; it can be seen from the substrate-time plot of Benzene that in about 8.5 h, 23 mg/l of substrate had been consumed. The degradation rate decreased till it stopped completely after 12 h. The Degradation Plot of Benzene is slightly similar to that o f Toluene.

In Microbial growth plot depicted in Figure 8, the Microbes underwent lag for about 3 h. This was followed by an exponential phase of 13 h. They began to die out by the 10th h.

Line weaver-Burk plot has been constructed, in Figure 9 in order to derive the various parameters for the Michaelis-Menten equation. The Equation $y = mx + c$ can be derived thus:

$y = 1.973 – 2.685$,
Slope $= 1.973$

$1/V_{max}$ or Intercept $c = - 2.685$, Therefore, $V_{max} =$ 0.3724mg/l per hour.

$K_m/V_{max.} = 1.973$, Therefore, $K_m = 0.7348$mg/l per hour.

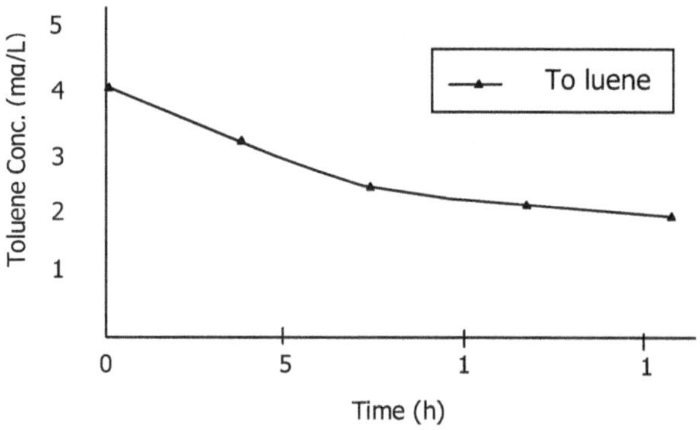

Figure 1. Toluene consumption with respect to time.

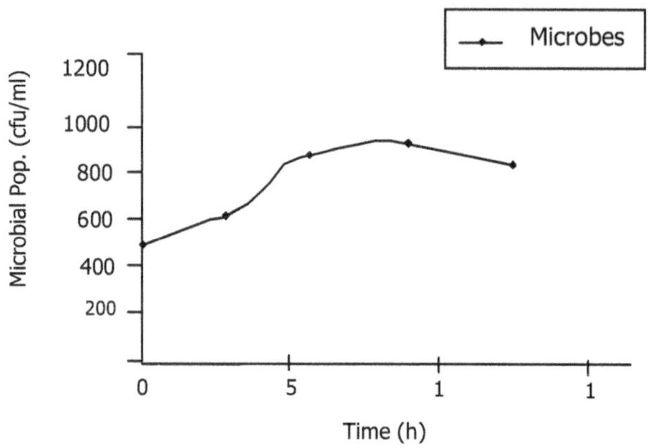

Figure 2. Microbial growth versus time.

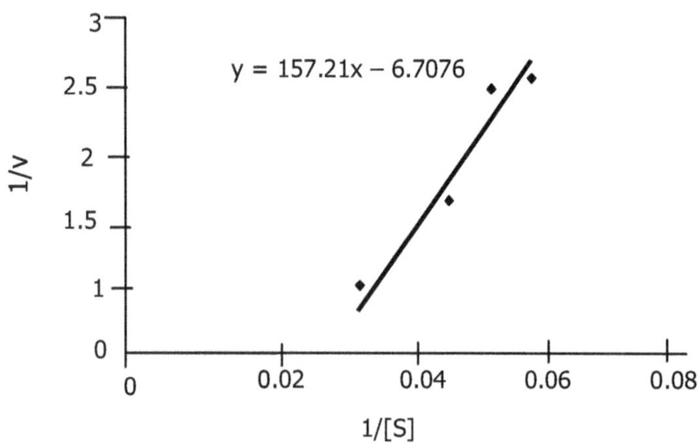

Figure 3. Line weaver-Burk plot of 1/V versus 1/S for fresh water medium.

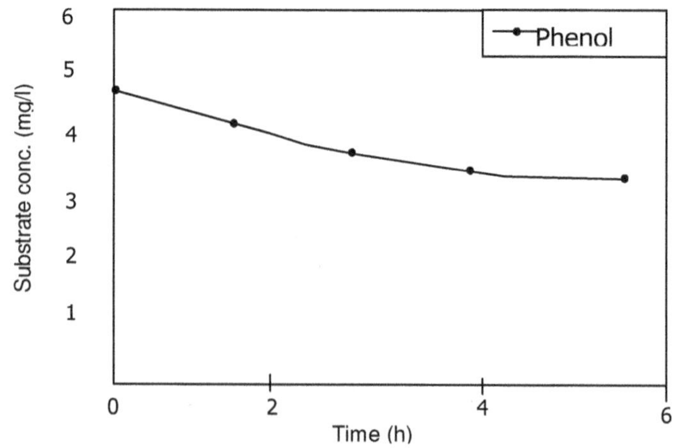

Figure 4. Phenol degradation plot of substrate concentration versus time for fresh water medium.

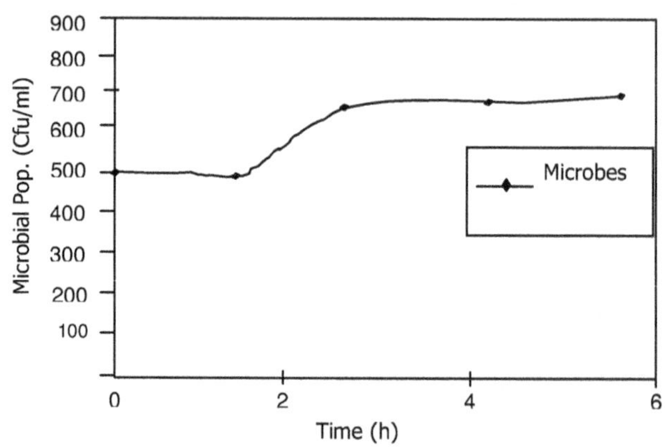

Figure 5. Microbial growth versus time for fresh water medium in phenol degradation.

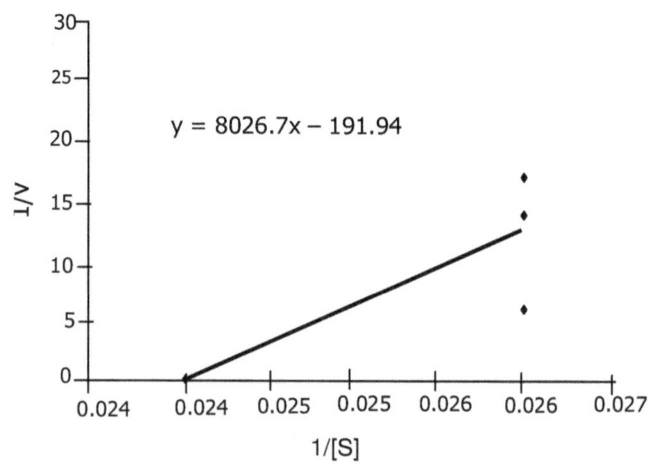

Figure 6. Line weaver-Burk plot of phenol in 1/V versus 1/S.

Conclusion

Predicting the rate of degradation of aromatic compounds requires a detailed understanding of the controlling processes and their dependence on some of the physicochemical parameters as well as environmental conditions. Simulation results were presented from the

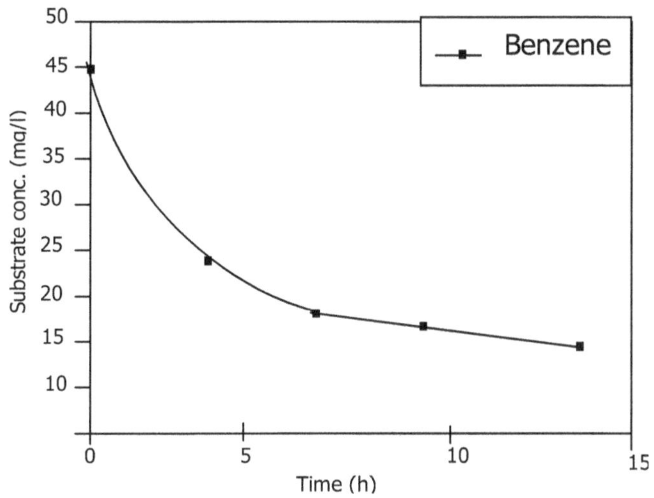

Figure 7. Benzene degradation plot of substrate concentration versus time for fresh water medium.

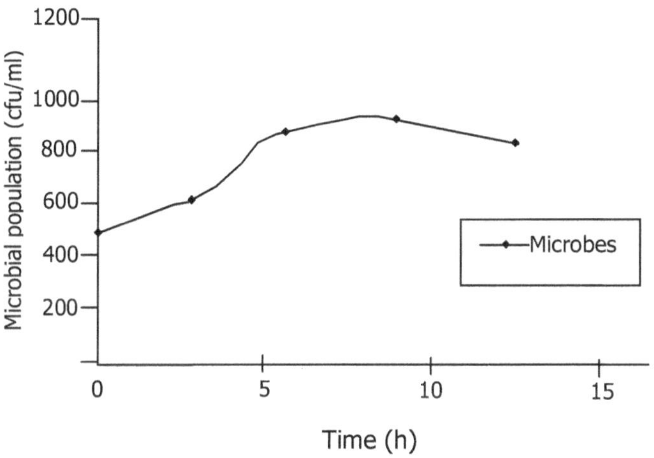

Figure 8. Microbial growth versus time for fresh water medium in benzene degradation

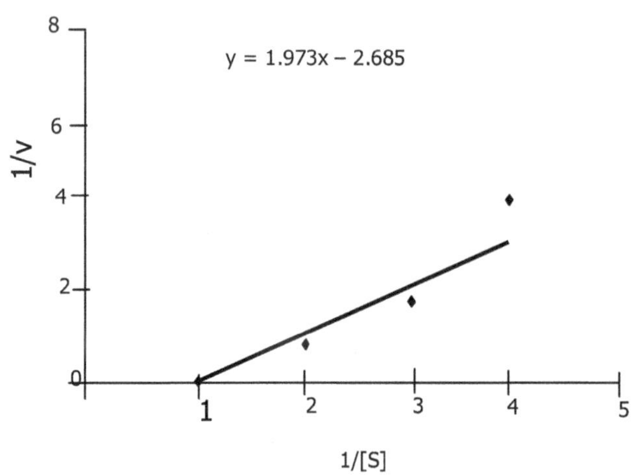

Figure 9. Line weaver-Burk plot of benzene in 1/V versus 1/S for fresh water medium.

model developed with was capable of addressing a wide range of some physicochemical conditions that influence the degradation of toluene, phenol and benzene in a fresh water medium in Niger Delta Area of Nigeria.

Our comparison of benzene, toluene and phenol degradation in a fresh water medium under different physicochemical parameters indicates that biodegradation of benzene and toluene was more effective than phenol. The physiochemical parameters play a major role in determining the rate of degradation of aromatic compounds in aqueous medium, since these pavements affect many biochemical processes, both in dry and wet season.

The results obtained in this paper were very promising since *Pseudomonas putida* was able to degrade the aromatic compounds in the fresh water medium.

This study has established that:

In the evaluation of the rate of degradation of aromatic compounds, the properties of the fresh water medium was considered as a major factors that influence the biodegradation process.

The microorganisms used in the investigation were found capable of degrading toluene, benzene and partially phenol.

It is considered by the author that the bioremediation or degradation of aromatic compounds e.g toluene, benzene and phenol as presented in this paper can be more effective if inhibiting agents are taking care of.

Isolation, identification and characteristics of heterotrophic and petroleum/aromatic hydrocarbon utilizing bacteria was presented in this paper with a view of possible organisms present in system.

Nomenclature

Y = Yield coefficient (g/cell/l)
ΔX = Change in biomass concentration (mg/l)
ΔS = Change in substrate concentration (mg/l)
Yx/s = Yield coefficient for biomass (g/cell/l)
Yp/s = Yield coefficient for products (g/cell/l)
S = Substrate concentration (mg/l)
Ya = Yield coefficient for the conversion of feed or substrate into the mass of cells (g/cell/ml).
m = Specific requirement for maintenance (mg/l)
dS = Change in substrate concentration (mg/l)
X = Biomass concentration (mg/l)
Rx = The rate of reaction of substrate (mg/l)
$\frac{dX}{dt}$ = Change in biomass concentration per unit time (mg/l/h)
μ = Biomass constant or specific growth rate (mg/l)
t = Time (h)
IS = Salinity inhibition constant (h^{-1})
KI = Substrate inhibition constant (mgl^{-1})

I*S = Constant depending on culture (h^{-1})
μm = Maximum specific growth rate (mg/l)
Sa = Salinity level (% g/cm^3)
ρ = Density of water medium (g/cm^3)
KS = Equilibrium constant of substrate (dimensionless)
Da = Constant for the salinity inhibition of maximum specific growth rate (h^{-1})
MToT = Total mass of substrate in the liquid phase in the bioreactor (g) or (cm^3)
Ma = Mass or volume of substrate in the gaseous phase in the bioreactor (g) or (cm^3)
R = Universal gas constant
T = Temperature
H = Enthaphy
Va = Volume of substrate in gaseous phase (cl^3)
VL = Volume of substrate in liquid phase (cl^3)
dP = Change in product concentration (mg/l)
δ = Constant of pH dependent (dimensionless)
ε = Constant of pH dependent (dimensionless)
SL = Substrate in liquid phase (dimensionless)
Ai = Positive constant (dimensionless)
KI = Positive constant (dimensionless)
τ = Cell age (h) (dimensionless)

REFERENCES

Adamczewska M, Siepak J, Gramowska H (2000). Studies of levels of polycyclic aromatic hydrocarbons in soils subjected to antropopressure in the city of Poznafi. Pol. J. Environ., 4: 305.

Aksu Z, Gonen F (2004). Biosorption of phenol by immobilized activated sludge in a continuous packed bed: Prediction of breakthrough curves. Process Biochem., 39(5): 599-613.

Alexievaa Z, Gerginova M, Zlateva P, Peneva N (2004). Comparison of growth kinetics and phenol metabolizing enzymes of *Trichosphoron Cutaneum* R57 and mutants with modified degradation abilities. Enzym. Microb. Technol., 34(3-4): 242-247.

Annadurai G, Juang RS, Lee DJ (2002). Microbiological degradation of phenol using mixed liquors of *Pseudomonas putida* and activated sludge. Waste Manag., 22(7): 703-710.

Bandhyopadhyay K, Das D, Bhattacharyya P, Maiti BR (2001). Reaction engineering studies on biodegradation of phenol by *Pseudomonas putida* MTCC 1194 immobilized on calcium alginate. Biochem. Eng. J., 8(3): 179-186.

Banerjee G (1997). Treatment of phenolic wastewater in RBC reactor. Water Res., 31(4): 705-714.

Belcher R, Nutten AJ, Macdonald AMG (1970). Quantitative Inorganic Analysis. 3rd edn. London: Burtterworths, p. 47.

Buitron G, Gonzalez LM, Lopez-Marin A (1998). Biodegradation of phenolic compounds by an acclimated activated sludge and isolated bacteria. Water Sci. Technol., 37(4-5): 371-378.

Chryplewicz-Kalwa M (1999). Characteristic of the rivers in the Wroclaw province in respect of the content of polycyclic aromatic hydrocarbons (in Polish). Gosp. Wodn., 3: 103.

Chung TP, Tseng HY, Juang RS (2003). Mass transfer effect and intermediate detection for phenol degradation in immobilized *Pseudomonas putida* systems. Process Biochem., 38(10): 1497-1507.

Dauta A (1982). Conditions de development du phytoplacton. Etude comparative du comportement de huit especes en culture: Role des nutriments, assimilation et stockage intracellulaire. Ann. Limnol., 18(3): 33-40.

De Voogt P, Van Hattum PB, Leonards P, Klamer JC, Govers H (1991). Bioconcentration of polycyclic aromatic hydrocarbons in the guppy (*Peocillia reticulate).* Aquat. Toxicol., 20: 169-194.

Djomo JE, Ferrioer V, Gauthier L, Zoll-Moreux C, Marty J (1995). Amphibian macronucleus test *in vivo*: Evaluation of the genotoxicity of some polycyclic aromatic hydrocarbons found in a crude oil. Mutagenesis, 10(3): 223-226.

Djomo JE, Garrigues P, Narbonne F (1996). Uptake and Depuration of polycyclic aromatic hydrocarbons from sediment by the zebrafish (*Brachydanio Rerio).* Environ. Toxicol. Chem., 15(7): 1177-1181.

Djomo JE, Dauta A, Ferrier V, Narbonne JF. Monkiedje A, Njine T, Garrigues P (2004). Toxic effects of some major polyarmatic hydrocarbons found in crude oil and aquatic sediments on *Scenedesmus subspicatus*. Water Res., 38: 1817-1821.

Egberongbe FOA, Nwilo PC, Badejo OT (2006). 'Oil Spill Disaster Monitoring along Nigerian Coastline', Disaster Preparedness and Management: Shaping the Change, Munich, Germanym, pp. 28-34.

Eisler R (1987). Polycyclic aromatic hydrocarbon hazards to fish, wildlife, and invertebrates: A synoptic review. US Fish Wildl. Serv. Biol. Rep., 85: 1-11.

Gernjak W, Krutzler T, Malato AGS, Caceres J, Bauer AR (2003). Fernandez-Alba, Photo-Fenton Treatment of water containing natural phenolic pollutants. Chemosphere, 50: 71-78.

Geyer H, Viswanathan R, Freitag D, Korte F (1981). Relationship between water solubility of organic chemicals and their bioaccumulation by the alga chlorella. Chemosphere, 10(11/12): 1307-1313.

Godjevargova T, Ivanova D, Alexieva Z, Dimova N (2003). Biodegradation of toxic organic components from industrial phenol production waste waters by free and immobilized *Trichosporon cutaneum* R57, Process Biochem., 38(6): 915-920.

Gonzalez G, Herrera MG, Garcia MT, Pena MM (2001). Biodegradation of phenol in a continuous process: comparative study of stirred tank and fluidized-bed bioreactors, Bioresour, Technol., 76(3): 245-251.

Halsall CJ, Barrie LA, Fellin P, Muir CG, Billeck BN, Lockhartl FYA (1997). Spatial and temporal variation of polycyclic aromatic hydrocarbons in the Arctic atmosphere. Environ. Sci. Technol., 31: 3593.

Hao OJ, Kim MH, Seagren EA, Kim H (2002). Kinetics of phenol and chlorophenol utilization by *Acinetobacter* species. Chemosphere, 46(6): 797-807.

Hidalgo A, Jaureguibeitia A, Prieto MB, Fernández CR, Serra JL, Llama MJ (2002). Biological treatment of phenolic industrial wastewaters by *Rhodococcus erythropolis* UPV-1. Enzyme Microb. Technol., 31(3): 221-226.

Hutchinson TC, Hellebust JA, Mackay D, Tam D, Kauss P (1979). Relationship of hydrocarbon solubility to toxicity in algae, cellular membrane effects, In: Oil Spill Conference American Petroleum Institute, Washington, DC, pp. 541-547.

IARC (1993). Monograph on the evaluation of carcinogenic risks of chemicals to man: Certain polycyclic hydrocarbons and heterocyclic compounds. Lyon, France, 3: 22-34.

Kibret M, Somitsch W, Robra KH (2000). Characterization of a phenol degrading mixed population by enzyme assay. Water Res., 34(4): 1127-1134.

Knightes CD, Peters CA (2006). Multisubstrate Biodegradation kinetics for binary and complex mixtures of polycyclic aromatic hydrocarbons. Environ. Toxicol. Chem., 25(7): 1746-1756.

Kochany J, Maguire RJ (1991). Abiotic transformations of polynuclear aromatic hydrocarbons and polynuclear aromatic nitrogen heterocycles in aquatic environments. Sc. Total Environ., 144: 17.

Kowalczyk D, Swietlik R, Dojlido J (2000). Polycyclic aromatic hydrocarbons in aquatic environment of the Radomka River basin. Field study in Poland. Environ. Prot. Eng., 26(4): 51.

Kreyzig E (1999). Advanced Engineering Mathematics. 8th edn. New York: John Wiley and Sons, Inc., pp.1–21.

Kumar A, Kumar S (2004). Biodegradation kinetics of phenol and catechol using *Pseudomonas putida* MTCC 1194. Biochem. Eng. J., 22: 151-159.

Kumaran P, Paruchuri YL (1997). Kinetics of phenol biotransformation. Water Res., 31(1): 11-22.

Lante A, Crapisi A, Krastanov A, Spettoli P (2000). Biodegradation of phenols by laccase immobilised in a membrane reactor. Process Biochem., 36(1-2): 51-50.

Lee FR, Gardner WS, Anderson JW, Blaylock JW, Barwell-Clarke J

(1978). Fate of polycyclic aromatic hydrocarbons in controlled ecosystem enclosures. Environ. Sci. Technol., 7: 832.

Lemaire P, Lemaire-Gony S, Berhaut J, Lafaurie M (1992). The uptake, metabolism and biological half-life of benzo(a)pyrene administered by force-feeding in sea bass (*Dicentrarchus labrax*). Ecotoxicol. Environ. Saf., 23: 244-251.

Levenspiel O (1999). Chemical Reaction Engineering. 3rd edn. New York: John Wiley and Sons, Inc., pp. 611-627.

Margesin R, Fonteyne PA, Redl B (2005). Low-temperature biodegradation of high amounts of phenol by *Rodococcus* spp. And basidiomycetous yeasts. Res. Microbiol., 156(1): 68-75.

Marrot B, Barrios MA, Moulin P, Roche N (2006). Biodegradation of high phenol concentration by activated sludge in an immersed membrane bioreactor. Biochem. Eng. J. 30: 174-183.

Mekenyan OG, Ankley GT, Veith GD, Call DJ (1994). ASARs for photoinduced: Acute lethality of polycyclic aromatic hydrocarbons to *Daphnia magna*. Chemosphere, 28(3): 567-582.

Mekenyan OG, Ankley GT, Veith GD, Call DJ (1995). QSARs for photoinduced toxicity of aromatic compounds. SAR QSAR Environ. Res., 4: 139-145.

Mill T, Babey WR, Lan BY, Baraze A (1981). Photolysis of polycyclic aromatic hydrocarbons in water. Chemosphere, 11/12: 1281.

Miller RM, Singer GM, Rosen JD, Bartha R (1988). Photolysis primes biodegradation of benzo(a)pyrene. Appl. Environ. Microbial., 58(5): 1724-1730.

Monteiro AMG, Boaventura AR Rodrigues AE (2000). Phenol Biodegradation by *Pseudomonas putida* DMS 548 in a batch reactor, Biochem. Eng. J., 6(1): 45-49.

Mordocco A, Kuek C, Jenkins R (1999). Continuous degradation of phenol at low concentration using immobilized *Pseudomonas putida*. Enzym. Microb. Technol., 25(6): 530-536.

Morrison RT, Boyd RN (1974). Organic Chemistry. 3rd edn. Boston: Okoh AI (2002) Assessment of the Potentials of some Bacterial Isolates for Application in the Bioremediation of Petroleum Hydrocarbon polluted soil. Ph.D Thesis. Obafemi Awolowo University, Ile-Ife, Nigeria, pp. 7-130.

Narbonne JG, Ribera D, Garrigues P, Lafaurie M, Romana A (1992). Different Pathways for the uptake of benzo(a)pyrene adsorbed to sediment by the mussel *Mytilus galloprovincialis*. Bull. Environ. Contam. Toxicol., 49: 150-156.

Neter J, Wasserman W, Kutner MH (1985). Applied linear statistical models. 2nd ed. Homewood, IL: Richard D. Irwin, pp. 171-174.

Newsted JL, Giesy JP (1987). Predictive models for photoinduced acute toxicity of polycyclic aromatic hydrocarbons to *Daphnia magna Strauss* (Cladocera, Crustacea). Environ. Toxicol. Chem., 6: 445-461.

Nuhoglu A, Yalcin B (2005). Modelling of Phenol Removal in a batch reactor. Process Biochem., 40(3-4): 1233-1239.

O'Brien PJ (1991). Molecular mechanisms of quinine cytotoxicity. Chem. Biol. Interact., 80: 1-41.

Okoh AI (2006). 'Biodegradation Alternative in the Cleanup of Petroleum Hydrocarbon Pollutants.' Biotechnol. Mol. Biol. Rev., 1(2): 38-50.

Pai SI, Hsu YL, Chong NM, Sheu CH (1995). Chem. Continuous degradation of phenol by *Rhodococus* sp. immobilized on granular activated carbon and in calcium alginate. Biorescour. Technol., 51(1): 37-42.

Park C, Marchland EA (2006). 'Modelling Salinity Inhibition Effects During Biodegradation of Perchlorate.' J. Appl. Microbiol., 101: 222-233.

Pazarlioglu NK, Telefoncu A (2005). Biodegradation of phenol by *Pseudomons putida* immobilized on activated pumice particles. Process Biochem., 40: 1807-1814.

Prasad R (2000). Petroleum Refining Technology. Nai Sarak Delhi: Khanna Publishers, pp. 346-349.

Prieto MB, Hidalgo A, Serra JI, Llama MJ(2002). Degradation of phenol by *Rhodococcus crythropolis* UPV-1 immobilized on Biolite in a packed-bed reactor. J. Biotechnol., 97: 1-11.

Reardon KF, Mosteller DC, Rogers J, DuTeau NM, Kim KH (2002). 'Biodegradation Kinetics of Aromatic Hydrocarbon Mixtures by Pure and Mixed Bacterial Cultures.' Environ. Health Perspect., 110(6): 1005-1010.

Richardson JF, Peacock DG (19991). "Biochemical Reactors and Process Control.' Coulson and Richardson Chemical Engineering: 3rd edn. Linacre House: Butterworth-Heinemann, 3: 330-340.

Sa CSA. Boaventura RAR (2001). Biodegradation of phenol by *Pseudomonas putida* DSM 548 in a trickling bed reactor. Biochem. Eng. J., 9(3): 211-219.

Sims RC, Overcash MR (1983). Fate of polynuclear aromatic compounds (PNAs) in soil-plant systems. Residue Rev., 88: 1-68.

Suietlik R, Kowalczyk D, Dojlido J (2002). Influence of selected physicochemical factors on the Degradation of PAHs in water. Polish J. Environ. Stud., 11(20): 165-169.

Swallow KC, Taghizadeh K, Weakland S, Plummer EF, Bussby WF, Lafleur AL (1995). Photooxidation of selected polycyclic aromatic hydrocarbons in aqueous organic media in the presence of Ti(IV)oxide. Int. J. Environ. Chem., 60: 113.

Thomas SD, LI QX (2000). Immunoaffinity chromatography for analysis of polycylic aromatic hydrocarbons in corals. Environ. Sci. Technol., 34: 2649.

Timur S, Pazarlioglu N, Piloton R Telefoncu A (2003). Detection of phenolic compounds by thick film sensors based on *Pseudomonas putida*. Talanta, 61(2): 87-93.

Ukpaka CP (2010). Investigation into the Rain Water quality of Ogba Community in Niger Delta Area of Nigeria. The Nigerian Academic Forum: Multidiscip. J., 18(2): 1-11.

Ukpaka CP (2004). Comparative Impact of Drilling Mud and Drill Cuttings on the Environment. Int. J. Sci. Tech. Nig., 2(3): 38-41.

Ukpaka CP (2004). Development of Model for Crude Oil Degradation in a Simplified Stream System. Int. J. Sci. Technol., 3(2): 34-37.

Ukpaka CP (2005). Biodegradation Kinetics for the production of carbon dioxide from natural aquatic ecosystem polluted with crude oil. J. Sci. Technol. Res., 4(3): 41-49.

Vysotskaya NA, Bortun LN (1984). The mechanism of radiation-induced hemolytic substitution of condensed aromatic hydrocarbons in aqueous solutions. Rad. Phys. Chem., 23(6): 731-738.

Wang SJ, Loh KC (1999). Modelling the role of metabolic intermediates in kinetics of phenol biodegradation. Enzym. Microb. Technol., 25: 177-184.

Watanabe K, Hino S, Takahashi N (1996). Responses of activated sludge to an increase in phenol loading. J. Ferment. Bioeng., 82(5): 522-524.

Zepp RG, Hoigne J, Bader H (1987). Nitrate-induced photooxidation of trace organic-chemical in water. Environ. Sci. Technol., 21(5): 443-450.

Zerbe J, Elbanowska H, Siepak J, Sobczynski T, Kabacinski M, Gramowska HH (1994). Metals and polycyclic aromatic hydrocarbons in the lakes water of the Wielkopolski National Park (In polish). Gosp. Wodn., 8: 174.

Zhu X, Venosa AD, Suidan MT, Lee K (2001). 'Guidelines for the Bioremediation of Marine Shorelines and Fresh Water Wetlands.' U.S. Environmental Protection Agency, Contract 68-c7-0057, pp. 1-163.

The protective effect of beta glucan against *Escherichia coli* infected mice via intraperitonial administration

Elufisan T. O.*, Oyelade Busayo and Oloke J. K.

National Centre for Technology Management, P.M.B. 012, Obafemi Awolowo University, Ile-Ife, Osun State, Nigeria.

Mice infected with *Escherichia coli* were protected against lethal peritonitis by the intra peritoneal administration 0.2 ml of 10 mg of poly-1,3,1,6-glucotriosyl-1,3,1,3-glucopyranose (PGG) glucan per gram body weight of animal 24 h prior to bacterial infection. This procedure employed in this study, is similar to that employ in a previous studies where rat model were protected with β-glucan against intra-abdominal sepsis, where the intramuscular doses of 10 ml to 10 g per animal 24 h and 4 h prior to surgical implantation of the bacterial inoculums reduced the early mortality associated with the peritonitis phase of this experimental disease process. In this study mice were protected with β-glucan against peritonist 24 h prior to intraperitoneal administration of bacteria inoculums. Quantitative cultures of the liver obtained from mice infected with *E.coli* showed significantly the liver of PGG glucan-treated infected animal has a reduced microbial load as compare to those infected and were not treated. The transient increase in survival rate of mice infected with *E.coli* that have been treated with PGG glucan- as compare with that of those not treated shows the significant effect of β glucan on *E.coli* infection.

Key words: Beta-glucan, *Escherichia coli,* poly-131-6-glucotriosyl-j31-3-glucopyranose (PGG) glucan, peritonitis, inoculums, intra-peritoneal, intramuscular

INTRODUCTION

The discovery of zymosan a crude insoluble extract from yeast cell wall, as a stimulant, that has an ability to stimulate the reticuloendothelial system initiate the identification of the immunological properties of yeast, (Riggi and DiLuzio, 1961; Ondoderk et al.,1992). Further study was able to identify the active moiety of the yeast cell wall as beta-glucan (Bacon et al., 1960; Phaff, 1963; Onderdonk et al., 1992).

Beta glucan had been described to be able to stimulate the immune system, owing to its ability to increase the plasma level and splenic level of Interleukin 1 and 2 in rat (Czop and Austen, 1985).

The immune modulating properties of β-glucan had been related to its ability to bind to specific receptor site of the human neutrophils and macrophages (Czop and Austen, 1985).

It has been shown that host defense responses related to the β-glucans include the activation of the alternative complement pathway, the release of lysosomal enzymes by monocytes, and the generation of leukotrienes by monocyte (Czop et al., 1978; Czop and Austen, 1985; Janusz et al., 1987). Immune modulators are known to exhibit their activities in animal by increasing the non specific component of the immune system such as the macrophages, and neutrophil, and can induce innate immune response by stimulating the maturation of the myeloid progenitor cells (Kounnikasis et al., 2003). This ability of immune modulators such as β-glucan to activate macrophages is considered the first line of defense against microorganism. Once the macrophages are activated they have the ability to stimulate the secondary

*Corresponding author. E-mail: ptemidayo@yahoo.com.

line of defense, which is the effective humoral and cellular response (Abass et al., 2000).

The ability of β-glucan to ensure microbial clearance and reduce mortality rate of infected animal through its activities on macrophages and neutrophils, (Kaiser et al., 1993) and also that not much have been said on the intra-peritoneal administration of neutrophil without using it as an adjuvant or supplement in a particular formulation led into this study.

This study also tries to investigate the effectiveness of a new concentration on the protective effect of β- glucan administration through the peritoneum.

MATERIALS AND METHODS

Animals

Fourteen-weeks old mice weighing between 22 and 24 g obtained from the central animal house University of Ibadan were used in this study.

The mice were maintained at a maximum of 5 mice per cage under standard laboratory condition. The mice were fed with pellets obtained from Ladokun feeds in Ibadan and water was supplied to the animal ad libtum.

Bacteria

E. coli; A multi-resistant strain of *E. coli* obtained from the medical microbiology laboratory of Baptist Medical centre Ogbomosho Oyo State Nigeria was used for this study. Strains were grown in nutrient agar and later in eosin methylene blue agar which is selective for *E. coli* .These were then subculture in nutrient after which the stock culture was prepared by harvesting the cell suspension from nutrient agar in normal saline. The stock was diluted and use in the protection study.

β-glucan

Beta-glucan was purified from the cell wall of *Saccharomyces cerevisae* by Professor J.K. Oloke of the Department of Pure and Applied Biology, Ladoke Akintola University of Technology, Ogbomoso, Oyo state, Nigeria and was made available for this study.

Protection study

To investigate the protective of the intra peritoneal administration of beta-glucan treatment 4 groups (n = 10) of mice were administered single intra-peritoneal dose of β glucan at 10 mg per 1000 mg body weight of mouse in 0.2 ml of saline. Saline was also administered to those that were used as control 24 h before the animals were challenged with 0.2 ml 0f the lethal dose (which is 0.2 ml 0f 4×10^{-2} cfu/ml of *E.coli)* of *E.coli by* intra-peritoneal administration.

The controls were not infected with *E. coli* but were only injected with 2 ml of saline solution.

Animals were observed twice daily for ten days within which the study last. The survival rate was recorded following the cause of infection and percentage survival was calculated using the ratio of the surviving animal each day to the total number of infected animals.

Microbial clearance

At the end of this study which is the 11th day following challenge with the lethal dose of *E.coli,* the surviving animal were humanely slaughtered, and the liver were harvested, weighed, macerated and homogenize in 20 ml nutrient broth in an aseptic condition. The liver homogenate were serially diluted and 0.1 ml of the serially diluted liver homogenate at a concentration of 0.2 ml of 1×10^{-7} cfu/ml of each homogenate was seeded into eosin methylene blue agar and incubated for 24 h at 25°C in other to evaluate the number of the colony forming units per gram liver tissue

RESULTS

A single dose of β-glucan, given intra- peritoneally increased the number of animals that survived by 40% from one (10%) in the control group to five in the intervention group (50%).Table 1 and Figure 1.

β-glucan was able to increase the survival rate in the intervention group because of its ability to stimulate the host innate immune response. The enhanced microbial killing as observed by significant reduction in the microbial bio-burden of the livers of the treated animal is shown in Table 2.

Statistical evaluation

Comparison of groups with regard to mortality was made by student's t test; comparisons of quantitative data were made with student's t test. The result of this study was found to be significant at a probability value of $p \leq 0.05$.

DISCUSSION

Previous studies have documented the nonspecific immune-modulatory properties of insoluble polysaccharides as zymosan (Riggi and DiLuzio, 1961; Williams et al., 1987; Reynolds et al., 1982; Cook et al., 1982 in animal test system). Also, Ondoker et al. (1992) were able to show the possibilities of using the solubilized form of beta glucan via percutaneous and trans thoracic administration and through intra muscular injection of the beta glucan to abrogate intra-abdominal sepsis.

The observation that yeast glucans could abrogate the deleterious effects of a variety of microbial pathogen prompted considerable interest in determining exactly how such compounds interact with the host immune system. The Moribund finding of a specific cell surface receptor for P-1, 3 glucans on macrophages and granulocytes suggested that the protective effects for the glucans are provided by such cells (Czop and Austen, 1985; Czop, 1986; Easson et al., 1990; Sherwood et al., 1987), possibly by cytokine production and release, improve phagocytic killing, or increased production of phagocytic cells. Because PGG glucans bind to cells with

Table 1. Percentage surviving mice following *E.coli* infection. These animals were observed for ten days and the percentage survival rate was recorded from the ratio of the mice that survive after infection to the total number of animal used.

Days	Percentage Survival			
	Infected animal with treatment (%)	Treated animal without infection (%)	Infected animal without treatment (%)	Animal without infection and treatment (%)
1	90	100	20	100
2	50	100	10	100
3	50	100	10	100
4	50	100	10	100
5	50	100	10	100
6	50	100	10	100
7	50	100	10	100
8	50	100	10	100
9	50	100	10	100
10	50	100	10	100

Table 2. Average microbial clearance rate in animal infected with *E.coli* the surviving animals were slaughtered and the liver harvested was homogenized in 20 ml nutrient broth to carry out colony counting in other to evaluate the microbial load in the liver (which is reported in colony forming unit).

Treatment	Cfu/g Liver
Infected Animal without Treatment	4.0×10^7
Infected Animal with Treatment	0.5×10^7

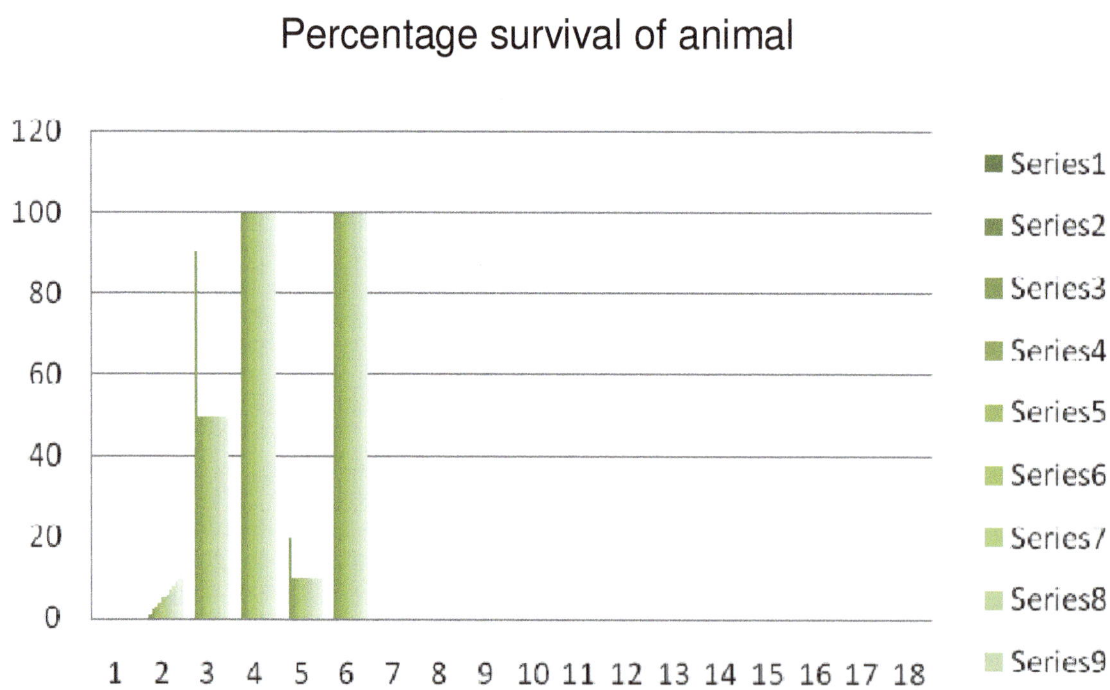

Figure 1. Histogram of the percentage rate of survival of animals after the completion of study. $p \leq 0.05$. Key 3 represent infected animal without treatment indicating how survival rate fell from 90 to 50%. 4 represent animals without infection but with treatment. 5 represents infected animal with treatment indicating how survival rate fell from 20 to 10% 6 rep animal without infection and without treatment.

an array of immune-modulatory activities, it is likely that the actual protection provoked by such materials is complex.

The protective effect of beta glucan is mediated through the stimulation of the microbicidal activities of the circulating blood cells of the innate immune system (Vetvicka et al., 1996).

The understanding of the ability of beta glucan to stimulate host immune defence and previous study that have been carried out on the efficacy of beta glucan prompted this study to investigate the protective effect of the intra peritoneal administration of beta glucan in an *in vivo* investigation.

The ability of beta glucan to significantly reduce the rate of death in infected animals as compare with other studies in which the effectiveness of beta glucan had been demonstrated in intramuscular administration further shows the ability of beta glucan to induce host immune cell in combating infectious diseases and like the result obtained by ondoker et al. (1992) soluble form beta glucan is also of significant important in combating microbial infection without any toxicity. The survival of the group of mice injected with beta glucan only further shows that the beta glucan in soluble form has no toxic effect and can be administered to animal directly instead of using it as an adjuvant.

Subsequent experiments had administered beta glucan to animals few hours before the introduction of infectious agents. This work however subject the animal to a 24 h treatment before exposure to infectious agents, during which there is possibilities of decrease in blood level of this protective agent. This therefore suggests two possibilities:

(1) This further establish the fact that the activities of beta glucan is based on its effect on the host immune system, and thus could possibly be a good agent in combating the continuous emergence of resistant disease condition.
(2) It also suggests the possibility of beta glucan to be used as a prophylactic agent against infection in the nearest future.

REFERENCES

Abass AK, Litchman AH, Pober JS (2008). Cellular and Molecular Immunology. 2nd edition Philadelphia, Pawb Sander Company,

Onderdonk ABR, Cisneros PH, Gary O. (1992). Anti-Infective Effect of Poly-13-6-Glucotriosyl-1,3-3-Glucopyranose Glucan *in vivo.* April 1992. J. Infect. Immun., 60(4): 1642-1647.

Bacon JSD, Farmer VC, Jones D, Taylor IF (1969). The glucan component of the cell wall of Baker's yeast (Saccharomycescerevisiae) considered in relation to its ultrastructure. Biochem. J., 114: 556-557.

Czop JK (1986). Phagocytosis of particulate activators of the alternative complement pathway: Effects of fibronectin. Adv. Immunol., 38: 361-398.

Czop JK, Austen KF(1985). A P-glucan inhibitable receptor on human monocytes: its identity with the phagocytic receptor for particulate activators of the alternative complement. Pathw. J. Immunol., 134: 2588-2593.

Czop JK, Austen KF (1985). Generation of leukotriene by human monocytes upon stimulation of their 13-glucan receptor during phagocytosis. Proc. Natl. Acad. Sci. USA, 82: 2751-2755.

Czop JK, Austen KF (1985). Properties of glucans that activate the human alternative complement pathway and interact with the human monocyte l-glucan receptor. J. Immunol., 135: 3388-3393.

Czop JK, Fearson DT, Austen KF (1978). Opsoninin dependent phagocytosis of activators of the alternative complement pathway by human monocytes. J. Immunol., 120: 11321138.

Easson DD, Ostroff Jr GR, James S (1990). Macrophage-targeted carbohydrate microcapsules for antigen and drug delivery. Abstr. Int. Congr. Infect. Dis., 58: 689.

Janusz MJ, Austen KF, Czop JK (1987). Lysosomal enzyme release from human monocytes by particulate activators is mediated by f-glucan inhibitable receptors. J. Immunol., 138: 3897-3901.

Kaizer AD, Kernodle BS (1998). Synergism between Poly- (1-6) beta – D- gluco- pyranosyl (1-3) – beta –d- glucopyranose.glucan and cefazolin in prophylaxis of *Staphylococcus infection* in a guinea pig model. Antimicrob. Agent Chemother., 42; 2449- 2451.

Kounnikasis B, Rose Monde MP, Brousseau G (2003). Anthrax protective effect of yeast beta, 1,3, glucan. MEDSCAP, Gen. Med., p. 5.

Phaff HJ (1963). Cell wall of yeasts. Annu. Rev. Microbiol., 17:15.

Riggi SJ, DiLuzio NR (1961). Identification of a reticuloendothelial stimulating agent in zymosan. Am. J. Physiol., 200: 297-300.

Sherwood ER, Williams DL, McNamee RB, Jones EL, Browdwe IW, DiLuzio NR (1987). Enhancement of interleukin-1 and interleukin-2 production of soluble glucan. Int. J. Immunopharmacol., 9: 261-267.

Multiplex polymerase chain reaction assay for the detection of *Salmonella enterica* serovars in shrimps in 4 h

K. Thirumalai Raj, G. Jeyasekaran*, R. Jeya Shakila, A. Jemila Thangarani and D. Sukumar

Department of Fish Processing Technology, Fisheries College and Research Institute, Tamil Nadu Veterinary and Animal Sciences University, Tuticorin 628 008, India.

A rapid and sensitive multiplex polymerase chain reaction (MPCR) based assay was developed for the detection of *Salmonella enterica* serovars such as Typhi (ATCC 122235), Paratyphi A (MTCC 735), Typhimurium (MTCC 98), Enteritidis (ATCC 13065), Weltevreden (MTCC 1169) Bovismorbificians (MTCC 1162), Brunei (MTCC 1168), Arizonae (MTCC 660) and Infantis (MTCC 1167) in shrimps within 4 h of pre-enrichment. The *Salmonella* genus specific gene of *himA* gene was selected and 16S-23S internal transcribed spacer region was used as an internal amplification control (IAC). The genomic DNA was extracted by using boiling and centrifugation method. Sensitivity of the assay was tested by artificially inoculating the shrimp homogenate with viable cells of *Salmonella*. The MPCR assay could detect up to 5 cells within 4 h of pre-enrichment. Amplification of DNA extracted from other bacterial pathogens viz. *Vibrio cholerae* (NICED 16582), *Escherichia coli* (ATCC 9637) and *Staphylococcus aureus* (ATCC 12598) yielded negative results. This MPCR assay provides specific, rapid and reliable results and allows for the cost effective detection of serovars of *S. enterica* in one reaction tube in mixed bacterial communities that are prevalent in shrimp products.

Key words: Multiplex polymerase chain reaction (MPCR), *Salmonella enteric*, *himA,* 16S-23S spacer region, 4 h assay.

INTRODUCTION

Salmonella is one of the major bacterial pathogens of food borne disease outbreaks. It has been classified into more than 2,000 serovars. Most of the human diseases caused by *Salmonella* have been associated with *Salmonella enterica* serotypes Typhi and Paratyphi. The most common contaminated foods associated with human salmonellosis include poultry, beef, pork, eggs, milk and seafood. The prevalence of *Salmonella* in Indian seafoods has been reported by few workers (Nambiar and Iyer, 1991; Hatta and Lakshmanaperumalsamy, 1997; Saroj et al., 2008). A zero tolerance has been prescribed for *Salmonella* in seafood for export trade (Liston et al., 1971). Among the *S. enterica* serovars, the most commonly implicated serovars in major outbreaks are Enteritidis, Anatum, Newport, Cerro, Montevideo, Infantis and Saintpaul (Perch et al., 2003). The incidence of *Salmonella* in fish and fishery products has also been reported from other countries including Japan (Saheki et al., 1989), Thailand (Rattagool et al., 1990), Hong Kong (Yam et al., 1999) and Spain (Martinez-Urtaza et al., 2003).

The ubiquitous distribution of *Salmonella* in the natural environment and its prevalence in many food types has made inspection of *Salmonella* in foods a mandatory requirement worldwide. Conventional methods that rely

*Corresponding author. E-mail: ttnjerosh99@bsnl.in.

Table 1. Primers used in the MPCR assay.

Gene name	Primer sequences	Annealing temperature (°C)	Product size (bp)	References
himA	F-CGT GCT CTG GAA AAC GGT GAG R-CGT GCT GTA ATA GGA ATA TCT TCA	60	123	Bej et al., 1994
ITS	F-TAT AGC CCC ATC GTG TAG TCA GAA C R-TGC GGC TGG ATC ACC TCC TT	58	312	Chiu et al., 2005

on phenotypic characterization often take several days for detection and identification of this pathogen. There are reports on the PCR detection of this pathogen in foods including seafood (Liu and Li, 2001; Sanathkumar et al., 2003; Jeyasekaran et al., 2010). A negative PCR results does not necessarily indicate that no template DNA was present in the sample. Inhibitory substances present in a sample may influence the outcome of the PCR by lowering or completely preventing the amplification (Lund and Madsen, 2006). A false-negative PCR result is a major concern to the food industry where PCR is being used for pathogen detection; therefore, internal standards or controls have to be included in the PCR to avoid false-negatives (Jones et al., 2000).

The non-amplification of target DNA and the amplification of non-target internal amplification control (IAC) when IAC is added to the PCR sample indicate that the IAC signal eliminates the possibility of false-negatives (Hoorfer et al., 2003). All the PCR based methods so far reported required about 12 to 24 h of pre-enrichment for the detection of *Salmonella* (Fach et al., 1999). The present study was aimed at developing a multiplex PCR that allowed the detection of *Salmonella* in food products by including 16S-23S spacer region as non target DNA (IAC) along with the *Salmonella* genus specific *himA* gene as a target DNA.

MATERIAL AND METHODS

Bacterial strains, media and reagents

Type cultures of *Salmonella* used in this work included *S. enterica* serovar Typhi (ATCC 122235), serovar Paratyphi A (MTCC 735), serovar Typhimurium (MTCC 98), serovar Enteritidis (ATCC 13065), serovar Weltevreden (MTCC 1169) serovar Bovismorbificians (MTCC 1162), serovar Brunei (MTCC 1168), serovar Arizonae (MTCC 660) and serovar Infantis (MTCC 1167). The other bacterial pathogens viz. *Vibrio cholerae* (NICED 16582), *Escherichia coli* (ATCC 9637) and *Staphylococcus aureus* (ATCC 12598) were used to check the specificity of the assay. The cultures were grown on Trypticase Soy broth (TSB) and incubated at 37°C for 18 to 24 h to obtain young culture prior to the extraction of DNA. Molecular grade water (Sartorius Stedim Biotech, Gottingen, Germany) was used for the preparation of buffers, chemicals and reagents. Media and their ingredients were purchased from Hi-Media Laboratories Pvt. Ltd, Mumbai, India.

Bacterial DNA extraction

The genomic DNA was extracted from young cultures (10^7 CFU/ml) by boiling- centrifugation method (Trkov and Avgustin, 1998). One milliliter aliquot of enrichment broth was centrifuged at 13,000xg for 10 min. The pellet was resuspended in 100 µl of molecular grade water, heated to 105°C in a dry bath (Bangalore Genei, Bangalore, India) for 10 min, cooled in ice and centrifuged at 10,000xg for 10 min. The supernatant was used for the PCR assay. The extracted genomic DNA was directly quantified in a Biophotometer (Eppendorf AG, Hamburg, Germany).

Primers

Two sets of primers belonging to *Salmonella* genus specific gene *himA* (Bej et al., 1994) and 16S-23S (Chiu et al., 2005) internal transcribed spacer region (ITS) were selected for this study. The IAC was incorporated into the PCR without impairing the amplification of *Salmonella* and without the loss of specificity and sensitivity on the detection of *Salmonella* in food products. All the primer sequences were synthesized by Ocimum Biosolutions Inc. Indianapolis, USA. The oligonucleotide primer sequences and their expected product sizes are listed in Table 1.

Uniplex PCR amplification and detection of *S. enterica*

The uniplex PCR for the detection of *Salmonella* using two sets of primers belonging to genus specific genes were performed in a Gradient Master Cycler (Eppendorf AG, Hamburg, Germany). All the PCR mixtures were handled in the Bio-Safety Cabinet Type II (Clean Air Systems, Chennai, India). The PCR amplification was performed in 25 µl of reaction mixture containing template DNA of *S. enterica* serovar Typhimurium (MTCC 98), 10X PCR buffer (100 mM Tris (pH 9.0), 500 mM KCl,15 mM $MgCl_2$ and 0.1% gelatin), 100 mM of each dNTP, 0.4 µg of forward and reverse primers of selected genes and 1.25 U of Taq DNA polymerase. The optimized uniplex PCR conditions consisted of initial denaturation at 94°C for 3 min. followed by denaturation at 94°C for 60 s; annealing temperature 60°C for *himA* gene and 58°C for 16S-23S internal transcribed spacer region (ITS) for 60 s; extension at 72°C for 90 s; for 35 cycles and final extension at 72°C for 3 min. The PCR products were run on 2% agarose gel containing ethidium bromide (5 mg/ml) in Tris Acetate EDTA buffer (TAE, pH 8.5) using submarine electrophoresis system (GE HealthCare Bio-Science Ltd., HongKong) and observed under UV Transilluminator using Gel Documentation System (Alpha Innotech, California, USA). The products were identified in comparison with the 100 bp DNA ladder (Real Biotech Corporation, Ohio, USA).

Optimization of multiplex PCR

Optimization was done by performing PCR in a gradient master cycler with different annealing temperatures. The MPCR conditions consisted of initial denaturation at 94°C for 3 min, followed by denaturation at 94°C for 60 s, annealing temperatures ranging from 57 to 63°C for 60 s; extension at 72°C for 90 s for 35 cycles; and final extension at 72°C for 3 min.

Multiplex PCR for *Salmonella*

With the optimized MPCR conditions, the multiplex PCR was performed for the selected serovars of *S. enterica* viz., Typhi (ATCC 122235), Paratyphi A (MTCC 735), Typhimurium (MTCC 98), Enteritidis (ATCC 13065), Weltevreden (MTCC 1169) Bovismorbificians (MTCC 1162), Brunei (MTCC 1168), Arizonae (MTCC 660) and Infantis (MTCC 1167). The optimized MPCR condition consisted of initial denaturation at 94°C for 3 min, followed by denaturation at 94°C for 60 s; annealing at 62°C for 60 s; extension at 72°C for 90 s; for 35 cycles and final extension at 72°C for 3 min. The amplified MPCR product was analyzed in 2% agarose gel containing ethidium bromide as described earlier.

Specificity of MPCR assay

The MPCR assay was performed with the other bacterial pathogens like *V. cholerae* (NICED 16582), *E. coli* (ATCC 9637) and *S. aureus* (ATCC 12598) to test for the specificity of the developed MPCR for *Salmonella*.

Detection limit of MPCR assay in shrimps

Shrimp homogenate (150 g) was sterilized at 100°C for 3 min. and then divided into six lots of 25 g and transferred into the sterile petriplates. Aliquots of 0.1 ml of tenfold serial dilution (10^{-2} to 10^{-7}) of an overnight grown culture of *S. enterica* serovar Typhimurium (MTCC 98) containing 1.00×10^7 CFU/ml was inoculated into 25 g of homogenized sample. About 100 µl of each dilution was plated on to BSA plates and incubated at 37°C for 48 h for enumerating the counts. The homogenized sample containing inoculum of *S. enterica* serovar Typhimurium (MTCC 98) was enriched with 225 ml of lactose broth (Hi-Media Laboratories, Mumbai, India) and incubated for 0, 2, 4, 6, and 8 h at 37°C. Uninoculated shrimp homogenate was used as a control. Immediately after the inoculation and after 2, 4, 6, and 8 h of pre-enrichment, 1.5 ml enriched sample was used to extract the genomic DNA and performed the PCR as described earlier. The experiment was independently conducted in triplicates.

Validation of MPCR assay

The seafood samples (10 Nos.) comprising crab, finfish and shrimp collected from the Fishing Harbour, Tuticorin, India were screened for the presence of *Salmonella* and clinical isolates (10 Nos.) of *Salmonella* received from Bose Clinical Laboratory, Madurai, India were checked for its confirmation with the developed method to validate the MPCR assay.

RESULTS

DNA extraction

The genomic DNA extracted by using the SDS-NaOH method had a DNA level of 85.6 ng/µl. The SDS-NaOH method easily extracted genomic DNA from bacterial cultures in less than 30 min (Trkov and Avgustin, 1998).

Uniplex PCR for *Salmonella*

The uniplex PCR performed using the genus specific *himA* gene and 16S-23S internal transcribed spacer region (ITS) of *Salmonella* produced an intense band of the expected product size (Figure 1). The product size of *himA* gene was 123 bp and 16S-23S internal transcribed spacer region (ITS) was 312 bp. The repeated PCR amplification of selected genes gave the reproducible results.

Optimization of multiplex PCR

The selected genes got amplified in all the annealing temperatures (57 to 63°C) tested (data not shown). Since the product yield was more at 62°C, it was selected as the optimum annealing temperature for the amplification of selected genes of *Salmonella*.

Development of MPCR

The *himA* and 16S-23S internal transcribed spacer region genes were found to be specific only for the genus *Salmonella* when they were tested with the optimized annealing temperature in the MPCR and got amplified for all the selected serovars of *S. enterica* viz., Typhi, Paratyphi A, Typhimurium, Enteritidis, Weltevreden, Bovismorbificians, Brunei, Arizonae and Infantis (Figure 2).

Sensitivity of detection of *Salmonella* in shrimps by MPCR

Even though the *Salmonella* genes were undetectable immediately after the inoculation in lactose broth (data not shown), the 225 ml homogenates inoculated with 10^{-2} and 10^{-3} dilutions of *Salmonella* equivalent to 1,06,753 and 20,184 cells showed the amplification of genus specific *himA* gene and 16S-23S internal transcribed spacer region after 2 h of enrichment (Figure 3). No amplification was observed in the lower dilutions. After 4 h of pre-enrichment, *Salmonella* genus specific *himA* gene and 16S-23S internal transcribed spacer region (ITS) were detectable even in the lower inoculum levels of 10^{-4}, 10^{-5}, 10^{-6} and 10^{-7} equivalent to 1,125, 246, 64 and 5 cells, respectively (Table 2). The degree of specificity and sensitivity of the 4 h MPCR assay was very significant and virtually detected *Salmonella* from shrimp homogenates even with 5 cells (Figure 4). Further pre-enrichment for 6 and 8 h also yielded the same

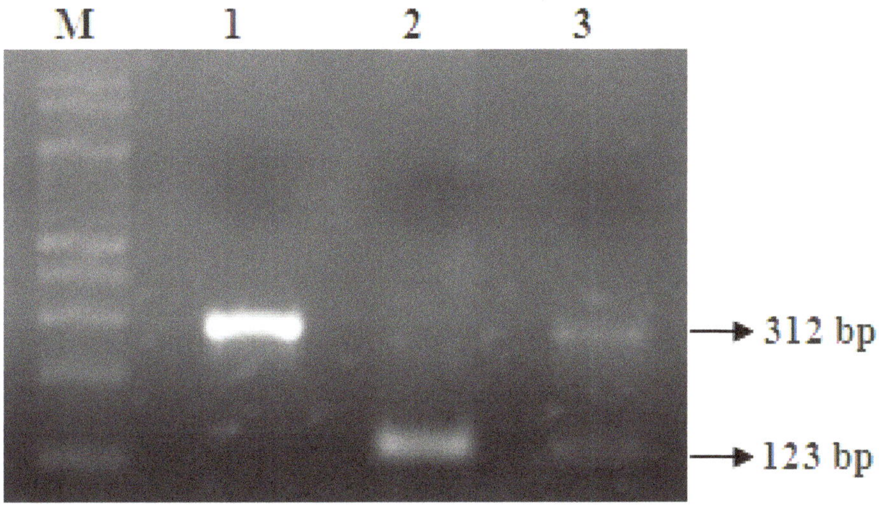

Figure 1. Ethidium bromide stained 2% agarose gel showing results of electrophoresis products of the PCR reaction for the detection of *S. enterica* serovar Typhimurium using genus specific *himA* and 16S-23S internal transcribed spacer regions (*ITS*) genes. Lane M: 100 bp DNA marker; lane 1: 16S-23S internal transcribed spacer region; lane 2: *himA*; lane 3: MPCR amplification of the selected two target genes.

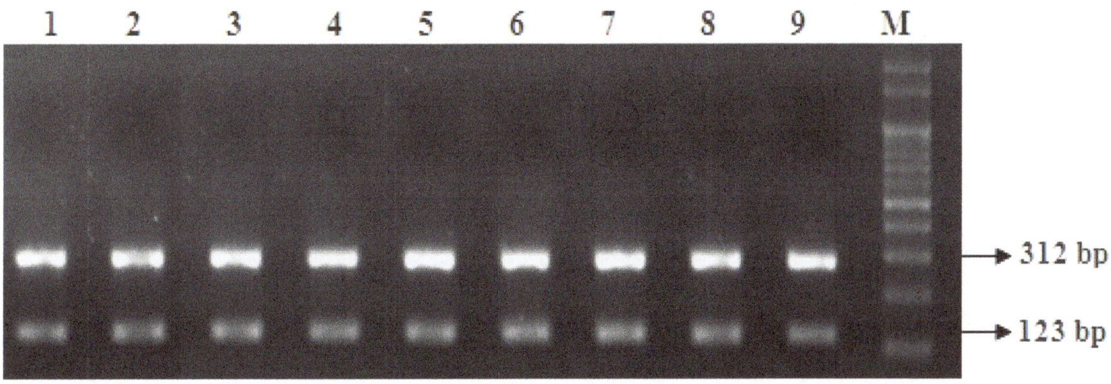

Figure 2. Ethidium bromide stained 2% agarose gel showing results of electrophoretic products of the MPCR reaction for the detection of *S. enterica* serovars using genus specific *himA* gene and 16S-23S spacer region. Lane 1: *S.* Typhi (ATCC 122235); lane 2: *S.* Paratyphi A (MTCC 735); lane 3: *S.* Typhimurium (MTCC 98); lane 4: *S.* Enteritidis (ATCC 13065); lane 5: *S.* Weltevreden (MTCC 1169); lane 6: *S.* Bovismorbificians (MTCC 1162); lane 7: *S.* Brunei (MTCC 1168); lane 8: *S.* Arizonae (MTCC 660); lane 9: *S.* Infantis (MTCC 1167); M: 100 bp DNA marker.

sensitivity. Hence, 4 h pre-enrichment was sufficient to detect the *Salmonella* in seafoods.

Validation of MPCR assay

The *Salmonella* genus specific *himA* gene and 16S-23S spacer region gave the positive results in 10% of the seafood samples screened for *Salmonella* and 70% of clinical isolates of *Salmonella* checked for confirmation.

DISCUSSION

The *himA* and *ITS* genes in this study were successfully used for the detection of *Salmonella* by PCR (Figure 1). Several uniplex PCR based methods for the specific detection of *S. enteric* in foods were earlier reported

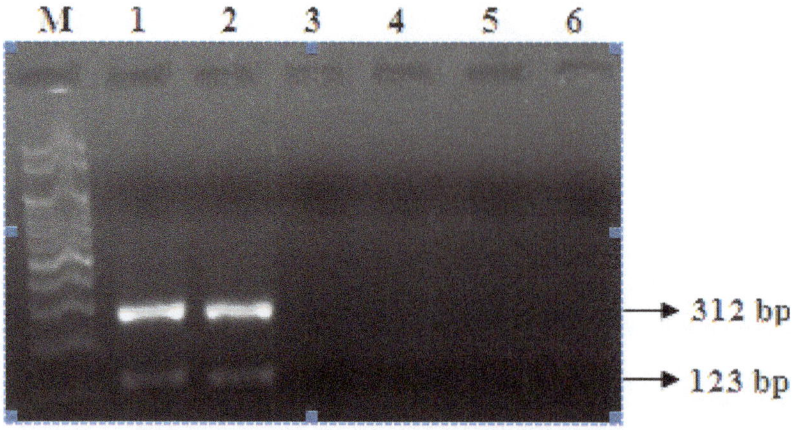

Figure 3. Agarose gel electrophoresis of the MPCR products obtained when testing the artificially contaminated shrimp muscle samples with *S. enterica* serovar Typhimurium (MTCC 98) using the genus specific *himA* gene and 16S-23S *ITS* region after 2 h of pre-enrichment. Lane M: 100 bp- DNA marker; Lane 1: 10^{-2}; lane 2: 10^{-3}; lane 3: 10^{-4}; lane 4: 10^{-5}; lane 5: 10^{-6}; lane 6: 10^{-7}.

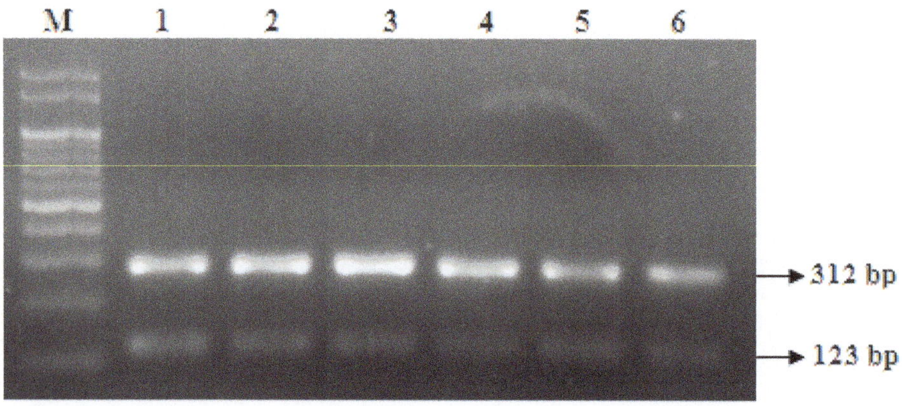

Figure 4. Agarose gel electrophoresis of the MPCR products obtained when testing the artificially contaminated shrimp muscle samples with *S. enterica* serovar Typhimurium (MTCC 98) using the genus specific *himA* gene and 16S-23S *ITS* region after 4 h of pre-enrichment. M: 100 bp DNA marker; Lane 1: 10^{-2}; lane 2: 10^{-3}; lane 3: 10^{-4}; lane 4: 10^{-5}; lane 5: 10^{-6}; lane 6: 10^{-7}.

Table 2. Detection of *S. enterica* serovar Typhimurium (MTCC 98) in artificially contaminated shrimps by MPCR assay.

Duration of pre-enrichment (in h)	Bacterial dilution					
	10^{-2}	10^{-3}	10^{-4}	10^{-5}	10^{-6}	10^{-7}
2	+	+	-	-	-	-
4	+	+	+	+	+	+
6	+	+	+	+	+	+
8	+	+	+	+	+	+

employing *invA* (Rahn et al., 1992), *hns* (Jones et al., 1993), *fimA* (Cohen et al., 1996), *histidine transport operon gene* (Cohen et al., 1993) and *hilA* (Pathmanathan et al., 2003).

It has been demonstrated that the detection of the genus *Salmonella* possessing two different genes by multiplex PCR with high specificity was possible in a relatively short period of time (Figure 2). This MPCR method therefore could be very useful in the diagnosis of different serovars of *S. enterica*, though some serovars lack any one of the conserved DNA sequences. Targeting multiple microbial pathogens simultaneously in a single PCR reaction was reported to be more time-efficient and cost effective (Brasher et al., 1998). When the specificity of the developed assay was checked, the other bacterial pathogens viz., *V. cholerae*, *E. coli* and *S. aureus* gave negative results.

The amplification of the internal amplification control (IAC) and genus specific genes in the inoculation studies at the 4 h pre-enrichment with lower inoculum levels without any mismatching indicated that the developed multiplex PCR assay was highly sensitive to detect the *Salmonella* (Figure 4 and Table 2). However, it has been earlier reported that *Salmonella* was detected at a level of 10^6 cfu per 25 g of food samples with a uniplex PCR-based commercial Kit after 18 h enrichment period (Fach et al., 1999). Kumar et al. (2008) has also found that *Salmonella* was detected by uniplex PCR in seafoods spiked with viable *Salmonella* cells to a level of 10^6 to 2 cfu per 25 g in 250 ml after 8 h pre-enrichment. Another report indicated the detection limit of optimized uniplex PCR using the *invA* gene as 2.60×10^4 cfu / ml (Moganedi et al., 2007). Incorporation of a 6 h non-selective pre-enrichment step further increased the detection limit to 26 cfu / ml. Similar results have been previously reported by other workers (Bej et al., 1994; Fratamico, 2003; Myint et al., 2006) who observed that at least 10^3 to 10^5 CFU / ml must be present to give positive results by uniplex PCR without a pre-enrichment step and that 1 to 10 CFU / ml could be detected after a pre-enrichment step. All the reported pre-enrichment methods were based on uniplex method and without the IAC.

Of the seafood samples tested for the validation of developed MPCR assay, 10% of samples were found to be positive for *Salmonella*. However, the MPCR method confirmed 70% of the clinical isolates of *Salmonella* and it showed that the method is more accurate and rapid than the conventional identification method. Our findings are supported by D'Aoust (1992), who observed that conventional methods show poor sensitivity and sometimes false-positives.

Based on the results, it can be concluded that the developed MPCR assay was highly specific, sensitive, rapid and reliable for the detection of serovars of *S. enterica* from seafood and clinical samples.

ACKNOWLEDGEMENTS

The financial assistance extended by the Department of Biotechnology, Govt. of India, New Delhi, India for conducting this project work is gratefully acknowledged. Authors wish to thank the Dean, Fisheries College and Research Institute, Tamil Nadu Veterinary and Animal Sciences University, Tuticorin, India for having provided necessary support for carrying out this work.

REFERENCES

Bej AK, Mahbubani MH, Boyce MJ, Atlas RM (1994). Detection of *Salmonella* spp. in oyster by PCR. Appl. Environ. Microbiol., 60: 368-373.

Brasher CW, DePaola A, Jones DD, Bej AK (1998). Detection of microbial pathogens in shellfish with Multiplex PCR. Curr. Microbiol., 37: 101-107.

Chiu TH, Chen TR, Hwang WZ, Tsen HY (2005). Sequencing of an internal transcribed spacer region of 16S-23S rRNA gene and designing of PCR primers for the detection of *Salmonella* spp. in food. Int. J. Food Microbiol., 97: 259-265.

Cohen JH, Mechanda MS, Lin W (1996). PCR amplification of the *fimA* sequence of *Salmonella typhimurium*, a specific method for detection of *Salmonella* spp. Appl. Environ. Microbiol., 62: 4303-4308.

Cohen ND, Neibergs LH, McGruder ED, Whitford HW, Behle RW, Ray PM, Hargis BM (1993). Genus specific detection of *Salmonella* using the polymerase chain reaction (PCR). J. Vet. Diagn. Investig., 5: 368-371.

D'Aoust JY (1992). Commercial diagnostics kits for the detection of foodborne *Salmonella*. In: Congress report *Salmonella* and salmonellosis, Ploufragan, France, pp. 9-19.

Fach P, Dilasser F, Grout J, Tache J (1999). Evaluation of a polymerase chain reaction – based test for detecting *Salmonella* spp. in food samples: Probelia *Salmonella* spp. J. Food Protect, 62: 1387-1393.

Fratamico PM (2003). Comparison of culture, polymerase chain reaction (PCR) TaqMan *Salmonella* and Transia card *Salmonella* assays for detection of *Salmonella* spp. in naturally-contaminated ground chicken, ground turkey and ground beef. Mol. Cell Probes, 17: 215-221.

Hatta AAM, Lakshmanaperumalsamy P (1997). Prevalence of *Salmonella* in fish and crustaceans from markets in Coimbatore, South India. Food Microbiol., 14: 111-114.

Hoorfer J, Cook N, Malorny B (2003). Making internal amplification control mandatory for diagnostic PCR. J. Clin. Microbiol., 41: 5385.

Jeyasekaran G, Raj KT, Shakila RJ, Thangarani AJ, Sukumar D, Jailani VAK (2010). Rapid detection of *Salmonella enterica* serovars by multiplex PCR. World J. Microbiol. Biotechnol. DOI10.1007/s11274-010-0538-9.

Jones DD, Law R, Bej AK (1993). Detection of *Salmonella* spp. in oysters using polymerase chain reaction (PCR) and gene probes. J. Food Sci., 58: 1191-1197.

Jones RN, Neale ML, Beattie B, Westmoreland D, Fox JD (2000). Development and application of a PCR-based method including an internal control for diagnosis of congenital cytomegalovirus infection. J. Clin. Microbiol., 38: 1-6.

Kumar R, Surendran PK, Thampuran N (2008). An eight-hour PCR-based technique for detection of *Salmonella* serovars in seafood. World J. Microbiol .Biotechnol., 24: 627-631.

Liston J, Matches JR, Baross J (1971). Survival and growth of pathogenic bacteria in seafoods. In: (Ed Kreuzer R) Fish Inspection and quality control. FAO, Rome, Italy, pp. 246-249.

Liu Y, Li Y (2001). Rapid detection of *Salmonella typhimurium* using immuno-optical sensing method. Sens. Autuators, B 72: 238-241.

Lund M, Madsen M (2006). Strategies for the inclusion of an internal amplification control in conventional and real time PCR detection of *Campylobacter* spp. in chicken fecal samples. Mol. Cell. Probes, 20: 92-99.

Martinez-Urtaza J, Saco M, Hernandez-Cordova G, Lonazo A, Garcia Martin O, Espinosa J (2003). Identification of *Salmonella* serovars isolated from live molluscan shellfish and their significance in the marine environment. J. Food Protect, 66: 226-232.

Moganedi KLM, Goyvae EMA, Ventor SN, Sidara MM (2007). Optimization of the PCR-*invA* primers for the detection of *Salmonella* in drinking water and surface waters following a pre-cultivation step. Water SA, 33: 195-202.

Myint MS, Johnson YJ, Tablante N, Heckert RA (2006). The effect of pre-enrichment protocol on the sensitivity and specificity of PCR for detection of naturally contaminated *Salmonella* in raw poultry compared to conventional culture. Food Microbiol., 23: 599-604.

Nambiar VN, Iyer KM (1991). Distribution of *Salmonella* serotypes in fish in retail trade in Cochin. Fish. Technol., 28: 33-37.

Pathmanathan SG, Cardona-Castro N, Sa´nchez-Jime´nez MM, Correa-Ochoa MM, Puthucheary SD, Thong KL (2003). Simple and rapid detection of *Salmonella* strains by direct PCR amplification of the *hilA* gene. J. Med. Microbiol., 52: 773–776.

Perch M, Braden CR, Bishop R, Fields P, Plikaytis R, Tauxe RV (2003). *Salmonella* surveillance summary, 2003. U.S Department of health and Human Services, Centers for Disease Control and Prevention, Atlanta, GA.

Rahn K, Degrandis DS, Clarke RC, Mcewen SA, Galan JE, Ginocchoio C, Curtiss R, Gyles CL (1992). Amplification of an *invA* gene sequence of *Salmonella typhimurium* by polymerase chain reaction as a specific method of detection of *Salmonella*. Mol. Cell. Probes, 6: 271-279.

Rattagool P, Wongchinda N, Sanghtong N (1990). *Salmonella* contamination in Thai shrimp. FAO Fishery Report. Supplement, FAO, Rome, 401: 18-23.

Saheki K, Kabayshi S, Habanishi T (1989). *Salmonella* contamination of eel culture ponds. Nippon Suisan Gakkai Shi, 55: 675-679.

Sanathkumar H, Sunil R, Venugopal MN, Karunasagar I, Karunasagar I (2003). Detection of *Salmonella* spp. in tropical seafood by polymerase chain reaction. Int. J. Food Microbiol., 88: 91-95.

Saroj SD, Shashidhar R, Karani M, Bandekar JR (2008). Rapid, sensitive, and validated method for detection of *Salmonella* in food by an enrichment broth culture- Nested PCR combination assay. Mol. Cell. Probes, 22: 201-206.

Trkov M, Avgustin G (1998). Specific detection of *Salmonella* spp. with molecular biological techniques. Animal Biotechnol., 72: 123-127.

Yam WC, Chan CY, Ho Bella SW, Tam TY, Kueh C, Lee T (1999). Abundance of clinical enteric bacterial pathogens in coastal waters and shellfish. Water Res., 56: 51-56.

Comparative efficacy of Rose Bengal plate test, standard tube agglutination test and Dot ELISA in immunological detection of antibodies to *Brucella abortus* in sera

Deepti Chachra, Hari Mohan Saxena*, Gurpreet Kaur and Mudit Chandra

Department of Veterinary Microbiology, College of Veterinary Science, Guru Angad Dev Veterinary and Animal Sciences University (GADVASU), Ludhiana, 141004 India.

Comparative efficacy of 3 serodiagnostic tests (RBPT, STAT and Dot ELISA) in detecting anti - *Brucella* antibodies in sera was evaluated on a total of 28 serum samples which included 18 samples from brucellosis suspected and 10 from normal healthy (brucellosis unaffected) cattle. Out of 18 sera from suspected cases, only 1 (5.55%) sample was found positive by STAT and 9 (50%) samples were positive by RBPT, whereas Dot ELISA could detect antibodies in all the 18 (100%) samples. Interestingly, RBPT could detect antibodies in 10 out of 17 (58.82%) samples found negative by STAT. The entire 9 samples positive by RBPT (100%) showed positive results with Dot ELISA also. Of the 9 RBPT negative samples, 11.25% showed positive and 88.88% showed negative results by STAT while all the 9 samples (100%) showed positive results with Dot ELISA. All the 10 sera from normal healthy animals were negative by RBPT, STAT and ELISA. Thus, Dot ELISA was found to be the most sensitive of the 3 tests used. It is, however, suggested that in order to get a fool proof diagnosis of *Brucella* infection, a combination of RBPT and Dot ELISA should be used, especially in case of samples found negative by either RBPT or STAT used alone or in combination.

Key words: Brucellosis, RBPT, STAT, Dot ELISA, *Brucella.*

INTRODUCTION

Brucellosis is an important and widely prevalent zoonotic disease of man and animals (cattle, buffaloes, sheep, goats, dogs etc) caused by *Brucella* organisms (Schelling et al., 2003). The common serological test used for the diagnosis of brucellosis is Rose Bengal plate test (RBPT) based on agglutination of colored particulate antigen (killed *Brucella* organisms) by the antibodies present in the patient's serum. Although it is a simple, cheap and effective test, the RBPT is generally considered to be less sensitive than other tests like standard tube agglutination test (STAT), complement fixation test (CFT) and enzyme linked immunosorbant assay (ELISA). ELISA has been claimed to be a good screening test whether used alone or in combination with the RBPT (Jacques et al., 1998).

However, few studies have been conducted on the comparative sensitivity of the 3 tests. In particular, the negative tests given by RBPT and STAT when used alone need to be further confirmed by other tests like ELISA, to avoid any possibility of wrong diagnosis owing to false negative reactions by these tests. The present study was therefore, undertaken to explore this aspect of serodiagnosis of brucellosis.

MATERIALS AND METHODS

Samples

A total of 28 serum samples from cattle including 18 from cattle suspected of brucellosis and 10 samples from brucellosis - free cattle from different parts of the Punjab state of India were included in the study. All the serum samples were subjected to the common serological tests RBPT, STAT and Dot ELISA, respectively. The

*Corresponding author. Email: hmsaxena@yahoo.com.

Table 1. Comparative sensitivity of RBPT, STAT and Dot ELISA.

S. no.	ELISA Comb and Lane no.	Sample no.	RBPT	STAT	ELISA
1	C1L2	J15	Positive	Negative	Positive
2	C1L8	J-7	Positive	1:20	Positive
3	C1L10	J18	Positive	1:20	Positive
4	C2L2	4-S	Positive	1:20	Positive
5	C2L3	4-C	Positive	Negative	Positive
6	C2L4	07-1222	Positive	Negative	Positive
7	C4L1	5-C	Positive	Negative	Positive
8	C4L6	07-1258	Positive	Negative	Positive
9	C4L9	47	Positive	Negative	Positive

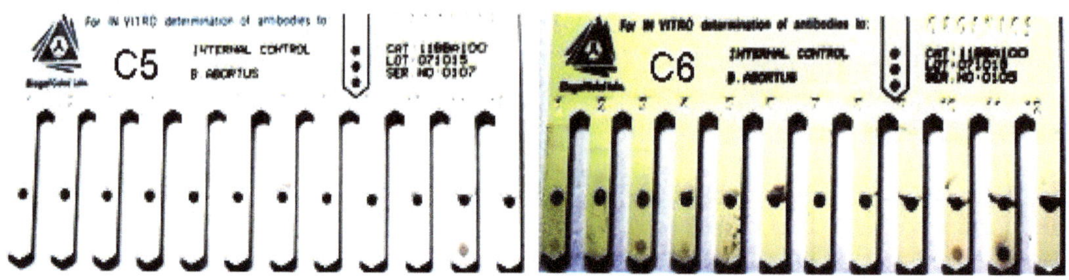

Figure 1. Dot ELISA of cattle sera for antibodies to *Brucella abortus.*

cattle included in the study had not been vaccinated against Brucellosis. Serum samples positive by all the 3 tests, that is, RBPT, standard tube agglutination test (STAT) and Dot ELISA were considered as positive controls. The 18 sera included in the study were selected out of 70 serum samples screened for Brucellosis. The 18 sera were positive by Dot ELISA but negative by RBPT and / or STAT. The animals positive by Dot ELISA were taken as positive since the results of Dot ELISA have been found to correlate well with the clinical picture and microbiological as well as molecular (PCR) analysis. However, the specimens could not be cultured for *Brucella* organisms.

In the absence of culture of *Brucella* organisms, the results of ELISA and PCR were considered as the gold standard.

Serological tests

Rose Bengal plate test (RBPT):

Equal volumes (10 µl of each) of RBPT colored antigen (IVRI, Izatnagar, India) and test serum were mixed on a clean glass slide (Morgan et al., 1978) with the help of a clean sterilized toothpick. The slide was observed after 1 min for the formation of clumps. The formation of clear clumps was considered a positive test while the absence of clear clumps was considered a negative reaction.

Standard tube agglutination test (STAT):

Plain antigen of *Brucella abortus* S99 (IVRI, Izatnagar) was used as per the method of Alton et al. (1975). Two fold serial dilutions (1:20 to 1:640) of the sera were prepared in phenol saline and equal quantity (0.5 ml) of antigen was added to each tube. All the tubes were incubated at 37°C for 24 h. The results were compared with the antigen control tube showing 50% agglutination. A titer of 1:40 or above was considered positive.

Dot ELISA:

ImmunoComb Bovine *Brucella* antibody test kit (Biogal-Galed Labs, Israel) was used for conducting Dot ELISA with the serum samples of cattle. The samples to be tested were mixed with the diluent in the first row of wells of a multichamber developing plate. The test spots on the comb were then incubated with the serum samples in the developing plate. The comb was then transferred to a well containing buffer to wash the unbound antibodies from the antigen spots. The comb was then allowed to react with an anti-cattle IgG alkaline phosphatase conjugate. After 2 more washes, the comb was moved to the last well. Color was developed by an enzymatic reaction. The intensity of color of the test spots corresponded directly to the antibody level in the test sample.

RESULTS

In the present study on 18 serum samples from cattle suspected of brucellosis, 9 (50%) samples were found to be positive while 9 (50%) samples were found to be negative by RBPT. Only 1 (5.55%) out of 18 samples was positive by STAT. All the 9 RBPT positive samples (100%) showed negative results by STAT. All the 18 samples (100%) showed positive results with Dot ELISA. (Table 1, Figure 1).

Among the RBPT negative samples, 8 out of 9 (88.88%) showed negative results by STAT while 1 out of 9 (11.11%) showed a titer of 1:40. However, all the samples negative by RBPT or STAT showed positive results with Dot ELISA. (Table 2, Figure 1).

All the serum samples from normal healthy cattle were negative by RBPT, STAT and Dot ELISA (Table 3, Figure 1).

Table 2. Detection of RBPT and STAT false negative cases by Dot ELISA.

S. no.	ELISA Comb and Lane no.	Sample no.	RBPT	STAT	ELISA
1	C1L7	5P	Negative	1:40	Positive
2	C1L9	J-4	Negative	1:20	Positive
3	C3L3	M-9	Negative	Negative	Positive
4	C3L10	M-18	Negative	Negative	Positive
5	C6L1	J-6	Negative	Negative	Positive
6	C6L3	J-19	Negative	Negative	Positive
7	C6L4	M-19	Negative	Negative	Positive
8	C6L5	M-20	Negative	Negative	Positive
9	C6L10	M-17	Negative	Negative	Positive

Table 3. Serological evaluation of sera from Brucellosis - free cattle.

S. no.	ELISA Comb and Lane no.	Sample no.	RBPT	STAT	Dot ELISA
1	C1L3	136P/J-26	Negative	Negative	Negative
2	C2L2	4	Negative	Negative	Negative
3	C3L7	M3	Negative	Negative	Negative
4	C4L5	A4	Negative	Negative	Negative
5	C5L3	D3	Negative	Negative	Negative
6	C5L6	2-P	Negative	Negative	Negative
7	C5L7	1-P	Negative	Negative	Negative
8	C5L8	6-P	Negative	Negative	Negative
9	C5L10	A-5	Negative	Negative	Negative
10	C6L2	J-16	Negative	Negative	Negative

DISCUSSION

The present study has shown that the commonly used conventional serodiagnostic tests for brucellosis, RBPT and STAT, may not be absolutely reliable. RBPT detected antibody in the sera of 50% of the animals suspected for brucellosis whereas, STAT could detect only 5.55% cases. Interestingly, RBPT, commonly believed to be cruder and less sensitive test could detect 58.82% samples negative by STAT. On the other hand, all the samples from suspected cattle shown to be negative by RBPT or STAT, were found positive by Dot ELISA.

Other workers have also reported comparable results with RBPT, STAT and ELISA in the serodiagnosis of brucellosis. Otlu et al. (2008) studied the seroprevalance of brucellosis in cattle and humans. Among the cattle sera, 32.92 and 34.64% were found positive by RBPT and SAT, respectively. Of the human samples, 13, 14.22 and 17.88% of the sera were found positive for brucellosis by RBPT, SAT and ELISA respectively. Kanani (2007) studied the seroprevalance of brucellosis by ELISA and compared its efficacy with RBPT and STAT. The tests revealed 5.94, 9.90 and 9.90% of the bulls positive for brucellosis by RBPT, STAT and ELISA, respectively. The sensitivity was reported to be 50% in case of RBPT and 62.5% in case of STAT considering ELISA as a gold standard test. However, Ofukwu et al. (2008) found a non - significant ($p < 0.05$) difference in the sensitivity for the RBPT and SAT tests. Jacques et al. (1998) assessed the efficacy of indirect ELISA in comparison with RBPT and CFT on sera from ewes infected with *Brucella melitensis*. The indirect ELISA was shown to be a good screening test to be used alone or in addition to RBPT. However, for an accurate and fool proof diagnosis of brucellosis, it is suggested on the basis of our findings that a combination of RBPT and ELISA should be followed in case of samples found negative by either RBPT or STAT used singly or in combination.

Conclusions

Comparative efficacy of 3 serodiagnostic tests (RBPT, STAT and Dot ELISA) in detecting anti - *Brucella* antibodies in sera was evaluated on a total of 28 serum samples which included 18 samples from brucellosis suspectted and 10 from normal healthy (brucellosis unaffected) cattle. Only 50% of the suspected samples were positive by RBPT, whereas Dot ELISA could detect antibodies in all the samples. RBPT could detect antibodies in 58.82% samples found negative by STAT while all the 9 RBPT negative samples showed positive results with Dot ELISA. It is, therefore, suggested that for an accurate and fool proof diagnosis of brucellosis, a combination of RBPT and ELISA should be followed in case of samples found negative by either RBPT or STAT used singly or in combination.

REFERENCES

Alton GG, Jones LM, Pietx DE (1975). Laboratory Techniques in Brucellosis. Second edition. FAO / WHO, Geneva, Monograph series no. 55.

Jacques I, Olivier–Bernardin V, Dubray G (1998). Efficacy of ELISA compared to conventional tests (RBPT and CFT) for the diagnosis of *Brucella melitensis* infection in sheep. Vet. Microbiol. 64(1): 61 – 73.

Kanani A (2007). Serological, cultural and molecular detection of Brucella infection in breeding bulls. PhD thesis, Anand Agricultural University, Anand.

Morgan WJB, Mackinnon DT, Gill KPW, Gower SGM, Norris PIW (1978). Brucellosis diagnosis: Standard Laboratory Techniques Report Series no. 1, Weybridge, England.

Ofukwu AR, Yohanna CA, Abuh HA (2008). Medicine on-line: Int. J. Med. http://www.priory.com/med/brucella.htm

Otlu S, Sahin M, Atabay HI, Unver A (2008). Serological investigations of Brucellosis in cattle, farmers and Veterinarians in the Kars district of Turkey. Acta vet. BRNO 77: 117-121.

Schelling E, Diguimbaye C, Daoud S, Nicolet J, Boerlin P, Tanner M, Zinsstag J (2003). Brucellosis and Q – fever seroprevalances of nomadic pastoralists and their livestock in Chad. Prev. Vet. Med. 61: 279-293.

Correlation and path coefficient studies in maize (*Zea mays* L.) genotypes under 40% soil moisture contents

Qurban Ali[1]*, Muhammad Hammad, Nadeem Tahir[1], Muhammad Ahsan[1], Shahzad Maqsood Ahmed Basra[2], Jehanzeb Farooq[1], Muhammad waseem[3] and Mehboob Elahi[3]

[1]Department of Plant Breeding and Genetics, University of Agriculture, Faisalabad, Pakistan.
[2]Department of Crop Physiology, University of Agriculture, Faisalabad, Pakistan.
[3]Department of Agronomy, University of Agriculture, Faisalabad, Pakistan.

The present study pertaining to the estimation of genetic parameters and characters association in forty maize genotypes was conducted in the glasshouse of the Department of Plant Breeding and Genetics, University of Agriculture, Faisalabad, during the crop season in February 2009 at 40% moisture level. The data for various quantitative characters were recorded, statistically analyzed. The path-coefficients were computed to estimate the contribution of individual characters to the fresh shoot length. Investigations regarding path coefficient showed that fresh root length had maximum direct effect on fresh shoot length at 40% moisture level followed by dry root weight, root density, leaf temperature and dry shoot weight. It was concluded that fresh root length, dry root weight, root density, leaf temperature and dry shoot weight are the characters which contribute largely to fresh shoot length of maize seedlings and selection can be made on the basis of these characters.

Key words: *Zea mays,* correlation, path coefficient, genotypic, phenotypic, Pakistan.

INTRODUCTION

Maize ranks third among the cereal crops worldwide after wheat and rice. In Pakistan, maize is grown on an area of 950 thousand hectares with total production of 3487 thousands tons (Anonymous, 2009, 2010) with an average yield of 2892 $kgha^{-1}$. It is consumed as food by human and feed for the livestock and poultry. It also fulfills the requirement of raw material in food, medicine and textile industries, which finally manufacture corn oil, corn flakes, dextrose, textile dyes etc. A plant may experience biotic and abiotic stresses in the field like diseases attack, water scarcity, water logging, salinity, high and low temperature extremes etc. either continuously or with some breaks at different times during the growing season (Trester and Bacic, 2005). Abiotic stresses limit crop productivity (Araus et al., 2002; Boyer, 1970).

*Corresponding author. E-mail: qurbanalisaim@yahoo.com.

Among various abiotic stresses drought is undoubtedly one of the worst natural enemy of life. It can occur in any region of the world, and can affect life from very basic personal inconvenience to nationwide. Drought can reduce crop yield, pasture deterioration and death of livestock. It strongly affects the production of cereals, and poses a serious threat to the food security of households. World food security is dependent on continuous crop improvement; in particular, the development of crops with increased tolerance to abiotic stresses especially drought and salinity (Denby and Gehring, 2005). The maize was grown at three levels of water availability (100, 75, or 60% of daily transpiration) during a period bracketing silking and at two plant densities (6 and 10 plants m^{-2}) without nutrient limitations to generate a range of levels of resource availability of water (Echarte and Tollenaar, 2006). A study for water stress at 3 growth stages before silking, at silking and during grain filling growth stages caused a

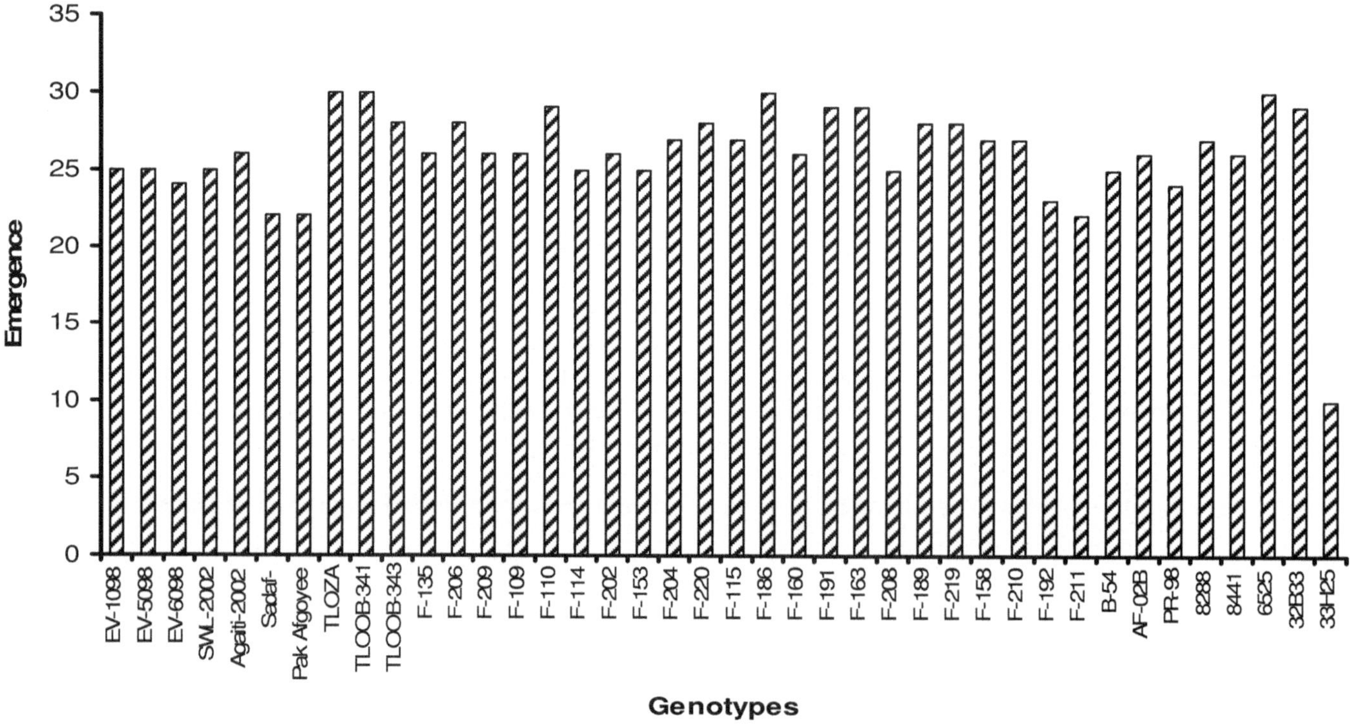

Figure 1. Emergence rate of different maize genotypes.

significant reduction in the different growth parameters studied at 90 days after planting as compared with the normal irrigation regime (Ghooshchi et al., 2008). Chlorophyll content measurements were performed on each leaf of several plants along the crop cycle (Moulin et al., 2009). Those measurements, as well as the surface measurements and the leaf insertion height measurements gave the vertical distribution of chlorophyll pigments within the canopy. The main objectives of the study were to identify the drought resistant maize genotypes. The study of traits which can help a plant breeder to select the better yield maize genotypes under drought stress conditions.

MATERIALS AND METHODS

The proposed study was carried out in the glasshouse of the Department of Plant Breeding and Genetics, University of Agriculture, Faisalabad. The experimental material consisted of 40 accessions namely: EV-1098, EV-5098, EV-6098, SWL-2002, Agaiti-2002, Sadaf, Pak Afgoyee, TLOZA, TLOOB-341, TLOOB-343, F-135, F-206, F-209, F-109, F-110, F-114, F-202, F-153, F-204, F-220, F-115, F-186, F-160, F-191, F-163, F-208, F-189, F-219, F-158, F-210, F-192, F-211, B-54, AF-02B, PR-98, 8288, 8441, 6525, 32B33 and 33H25.

These accessions were sown in polythene bags (18 × 9 cm) filled with sandy loam soil (pH 7.8 and EC 1.7 dS m^{-1}) in the glasshouse of Department of Plant Breeding and Genetics, University of Agriculture, Faisalabad. Field capacity of the soil was determined before sowing. Two seeds per polythene bag were sown and thinned up one healthy seedling after emergence. After 21 days of sowing the stress levels that is, 40% of the field capacity was applied. All the recommended agronomic and cultural practices were carried out. According to Jamieson et al. (1995) water requirement of maize at the time of tesseling is 135 mm/month (4.5 mm/day) and this requirement may increase up to 195 mm/month (6.5 mm/day) during hot windy conditions. The moisture level was maintained by volume on alternate days by using moisture meter (ΔT-NH2, Cambridge, England). The data was recorded for traits, chlorophyll contents (Chl. C), fresh shoot length (FSL), fresh root length (FRL), dry shoot weight (DSW), dry root weight (DRW), root density (RD), leaf temperature (LT), survival rate (SR), root-shoot length ratio (RSLR) and root-shoot weight ratio (RSWR). The chlorophyll contents were calculated by using spectrophotometer (T-60) after crushing and dipping the sample leaves from each replication of each accession in 80% acetone at 663 nm and 645 nm OD's value according to the method of Bwinsma (1963). The data was analyzed statistically using analysis of variance technique (Steel et al., 1997) and Duncan multiple range (DMR) test at 1% significance level was used to compare the treatments means. Genotypic and phenotypic correlations will be calculated to observe the association between different traits (Kwon and Torrie, 1964). Path coefficient analysis will be performed (Dewey and Lu, 1959) to assess the direct and indirect effects on fresh shoot length using genotypic correlations where association of all the aforementioned traits were calculated by keeping one at a time as response variable and other contributing traits as causal variables.

RESULTS AND DISCUSSION

It is clear from the graph (Figure 1) that the lines TLOZA, TLOOB-341, F-110, F-168, 6525 and 32B33 have greater value of emergence rate which shows the greater ability to tolerate the drought stress conditions while the

Table 1. Estimates of genetic components.

Traits	Genotypic coefficient variation (%)	Phenotypic coefficient variation (%)	Standard deviation	Genotypic variance	Phenotypic variance	Broad sense heritability h^2 BS (%)
Leaf temperature	20.713	22.74	1.554	17.636	21.257	83.0
Chlorophyll contents	135.41	135.547	0.017	0.208	0.209	99.80
Fresh shoot length	18.623	20.188	1.817	28.251	33.201	85.10
Root density	2.0204	2.309	0.192	0.565	0.62	91.10
Fresh root length	22.089	28.555	0.058	0.007	0.013	59.80
Root-shoot length ratio	34.29	40.612	0.579	1.248	1.751	71.30
Survival rate	40.642	42.435	4.872	394.792	430.403	91.70
Dry shoot weight	24.239	29.776	0.72	0.015	0.023	66.30
Dry root weight	20.205	30.544	0.254	0.075	0.172	43.80
Root-shoot weight ratio	20.418	23.256	0.066	0.022	0.028	77.10

genotype 33H25 has lowest value which means it has very less ability to tolerate the drought stress conditions (Ghooshch, et al., 2008). The genetic para meters as indicated in Table 1, it is clear that the higher genotypic and phenotypic variances SR, LT, and FSL as 394.792, 17.636, 28.251 and 430.403, 21.257, and 33.201 respectively, while lowest for FRL and RSWR 0.007, 0.013 and 0.022 and 0.023 respectively. The highest genotypic and phenotypic coefficients of variability for Chl.C and SR as 135.41, 40.642, 135.547 and 42.435% respectively, while lowest for RD as 2.0204 and 2.309 respectively. The highest value for standard deviation was for SR as 4.872 and lowest for Chl.C as 0.017. The higher values of genotypic variance and genotypic coefficient variance indicated that these traits can be use for selecting higher yielding maize genotypes. The same results were by Ojo et al. (2006). The highest heritability values were found for Chl.C, SR, and RD as 99.80, 91.70 and 91.10% respectively while DRW showed lowest heritability value as 43.80%. The higher values of heritabilities indicated that selection can be made on the bases of these traits (Dasgupta et al., 1992). The correlation coefficients (in Tables 2 and 3) a positive and significant genotypic and phenotypic correlation of Chl.C was found for RD and SR but negatively correlated with LT. The same results were by Veronica et al. (2009). The DRW and DSW were positively and significantly correlated with all traits except RSLR and FRL at genotypic and phenotypic levels but DSW was negatively and significantly correlated with all traits except RSWR at genotypic and phenoltypic levels. The FRL was positively and significantly correlated with all traits except RSLR, Chl.C and DRW at genotypic and phenotypic levels but FRL was negatively and significantly correlated with LT and RSWR at genotypic and phenotypic levels. Similar results were by Ojo et al. (2006) and Malik et al. (2005). The LT was positively and significantly correlated with DRW, DSW, RSWR and FSL at genotypic and phenotypic levels but LT was negatively and significantly correlated with traits except RSLR, SR, Chl.C and RD at genotypic and phenotypic levels. The same results were by Ojo et al. (2006) and Yousuf and Saleem (2001). The RD was positively and significantly correlated with all traits except RSWR and LT at genotypic and phenotypic which were negatively and significantly correlated RD. The RSLR was positively and significantly correlated with traits RSLR and SR at genotypic and phenotypic levels but RSLR was negatively and significantly correlated with RSWR, FSL and LT at genotypic and phenotypic levels. The other traits were non-significantly correlated with RSLR. Similar results were by Xu et al. (2007). The RSWR was positively and significantly correlated with traits LT at genotypic and phenotypic levels but RSLR, DSW, FRL, and RD were negatively and significantly correlated with RSWR at genotypic and phenotypic levels. The genotypic correlation of RSWR was significant for DRW but

Table 2. Genotypic correlation for different traits.

Variables	Chl.C	DRW	DSW	FRL	LT	RD	RSLR	RSWR	SR
DRW	0.0548								
DSW	0.1863	0.663 **							
FRL	0.1729	0.0408	0.2696						
LT	-0.6151 **	0.1471 **	0.1247 **	-0.1195 **					
RD	0.858 **	0.2034 **	0.2559 **	0.3389 *	-0.7781 **				
RSLR	0.1416	-0.3084	-.1954	0.6737 **	-0.1698 **	0.2153 **			
RSWR	0.1182	0.5257 **	-.3386 *	-0.2712 *	0.0974 *	- 0.0408 *	-0.0969 *		
SR	0.7304 **	0.1495 *	.2465 **	0.2633 **	-0.2719 **	0.6984 **	0.1705 **	-0.0309	
FSL	0.0317	0.3756**	.5456 **	0.3708 *	0.0903 *	8.86E-02 *	-0.4513 **	0.1862 **	0.0408 **

DRW = Dry root weight, DSW = Dry shoot weight, FRL = Fresh root length, LT = Leaf temperature, Chl.C = Chlorophyll contents, RD = Root density, RSLR = Root-shoot length ratio, RSWR = Root-shoot weight ratio, SR = Survival rate and FSL = Fresh shoot length, * = Significant, ** = Highly significant.

Table 3. Phenotypic correlation for different traits.

Variables	Chl.C	DRW	DSW	FRL	LT	RD	RSLR	RSWR	SR
DRW	0.0410								
DSW	0.1388	0.5100**							
FRL	0.1563	0.0491	0.2354						
LT	-0.5869**	0.1435**	0.1198**	-0.0936**					
RD	0.7211**	0.1833**	0.1206**	0.2428**		-0.6183**			
RSLR	0.1246	-0.2277	-0.1851	0.6372**	-0.1233**	0.1471**			
RSWR	-0.0721	0.428**	-0.4494**	-0.1622**	0.029*	- 0.0322*	-0.0294**		
SR	0.6956**	0.1589**	0.1695**	0.2258**	-0.2523**	0.5852**	0.1345**	-0.0178**	
FSL	0.0261	0.3152 *	0.3994*	0.3293*	0.0814*	0.0991*	-0.4346**	0.1536 **	0.0523**

DRW = Dry root weight, DSW = Dry shoot weight, FRL = Fresh root length, LT = Leaf temperature, Chl.C = Chlorophyll contents, RD = Root density, RSLR = Root-shoot length ratio, RSWR = Root-shoot weight ratio, SR = Survival rate and FSL = Fresh shoot length, * = Significant, ** = Highly significant.

but non-significant at phenotypic level. The SR was positively and significantly correlated with all traits except LT and RSWR at genotypic and phenotypic levels but SR was negatively and non-significantly correlated with RSWR at genotypic. Similar results were found by Malik et al. (2005). The FSL was positively and significantly correlated with all traits at genotypic and phenotypic levels but RSLR was negatively and

Table 4. Direct (In parenthesis) and indirect effect of various traits on FSL.

Variables	Chl.C	DRW	DSW	FRL	LT	RD	RSLR	RSWR	SR
Chl.C	(0.2892)	0.0371	-0.1205	0.1291	-0.027	-0.1916	-0.1762	0.0614	-0.06
DRW	0.0158	(0.6778)	-0.4287	0.0517	0.0065	-0.0454	0.3836	-0.2734	-0.0123
DSW	0.0539	0.4494	(-0.6466)	0.3416	0.0055	-0.0571	0.2431	0.1761	-0.0202
FRL	0.05	0.0277	-0.1743	(1.2671)	-0.0052	-0.0757	-0.8381	0.1411	-0.0216
LT	-0.1779	0.099	-0.0806	-0.1514	(0.0439)	0.1737	0.2112	-0.0506	0.0223
RD	0.2482	0.1379	-0.1654	0.4295	-0.0341	(-0.2233)	-0.2678	0.0212	-0.0574
RSLR	0.041	-0209	0.1264	0.8536	-0.0074	-0.0481	(-1.2441)	0.0504	-0.014
RSWR	-0.0342	0.3563	0.2189	-0.3437	0.0043	0.0091	0.1206	(-0.5201)	0.0025
SR	0.2113	0.1013	-0.1594	0.3336	-0.0119	-0.1559	-0.2121	0.0161	(-0.0821)

DRW = Dry root weight, DSW = Dry shoot weight, FRL = Fresh root length, LT = Leaf temperature, Chl.C = Chlorophyll contents, RD = Root density, RSLR = Root-shoot length ratio, RSWR = Root-shoot weight ratio, SR = Survival rate and FSL = Fresh shoot length.

significantly correlated with FSL at genotypic and phenotypic levels.

The genotypic and phenotypic correlation of FSL was non-significant for Chl.C.

A perusal of the data given in Table 4 indicates that the genotypic correlation between leaf temperature and fresh shoot length was positive (0.0903). The direct effect of leaf temperature on fresh shoot length was positive (0.0439) whereas FSL has negative indirect effects through Chl. C, DSW, FRL, RSWR while DRW, RD and RSL ratio have positive indirect effects on FSL. The genotypic correlation between chlorophyll contents and fresh shoot length was positive (0.0317). The direct effect of Chl. C on fresh shoot length was positive (0.2892) whereas Chl. C has negative indirect effects through DSW, LT, RD ,SR, RSL ratio while DRW, FRL and RSW ratio have positive indirect effects through Chl.C on FSL. Similar results were found by Boyer and Westgate (2004) and Asrar-ur-Rehman et al. (2007). The genotypic correlation between root density and fresh shoot length was positive (0.088). The direct effect of root density on fresh shoot length was negative (-0.2233) whereas root density has negative indirect effects through DSW, LT, SR, RSL ratio while others have positive indirect effects through root density on FSL.

The genotypic correlation between fresh root length and fresh shoot length was positive (0.3708). The direct effect of fresh root length on fresh shoot length was positive (1.2671) whereas fresh root length has negative indirect effects through DSW, LT, RD, SR, RSL ratio while others have positive indirect effects through fresh root length on FSL. The genotypic correlation between root-shoot length ratio and fresh shoot length was negative (-0.4513). The direct effect root-shoot length ratio on fresh shoot length was higher but negative (-1.2441) whereas root-shoot length ratio has negative indirect effects through DRW, LT, RD and SR while others have positive indirect effects through root-shoot length ratio on FSL. The genotypic correlation between survival rate and fresh shoot length was positive (0.0408). The direct effect of survival rate on fresh shoot length was negative (-0.0821) whereas survival rate has negative indirect effects through DSW, LT, RD and FRL while others have positive indirect effects through survival rate on FSL. The genotypic correlation between root dry weight and fresh shoot length was positive (0.3756). The direct effect of root dry weight on fresh shoot length was positive (0.6778) whereas root dry weight has negative indirect effects through DSW, RD, SR and RSL ratio while others have positive indirect effects through root dry weight on FSL. Similar findings were by Aslam and Tahir (2003) and Xu et al. (2007).The genotypic correlation between shoot dry weight and fresh shoot length was positive (0.5456). The direct effect of shoot dry weight on fresh shoot length was negative (-0.6466) whereas shoot dry weight has positive indirect effects through all traits except RD and SR while others have negative indirect effects through shoot dry weight on FSL, genotypic correlation between root-shoot weight ratio and fresh shoot length was negative (-0.1862). The direct effect of root-shoot weight ratio on fresh shoot length was negative (-0.5201) whereas root-shoot weight ratio has negative indirect effects through all traits except Chl.C and FRL while others have positive indirect effects through shoot dry weight on FSL.

Conclusions

The traits dry root weight, dry shoot weight, root density, survival rate and fresh shoot length were positively correlated at genotypic level could be used for the selection of batter yielding lines under drought stress conditions. Investigations regarding path coefficient showed that fresh root length had maximum direct effect on fresh shoot length followed by dry root weight, root density, leaf temperature and dry shoot weight at three moisture levels. It was concluded that fresh root length, dry root weight, root density, leaf temperature and dry shoot weight are the characters which contribute largely to fresh shoot length of maize seedlings and selection can be made on the basis of these characters.

REFERENCES

Anonymous (2009-2010).Economic Survey of Pakistan. Govt. of Pakistan, Finance and Economic Affairs Division, Islamabad, Pakistan, p. 21.

Araus J, Slafer GA, Reynolds MP, Toyo C (2002). Plant breeding and drought in C3 cereals, what should we breed for? Ann. Bot., 89: 925-940.

Aslam M, Tahir MHN (2003). Correlation and Path Coefficient Analysis of Different Morpho-Physiological Traits of Maize Inbreds under Water Stress Conditions. Inter. J. Agric. Biol. 4: 446–448.

Asrar-ur-Rehman SU, Saleem M, Subhani GM (2007).Correlation and path coefficient analysis in Maize (*zea mays l.*). J. Agric. Res., 45: 177-183.

Boyer JS (1970). Leaf enlargement and metabolic rates in corn, soybean, and sunflower at various leaf water potential. Plant Physiol., 46: 233-235.

Boyer JS, Westgate ME (2004). Grain yields with limited water. J. Exper. Bot., 55: 2385–2394.

Denby K, Gehring C(2005). Engineering drought and salinity tolerance in plants: lessons from genome-wide expression profiling in Arabidopsis. Trends in Biotech., 23: 547-552.

Dasgupta T, Islam MO, Gayen P(1992).Genetic variability and analysis of yield components in chickpea (*Cicer arietinum* L.). Annals Agric. Res., 13: 157-160 [Pl. Br. Abst. 64(7): 7129; 1994].

Dewey DR, Lu KH (1959). A correlation and path coefficient analysis of components of crested wheatgrass seed production. Agron. J. 51: 515-518.

Echarte L, Tollenaar M (2006).Kernel set in maize hybrids and their inbred lines exposed to stress. Crop Sci., 46: 870–878.

Ghooshchi F, Seilsepour M,Jafari P (2008). Effects of water stress on yield and some agronomic traits of maize [SC 301]. American-Eurasian J. Agric. Environ. Sci., 4: 302-305.

Kwon SH, Torrie JH(1964). Heritability and interrelationship of two soybean (*Glycine max* L.) populations. Crop Sci., 4: 196-198.

Malik HN, Malik SI, Hussain M, Chughtai SUR, Javed HI (2005). Genetic Correlation among Various Quantitative Characters in Maize (*Zea mays* L.) Hybrids. J. Agric. Social Sci., 3:262–265.

Moulin S, Baret F, Bruguier N, Bataille C (2009). Assessing the vertical distribution of leaf chlorophyll content in a maize crop. INRA - Unite Climate, Soil, Environment (CSE), pp. 7803-7929.

Ojo DK,Omikunle OA,Oduwaye OA,Ajala MO, Mao SA (2006). Heritability, Character Correlation and Path Coefficient Analysis among Six Inbred-Lines of Maize *(Zea mays* L.). World J. Agric. Sci., 2: 352-358.

Steel RGD, Torrie JH, Dicky DA (1997).Principles and procedures of Statistics. A biometrical Approach 3rd Ed. McGraw Hill Book Co. Inc. New Yark, pp. 400-428.

Trester M, Bacic A (2005).Abiotic stress tolerance in grasses, From model plants to crop plants. Plant Physiol., 137: 791-793.

Veronica C, Gitelsona A,Schepersca J(2009). Non –destructive determination of maize leaf and canopy chlorophyll content. J. Pl. Physiol., 166: 157-167.

Xu H, Lu J, Zhou X (2007). Drought resistance of regenerative plants from hairy root cultures in maize under water stress. Nongye Gongcheng Xuebao/Trans. Chin. Soc. Agric. Eng., 23: 19-23.

Yousuf M, Saleem M(2001).Correlation Analysis of S1 Families of Maize for Grain Yield and its Components. Inter. J. Agric. Biol. 4: 387–388.

18

Brevibacillus parabrevis, Acinetobacter baumannii and Pseudomonas citronellolis - Potential candidates for biodegradation of low density polyethylene (LDPE)

R. Pramila[1], Kesavaram Padmavathy[2], K. Vijaya Ramesh[1]* and Krishnan Mahalakshmi[2]

[1]Department of Plant Biology and Plant Biotechnology, Quaid-e-Millath Government College for Women, Chennai 600 002, Tamil Nadu South India.
[2]Department of Microbiology, Sree Balaji Dental College and Hospital, Chennai 600 100, Tamil Nadu, South India.

Low density polyethylene is one of the polymers that is up till date nearly impossible to be degraded safely. Biodegradation is the safest method of breakdown that possibly leaves behind less toxic residue and shows potentials of bio-geo chemical cycling of the substrate. Considerable amount of work has been carried out in this area, but most of them are associated with blended low density polyethylene (LDPE). Previous reports also rely on host consortia for degradation. Various techniques have been implicated in designing this kind of study. BATH test, to evaluate the hydrophobicity of the isolates was performed apart from the calculation of generation time of the isolates in LDPE incorporated growth medium. Biofilm formation was also quantified by using protein estimation technique. Our findings corroborated previous findings in most of the techniques, but in Sturm test, a technique to evaluate concentration of carbon dioxide, a final product of biodegradation of LDPE, we have taken dissolved CO_2 (that is, CO_2 present in the soluble form in the growth medium, apart from the gaseous CO_2 collected in the KOH tube) also in consideration which is not reported by any of the previous workers. The current article investigated the biodegradation ability of bacteria isolated from a municipal landfill area near Pallikaranai, Chennai, South India. The bacteria were subjected to growth in a medium containing LDPE as the sole carbon source with and without a nitrogen source. Four bacterial species were isolated. According to the *16S rRNA* gene sequences they were identified as *Brevibacillus parabrevis* (PL-1), *Acinetobacter baumannii* (PL-2, PL-3) and *Pseudomonas citronellolis* (PL-4). Bacterial adhesion to hydrocarbon (BATH test) was done to determine the bacterial hydrophobicity. Bacterial biomass was quantified to estimate the population density of the biofilm. This work clearly identifies with our objective to find a right microbe to degrade the resistant LDPE by giving out promising results from the present study.

Key words: Biodegradation, low density polyethylene (LDPE), bacterial adhesion to hydrocarbon (BATH test), sturm test, *Pseudomonas*, *Acinetobacter*, *Brevibacillus*.

INTRODUCTION

Plastics are polymers that consist of monomers linked together by chemical bonds (Shah et al., 2009). The polymers include, polyethylene, polypropylene, polystyrene, polyurethane, nylon etc., Polyethylene is a thermoplastic polymer produced by combining monomers of ethylene. Low density polyethylene is a thermoplastic made from petroleum (Shah et al., 2009). Plastic bags do bring a lot of convenience to people's life, but at the same time, it also generates long term harms. When the plastic products mix in the soil and accumulate continuously, they affect the absorption of nutrients and water by the crops, thus reducing their output. The littering also reduces rate of rainwater percolation. Biodegradation is a process where complex organic molecule breaks into

*Corresponding author. E-mail: ramesh.vijaya67@gmail.com

simple molecules as a result of the action of micro-organisms like bacteria, fungi, or algae. Biodegradation of the materials has to be scientifically measurable. Since most biodegradation produces CO_2 as a by-product, usually this is measured by the amount of CO_2 produced after degradation of carbon containing compounds. The following strategies are used to assess and monitor the biodegradation of the polymer:

1. Accumulation of biomass (experimentally determine the growth rate of micro organisms with the polymer as the sole carbon source).
2. Oxygen uptake rate.
3. CO_2 evolution rate.
4. Surface changes.
5. Changes in the mechanical and physical properties of the polymer.

Biodegradation of polyethylene has been studied extensively earlier (Pometto et al., 1992) but the results were based on PE blend with starch (Pometto et al., 1993; Breslin, 1993).

Fourst et al. (1997) have reported the biodegradation of LDPE/cellulose blends by cellulose blends by common fungi. El-shafei et al. (1998) have reported the biodegradation of disposable polyethylene by fungi and *Streptomyces* sp. Yamada-Onodera et al. (2001) have reported the degradability of polyethylene by *Penicillium simplicissimum.* Kathiresan (2003) has reported isolating fungi from the mangrove soil which has the potential to degrade polyethylene materials. Watanabe et al., (2009) isolated and identified three types of low density polyethylene (LDPE) degrading microbes *Bacillus circulans*, *Bacillus brevies* and *Bacillus sphaecicus* by soil burial method. Sindhuja et al. (2011) have reported degradation of LDPE by marine fungi.

Biodegradation of polyethylene as revealed by structural changes have been studied by many workers using Scanning Electron Microscopy (SEM). (Raghavan et al.,1992; Weiland et al.,1995; Bonhomme et al. 2003; Gilan et al., 2004; Aamer, 2007; Zahra et al., 2010; Mumtaz et al., 2010; Pramila and Vijaya, 2011) The previous findings do have some bearing about microbial remediation strategies but the onus lies on the time taken for the degradation. The work has proceeded towards isolating microbial species that can degrade LDPE in lesser time. Reports of Sturm test done in this study reveals some amount of evidence in this line.

MATERIALS AND METHODS

Preparation of LDPE powder

LDPE sheets were cut into small bits and immersed in xylene and boiled for 15 min. Xylene dissolved the LDPE and the residue was crushed while it was warm by hand with help of gloves. The LDPE powder so obtained was washed with ethanol to remove residual xylene and allowed to evaporate (approximately 2 to 3 h) to remove ethanol. The powder was dried in hot air oven at 60°C overnight. The LDPE powder was stored in closed containers in room temperature.

Isolation of microorganisms

Soil sample was collected from the municipal solid waste landfill area, Pallikaranai, Chennai Tamil Nadu, South India. To isolate bacteria, 5 g of soil sample and 500 mg LDPE powder was added in 100 ml sterilized synthetic medium (SM) and incubated at 37°C for 24 h. The organisms grew by utilizing LDPE as carbon source. Bacteria isolated were obtained as pure cultures.

Determination of growth rate of bacteria colonized on LDPE

250 mg of LDPE powder was added to overnight culture inoculated into 50 ml SM and shaken. OD values were taken every 30 min at 500 nm using UV-VIS spectrophotometer (Model SL-159). From the values obtained for period of 5 h, generation time was calculated. SM without culture served as a control.

Determination of bacterial hydrophobicity

Bacterial cell surface hydrophobicity was determined by the BATH test (Bacterial adhesion to hydrocarbon). 24 h culture (5 ml) in nutrient broth was centrifuged at 10,000 rpm for 15 min and washed twice with Phosphate-Urea-Magnesium (PUM) buffer. Supernatant was discarded and the pellet was re-suspended in PUM buffer. Absorbance of the suspension was measured at 400 nm using UV-VIS spectrophotometer (Model SL-159). 0.2 ml of hexadecane was added to the suspension and shaken for 20 min. Test tubes were kept undisturbed for 5 min, which resulted in separation of two phases into organic and aqueous. Absorbance of aqueous layer was measured at 400 nm using UV-VIS spectrophotometer (Model SL-159). Culture free buffer served as the blank.

Estimation of bacterial biomass colonizing the LDPE

LDPE films of similar size and weight were used for the colonizing studies. The LDPE sheets were disinfected with 70% ethanol for 30 min and transferred to sterile distilled water for 10 min. The disinfected LDPE sheets were used for further studies. Biofilm formed on the polyethylene surface was quantified by protein estimation.

Culture of 24 h was inoculated in 100 ml SM. 3 pieces of disinfected LDPE sheets were transferred to the medium. The conical flasks were kept over magnetic stirrer at 37°C. One polyethylene film was taken out after every 2 days, boiled with 5 ml of 0.5 mol^{-1} NaOH and the suspension was centrifuged at 10,000 rpm for 15 min. Supernatant was kept aside and the pellet was subjected to the same procedure. The two supernatants were combined and the protein concentration was determined according to Lowry et al. (1951).

Quantification of biofilm

Bacterial culture of 24 h was inoculated in 100 ml SM. 4 pieces of disinfected LDPE sheets were transferred to the medium and incubated at 37°C. After 2 days, one LDPE film was taken out. 10 ml of 95% ethanol was added to the LDPE sheet and shaken vigorously for 10 min to remove biofilm from the sheets. Absorbance was taken at 540 nm using the UV-VIS spectrophotometer. 95% ethanol served as blank.

Table 1. Generation time of the bacterial isolates

Name of the isolate	Generation time in min
PL- 1	167
PL- 2	79
PL- 3	80
PL- 4	28

CO_2 evolution test (Sturm test)

Capacity autoclavable plastic containers of 100 ml were used for the study. A separate set up was kept for control that was maintained un-inoculated. Briefly, sterile CO_2 free air was passed into the container having inoculated synthetic medium, supplemented with LDPE powder. After the stipulated time [48 h for bacteria], potassium hydroxide (KOH) solution [1 M] that had trapped the CO_2 liberated by the inoculants, [after utilization of LDPE the sole carbon source] was gravimetrically quantified for test as well as control. The dissolved carbon dioxide present in the medium was also estimated using titration method. Briefly, sample (25 ml) was taken in a conical flask and 0.05 ml of 0.1 N sodium thiosulphate solution was added. After the addition of 2 drops of methyl orange indicator, this solution was titrated against 0.02 N sodium hydroxide solution. End point was the change in color from orange red to yellow. Following this, two drops of phenolphthalein indicator was added and titration continued till a pink color developed. Volumes of the titrant used were noted and the amount of CO_2 calculated using the formula:

$$\frac{A \times B \times 50 \times 1000}{V}$$

Where,
A = ml of NaOH titrant
B = normality of NaOH
V = ml of the sample

Identification of bacteria

To perform the morphological and biochemical studies on the isolated bacteria for identification, the cultures were streaked onto nutrient agar (NA) plates and Hi Media Biochemical Kit for identification of Gram negative bacteria K22. For the morphological identification following parameters were taken into consideration:

1. Gram character.
2. Motility.
3. Basic biochemical utilization tests like Catalase and Oxidase.
4. Other biochemical tests like carbon, amino acids, nitrate, other enzyme production etc.

16S rRNA gene sequencing

Genomic DNA was extracted by boiling-lysis method and stored at – 80°C until PCR assay. Amplification of *16S rRNA* was carried out using previous described primers

Forward – CCGAATTCGTCGACAACAGAGTTTGATCCTGGCTCAG,
Reverse -CCCGGGATCCAAGCTTAAGGAGGTGATCCAGCC) (Weisburg et al., 1991).

PCR was carried out in a 50 µl reaction volume which contained 5 µl of the template, 10 pM of each primer (Sigma Aldrich), 0.25 mM of each dNTP (Bangalore Genei, India), 2U of *Taq* polymerase (Bangalore Genei, India).The thermocycling conditions included, an initial denaturation step at 94°C, 3 min, followed by 35 cycles of denaturation at 94°C, 30 s, annealing at 65°C, 30 s and extension at 72°C, 1 min and a final extension at 72°C, 7 min.

The amplified PCR products were resolved in 1% agarose gel in TBE. The amplicons were visualized in a UV transilluminator and gel documentation was carried out (BioRad, Hercules, CA). DNA sequencing was performed to identity the amplified PCR products using applied biosystem (ABI) 3130 Genetic Analyser with ABI PRISM BigDye Terminators V 3.1. The gene sequences were compared with sequences in the GenBank database by using BLAST program. The sequences were deposited in the GenBank database.

RESULTS

Bacterial generation time

Experiments performed to calculate generation time (Table 1) of each of the bacterial isolates in SM (synthetic medium) with LDPE as a carbon source revealed that isolate PL-4 had doubling time of 28 min, culture PL–2 and PL–3 had a generation time of 79 min and 80 min respectively, followed by PL-1 that showed a doubling time of 167 min.

BATH test

BATH test (Table 2) of the bacterial culture revealed some level of hydrophobicity in PL-2 and PL-3 followed by PL -1 and PL-4

Bacterial biomass colonized on LDPE

Bacterial colonization was measured as extractable protein of the polyethylene. Isolate PL-1 showed considerable increase in protein concentration than PL-2 & PL-3 and PL-4 (Table 3)

Bio-film quantification

The bacterial isolates PL-1, PL-2 and PL-3 rose during 2nd day and remained stable for 2 additional days,

Table 2. BATH test.

Name of the isolate	OD value before adding hexadecane	OD value after adding hexadecane
PL- 1	0.432	0.28
PL- 2	0.86	0.35
PL- 3	0.87	0.35
PL- 4	0.40	0.380

Table 3. Measurement of bacterial colonization (490 nm).

Name of the isolate	Protein concentration (μg mg^{-1})	
	After 2 days	After 6 days
PL- 1	15	3
PL- 2	17	19
PL- 3	17	20
PL- 4	3	9

Table 4. Quantification of bacterial bio-film.

Name of the isolate	OD value after 2 days	OD value after 4 days	OD value after 6 days
PL- 1	0.020	0.023	0.008
PL- 2	0.008	0.010	0.006
PL- 3	0.008	0.010	0.005
PL- 4	0.055	0.008	0.017

Table 5. Sturm test for bacteria (48 h).

Name of the Isolate	Amount of CO_2 (g l^{-1})
PL- 1	0. 7042
PL- 2	1. 0603
PL- 3	1. 0604
PL- 4	0. 5706

followed by a gradual decrease in the 6th day whereas PL-4 showed a gradual decrease from the 4th day onwards (Table 4)

Sturm test

The value for CO_2 evolution from the degradation of LDPE sample by bacterial isolates (Table 5) shows that PL-2 and PL-3 degraded LDPE with a higher efficiency than the other isolates.

Identification of bacterial strains

Based on the *16S rRNA* gene sequencing, the bacterial isolates were identified as *Brevibacillus parabrevis* (PL-1), *Acinetobacter baumannii* (PL-2), *Acinetobacter baumannii* (PL-3) and *Pseudomonas citronellolis* (PL-4).

The DNA sequences of these bacterial isolates were deposited in the GenBank- Genetic sequence database at the National Center for Biotechnical Information (NCBI) under the following accession numbers JQ304812 (strain PL1), JQ304813 (strain PL2), JQ294033 (strain PL3), JQ304811 (strain PL4).

DISCUSSION

The list of pollutants which pose environmental problems and health hazard and are tough for biodegradation, is a long one. Present study ealt with the isolation of polyethylene degrading micro organisms from the municipal landfill soil.

Low density polyethylene films were used in this study. The micro organisms with ability to degrade LDPE were isolated in synthetic medium supplemented with LDPE powder and these micro organisms used for degradation study. Several methods were employed to monitor the biodegradation of polyethylene.

Rosenberg et al. (1980) have described BATH test to estimate the bacterial cell surface hydrophobicity that can be directly related to the ability to form an effective biofilm over any hydrophobic surfaces. This test was followed by Hadad et al. (2005) where the results showed lower reduction in turbidity of the bacterial suspension. Our results of BATH test revealed some levels of decrease in hydrophobicity of strains PL-2, PL-3 and PL-1 followed by PL-4.

Estimating bacterial biomass density in the biofilm is a difficult task since the bacterial cells are strongly attached to the polyethylene surface. We had followed the technique of (Hadad et al., 2005) and the results proved that the bacterial biomass of isolate PL-4 increased from 2nd day to 6th day of incubation followed by PL-1 and PL-2 and PL-3.

We had followed the technique of O'Toole et al (2000) to measure the absorbance of the biofilm extracted with ethanol. There was a mild increase in the OD value of the isolates PL-1, PL–2 and PL-3 from 2nd to 4th day but decreased in 6th day whereas PL-4 isolate showed a poor formation of biofilm, but interestingly, PL-4 isolate showed a doubling time of about 28 min, indicating the fact that biofilm formation cannot be relied upon totally for deciding the rate of degradation as also observed by Hadad et al. (2005).

Sturm test is commonly employed for evaluation of the biodegradability of polymer materials. We used a Modified Sturm test for the measurement of CO_2 both in gaseous and dissolved form in SM. Incidentally, all previous reports did not take into account the dissolved CO_2 during Sturm Test evaluation (Shah et al., 2007; Muhammad et al., 2009; Shah et al., 2009)

The CO_2 evolution test gave a valid data about the degradation rate by the bacterial isolates. CO_2 evolved by isolate PL-2 and PL-3 was measured to be 1.0603 g l^{-1} followed by isolate PL-1 that evolved 0.7042 g l^{-1} and isolate PL-4 which gave out 0.5706 g l^{-1} of measurable CO_2. Shah et al. (2007) performed this test with a consortium of bacterial and fungal isolates and he reported a CO_2 concentration of 1.85 g l^{-1} for 4 weeks. Studies done by Muhammad Ishtiaq Ali et al. (2009) reported a concentration of about 10 g l^{-1} of CO_2 after a period of 30 days. Shah et al. (2009) also reported CO_2 concentration of about 1.85 g l^{-1} after a 30 day period of growth of *Fusarium* sp. on LDPE films. On comparison with the above reports, our studies have revealed a much higher concentration of CO_2 for our isolates and this shows that there are potential abilities in these microbes that could be tapped to draw out a biodegradation strategy. Our Sturm Test study is unique in the fact that we have worked with individual organisms rather than consortia as reported by many workers.

Conclusion

LDPE can be biodegradable if the right microorganism is isolated. We have proved that the hydrophobic LDPE film can act as a substratum for some groups of micro-organisms which formed a biofilm on the LDPE film. The isolates also grew on minimal medium containing only LDPE in the powdered form as the carbon source even without any nitrogen source. We have also proved that the LDPE can be totally degraded into carbon dioxide which brings us closer to the fulfillment of the objective of isolating a microorganism that can completely degrade the recalcitrant polyethylene if the right conditions are provided.

ACKNOWLEDGEMENTS

The authors are grateful to the University Grants Commission (F.MRP-3475/10 LINKNO: 345 dated August, 2010) for providing financial assistance for the completion of this work.

REFERENCES

Ali Shah A, Fariha H, Abdul H, Javed IA (2009). Isolation of *Fusarium* sp. AF4 from sewage sludge, with the ability to adhere the surface of polyethylene. Afr. J. Microb. Res., 3: 658-663.

Bonhomme S, Cuer A, Delort M, Lemaire J, Sancelme M, Scott C (2003). Environmental biodegradation of polyethylene. Poly. Degrad. Stabil., 81: 441-452.

Breslin VT (1993). Degradation of starch – plastic composites in a municipal solid waste landfill. J. Environ. Poly. Degrad., 1: 127-141.

El-Shafei HA, El-Nasser NH, Kansoh AL, Ali AM (1998). Biodegradation of disposable polyethylene by fungi and *Streptomyces* species. Poly. Degrad. Stabil., 62: 361-365.

Fourst CJ, Mills RR, Haas T, Castevans R, Ottenbrite R (1997). Biodegradation of LDPE/Cellulose blends by common fungi. Macromol. Symp., 123: 1-8.

Gilan I, Hadar Y, Sivan A (2004). Colonization, biofilm formation and biodegradation of polyethylene by a strain of *Rhodococcus ruber*. Appl. Microb. Biotech., 65: 97-104.

Hadad D, Geresh S, Sivan A (2005). Biodegradation of polyethylene by the thermophilic bacterium *Brevibacillus borstelensis*. J. Appl. Microb., 98:1093-1100.

Kathiresan K (2003). Polythene and plastic degrading microbes in an Indian Mangrove soil. Rev. Biol. Trop., 51: 629-633.

Lowry O, Rosebrough NI, Farr AL, Randall RJ (1951). Protein measurement with the Folin Phenol reagent. J. Biol. Chem., 193: 265-275.

Muhammad Ishtiaq Ali, Qaiser P, Bashir A, Imran J, Raja R, Saadia A, Naima A, Pir BG, Safia A, Abdul H (2009). Studies on biodegradation of cellulose blended polyvinyl chloride films. Inter. J. Agri. Biol., 11: 577-580.

Mumtaz T, Khan MR, Hassan MA (2010). Study of environmental biodegradation of LDPE films in soil using optical and scanning electron microscopy. Micron, 41: 430-438.

O'Toole G,Kaplan HB, Kolter R (2000). Biofilm formation as microbial development. Ann. Rev. Microb., 54: 49-79.

Pometto AL, Lee BT, Johnson KE (1992). Production of an extracellular polyethylene-degrading enzymes by *Streptomyces* species. Appl. Environ. Microb., 58:731-733.
Pramila R, Vijaya RK (2011). Biodegradation of low density polyethylene (LDPE) by fungi isolated from marine water – a SEM analysis. Afri. J. Microbiol. Res., 5: 5013-5018.
Rosenberg M, Gutnik D, Rosenberg E (1980). Adherence of bacteria to hydrocarbons: a simple method for measuring cell surface hydrophobicity. FEMS. Microbiol. Lett., 9:29-33.
Shah AA, Hassan F, Hameed A, Ahmed S (2009). Biological degradation of plastic – a comprehensive review. Biotech. Adv., 26: 246-265.
Sindhuja P, Padmapriya M, Pramila R, Vijaya RK (2011). Biodegradation of Low Density Polyethylene (LDPE) by Fungi isolated from marine water. Res. J. Biol. Sci., 6:141-145.
Watanabe T, Ohtake Y, Asabe H, Murakami N, Furukawa M (2009). Biodegradability and degrading microbes of low density polyethylene. J. App. Poly. Sci., 111: 551-559.
Weiland M, Daro A, David C (1995). Biodegradation of thermally oxidized polyethylene. Poly. Degrad. Stab., 48:275-289.
Weisburg WG, Barns SM, Pelletier D, Lane DJ (1991). 16S Ribosomal DNA amplification for phylogenetic Study. J. Bact., 173: 697-703.
Yamada Onodera K, Mukumoto H, Katsuyaya Y, Saiganji A, Tani Y (2001). Degradation of polyethylene by a fungus, *Penicillium simplicissimum*. YK. Poly. Degrad. Stab., 72:323-327.
Zahra S, Shojaosadati S A, Mohammad-Taheri M, Nosrati M (2010). Biodegradation of low density polyethylene (LDPE) by isolated fungi in solid waste medium. Waste Manage., 30:396-401.

Bacillus megaterium, a new pathogen on lupine plants in Egypt

Abdel-Monaim M. F.[1]*, Gabr M. R.[2], El-Gantiry S. M.[1], Shaat M. N.[2] and El-Bana A. A.[2]

[1]Legume Diseases Research Department, Plant Pathology Research Institute, Agricultural Research Center, Giza, Egypt.
[2]Plant Pathology Department, Faculty of Agriculture, El-Minia University, Minia, Egypt.

The causal agents of bacterial blight in lupine (*Lupinis termis*) were isolated from leaves displaying symptoms in Minia governorate, Egypt. The pathogens were characterized by biochemical and physiological tests, and identified as *Bacillus megaterium*. Tissue extracts prepared from experimentally diseased shoot systems showed great pectolytic and cellulolytic activities while healthy tissue extracts of both organs showed slight activities of the enzymes. In varietal response test, four lupine cultivars that is, Australian, Balady, Giza 1 and Giza 2 were tested for their susceptibility to all tested *B. megaterium* isolates and the cultivar Balady was most sensitive to all isolates. However, isolates B2 and B6 were more virulent than the others. Two isolates of *B. megaterium* isolates that caused blight on lupine plants were subsequently tested on other foliar plants representatives of 17 plant species. Both *B. megaterium* isolates could not infect any of the tested plants but produced small necrotic spots on faba bean leaves. These results indicate the host specificity of this bacterium towards its host plant and accordingly justify the suggested forma specials *lupini* to be given to the lupine bacterial pathogen *B. megaterium*. According to the literature review, this is the first report on the occurrence of *B. megaterium* as a causal agent of leaf blight of lupine plants in Egypt.

Key words: *Bacillus megaterium*, lupine, pectolytic and cellulolytic activities, cultivars.

INTRODUCTION

Lupine (*Lupinus termis* Forsk) is one of the oldest field crops grown in Egypt. Lupine uses as fodder crop and green manure for sandy and poor soils to reclaim new lands. The seeds of lupine contain great ratio of proteins, fibers and carbohydrates. It is also used for medical and industrial purposes (Maknickiene, 2001).

Lupine plants are prone to many diseases which attack the foliar parts causing downy mildew, rust and leaf blight diseases (Paulitz and Cote, 1991; Yang and Sweetingham, 2002). Such diseases are destructive and cause great loss in seed yield and quality (Osman et al., 1986; Muller et al., 1999). Hosford (1982) found that a form of *Bacillus megaterium* causes a white to very light tan blotching and streaking of wheat leaves. The name *B. megaterium* de Bary 1884 pv. Cerealis pv. Nov. is proposed for this pathovar; the halopathotype strain is WB 28 (Dye et al., 1980). Plant diseases caused by plant pathogens are a complicated process because a number of factors play a part. However, direct involvements of pectic and cellulitic enzymes produced by the pathogen in pathogensis were reported (Gaber et al., 1990; Walker et al., 1994).

The objective of this investigation is to isolate and identify the pathogenic agent that caused leaf blight of lupine plant at Minia governorate and test the varietal response to infection with *B. megaterium,* furthermore, the cell wall degradation enzymes in pathogenesis were

*Corresponding author. E-mail: fowzy_2008@yahoo.com.

Figure 1. Naturally diseased lupine foliage showing blighted leaves and stem. The affected tissues became winked, weathered and turned grayish – brown.

discussed.

MATERIALS AND METHODS

Isolation of pathogenic organisms

Diseased lupine plants showing spotting and blight symptoms (Figure 1) were collected from different fields located at Minia Governorate for isolation. The infected stems and leaves were first separated from the plants and washed thoroughly using running tap water then cut into small parts, surface sterilized using 0.1% mercuric chloride solution for 2 min then washed several times in sterilized distilled water and maceration in sterilized mortar containing small amounts of sterilized distilled water, left for 15 min. Loopfulls of the obtained homogenates were streaked onto nutrient sucrose agar medium in Petri dishes. All the inoculated plates were incubated at about 25±2°C and examined daily for bacterial and fungal growth.

To obtain pure cultures of the developing growth: in case of bacteria, the developing growths were further streaked onto the same medium and subcultures were prepared from single separated colonies and maintained on nutrient sucrose agar slopes and stored at 5°C for future studies.

Pathogenicity tests

The purified bacterial isolates secured from diseased lupine plant organs were tested for their pathogenicity on healthy lupine plants cv. Balady grown in pots (25 cm in-diameter) containing sterilized soil and sown with disinfested seeds. Soil sterilization was carried out using formalin solution 5% whereas the seeds were disinfested in 0.1% mercuric chloride for 2 min then washed several times with tap water and sown in the prepared pots (10 seeds/pot). The disinfested soil before sowing the seed was left to aeration for 3 weeks to get rid of the chemical remains. The inocula of the purified bacterial isolates was prepared by streaking the secured purified isolates onto glucose nutrient agar in plates (3 plates/ isolate) and incubated for 48 h. at 25 ±2°C. The developing growth of each isolate was suspended in appropriate amount of water to obtain inocula of 1-1.5 × 10^8 colony– forming units (CFU) /ml (Shoeib, 1997). Foliage of 30 day old lupine plants (cv. Balady) grown in pots was sprayed with the prepared bacterial inocula by hand automizer. Each isolate was used to inoculate 30 days old lupine plants grown in 3 pots as replicates. Check treatment was sprayed with water. Inoculated and uninoculated plants were enclosed within polyethylene bags for 24 h. to maintain adequate humidity. All plants were placed in the greenhouse for 2 weeks, irrigated as needed and their foliage was carefully observed for disease development and the percentage of blighted leaf area was assayed. After the incubation period, disease severity was evaluated using the disease index (DI) according to Abdel-Monaim (2008) as follows: 0 = no symptoms, 1=1–9%, 2=10–24%, 3=25-49%, 4=50-74%, 5=75-100% of blighted leaf area. The mean of disease index (DI) and disease severity index (DSI) for each replicate was calculated by the formula suggested by Liu et al. (1995) and calculated as follows:

$$DSI = \frac{\sum d}{d\ max\ X\ n} \times 100$$

Wher d is the disease rating of each plant, and d max is the maximum disease rating and n is the total number of plants examined in each replicate. Reisolation was carried out from some of the experimentally diseased plants to fulfill Kock's postulations and the developing bacterial colony was compared with the original isolates.

Identification of the causal organisms

Identification of the bacterial isolates was performed according to Bergey's manual of systematic bacteriology (Sneath et al., 1986).

Assessment of some hydrolytic enzymes that is, cellulase and pectinase in diseased and healthy lupine tissues

Assessment of pectinase and cellulase enzymes were assayed in tissue extracted prepared from diseased and healthy shoot system taken from the subjected plants during the pathogenicity test.

Preparation of tissue extracts

Half gram of leaf tissues (either healthy and/or infected) were existed and separately macerated with clean mortar and pestle containing 5 ml of 0.05 M phosphate buffer (pH 6). The homogenated tissue extracts were filtered though several layers of cheese cloth, cooled to temperature near zero then centrifuged at 5000 rpm for 20 min. The clarified enzyme preparations of diseased and healthy tissues were directly subjected to the viscometrical assessment.

Assessment of pectic enzymes

Assessment of pectic enzyme was employed viscometrically according to the method of Mahadevan and Sridhar (1982). This was carried out by measuring the reduction in viscosity of the reaction mixtures containing 2 ml of crude enzyme preparations (tissue extract), 5 ml, 1.5% citrus pectin solution in 0.1 M phosphate buffer at pH 5 and/or 8 adjusted by 0.3 M NaOH or HCl. The reaction mixtures were incubated at 28°C and the loss in viscosity of the mixture was measured after 90 min against blank containing boiled inactivated extracts instead of the active ones. Reduction in viscosity of the substrate was calculated using Fenske- Ostwan Viscosimeter according to the formula:

$$V = \frac{T_0 - T}{T_0 - T\,H_2O} \times 100$$

where V= percent of loss in viscosity, T_0=Flow time in seconds at of blank (boiled enzyme), T = Flow time in seconds after incubation (90 min), T H_2O=Flow time of distilled water.

Cellulase assessments

Cellulase activity was assayed viscometrically in mixtures containing 5 ml and 1.5% carboxymethylcellulose (CMC) in 0.05 M phosphate buffer at pH 6 mixed with 2 ml crude enzyme preparations. The mixtures were incubated at 28°C and the percentage loss in viscosity was estimated after 90 min against control containing heat inactivated tissue extracts instead of the active ones.

Varietal response

Response of four lupine cultivars namely cvs. Balady, Australian, Giza 1 and Giza 2 to infection with the infection isolates were investigated using some selected bacterial isolates which showed high pathogenic property through the pathogenicity test.

The tested cultivars were grown in pots containing sterile soil and the inocula of the bacterial isolates were prepared and applied similarly as was done in the pathogenicity test. Data were recorded after 15 days as mentioned above.

Host range

The pathogenicity of two bacterial isolates (isolates B2 and B6) were further tested against other plants selected to be representatives of different plant species presented in Table 1. Seeds of the tested plant species were disinfested and sown in pots containing sterile soil (10 seeds / pot). There were three pots per treatment. After 45 days of sowing, the plants were inoculated by spraying their foliage with the inocula of the pathogen bacterial isolates to be tested prepared as described before in pathogenicity test. Control plants were sprayed with tap water. Disease severity index (DSI) was recorded 15 days after inoculation.

Statistical analysis

All experiments were performed twice. Standard deviation (SD) was used for analysis of variances as described by Little and Hills (1972).

RESULTS

Isolation of the causal organisms

Isolation trials carried out from diseased specimens taken from lupine or spotted/ blighted foliar on sucrose nutrient agar media resulted in development of 6 bacterial isolates that was purified and kept in agar slants.

Pathogenicity tests

Lupine plants inoculated with purified bacterial isolates behaved aggressively towards the inoculated plants. The inoculated plants soon became soft discolored grayish-brown bent over and eventually collapsed (Figure 2). Also stem inoculated lupine plant through artificial wound inoculated by an effective *B. megaterium* isolate (B2) causing discoloration and softening to the surrounding tissues (Figure 3). The infection extended to the adhering leaf. Data present in Figure 4 show that bacterial isolate B2 recoded the highest disease severity (100%) followed by isolate B6 (96%), while isolate B3 recoded the lowest ones (84%).

Identification of the causal organisms

The morphological and physiological activities performed on six bacterial lupine isolates that proved pathogenic on

Table 1. List of plant species tested for their reaction to lupine foliar pathogens.

Scientific name	Common name	Variety
Family: Leguminaceae		
Glycine max, Merr.	Soybean	Clark
Medicago sativa, L.	Alfalfa	Giza 1
Phaneolus vulgaris, L.	Bean	Giza 6
Vicia faba, L.	Faba bean	Giza 3
Pisum sativum, L.	Pea	Master peas
Lupinus termis, Forsk.	Lupine	Balady
Lens esculenta, L.	Lentil	Giza 4
Trifolium alexandrinum, L.	Egyptian clover	Giza 1
Trigonellafoenum graecum, L.	Ferugreek	Balady
Vigna unguiculata, L.	Cowpea	Cream 7
Cicer arietinum, L.	Chickpea	Giza 3
Medicago playmorpha,L.	Medic	Balady
Melilotus indica,L.	India melilot	Balady
Family: Garmineae		
Hordeum vulgare, L.	Barley	Giza 125
Triticum sativum, L.	Wheat	Giza 168
Family: Compositae		
Heliathus amnuus, L.	Sunflower	Giza 1
Family: Asteraceae		
Xanthium pungens, Wallr	Cocklebur	Balady
Family: Chenopodiaceae		
Spinacia oleracea	Spinach	Balady

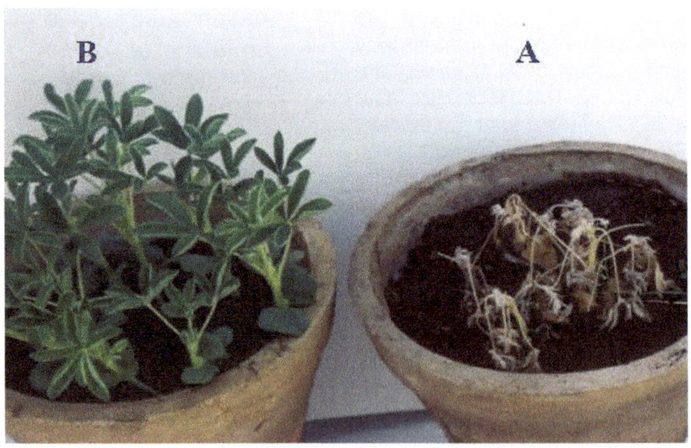

Figure 2. A: Severity attacked lupine plants by *B. megaterium* isolate B2. The attacked plants turned grayish – brown, reduced in size, bent over and collapsed. B: Control healthy plants.

their host and presented in Table 2 show the following: the isolates growth on glucose medium had creamy smooth colonies tended to be yellow on ageing, cells were Gram-positive, rods, spore formers, motile. Sporangia were not swollen within the vegetative cell. They were strict aerobes that could grow in broth medium amended with 5% NaCl, catalase positive, oxidase negative, V.P. negative. Negative results were recorded in each of starch hydrolysis, Indol and H_2S production. Nitrate was not reduced to nitrite expect in case of isolate B5.

All isolates produced acid from glucose, glycerol,

Figure 3. Stem inoculated lupine plant through artificial wound inoculated by an effective *B. megaterium* isolate (B2) causing discoloration and softening to the surrounding tissues. The infection extended to the adhering leaf.

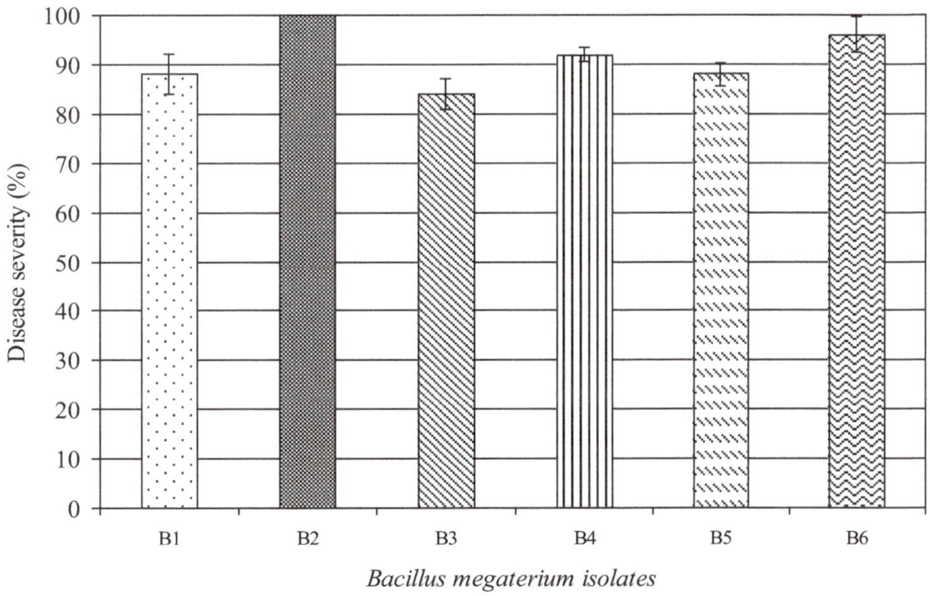

Figure 4. Pathogenicity test of bacterial isolates obtained from naturally diseased lupine foliar cv. Balady. Mean ± SDs for three replicates per treatment are shown.

maltose, raffinose and xylose but not from sorbitol or dextrin. Acid was also produced from sucrose, fructose and mannitol by the growing isolates except isolate B4. The isolates except isolate B5 gave negative results in gluconate and arginine dehydrolyzate tests. All the isolates utilized citrate and fumarate as a sole carbon source but not benzoate or tartarate. However, they varied in utilization of each of malonate, acetate and

Table 2. Morphological, physiological and biochemical characters of *B. megaterium* isolated from lupine foliage plants.

Characters	Reaction					
	B1	B2	B3	B4	B5	B6
Cell shape	Rod	Rod	Rod	Rod	Rod	Rod
Motility	+	+	+	+	+	+
Gram reaction	+	+	+	+	+	+
Spore forming	+	+	+	+	+	+
Growth at 45°C	+	+	+	+	+	+
Aerobiosis	Strict aerobic	Strict aerobic	Strict aerobic	Strict aerobic	Strict aerobic	Strict aerobic
Starch hydrolysis	-	-	-	-	-	-
Nitrate reduction to nitrite	-	-	-	-	+	-
Indole formation	-	-	-	-	-	-
V.P. test	-	-	-	-	-	-
Gelatin liquefaction	+	+	-	+	-	+
Oxidase	-	-	-	(+)	-	-
Catalase	+	+	+	+	+	+
Casein hydrolysis	-	+	-	+	-	+
Growth in						
5% NaCl	+	+	+	+	+	+
7% Nacl	-	-	-	-	-	-
Urease	-	-	-	+	-	-
Rot potato slices	-	+	-	-	-	+
Acid from:						
Glucose	+	+	+	+	+	+
Succrose	+	+	+	-	+	+
Fractose	+	+	+	-	+	+
Mannitol	+	+	+	-	+	+
Xylose	+	+	(+)	(+)	+	+
Maltose	+	+	+	+	+	+
Arabinose	-	-	+	+	-	-
Galactose	+	+	-	-	+	+
Glycerol	+	+	+	(+)	+	+
Lactose	+	-	-	-	+	-
Raffinose	+	(+)	(+)	(+)	+	+
Sorbitol	-	-	-	-	-	-
Mannose	-	+	+	-	+	+
Dextrin	-	-	-	-	-	-
Hypersensitive reaction tobacco leaves	-	-	-	-	-	-
H_2S production	-	-	-	-	-	-
2- Ketogluconate test	-	-	-	-	+	-
Lipolatic activity (tween 80)	-	+	+	+	-	+
Aesculin hydrolysis	+	+	+	+	+	+
Pectate hydrolysis	-	-	-	-	-	-
Utilization of:						
Malonate	-	-	-	(+)	+	-
Citrate	+	+	(+)	+	+	(+)
Acetate	(+)	-	-	(+)	+	-
Benzoute	-	-	-	-	-	-
Tartrate	-	-	-	-	-	-
Fumarate	+	+	+	+	+	+
Succinate	-	-	-	+	+	(+)

+ = Positive reaction, - = Negative reaction, (+) = Weak reaction.

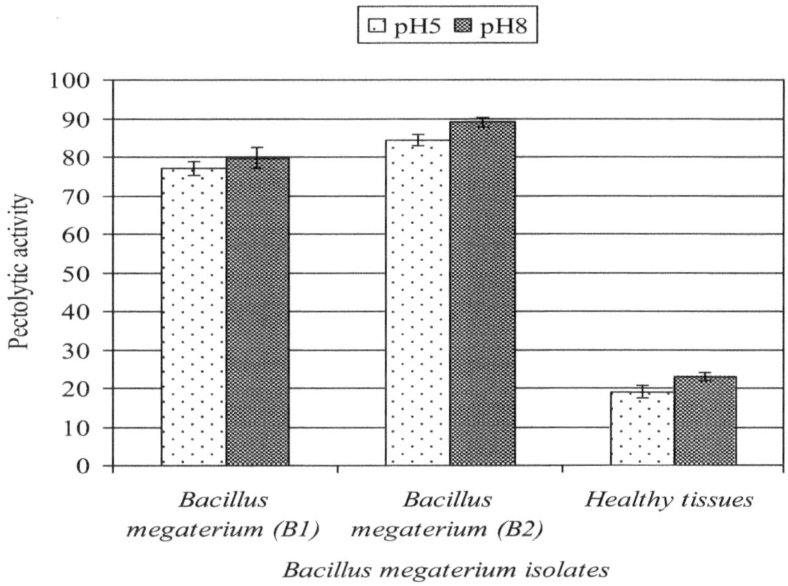

Figure 5. Pectolytic activity of extracts prepared from diseased foliar tissues infected with *B. megaterium* isolates B2 and B6 compared with healthy tissue extract. Mean ± SDs for three replicates per treatment are shown.

succenate.

According to Bergey's Manual of Systematic Bacteriology, 1986, these isolates resembles *B. megaterium* in the majority of the determinative tests except in starch hydrolysis hence it could be identified as pathogenic strain of the bacterium *B. megaterium.*

Assessment of pectolytic enzymes activity in diseased shoot and root system

Assessment of pectolytic enzymes secreted in diseased foliage by *B. megaterium* (B2 and B6) was assayed viscometrically. Great loss in viscosity of the pectin solution (enzyme substrate) was recorded according to the effect of the pectolytic enzyme(s) secreted in diseased tissues by bacterial isolates (Figure 5). While, the loss in viscosity took place at the two hydrogen ion concentrations (pH 5 and 8), the activity was higher at pH 8 in case of tissue extracts prepared from diseased shoot systems. The average of pectolytic activity means were 77.0, 79.7% and 84.3, 89.0% at two pHs due to the *B. megaterium* isolates B2 and B6, respectively and the enzyme activity was slightly higher at pH 8.

Assessment of cellulolytic enzyme activity

Diseased shoot system extracts showed marked cellulase activity when assayed at pH 6. Data present in Figure 6 show that the enzyme activity was 74.0 and 81.0% in case of the two bacterial infection (*B .megaterium* isolates B2 and B6) compared with 35.0% cellulytic activity in healthy tissues.

Varietal response

Foliage of the aforementioned cultivars was tested for their susceptibility towards six *B. megaterium* isolates. Data present in Figure 7 show that all bacterial isolates proved pathogenic towards the four cultivars being Balady cultivar the most susceptible. However, isolates B2 and B6 were more aggressive pathogen towards all tested cultivars especially Balady causing 94.5 and 97.7% disease severity. On the other hand, the Australian cultivar despite its susceptibility to the bacterial isolates, isolate B4 was less virulent on this cultivar.

Hast range

Effective isolates of *B. megaterium* isolates B2 and B6 which proved highly pathogenic on lupine plants were subsequently tested on other plant foliages representatives of 18 plant species as mentioned before to inquire about their host range. Both *B. megaterium* isolates B2 and B6 used in this study were avirulent to all inoculated plants, however, produced only small necrotic spots on faba bean leaves of the inoculated plants (Data not shown).

DISCUSSION

Diseased lupine plants showing foliar blight symptoms

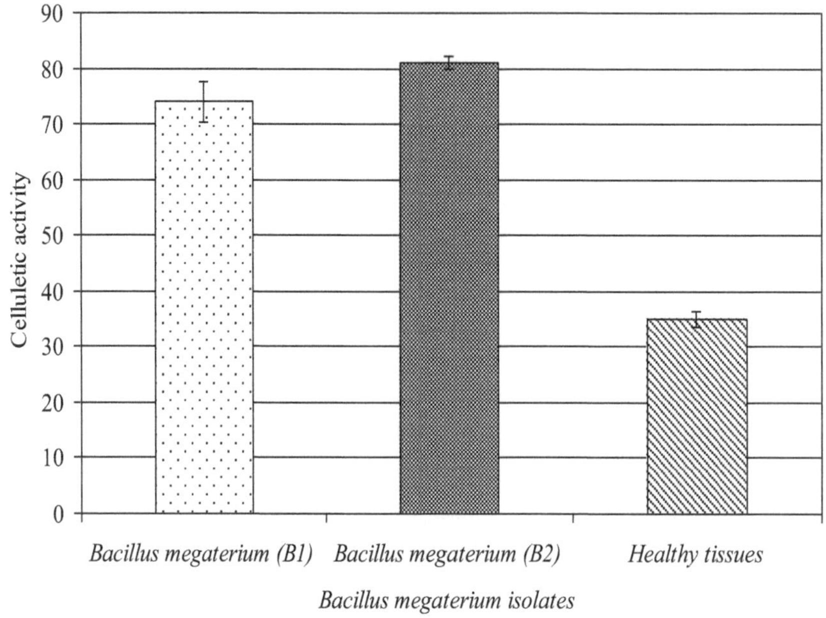

Figure 6. Cellulolytic activity of extracts prepared from diseased lupine foliar tissues infected with *B. megaterium* isolates B2 and B6 compared healthy tissue extract. Mean ± SDs for three replicates per treatment are shown.

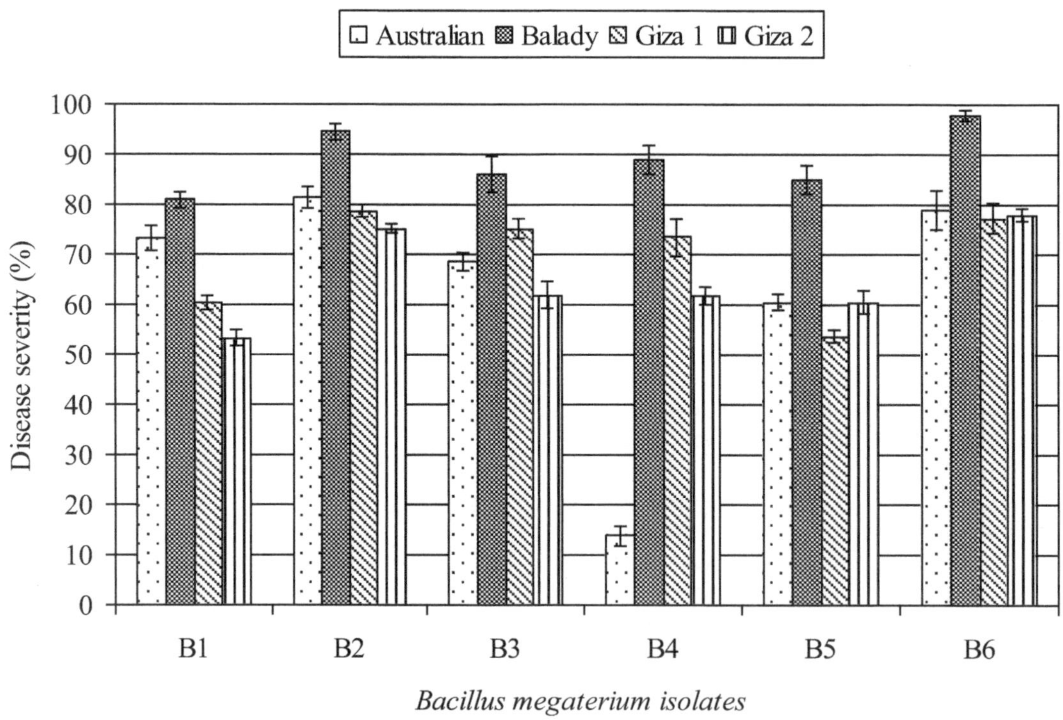

Figure 7. Cultivars response to lupine pathogenic *B. megaterium* isolates on lupine foliage. Mean ± SDs for three replicates per treatment are shown.

were collected from El- Minia Governorate for isolation trials. From diseased lupine foliage, the recovered bacterial isolates proved virulent on lupine plants in the pathogenicity test and incited symptoms on the subjected plants. The infected plants soon became blighted, turned grayish- brown and rather soft. The diseased plants bent over and eventually collapsed in a shorter period.

The published reports concerning lupine foliar diseases is rather limited as compared with those reported for root diseases. Kalis- Kuznia et al. (1991); Yang et al. (1996), Bolland et al. (2001) and Yang and Sweetingham (2002) reported the fungus *Pleiochaeta setosa* to cause dark brown and sunken lesions on leaves and stems of lupine. Paulitz and Cote (1991) reported *Phoma* sp to incite reddish brown lesions on the lower stems of lupine. Muller et al. (1999) found lupine anthracnose caused by *Colletotrichum* sp has been a major cause of yield losses in Germany.

No bacterial pathogens were reported to be isolated from diseased lupine foliage. Hence the obtained bacterial isolates during the present work could be regarded as hitherto unrecorded bacterial pathogen of lupine. The bacterial isolates were all similar to each other and were spore formers, aerobic, Gram positive motile. The determinative tests were carried out on these isolates and according to Bergey's Manual of Systematic Bacteriology, 1986, the bacteria could be identified as *B. megaterium* according to the majority of the tested characters except starch hydrolysis, nevertheless deviation in one character does not exclude the bacterium from a given species.

It is worth noting that Hosford (1982) isolated a spore forming bacterium causing white blotch to wheat in North Dakota. The author identified the bacterium as *B. megaterium* pv. *cereals*. Also, Ismail (2004) in Egypt isolated *B. megaterium* from affected sunflower leaves. The bacterium could incite black spots on sunflower leaves when experimentally inoculated.

Tissue extracts prepared from experimentally diseased shoot systems showed great pectolytic and cellulolytic activities while healthy tissue extracts showed slight activities of the enzymes. It has long been known that some plant pathogens secreted tissue macerating enzyme that is, pectinase and cellulose in the invaded tissues during the pathogenesis processes. These enzymes cause disintegration and collapse to the subjected tissues, hence facilitates the development of disease processes (Saleh and Stead, 2003; Ismail et al., 2012).

In varietal response test four lupine cultivars namely cvs. Australian, Balady, Giza 1 and Giza 2 were tested for their susceptibility *B. megaterium*. All the tested cultivars were susceptible to all tested *B. megaterium* isolates. The six bacterial isolates similarity attacked the four tested cultivars being cv. Balady the more sensitive cultivar.

Infective two *B. megaterium* isolates were subsequently tested for their virulence on other foliar plants representatives of 17 plant species. None of the inoculated plants were affected with any of the *B. megaterium* isolates except production of small and limited necrotic spots on inoculated faba bean leaves. These results indicate the host specificity of this bacterium towards its host plant and accordingly justify the suggested forma specials *lupini* to be given to the lupine bacterial pathogen *B. megaterium*.

REFERENCES

Abdel-Monaim MF (2008). Pathological Studies of Foliar and Root Diseases of Lupine with Special Reference to Induced Resistance. Ph. D. Thesis, Faculty of Agriculture, Minia University.

Bolland MDA, Sweetingham MW, Jarvis RJ (2001). Effect of *Pleiochaeta setosa* on field responses of *Lupinus angustifolius* and *L. luteus* to applications of phosphorus. Australian J. Exp. Agric. 41(4):549-556.

Gaber MR, Saleh OI, Abo El-Fotouh E (1990). Pectolytic, cellulolytic and proteolytic activities of two isolates of *Botryodiplodia theobramae.* Annals Agric. Sci. Fac. Agric. Ain Shams Univ. Cairo, Egypt 35(1):445-457.

Hosford RM Jr (1982). White blotch in wheat incited by *Bacillus megaterium* pv. *cerealis*. Phytopathol. 72:1453-1459.

Ismail ME, Abdel-Monaim MF, Mostafa YM (2012). Identification and pathogenicity of phytopathogenic bacteria associated with soft rot disease of girasole tuber in Egypt. J. Bacteriol. Res. 4(1):1-8.

Ismail ME (2004). Bacterial blight of sunflower plants in Egypt. Assuit J. Agric. Sci. 35(4):283:296.

Kalis-Kuznia RA, Stewart EL, Meronuck RA (1991). First record and notes on *Pleiochaeta setosa* in Minnesota. Mycol. 83(6):826-828.

Little TM, Hills FJ (1972). Statistical Methods in Agric. Res. USD, California, USA. pp. 9-27.

Liu L, Kloepper JW, Tuzun S (1995). Introduction of systemic resistance in cucumber against fusarium wilt by plant growth-promoting rhizobacteria. Plytopathol. 85:695-698.

Mahadevan A, Sridhar R (1982). Methods in Physiological Plant Pathology, Sivakami publications, India. pp.119.

Muller C, Steinbach P, Brother H (1999). Detection of *Colletotrichum* sp. on seed samples of lupine. Gesunde Pflanzen, 51(5):150-154.

Osman AR, Fahim MM, Sahab AF, Abdel- Kader MM (1986). Losses due to wilt of lupine. Egypt. J. Phytopathol. 15:27-34.

Paulitz TC, Cote E (1991). First report of *Phoma* spp. on white lupine in North America. Plant Dis. 75:862.

Saleh OI, Stead D (2003). Bacterial soft rot disease of pea in Egypt. Integrated control in protected crops, Mediterranean climate IOBC WPRS Bulletin. Editura Ceres, Bucharst, Romania, 26(10):115-120.

Shoeib Alia A (1997). Inhibitory effect of *Pseudomonas fluorescens* on bacterial growth and pathogenicity of *Burkholderia (=Pseudomonas) solanacearum Agrobacterium tumefaciens*. 8th Con. of the Egyptian Phytopathol. Sci. Cairo:403-415.

Sneath PHA, Mair NS, Sharpe ME, Holt JG (1986). Bergey's Manual of Systematic Bacteriol. Williams Wilkins Co., Baltimore. p. 1123.

Walker DS, Reeves PJ, Salmond GP (1994). The major secreted cellulose, CelV, of *Erwinia carotovora* sub sp. *carotovora* is an important soft rot virulence factor. MPMI 7(3):425-431.

Yang HA, Sweetingham MW (2002). Variation in morphology and pathogenicity of *Pleiochaeta setosa* isolates from *Lupinus* spp. and other legupms. Australian Plant Pathol. 31(3):273-280.

Yang HA, Sweetingham MW, Cowling WA (1996). The leaf infection process and resistance to *Pleiochaeta setosa* in three lupine species. Australian J. Agric. Res. 47(5):787-799.

In vitro susceptibility of some uropathogens and a comparative assessment of antibacterial activities of local and imported multodiscs

Olajuyigbe Olufunmiso O.[1,2]* and Adeoye Oluwaseun[2]

[1]Phytomedicine Research Centre, University of Fort Hare, Alice 5700, South Africa.
[2]Biosciences and Biotechnology Department, Babcock University, P. M. B. 21244, Ikeja, Lagos, Nigeria.

The *in vitro* antimicrobial susceptibility of recently isolated uropathogens was investigated using two locally produced brands of antibiotics and one imported brand, in a comparative study to determine their degree of effectiveness and the susceptibility profiles of these uropathogens. Seventy eight (78) bacterial strains containing 12 different species of both Gram negative and Gram positive bacteria were isolated and investigated. *Escherichia coli* was identified as the leading cause of urinary tract infections being the most isolated uropathogen. The activity of FD was comparable with that of AB (imported) which had the most effective antibacterial activities while those contained in JD were the least effective. Of the 18 different antibiotics employed, fluoroquinolones were the most effective antibiotics against all the bacterial isolates, followed by gentamicin > augmentin > nalidixic acid > nitrofurantoin > chloramphenicol while other antibiotics exhibited varying degree of activities on the bacterial isolates. It was therefore concluded that some locally manufactured antibiotics are as effective as imported brands while fluoroquinolones, augumentin, nitrofurantoin, gentamicin and nalidixic acid could be considered for first-line therapy in UTIs.

Key words: Uropathogens, UTIs, susceptibility, antibiotics, multodiscs.

INTRODUCTION

Urinary tract infections (UTIs) are one of the most common bacterial infections in humans both in the community and hospital setting (Tice, 1999). It accounts for 7 million consultations and >1 million hospital admissions a year in the USA (Stamm and Hooton, 1993). It is one of the most common bacterial infections encountered by clinicians in developing countries (Tessema et al., 2007) and its isolates are much more frequent in females than in males (Hunjak et al., 2007). This may be due to anatomical predisposition or urothelial mucosa adherence to the mucopolysaccharide lining or other host factors (Schaeffer et al., 2001). Of all pathogens from patients with simple UTI, *Escherichia coli* is the most common cause of both complicated and uncomplicated urinary tract infections (UTIs) (Yüksel et al., 2006; Yamamoto, 2007; Tessema et al., 2007) with *Enterococcus* spp., *Pseudomonas aeruginosa, Proteus mirabilis, Acinetobacter baumanii, Citrobacter* spp., *Serratia* spp., coagulase-negative staphylococci and *Klebsiella* spp being the next most frequently encountered species (Rafal'skiĭ et al., 2006).

Uncomplicated urinary tract infection (UTI) caused by uropathogenic *E. coli* (UPEC) represents a prevalent and potentially severe infectious disease (Hagan and Mobley, 2007). While the treatment and management of uncomplicated urinary tract infections is important (Jackson, 2007), management has become more complicated in the last decade due to the trend toward increasing antimicrobial resistance to ampicillin and trimethoprim/sulfamethoxazole (TMP/SMX) (Gobernado et al., 2007), ciprofloxacin, gentamicin and ceftriaxone (Laupland et al., 2007). In order to determine the extent of drug-resistance among Enterobacteriaceae isolated from urinary tract infection, *in vitro* experiments becomes essentially important as the emergence of enterococci with alarming rates of resistance concomitantly to penicillins and

*Corresponding author. E-mail: funmijuyigbe12@yahoo.com

aminoglycosides highlights the need for a more rational and restricted use of antimicrobials. In order to minimize the selection and spread of such strains, an early detection of these problem pathogens is also important for preventing any treatment failures (Miskeen and Deodhar, 2002).

Also, the knowledge of the sensitivity of uropathogens to antimicrobials can help to initiate empirical therapy of urinary tract infections (Sanchez et al., 2004). In view of the fact that *in vitro* susceptibility studies have indicated that a significantly high proportion of the urinary *E. coli* isolates has already developed resistance to the currently prescribed empirical antibiotics, namely the fluoroquinolones, a transition in empirical therapy appears imminent. Although antimicrobial resistance is a global concern (Blondeau and Vaughan, 2000), antimicrobial resistance among Enterobacteriaceae isolated from intra-abdominal infections is a problem (Baquero et al., 2006). Worldwide data showed that there is increasing resistance among urinary tract pathogens to conventional drugs (Hryniewicz et al., 2001). Multiple drug resistance (MDR) to β-lactams, aminoglycosides and quinolones mediated through R plasmids among Gram-negative bacteria has become a major nosocomial problem worldwide (Babinchak et al., 2005). Antimicrobial treatment of nosocomial infections caused by these bacteria is compromised (Barrett, 2005). The ability of both nosocomial and community-acquired pathogens to develop resistance to powerful broad-spectrum agents, presents a great challenge for prescribing patterns and development of new drugs relatively resistant to inactivation.

Hence, with the emergence of antimicrobial resistance in bacteria as a global problem, national and international surveillance programmes have been developed to monitor resistance (Felmingham, 2002). While surveillance of Enterobacteriaceae monitors changes in antimicrobial susceptibility and prevalence of isolates resistant to multiple classes of antimicrobial agents (Karlowsky et al., 2003), selection of oral antibiotics for the management of patients with infections should be based on knowledge of the susceptibility patterns of these isolates (Murray, 1991). An early detection of these problematic uropathogens will help in preventing any treatment failure (Ishikawa et al., 2004).

To this end, this study was designed to investigate the antibiotic susceptibility patterns of recently isolated community-acquired urinary tract pathogens from some teaching hospitals in South-Western Nigeria, using two locally manufactured antibiotics and one imported antibiotics sold in Nigerian markets. Analysis of these antimicrobial data will provide information for comparison with national trends, allow the rational selection of antibiotics for empiric treatment of UTIs in this country and compare the efficacy of locally manufactured antibiotics with the imported multodisk sold in Nigerian markets.

MATERIALS AND METHODS

Specimen collection

Freshly voided midstream urine specimens were collected aseptically from some patients who attended three teaching hospitals in South Western Nigeria, either as inpatient or out patient, with symptoms suggestive of UTIs (Savas et al., 2006; Santo et al., 2007). All patients had clinical evidence of urinary tract infections, as determined by the treating physician. Only a single positive culture per patient was included in the analysis. These patients did not include those who were on antibiotics a week before the samples were collected. The urine samples were collected into labeled 20 ml calibrated sterile bottles containing boric acid (0.2 mg) added to prevent the growth of bacteria in the urine. All patients were instructed on how to collect the urine samples aseptically. They were advised to take the samples to the laboratory immediately for culturing.

Bacteriological analysis

In the hospital laboratory, each well mixed urine sample (5 µl) was inoculated on McConkey agar (Oxoid), blood agar (Oxoid), and cysteine lactose electrolyte deficient agar (CLED, International Diagnostic Group). The inoculum on each plate was streaked out for discrete colonies with a wire loop following standard procedures (Cheesborough, 2006). The culture plates were incubated at 37°C for 24 h and observed for growth colonies. All the bacteria were isolated and identified using morphological, microscopy and biochemical tests following standard procedures described by Cowan and Steel (1974) and Cheesborough (2006).

Antibiotic susceptibility testing

The antibiotic susceptibility tests were performed by disc diffusion technique using three different commercially available discs on Mueller Hinton agar plates. Susceptibility testing was performed by using a standard agar dilution technique (Washington and Sutter, 1980) with Mueller Hinton agar (Lab. M; International Diagnostic Group Plc., Lancashire, UK) which is a susceptibility test medium recommended by the National Committee for Clinical Laboratory Standards, (NCCLS) (Crider and Colby, 1985) because of its low content of inhibitory substances. 100 µl (approximately 106 cfu/ml) of overnight broth culture of each test organism was dispersed into 20 ml volumes of molten Mueller Hinton Agar prepared according to manufacturer's instruction, swirled gently to ensure an even distribution of inoculums, poured into sterile Petri dishes and allowed to set.

Each set of antibiotic discs was aseptically dispensed onto the surface of the inoculated agar plate, pressed down to ensure complete contact with the agar surface while the plates were inverted and incubated at 35°C for 24 h within 15 min after the discs were applied. The assessment of antibacterial activities was based on measurement of the diameter of the zones of inhibitions to the nearest millimeter with calibrated transparent meter rule held on the back of the inverted Petri plates. The different multodiscs were identified as AB (imported), FD and JD. Each multodisc respectively contained 12, 13 and 15 different antibiotics as shown in Table 1.

RESULTS

From the study, *E. coli* was the most isolated (34.6%), followed by *Proteus morganella* (10.3%), *Enterobacter*

Table 1. Different multodiscs from different manufacturers and their various concentrations.

Drug	AB	FD	JD	Concentration (µg)
Amoxicillin	AMX	AMX	na	25
Ofloxacin	OFL	OFL	OFL	5
Ciprofloxacin	CPX	CPX	CPX	10
Gentamicin	GEN	GEN	GEN	10
Chloramphenicol	CHL	CHL	CHL	30
Cotrimoxazole	COT	COT	COT	25
Erythromycin	ERY	ERY	ERY	5
Nitrofurantoin	NIT	NIT	NIT	300
Augmentin	AUG	AUG	AUG	30
Tetracycline	TET	TET	TET	30
Nalixidic acid	NAL	na	na	25
Cloxacillin	CXC	na	na	5
Pefloxacin	PFX	PFX	na	5
Streptomycin	STR	STR	STR	10
Ceftriazone	na	CEF	CEF	30
Ampiclox	na	na	APX	30
Lincomycin	na	na	LIN	30
Cephalexin	na	na	COX	15
Ampicillin	na	na	PN	15

JD Gram – negative: LN – na; DOM – na; ED – July, 2011JD Gram – positive: LN – na; DOM – na; ED – July, 2011; FD Gram – negative: LN – 0303; DOM – na; ED – Oct, 2011; FD Gram – positive: LN – 0461; DOM – na; ED – Oct, 2011; AB Gram – negative: LN– JB07/BP; DOM – na; ED – December, 2011; AB Gram – positive: LN – JB04/P; DOM – na; ED – December, 2011; Key: na – Not available; LN – Lot number; DOM – date of manufacture; ED – Expiry date.

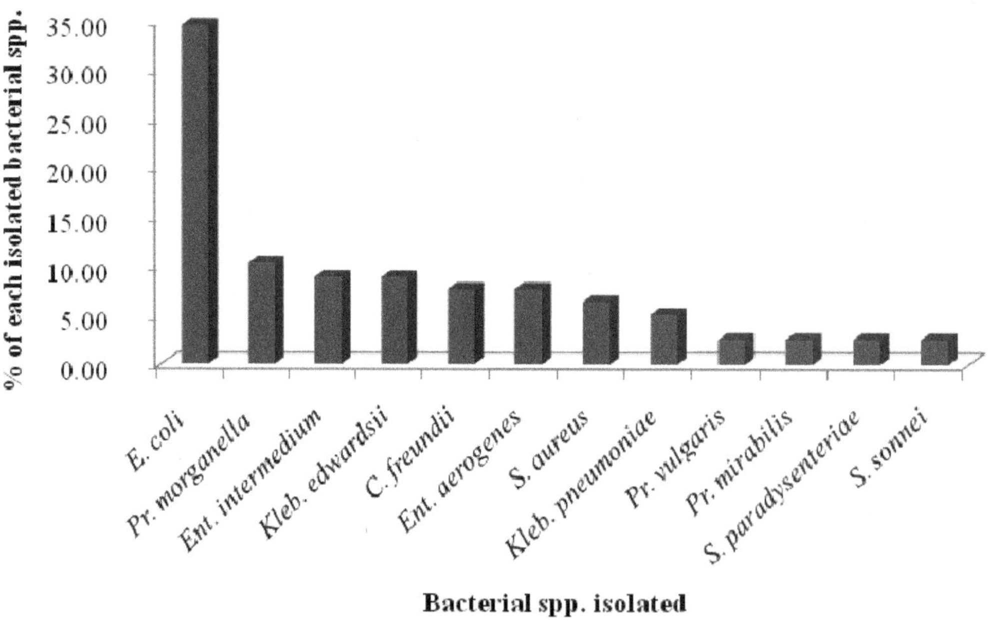

Figure 1. Percentage and type of each bacterial species isolated.

intermedium and *Klebsiella edwardsii* (8.9%), *Citrobacter freundii* and *Enterobacter aerogenes* (7.7%), *Staphylococcus aureus* (6.4%), *Klebsiella pneumoniae* (5.1%) as well as *Proteus vulgaris, P. mirabilis, Shigella paradysenteriae and Shigella sonnei* (2.6%) which were the least isolated as shown in Figure 1. The clinical

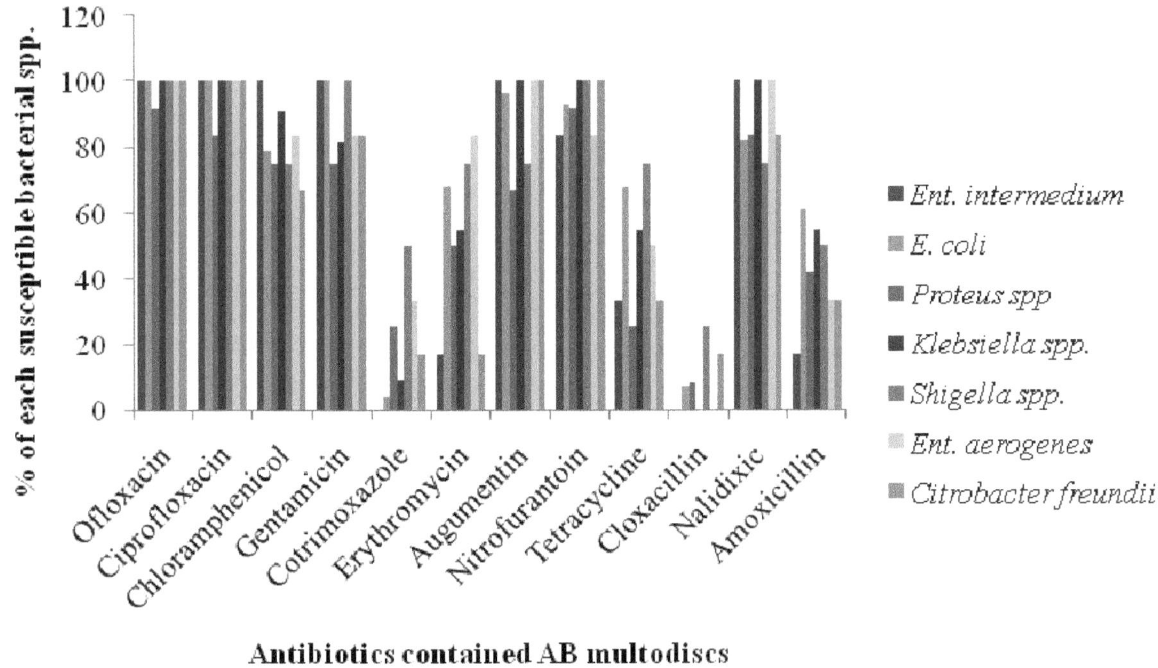

Figure 2. Percentages of each bacterial spp. susceptible to each antibiotic disc contained in AB multodisc.

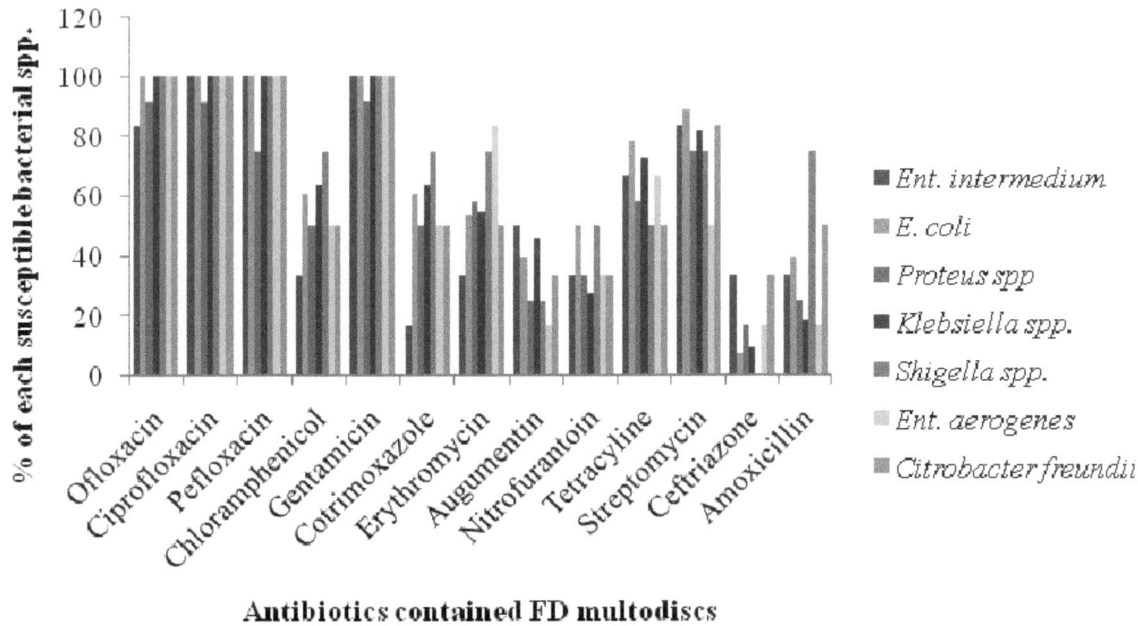

Figure 3. Percentages of each bacterial spp. susceptible to each antibiotic disc contained in FD multidisc.

isolates were susceptible to the routinely prescribed antibiotics. Percentages of different bacterial spp. susceptible to different antibiotic discs contained in the different brands of multodisc are as shown in Figures 2 to 4.

From the bacterial susceptibility profiles presented in Figures 2 to 4, the antibiotics contained in the AB were most effective against all the bacterial isolates, followed by the antibiotics contained in FD, while those contained in JD were the least effective. Fluoroquinolones were the most effective antibiotics against all the bacterial isolates, followed by Gentamicin > Augmentin > Nalidixic acid > Nitrofurantoin > Chloramphenicol while other antibiotics exhibited varying degrees of antibacterial activities on the

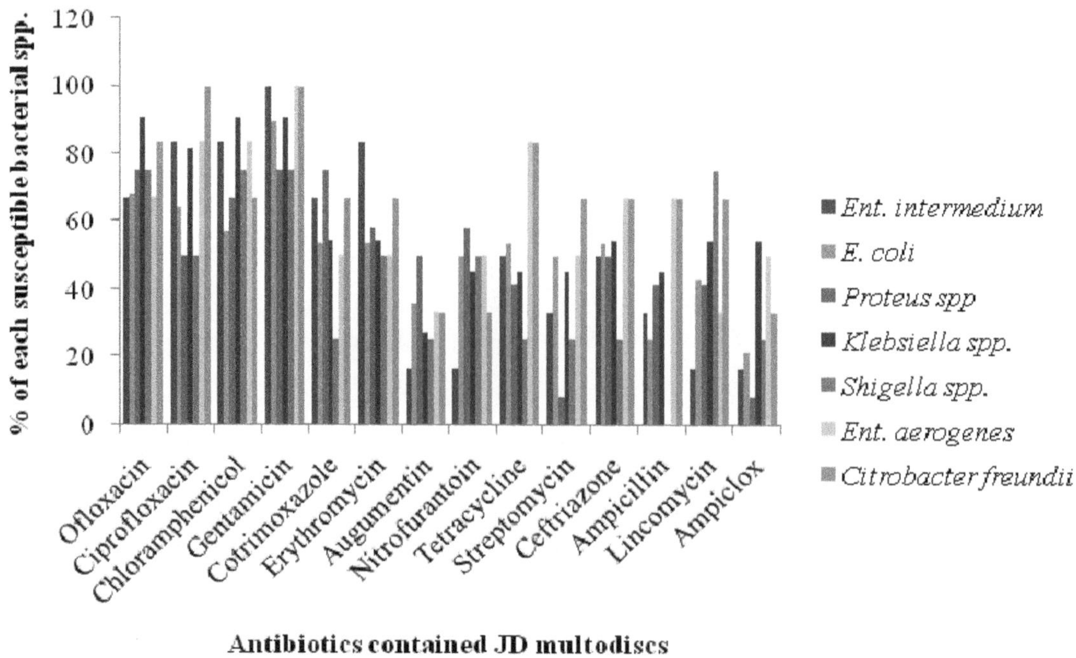

Figure 4. Percentages of each bacterial spp. susceptible to each antibiotic disc contained in JD multodisc.

isolates. With the exception of *Proteus* spp. not 100% susceptible to any of the antibiotics, many isolates were 100% susceptible to 3 to 4 different antibiotics, such as fluoroquinolones, augumentin, nitrofurantoin, gentamicin and nalidixic acid, from both AB and FD multodiscs while only few of the bacteria were 100% susceptible to gentamicin only in JD. In addition to being susceptible to fluoroquinolones, *E. intermedium* strains were 100% susceptible to chloramphenicol, gentamicin, augumentin and nalidixic acid, nitrofurantoin (83.3%), tetracycline (33.3%), amoxicillin and erythromycin (16.7%) while they were highly resistant to both cotrimoxazole and cephalexin. *E. coli* strains were 100% susceptible to gentamicin followed by augumentin (96.4%), nitrofurantoin (92.9%), nalidixic acid (82.1%), chloramphenicol (78.6%), tetracycline and erythromycin (67.9%), amoxicillin (60.7%) and cotrimoxazole (3.6%). *Klebsiella* spp. were 100% susceptible to augumentin, nitrofurantoin and nalidixic acid, chloramphenicol (90.9%), gentamicin (81.8%), erythromycin, tetracycline and amoxicillin (54.5%) and cotrimoxazole (9.1%).

Enterobacter aerogenes were 100% susceptible to augumentin and nalidixic acid, chloramphenicol, gentamicin, erythromycin and nitrofurantoin (83.3%), tetracycline (50%), cotrimoxazole and amoxicillin (33.3%). *C. freundii* were 100% susceptible to augumentin and nitrofurantoin, nalidixic acid and gentamcin (83.3%), chloramphenicol (66.7%), tetracycline and amoxicillin (33.3%), cotrimoxazole, erythromycin (16.7%). While augumentin was the most effective against *Proteus* spp. Having 96.4% susceptible, *Proteus* spp. were (91.7%) susceptible to ofloxacin and nitrofurantoin, nalidixic acid and ciprofloxacin (83.3%), chloramphenicol (78.6%), gentamicin (75%), erythromycin (50%), amoxicillin (41.7%), tetracycline and cotrimoxazole (25%), and cephalexin (8.3%). *Shigella* spp. was 100% susceptible to gentamicin and nitrofurantoin, erythromycin, chloramphenicol, augumentin, tetracycline and nalidixic acid (75%), cotrimoxazole and amoxicillin (50%) as well as cephalexin (25%). Apart from being 100% susceptible to fluoroquinolones, gentamicin, augumentin, nitrofurantoin and nalidixic acid, most of these bacteria exhibited multidrug resistance to all other antibiotics used in this study, while cephalexin was totally ineffective against all the tested bacterial species.

DISCUSSION

As a result of the fact that most UTIs are treated empirically, the selection of an antimicrobial agent is determined by the most likely pathogen and its expected susceptibility pattern. Monitoring antibiotic susceptibility patterns of uropathogens at a local level will yield important information regarding emerging problems of antibiotic resistance and provide assistance in managing empirical therapy.

In this study, the most common organisms were *E. coli* (34.6%), *Enterobacter* spp (16.7%), *Proteus* spp (15.4%) and *Klebsiella* spp (14.1%) indicating that *E. coli* is the most common cause of UTIs. This result is in agreement with reports from earlier investigators (Henry et al., 1998; NCCLS, 1998; Xu et al., 1999; Tice, 1999; Blondeau et al., 1999; Zhanel et al., 2005) and contrary to the report of Gruneberg (1994) who reported that *E. coli*, as the leading cause of uropathogen, was being replaced by other members of the Enterobacteriacea and Enterococci. Treatment of UTIs is a major community indication of antibiotic usage (Fihn, 2003; Hooton, 2003). Fluoroquinolones (Fihn, 2003; Hooton, 2003), trimethoprim-sulphamethoxazole (cotrimoxazole) or trimethoprim alone (Warren et al., 1999; Hooton and Stamm, 1997; Nicolle, 2002), ampicillin or amoxicillin (Gupta et al., 2001) and nitrofurantoin (Nicolle, 2002), have been implicated as being frequently used for the treatment of UTIs. While resistance to nitrofurantoin among *E. coli* from UTIs remain low despite more than 50 years widespread use of the drug (Kahlmeter, 2000; Mazzulli et al., 2001), resistance to nalidixic acid (Kahlmeter, 2000), cotrimoxazole (Mclssac et al., 2004) have been reported. Cotrimoxazole prescriptions for UTI have declined while fluoroquinolone prescriptions have increased (Gupta et al., 2001).

This study indicated that fluoroquinolones, augumentin, nitrofurantoin, gentamicin and nalidixic acid were active against all the isolates and could be used as first-line therapeutic agents in UTIs. The use of cotrimoxazole, ceftriazone, clindamycin, ampicillin, cephalexin and cloxacillin should be discouraged as they were ineffective against the isolates. The observed high rate of resistance to these ineffective antibiotics may be a reflection of the previous exposure of the isolates to them and acquisition of resistant genes. The level of susceptibility of *E. coli* and the varied degree of susceptibility of other enterobacteriaceae to the effective antibiotics was found to be comparable with results from other investigators (Cunney et al., 1992; Jones et al., 1999; Vromen et al., 1999; Fluit et al., 2000). The susceptibility of the uropathogens to fluoroquinolones may, however, indicate that virulent strains might be less resistant to antimicrobials than strains causing only colonization or lower tract UTI as previously observed by Komp et al. (2005) and Roos and Klemm (2006).

Although the activities of two locally manufactured multodiscs were compared against an imported multodiscs AB, the obtained result indicated that FD produced a reasonably comparable result with the AB while JD was the least effective. The observed disparity could have resulted from differences in production techniques, location, quality of raw materials used during production and the shelf lives of these products. It will, thus, be worthwhile that local and international manufacturers produce multodiscs that will contain the same number, quality and types of antibiotics. They should also establish and maintain good quality standards in their productions.

In conclusion, this study indicates that some antibiotics commonly used in UTI treatments are still effective, particularly in both community and hospital infections. These may be of immense value for use to determine drugs of choice in the treatment of UTIs prior to outcome of laboratory investigations while fluoroquinolones, augumentin, nitrofurantoin, gentamicin and nalidixic acid could be considered for first-line therapy for UTIs, in agreement with previous reports (Stamm, 2002; Cunha, 2006; Nicolle et al., 2006). Although there are some others "old antibiotics" with a role that may be underestimated for UTIs (Honderlick et al., 2006), prudent and rationale use of antibiotics must encourage prescribing fluoroquinolones and other indicated antibiotics parsimoniously for uncomplicated UTIs.

REFERENCES

Babinchak T, Ellis-Grosse E, Dartois N, Rose GM, Loh E, Tigecycline 301 Study Group, Tigecycline 306 Study Group (2005). The efficacy and safety of tigecycline for the treatment of complicated intra-abdominal infections: analysis of pooled clinical trial data. Clin. Infect. Dis., 41(Suppl 5): S354-367.

Baquero F, Cercenado E, Cisterna R, de la Rosa M, Garcia-Rodriguez JA, Gobernado M, Perez JL, Manchado P, Martin R, Pascual A, Picazo J, Prats G, Rubio C, Snyder TA, Sanz-Rodriguez C (2006). Patterns of susceptibility to antibiotics of Enterobacteriaceae causing intra-abdominal infection in Spain: SMART 2003 study outcomes. Rev. Esp. Quimioter., 19(1): 51- 59.

Barrett JF (2005). MRSA—what is it, and how do we deal with the problem? Expert Opin. Ther. Target., 9: 253-265.

Blondeau JM, Vaughan D (2000). In vitro activity of 19 antimicrobial agents against 3513 nosocomial pathogens collected from 48 Canadian medical centres. The Canadian Antimicrobial Study Group. Division of Clinical Microbiology. Int. J. Antimicrob. Agents, 15(3): 213-219.

Cheesborough M (2006). Medical Laboratory Manual for Tropical Countries, II Micrbiology (ELBS), Butterworth, Kent, U.K., pp. 23-78.

Cowan SJ, Steel KJ (1974). Cowan and Steel manual for identification of medical bacteria, 2nd edn Cambridge University Press, London, pp. 176-232.

Crider SR, Colby SD (1985). Susceptibility of Enterococci to Trimethoprim and Trimethoprim-Sulfamethoxazole. Antimicrob. Agents Chemother., 27(1): 71-75.

Cunha BA (2006). New uses for older antibiotics: nitrofurantoin, amikacin, colistin, polymyxin B, doxycycline, and minocycline revisited. Med. Clin. N. Am., 90: 1089-1107.

Felmingham D (2002). The need for antimicrobial resistance surveillance: Review. J Antimicrob Chemother., 50 (Suppl. S1): 1-7.

Fihn SD (2003). Clinical practice. Acute uncomplicated urinary tract infection in women. N. Engl. J. Med., 349: 259-266.

Fluit AC, Jones ME, Schmitz FJ, Acar J, Gupta R, Verhoef J (2000). Antimicrobial resistance among urinary tract infection (UTI) isolates in Europe: Results from the SENTRY Antimicrobial Surveillance Program 1997. Antonie van Leeuwenhoek, 77(2): 147-152.

Gobernado M, Valdés L, Alós JI, García-Rey C, Dal-Ré R, García-de-Lomas J (2007). Antimicrobial susceptibility of clinical *Escherichia coli* isolates from uncomplicated cystitis in women over a 1-year period in Spain. Rev. Esp. Quimioter., 20(1):68-76.

Gupta K, Hooton TM, Stamm WE (2001). Increasing antimicrobial resistance and the management of uncomplicated community-acquired urinary tract infections. Ann. Intern. Med., 135: 41-50.

Hagan EC, Mobley HL (2007). Uropathogenic *Escherichia coli* outer membrane antigens expressed during urinary tract infection. Infect. Immun., 75(8): 3941-3949.

Honderlick P, Cahen P, Gravisse J, Vignon D (2006). Uncomplicated urinary tract infections, what about fosfomycin and nitrofurantoin in 2006. Pathol. Biol., 54(8-9): 462-466.

Hryniewicz K, Szczypa K, Sulikowska A, Jankowski K, Betlejewska K, Hryniewicz W (2001). Antibiotic susceptibility of bacterial strains

isolated from urinary tract infections in Poland. J. Antimicrob. Chemother., 47(6): 773-780.
Hunjak B, Pristas I, Stevanović R (2007). Uropathogens and antimicrobial susceptibility in outpatients. Acta Med. Croatica, 61(1):111-5.
Ishikawa K, Miyakawa S, Tanaka T, Naide Y, Shiroki R, Hoshinaga K (2004). The trend and susceptibility to antibacterial agents of enterococcus species from urinary tract infections. Nippon Hinyokika Gakkai Zasshi., 95(1): 25-34.
Jackson MA (2007). Evidence-based practice for evaluation and management of female urinary tract infection. Urol. Nurs., 27(2): 133-136.
Jones RN, Kugler KC, Pfaller MA, Winokur PL (1999). The SENTRY Surveillance Group, North America. Characteristics of pathogens causing urinary tract infections in hospitals in North America: Results from the SENTRY Antimicrobial Surveillance Program, 1997. Diagn. Microbiol. Infect. Dis., 35(1): 55-63.
Karlowsky JA, Jones ME, Thornsberry C, Friedland IR, Sahm DF (2003). Trends in antimicrobial susceptibilities among Enterobacteriaceae isolated from hospitalized patients in the United States from 1998 to 2001. Antimicrob. Agents Chemother., 47(5): 1672-1680.
Laupland KB, Ross T, Pitout JD, Church DL, Gregson DB (2007). Investigation of sources of potential bias in laboratory surveillance for anti-microbial resistance. Clin Invest. Med., 30(4): E159-166.
Miskeen PA, Deodhar L (2002). Antimicrobial susceptibility pattern of *Enterococcus* species from urinary tract infections. J. Assoc. Phys. Ind., 50: 378-381.
Murray PR (1991). Antimicrobial activity of seven oral antibiotics against selected community- and hospital-acquired pathogens. Clin. Ther., 13(32): 224-231.
Nicolle L, Anderson PA, Conly J, Mainprize TC, Meuser J, Nickel JC, Senikas VM, Zhanel GG (2006). Uncomplicated urinary tract infection in women. Current practice and the effect of antibiotic resistance on empiric treatment. Can. Fam. Phys., 52: 612-618.
Nicolle LE (2002). Urinary tract infection: traditional pharmacologic therapies. Am J Med., 113(Suppl 1A): 35-44.
Rafal'skiĭ VV, Strachunskiĭ LS, Babkin PA, Valenskaia VS, Gabbasova LA, Dmitrieva OB, Emel'ianova IV, Krupin VN, Malev IV, Petrov SB, RokhlikovIM, Furletova NM, Khaĭrullov AS (2006). Resistance of causative agents of uncomplicated urinary tract infections in Russia. Urol., (5): 34-7.
Roos V, Klemm P (2006). Global gene expression profiling of the asymptomatic bacteriuria *Escherichia coli* strain 83972 in the human urinary tract. Infect. Immun., 74: 3565-3575.
Sanchez MJM, Guillan MC, Fuster FC, Lopez MR, Jimenez RM, Garcia AJ (2004). Antimicrobial susceptibility of *Escherichia coli* isolated from bacteriurias in Bierzo health area during 2003. Actas Urol. Esp., 28(8): 588-593.
Santo E, Salvador MM, Marin JM (2007). Multidrug-resistant urinary tract isolates of *Escherichia coli* from ribeirao preto, Sao Paulo. Braz. J. Infect. Dis. 11(6): 1-5.
Savas L, Guvel S, OnlenY, Savas N, Duran N (2006). Nosocomial urinary tract infections: microorganisms, antibiotic sensitivities and risk factors. West Ind. Med. J. 55(3): 1-9.
Schaeffer AJ, Rajan N, Cao Q, Anderson BE, Pruden DL, Sensibar J, Duncan JL (2001). Host pathogenesis in urinary tract infection. Int. J. Antimicrob. Agents, 17: 245-251.
Stamm WE, Hooton TM (1993). Management of urinary tract infections in adults. New Eng. J. Med., 329(18): 1328-1334.
Tessema B, Kassu A, Mulu A, Yismaw G (2007). Predominant isolates of urinary tract pathogens and their antimicrobial susceptibility patterns in Gondar University Teaching Hospital, Northwest Ethiopia. Ethiop. Med. J., 45(1): 61-67.
Tice AD (1999). Short course therapy of acute cystitis: a brief review of therapeutic strategies. J. Antimicrob. Chemother., 43(Suppl. A): 85-93.
Warren JW, Abrutyn E, Hebel JR, Johnson JR, Schaeffer AJ, Stamm WE (1999). Guidelines for antimicrobial treatment of uncomplicated acute bacterial cystitis and acute pyelonephritis in women. Infectious Diseases Society of America (IDSA). Clin. Infect. Dis., 29: 745-758.
Washington JA (II), Sutter VL (1980). Dilution susceptibility test: agar and macro broth dilution procedures. In E. H. Lennette, A. Balows, W. J. Hausler, Jr., and J. P. Truant (ed.), Manual of clinical microbiology, 3rd ed. American Society for Microbiology, Washington, D. C., pp. 453-458.
Xu Y, Chen M, Biedenbach DJ, Deshpande LM, Jones RN (1999). Evaluation of the in vitro antimicrobial activity of cefepime compared to other broad-spectrum β lactams tested against recent clinical isolates from 10 Chinese hospitals. Diagn. Microbiol. Infect. Dis., 35(2): 135-142.
Yamamoto S (2007). Molecular epidemiology of uropathogenic *Escherichia coli*. J. Infect. Chemother., 13(2): 68-73.
Yüksel S, Oztürk B, Kavaz A, Ozçakar ZB, Acar B, Güriz H, Aysev D, Ekim M, Yalçinkaya F (2006). Antibiotic resistance of urinary tract pathogens and evaluation of empirical treatment in Turkish children with urinary tract infections. Int. J. Antimicrob. Agents, 28(5): 413-416.
Zhanel GG, Hisanaga TL, Laing NM, DeCorby MR, Nichol KA, Palatnik LP, Johnson J, Noreddin A, Harding GK, Nicolle LE, Hoban DJ (2005). Antibiotic resistance in outpatient urinary isolates: Final results from the North American Urinary Tract Infection Collaborative Alliance (NAUTICA). Int. J. Antimicrob. Agents, 26: 380-388.

Effects of Mn^{2+} levels on the resistance properties of *Bacillus cereus* spores

Lawrence A. Klobutcher[1], Elena Gaidamakova[2], Michael J. Daly[2], Barbara Setlow[1], and Peter Setlow[1]*

[1]Department of Molecular, Microbial and Structural Biology, University of Connecticut Health Center, Farmington, CT 06030-3305, USA.
[2]Department of Pathology, School of Medicine, Uniformed Services University of the Health Sciences, Bethesda, MD 20814, USA.

In some *Bacillus* species, manganese levels influence the resistance properties of spores. To determine if this was true for *Bacillus cereus*, bacteria were sporulated with different $MnCl_2$ concentrations resulting in spores with 30-fold differences in core Mn^{2+} levels. Spores with different Mn^{2+} levels displayed no differences in resistance to dry heat, UV radiation, γ-radiation, or hydrogen peroxide. However, spores with the lowest Mn^{2+} level were less resistant to wet heat. Overall, Mn^{2+} levels were not a major factor in *B. cereus* spore resistance, and this suggests that this will also be true for the closely related *B. anthracis* spores.

Key words: Manganese (Mn^{2+}), spores, spore resistance, γ-radiation, *Bacillus*, *Deinococcus*.

INTRODUCTION

Spores of *Bacillus* species are extremely resistant to a variety of stress factors including wet or dry heat, UV or γ-radiation, or toxic chemicals, including oxidizing agents such as hydrogen peroxide (H_2O_2) (Setlow and Johnson, 2012). Spore resistance to these agents is due to a number of factors including the proteinaceous spore coats, the low hydration level of the spore core, the high levels of pyridine-2,6-dicarboxylic acid (dipicolinic acid (DPA)) in the spore core, the saturation of the spore genome with the α/β-type small, acid-soluble proteins (SASP), and DNA repair during spore outgrowth. A number of reports indicate that some agents, in particular UVA radiation and γ-radiation, kill growing cells of many aerobic organisms by generating reactive oxygen species (ROS) that damage macromolecules, including proteins and nucleic acids (Avery, 2011; Daly, 2012). Interestingly, elevated Mn levels are associated with increased resistance of growing cells of many bacteria and archea to UVA and γ-radiation, perhaps due to non-enzymatic detoxification of ROS by Mn^{2+} containing small molecules (McEwan, 2009; Daly, 2012).

Previous work has also shown that spore Mn levels have significant effects on the resistance properties of *Bacillus megaterium* spores, with elevated Mn levels associated with increased spore resistance to wet or dry heat, UVC radiation, and hydrogen peroxide, although, Mn levels had no effect on spore γ-radiation resistance (Donnellan and Stafford, 1968; Aoki and Slepecky, 1974; Ghosh et al., 2011). Elevated Mn levels are also associated with elevated resistance of *Bacillus fastidiosus* spores to wet heat (Aoki and Slepecky, 1973). In contrast, *Bacillus subtilis* spores with over a 200-fold range of protoplast Mn levels exhibited no significant differences in resistance to wet or dry heat, γ-radiation or hydrogen peroxide, although, spores with low Mn levels were less resistant to UVC radiation than high Mn spores (Ghosh et al., 2011; Granger et al., 2011). Given the increasing use of γ-radiation as a means for sterilization of foodstuffs and the importance of *Bacillus cereus* as a food-borne pathogen (Setlow and Johnson, 2012), it

*Corresponding author. E-mail: setlow@nso2.uchc.edu.

Table 1. Levels of Mn^{2+} in spores prepared in media with different amounts of added $MnCl_2$*.

[$MnCl_2$] Added to sporulation medium (µmol l^{-1})	Mn^{2+} content of spores (µg gm^{-1} dry weight)
0.5	73
1	485
10	1709
100	2311

**B. cereus* was sporulated at 30°C in Ellar's modified liquid sporulation medium with different $MnCl_2$ concentrations added (Stewart et al., 1981), and the spores were then purified and treated with EDTA (Ghosh et al., 2011; Granger et al., 2011). Spore Mn^{2+} and DPA levels (data not shown) were determined as described (Yi and Setlow, 2010; Ghosh et al., 2011; Granger et al., 2011).

seemed worthwhile to examine the effects of *B. cereus* spore Mn levels on these spores' resistance properties. In addition, given the close relatedness between *B. cereus* and *Bacillus anthracis* (Priest et al., 2004), results with *B. cereus* spores will also likely be applicable to *B. anthracis* spores.

RESULTS AND DISCUSSION

Preparation of *B. cereus* spores with varying levels of Mn

B. cereus was able to form spores in the presence of 0.5 to 100 µmol L^{-1} added $MnCl_2$, resulting in an ~30-fold increase in spore Mn levels in EDTA-washed spores at the highest $MnCl_2$ levels employed (Table 1). As found with spores of *B. megaterium* and *B. subtilis* (Ghosh et al., 2011; Granger et al., 2011), EDTA treatment removed < 2% of Mn from spores prepared with $\leq$ 10 µmol L^{-1} Mn, and only ~ 10% from spores prepared with 100 µmol L^{-1} Mn. These results suggest that ≥ 90% of Mn incorporated into spores prepared with $\leq$ 100 µM Mn is in the spore core, as EDTA should remove all Mn, except for that present in the spore core where most divalent cations are chelated with DPA. Note that even the maximal level of Mn found in spores will chelate < 5% of spore DPA, and DPA levels in spores made with various Mn levels were all within 7% of each other.

Resistance properties of spores with different Mn levels

Analysis of the resistance of spores containing different Mn levels, with the exception of γ-radiation resistance, was carried out at least three times, each time using at least duplicate measurements of spore viability, and always with essentially identical results. Spore treatment with γ-radiation was carried out only once, but the irradiated samples were analyzed at least three times, each using at least duplicate measurements of spore viability, again with essentially identical results. One representative measurement of the resistance of spores prepared with different Mn levels to various agents is shown in Figure 1. In these measurements, values for spore viability are $\leq \pm$ 20%. These results indicate that over a 30-fold range, Mn levels play no notable role in *B. cereus* spore resistance to dry heat, UVC radiation, γ-radiation or H_2O_2 (Figure 1a to d). However, spores with the lowest Mn levels did exhibit lower resistance to wet heat (Figure 1e).

Compared to other *Bacillus* species where the role of Mn on multiple resistance properties has been well-studied, *B. cereus* is more similar to *B. subtilis,* in that, only a single resistance property is affected by Mn (wet heat and UVC radiation, respectively). Why might these species differ in regard to the effects of Mn compared to *B. megaterium*, where multiple resistance properties are sensitive to Mn levels? One potential explanation is that the resistance properties of the different species are related to the number of chromosomes in the spores, since *B. subtilis* and *B. cereus* spores are monogenomic, while *B. megaterium* spores are digenomic (Hauser and Karamata, 1992). Multiple chromosomes in spores would allow for recombinational repair of DNA damage early in spore outgrowth, and it is possible that a component of this process is particularly sensitive to ROS. However, this scenario seems unlikely, as a number of reports indicate that wet heat and H_2O_2 kill spores of *Bacillus* species by damage to proteins, and not by DNA damage (Palop et al., 1998; Coleman et al., 2007, 2010; Setlow and Johnson, 2012). Moreover, even in the case of γ-irradiation there is uncertainty in regard to the type, level and possible importance of DNA damage in spore killing. The level of γ-irradiation-induced DNA damage would be expected to be significantly less in *Bacillus* spores compared to vegetative cells, since the core water content of spores is extremely low (Setlow and Johnson, 2012). Indeed, because of spores' low core water content, spore DNA is predicted to be significantly more resistant to double strand break (DSB) formation by γ-radiation, and Mn-complexes that act as antioxidants may give only minimal protection against γ-radiation (Gaidamakova et al., 2012). As a result, DNA may not even be the lethal target for ROS generated by oxidative stress caused by γ-radiation and other agents (Daly,

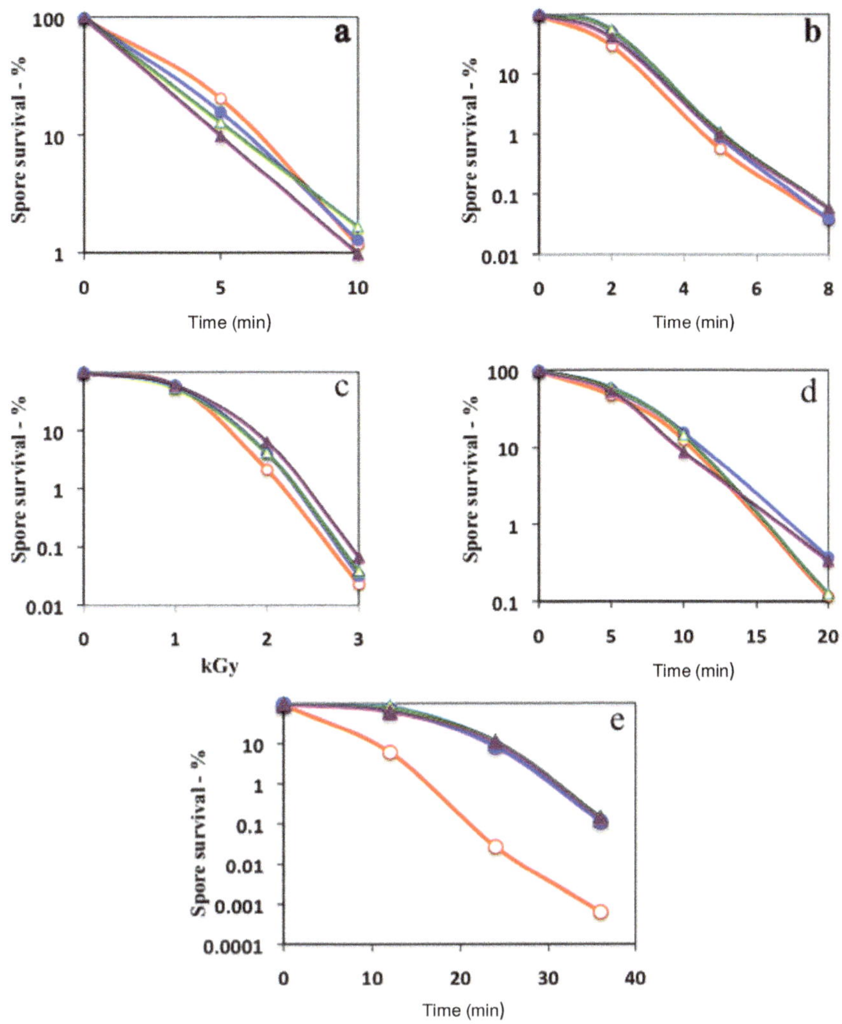

Figure 1. Resistance properties of *B. cereus* spores with different Mn contents. Plots illustrating survival of *B. cereus* spores prepared with varying $MnCl_2$ concentrations following treatment with (a) dry heat at 120°C, (b) UVC radiation at $5x10^{-4}$ J min^{-1} cm^{-2}, (c) varying doses of γ-radiation measured in kiloGrays (kGy), (d) H_2O_2 (5% in 25 mM KPO_4 buffer (pH 7.4) at 23°C) and, (e) wet heat at 87°C are shown. All values were ≤ +/- 20%, and all resistance measurements were made at least twice, with at least duplicate determinations at each time point. The symbols denoting the $MnCl_2$ added to the sporulation medium are: (○) 0.5 µmol l^{-1}; (●) 1 µmol l^{-1}; (△) 10 µmol l^{-1}; and (▲) 100 µmol l^{-1}. Spores of *B. cereus* T (originally obtained from H.O. Halvorson) were prepared at 37°C in liquid medium with different levels of added $MnCl_2$, and harvested and purified as described (Stewart et al., 1981; Ghosh and Setlow, 2010). These spores were free (> 98%) of growing or sporulating cells, germinated spores and cell debris as determined by phase contrast microscopy. Prior to analyses of spore resistance, spore preparations were incubated for 1 h with 10 mM EDTA at 4°C, and then washed thoroughly with water and stored in water at 4°C protected from light. Measurements of spore killing by wet heat, dry heat, UVC radiation, γ-radiation in liquid, or hydrogen peroxide were all carried out as described (Ghosh et al., 2011; Granger et al., 2011).

2012).

Given the above facts, it is not clear why changes in Mn levels would have such different effects on the resistance of spores of *B. cereus*, *B. megaterium* and *B. subtilis* to wet or dry heat, UVC radiation or hydrogen peroxide. The differences in these effects is most striking in examining wet heat resistance, as a low Mn level has no effect on *B. subtilis* spore wet heat resistance, decreases *B. cereus* spore wet heat resistance slightly, and decreases *B. megaterium* wet heat resistance markedly (Figure 1e; Ghosh et al., 2011; Granger et al., 2011). Why might Mn levels alter or not alter spores' wet heat resistance? While there is no definitive answer for this question, we suggest the following, and highly speculative, scenario. Since wet heat killing of spores of *Bacillus* species is most likely by protein damage (Coleman et al., 2007, 2010), perhaps in *B. megaterium* spores increasing Mn levels protect some key protein whose inactivation by wet heat results in spore death. In *B. subtilis* spores this same protein might be more resistant to wet heat, such that it is wet heat damage to another protein, and one whose stability is insensitive to Mn levels that results in spore death. In *B. cereus* spores, perhaps cumulative wet heat damage to several proteins causes spore death, and Mn levels affect the stability of only one of these proteins.

Two other points are also worth noting on the effects of Mn levels on spore resistance. First, given the close relatedness between *B. cereus* and *B. anthracis* (Priest et al., 2004), it seems likely that Mn levels will also have minimal effects on the resistance of *B. anthracis* spores. Given the interest in the killing and resistance of *B. anthracis* spores, this is a conclusion with significant applied importance. Second, usual media for spore preparation are invariably supplemented with $MnCl_2$, and with concentrations ≥ 10 µmol l^{-1}. At least with spores of *B. cereus* and *B. subtilis*, this Mn concentration results in spores with maximal resistance to all agents examined. Consequently, one does not have to worry that sporulation in media with extremely high Mn levels will give spores of these species with abnormally elevated resistance.

ACKNOWLEDGEMENTS

This work was supported by grants from the Army Research Office (P.S.) and the Air Force Office of Scientific Research [FA9550-07-1-0218] (M.J.D.).

REFERENCES

Aoki H, Slepecky RA (1973). Inducement of a heat-shock requirement for germination and production of increased heat resistance in *Bacillus fastidiosus* spores by manganous ions. J. Bacteriol. 114:137-143.

Aoki H, Slepecky RA (1974). The formation of *Bacillus megaterium* spores having increased heat and radiation resistance and variable heat shock requirements due to manganous ions. In Spore Research 1973 ed. Barker AN, Gould, GW, Wolf J, Academic Press, London, pp. 93-102.

Avery SV (2011). Molecular targets of oxidative stress. Biochem. J. 434:201-210.

Coleman WH, Chen D, Li YQ, Setlow, P (2007). How moist heat kills spores of *Bacillus subtilis*. J. Bacteriol. 189:8458-8466.

Coleman WH, Zhang P, Li YQ, Setlow P (2010). Mechanism of killing of spores of *Bacillus cereus* and *Bacillus megaterium* by wet heat. Lett. Appl. Microbiol. 50:507-514.

Daly MJ (2012) Death by protein damage in irradiated cells. DNA Repair 11:12-21.

Donnellan JE Jr, Stafford RS (1968). The ultraviolet photochemistry and photobiology of vegetative cells and spores of *Bacillus megaterium*. Biophys. J. 8:17-27.

Gaidamakova EK, Myles IA, McDaniel DP, Fowler CJ, Valdez PA, Naik S, Gayen M, Gupta P, Sharma A, Glass PJ, Maheshwari RK, Datta SK, Daly MJ (2012). Preserving immunogenicity of lethally irradiated viral and bacterial vaccine epitopes using a radio-protective Mn^{2+}-peptide complex from *Deinococcus*. Cell Host Microbe 12:117-124.

Ghosh S, Ramirez-Peralta A, Gaidamakova E, Zhang P, Li Y-Q, Daly MJ, Setlow P (2011). Effects of Mn levels on resistance of *Bacillus megaterium* spores to heat, radiation and hydrogen peroxide. J. Appl. Microbiol. 111:663-670.

Ghosh S, Setlow P (2010) The preparation, germination properties and stability of superdormant spores of *Bacillus cereus*. J. Appl. Microbiol. 108:582-590.

Granger AC, Gaidamakova EK, Matrosova VY, Daly MJ, Setlow P (2011). Effects of levels of Mn and Fe on *Bacillus subtilis* spore resistance, and effects of Mn^{2+}, other divalent cations, orthophosphate, and dipicolinic acid on resistance of a protein to ionizing radiation. Appl. Environ. Microbiol. 77:32-40.

Hauser PM, Karamata D (1992). A method for the determination of bacterial spore DNA content based on isotopic labeling, spore germination and diphenylamine assay; ploidy of spores of several *Bacillus* species. Biochimie 74:723-733.

McEwan AG (2009). New insights into the protective effect of manganese against oxidative stress. Mol. Microbiol. 72:812-814.

Palop A, Rutherford GC, Marquis RE (1998). Inactivation of enzymes within spores of *Bacillus megaterium* ATCC 19213 by hydroperoxides. Can. J. Microbiol. 44:465-470.

Priest FG, Barker M, Baillie LWJ, Holmes EC, Maiden MCJ (2004). Population structure and evolution of the *Bacillus cereus* group. J. Bacteriol. 186:7959-7970.

Setlow P, Johnson EA (2012). Spores and their significance. In Food Microbiology, Fundamentals and Frontiers, 4th edition (Doyle MP, Buchanan R, eds.), p. 45-79. ASM Press, Washington, DC.

Stewart GSAB, Johnstone K, Hagelberg E, Ellar DJ (1981). Commitment of bacterial spores to germinate. Biochem. J. 198:101-106.

Yi X, Setlow P (2010). Studies of the commitment step in the germination of spores of *Bacillus* species. J. Bacteriol. 192:3424-3433.

Characterization of antibiotic resistance in environmental enteric pathogens from Kibera slum in Nairobi-Kenya

Muhonja Christabel[1]*, Nancy Budambula[1], John Kiiru[2] and Samuel Kariuki[2]

[1]Jomo Kenyatta University of Agriculture and Technology, P. O. Box 62000-00200, Nairobi, Kenya.
[2]Centre for Microbiology Research, Kenya Medical Research Institute. P. O. BOX 54840-00200, Nairobi, Kenya.

Kibera slum is characterized by poor sanitation hence the frequent outbreaks of diarrheal diseases. Emergence of antibiotic resistance by diarrhoea-causing bacteria is an inevitable challenge in the area. Diarrhea-causing bacteria were isolated from water, soil, vegetables, meat and dry foods samples and their antibiotic susceptibility was characterized. 237 samples were aseptically collected and analyzed. Morphological and biochemical characterization was done using Bergey's manual of determinative bacteriology as a reference. 174 *Escherichia coli*, 8 *Salmonella* and 6 *Shigella* isolates were identified. Drug susceptibility of the isolates was done using disk diffusion method where 9 antibiotics from 5 classes of antibiotics were used. The frequencies of resistant isolates to antibiotics were as follows: ampicillin (56.79%), trimethoprim+ sulphamethoxazole (29.63%), augmentin (27.16%), tetracycline (18.52%), streptomycin (13.57%), chloramphenicol (7.41%), nalidixic acid (4.94%), gentamycin (2.47%) and ciprofloxacin (0%). Polymerase chain reaction was done to amplify the antibiotic determinants *tet-A, sul 1* and *dfrb1.* Gel electrophoresis revealed presence of *tet-A* gene and *sul 1* genes but absence of *dfrb1* in 15 isolates that were resistant to tetracycline and SXT. Plasmid extraction and profiling was done and plasmid sizes compared to plasmids of *E. coli* V517 and 39R861. Large and small plasmids were present in the isolates with *tet-A* and *sul 1* genes. Plasmid of 63 bp was present in all isolates. Conjugation experiments confirmed 100% transfer of 63 bp plasmid and 90% complete phenotype in all cases. This study showed presence of contamination of the area by antibiotic resistant diarrhea-causing pathogens, hence proper hygiene procedures and adherence to correct drug prescriptions need to be reinforced.

Key words: Characterization, antibiotic resistance, susceptibility, polymerase chain reaction, transconjugants, phenotype, plasmids.

INTRODUCTION

Worldwide, an estimated one billion people live in informal urban settlements or "slums" (UN-Habitat, 2006). These settlements pose grave threats to the health of their inhabitants, stemming from poor-quality housing, lack of infrastructure and minimal access to refuse collection, health care or other essential services).

A study conducted in Kibera sub location in Nairobi showed that coliform levels for all the water samples tested was above the recommended limit. The mean fecal coliform count in outhouse water was 93/100 ml and 103.4/100 ml for in-house water, thus the water was potentially highly contaminated with enteric pathogens (Chemuliti et al., 2010). According to the Kenyan Ministry of Health Report (2003), diarrhea is the leading cause of death in the Kibera slum followed by malaria and respiratory diseases.

*Corresponding author. E-mail: muhosh@gmail.com.

Table 1. Antimicrobial families, genetic markers and primer sequences for resistance genes tested.

Antimicrobial family	Genetic marker	Primer sequence (Forward and Reverse)	Amplicon size (bp)	Genbank accession no	Reference
Tetracycline	*Tet(A)*	GTGAAACCCAACATACCCC (F) GAAGGCAAGCAGGATGTAG(R)	888	X00006	J. Harel
Trimethoprim	*Dhfrlb*	AGTATCATTGATAGCTGCG (F) GTAGTGCGCGAAGCGAAC (R)	517	DQ388123.1	J. Harel
Sulfonamides	*sul1*	TTCGGCATTCTGAATCTCAC(F) ATGATCTAACCCTCGGTCTC(R)	822	X12869	R. C. Levesque

According to Murray et al. (1999), the pathogenic enteric bacteria include organisms such as *Salmonella, Shigella, Vibrio* and *Escherichia coli.* Until the 1940s only *Salmonella* and *Shigella* were considered as gastrointestinal pathogens of medical importance in the family *Enterobacteriaceae.* It is now a well-established fact that *E. coli* is a significant cause of diarrheal illnesses both in infants and adults in many parts of the world (Kinge et al., 2010; Schierack et al., 2009; Kreig and Holt, 1984). Data on clinical isolates is plenty while less attention has been given to environmental isolation of these enteric pathogens (Sharma et al., 2010). Samples from the environment such as water, soil, meat, vegetables and dry foods are probable reservoirs of these pathogens. Studying antimicrobial resistance in humans and animals is important in order to detect changes in patterns of resistance, implement control measures on the use of antimicrobial agents, and to prevent the spread of multidrug-resistant strains of bacteria (Duijkeren et al., 2003).

The progressive increase in antimicrobial resistance among enteric pathogens particularly *Shigella, Vibrio cholerae,* Enteropathogenic *E. coli, Salmonella typhi* and *Salmonella enteriditis* species are becoming a critical concern worldwide, particularly in the developing world where there are high rates of diarrheal diseases which are associated with mortality (Sang et al., 2011; Fricke et al., 2008; Kumar et al., 2008). Surveillance data for antibiotic resistance are necessary to define or update guidelines for empirical treatment, as well as a guide for appropriate drug supplies.

MATERIALS AND METHODS

Characterization and identification of bacteria

A total of two hundred and twenty samples comprising of meat, water, vegetables, soil and dry food were aseptically collected from Gatwekera and transported to the laboratory in a cool box for analysis within 1 h of collection between 10th October 2010 and 1st February 2011. Meat was collected from 15 retail butcheries, vegetables were from 14 open air markets, soil and dry foods were also sampled from these open air markets. The samples were enriched in selenite-F broth overnight, before they were incubated on MacConkey and XLD agar for morphological characterization and subsequently biochemical characterization of the individual colonies. Biochemical tests were done by the use of TSI and API 20 E kit. Identification was done with reference to Bergey's manual of determinative bacteriology.

Antibiotic susceptibility testing

Eighty isolates were selected randomly from all sample types and subjected to antibiotic susceptibility testing, using the method described by Bauer et al. (1966). Cultures were tested for sensitivity to different antimicrobials selected from five classes of commonly used antibiotics namely, amynoglycosides, tetracycline, beta-lactams, sulphonamides, quinolones and chloramphenicol. Quality control was provided for by *E. coli* ATCC 25922. The following antibiotic disks were used: ampicillin (10 μg), augmentin (30 μg), streptomycin (10 μg), chloramphenicol (30 μg), ciprofloxacin (5 μg), nalidixic acid (30 μg), tetracycline (30 μg), trimethoprim + sulfamethoxazole (25 μg) and gentamicin (10 μg).

PCR amplification of antibiotic resistance genes

The polymerase chain reaction for genes encoding resistance to Sulfonamides, Tetracycline and Trimethoprim was done according to Sambrook and Russel (2007) with some modifications. PCR amplification was performed with 2 μl of the DNA template. The PCR conditions were 30 cycles of amplification at a denaturation temperature of 95°C for 1 min, at an annealing temperature of 55°C for 1 min and an extension temperature of 72°C for 3 min. This step was followed by a final extension at 72°C for 10 min. PCR products were resolved on 1% agarose gels, stained with ethidium bromide, visualized under UV light and photographed. The primer sets for the genes for antibiotic resistance that were targeted are shown in Table 1.

Plasmid extraction

Plasmid extraction and profiling for the antibiotic resistant strains were done by the alkaline lysis technique (Birnboim and Doly, 1979). Plasmid sizes were determined by comparison with plasmids of known molecular sizes from *E. coli* strain V517 which bears plasmids of molecular weight 35.8, 4.8, 3.7, 2.6, 2.0, 1.8, 1.4 MDa and strain 39R861 which has plasmids of molecular weight 98, 42, 24, and 4.6 MDa on 1% horizontal agarose gels.

Table 2. A summary of the biochemical reactions carried out on the four different isolates that were identified in this study.

Biochemical reaction (test)	Isolate identity			
	E. coli	*Salmonella typhymurium*	*Salmonella typhi*	*Shigella flexneri*
TSI	A/Ag	K/A	K/A	K/A
H_2S (TSI)	–	+	+	–
Motility	+	+	+	–
Indole production	+	–	–	–
Urease activity	–	–	–	–
Cirate utilization	–	+	–	–
ONPG	+	–	–	–
LDC	+	+	+	–
ODC	–	+	–	–
Mannitol	+	+	+	+
Sorbitol	+	+	+	–
Rhamnose	+	+	–	–
Melibiose	+	+	+	–
Arabinose	+	+	–	–

TSI, Triple sugar iron; A/Ag, acid slant and butt + gas; K/A, alkaline slant, acid butt, no gas; H_2S, hydrogen sulfide gas; LDC, lysine decarboxylase; ONPG, ortho-nitrophenyl-beta-D galactopyranoside; ODC, ornithine decarboxylase.

Conjugation experiments

Attempts to transfer antibiotic resistance determinants from the resistant bacteria to susceptible recipients were done (Finlay and Falkow, 1988). In each case, appropriate antibiotics were used to select for transconjugants on Mueller Hinton agar. The mating assays were carried out using the nalidixic acid-resistant *E. coli* K12 strain as the recipient. In order to confirm the transfer of the resistance phenotype from donor to recipient, antibiotic susceptibility of the transconjugants was performed using the same panel of antibiotic disks used for the donor and similarly, plasmid extraction and profiling of transconjugants was done to ascertain transfer of R-plasmids.

Data management and analysis

Data was recorded in lab workbooks and later entered in excel worksheets and then analyzed using SPSS package. The output of the analysis was presented in form of frequency tables, bar charts and gel photographs.

RESULTS

Characterization and identification of isolates

After morphological and biochemical characterization of the isolates, 174 *E. coli*, 8 *Salmonella* and 6 *Shigella* were identified. The biochemical reactions that led to identification of the isolates under this study was summarized and presented in Table 2.

Sample source contamination

The highest frequency of contamination was seen in dry foods (93.75%), meat (92.68%), soil (88.90%) while the least contaminated sources were water (76.10%) and vegetables (60.71%) as shown in Table 3.

Antibiotic susceptibility outcomes

The highest frequencies of resistant isolates to various drugs were as follows: ampicillin (56.8%), sulphamethoxazole-trimethoprim (29.63%), augmentin (27.60%) and tetracycline (18.52%). It was also noted that the frequency of resistance to specific antibiotics was not dependent on the sample type from which the isolate was sourced. There was no trend of variation in the number of resistant isolates from certain sample types from one antibiotic to another. This is shown in Table 4.

Out of the 80 isolates that were screened for antibiotic susceptibility, 18.75% showed multiple antibiotic resistance, they resisted 3 or more antibiotics while 81.25% were not multiple antibiotic resistant. The resistance frequencies versus number of drugs to which they were resistant was as follows: 0 drugs (17.50%), 1 drug (42.50%), 2 drugs (21.25%), 3 drugs (8.75%), 4 drugs (3.75%), 5 drugs (2.50%), 6 drugs (2.50%), 7 drugs (1.25%), 8 drugs (0.00%) and 9 drugs (0.00%).

Molecular characterization of the antibiotic resistance

DNA from twenty four isolates that were resistant or gave intermediate outcomes to tetracycline and SXT was subjected to PCR with primers *sul* 1 and *tet A*. The 24 isolates were grouped into two groups of 12 each and DNA from each group was taken through two monoplex

Table 3. The frequency of sample source contamination for the five sources from which the samples were collected.

Source	Samples collected	Contaminated samples	Frequency of contamination (%)
Dry foods	32	30	93.75
Meat	41	38	92.68
Soil	45	40	88.90
Vegetables	56	34	60.71
Water	46	35	76.10
Total	220	180	256.30

Table 4. Frequencies of the resistant isolates obtained from the various environmental samples collected from Gatwekera village of Kibera slums to nine different antibiotics.

Drug	% of resistant isolates (N=80)	Source	*E. coli*	*Salmonella* spp	*Shigella* spp
CIP	0.00% (n=0)	Soil	0	*Salmonella* spp (0)	*Shigella* spp (0)
		Meat	0	*Salmonella* spp (0)	*Shigella* spp (0)
		Water	1	*Salmonella* spp (0)	*Shigella* spp (0)
		Vegetables	0	*Salmonella* spp (0)	*Shigella* spp (0)
		Fish	0	*Salmonella* spp (0)	*Shigella* spp (0)
AMP	56.25% (n=45)	Soil	7	*Salmonella typhymurium*(2)	*Shigella* spp (0)
		Meat	5	*Salmonella typhi* (1)	*Shigella* spp (0)
		Water	11	*Salmonella* spp (0)	*Shigella flexneri* (1)
		Vegetables	11	*Salmonella typhymurium*(2)	*Shigella* spp (0)
		Fish	5	*Salmonella* spp (0)	*Shigella* spp(0)
SXT	30.00% (n=24)	Soil	6	*Salmonella* spp (0)	*Shigella* spp(0)
		Meat	4	*Salmonella* spp (0)	*Shigella* spp(0)
		Water	8	*Salmonella typhi* (1)	*Shigella* spp(0)
		Vegetables	5	*Salmonella* spp (0)	*Shigella* spp(0)
		Fish	0	*Salmonella* spp (0)	*Shigella* spp (0)
S	15.00%(n=12)	Soil	0	*Salmonella* spp (0)	*Shigella* spp(0)
		Meat	2	*Salmonella typhymurium* (1)	*Shigella* spp(0)
		Water	2	*Salmonella typhymurium* (1)	*Shigella* spp(0)
		Vegetables	3	*Salmonella typhymurium* (2)	*Shigella* spp(0)
		Fish	1	*Salmonella* spp(0)	*Shigella* spp(0)
AMC	27.50% (n=22)	Soil	4	*Salmonella typhymurium* (1)	*Shigella spp*(0)
		Meat	3	*Salmonella* spp(0)	*Shigella spp*(0)
		Water	6	*Salmonella typhymurium* (1)	*Shigella spp*(0)
		Vegetables	5	*Salmonella typhymurium* (1)	*Shigella* spp(1)
		Fish	1	*Salmonella* spp(0)	*Shigella spp*(0)
TET	18.75%(=15)	Soil	2	*Salmonella* spp(0)	*Shigella* spp(1)
		Meat	3	*Salmonella typhymurium* (1)	*Shigella* spp(0)
		Water	4	*Salmonella typhymurium* (1)	*Shigella spp*(0)
		Vegetables	1	*Salmonella spp*(0)	*Shigella* spp(0)
		Fish	1	*Salmonella typhymurium* (1)	*Shigella* spp(0)
NA	3.75% (n=3)	Soil	0	*Salmonella* spp(0)	*Shigella* spp(0)

Table 4. Continued

		Meat	1	*Salmonella* spp(0)	*Shigella* spp(0)
		Water	0	*Salmonella* spp(0)	*Shigella* spp(0)
		Vegetables	0	*Salmonella* spp(0)	*Shigella* spp(1)
		Fish	1	*Salmonella* spp(0)	*Shigella* spp(0)
C	7.50%(n=6)	Soil	1	*Salmonella* spp(0)	*Shigella* spp(0)
		Meat	1	*Salmonella* spp(0)	*Shigella* spp(0)
		Water	3	*Salmonella* spp(0)	*Shigella* spp(0)
		Vegetables	1	*Salmonella* spp(0)	*Shigella* spp(0)
		Fish	0	*Salmonella* spp(0)	*Shigella* spp(0)
CN	2.50%(n=2)	Soil	0	*Salmonella* spp(0)	*Shigella* spp(0)
		Meat	1	*Salmonella* spp(0)	*Shigella* spp(0)
		Water	0	*Salmonella* spp(0)	*Shigella* spp(0)
		Vegetables	0	*Salmonella typhymurium* (1)	*Shigella* spp(0)
		Fish	0	*Salmonella* spp(0)	*Shigella* spp(0)

AMC, Amoxicillin; C, Chloramphenicol; Te, Tetracycline; Amp, Ampicillin; SXT, Sulphamethoxazole+Trimethoprim, S, Streptomycin; NA, Nalidixic acid; CIP, Ciprofloxacin; CN, Gentamycin.

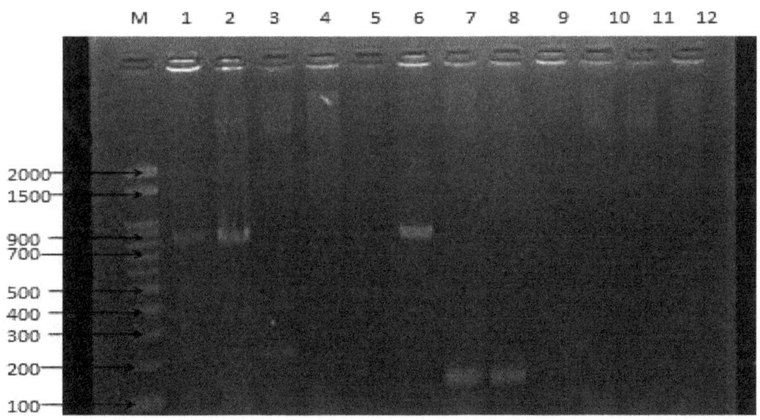

Plate 1. *Sul 1* genes from 12 isolates which were resistant to sulphamethoxazole + trimethopim. Lanes 7 and 4 are for *Salmonella typhimurium* while lanes 2, 3, 5, 6, 8, 9, 10, 11 and 12 are *E. coli*. M is the molecular marker, Hyper ladder (band sizes are in base pairs).

PCRs using the two antibiotic determinants.

In Plate 1, DNA from 12 isolates was amplified for the gene *sul 1.*Only lanes 1, 2 and 6 were positive for the gene encoding sulphonamides resistance (*sul 1)* while lanes 3, 4, 5, 7, 8, 9, 10, 11 and 12 were negative for the gene.

In Plate 2, DNA from 11 isolates was amplified for the gene *Tet A.* Lanes 2, 3 and 4 were positive for the gene encoding tetracycline resistance; *tet A* while lanes 1, 5, 6, 7, 8, 9, 10, 11 and 12 were negative for the gene.

In Plate 3, DNA from 12 isolates was amplified for the gene *Tet A.* Lanes 5, 6, 7, 8, 9, 10, 11, and 12 were positive for the gene encoding tetracycline resistance; *tet A* while lanes 1, 2, 3 and 4 were negative for the gene.

In Plate 4, DNA from 12 isolates was amplified for the gene *sul 1.*Only lanes 9, 10, 11 and 12 were negative for the gene encoding sulphonamides resistance (*sul 1)* while lanes 1, 2, 3, 4, 5, 6, 7 and 8 were positive for the gene.

Fifteen isolates that previously were positive for genetic determinants *sul* 1 and *tet A* or one of the two exhibited the presence of at least one plasmid. All of the fifteen isolates had the plasmid of size 43 bp. This is shown in Plate 5.

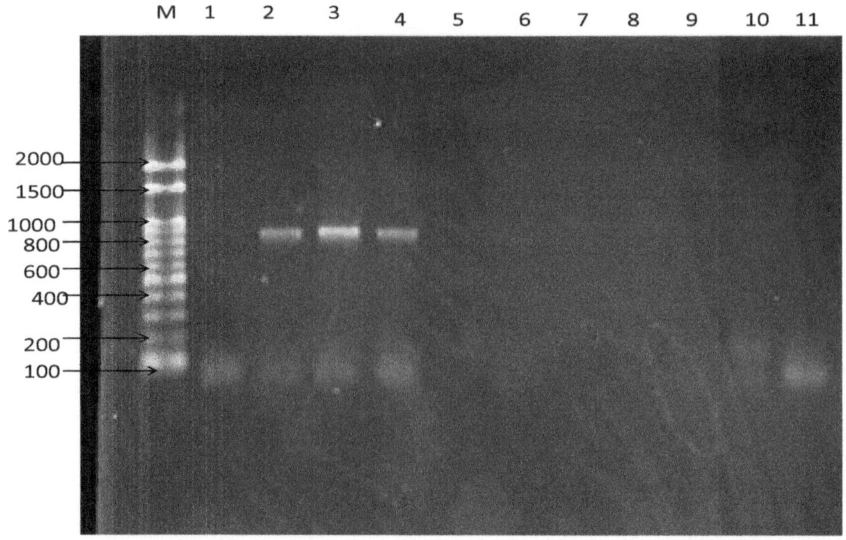

Plate 2. *Tet A* genes for 11 isolates which were resistant to tetracycline. Lanes 7 and 4 are for *Salmonella typhymurium*. 1, 2, 3, 5, 6, 8, 9, 10, 11 while 12 are for *E. coli*. M is the molecular marker, Hyper ladder (band sizes are in base pairs).

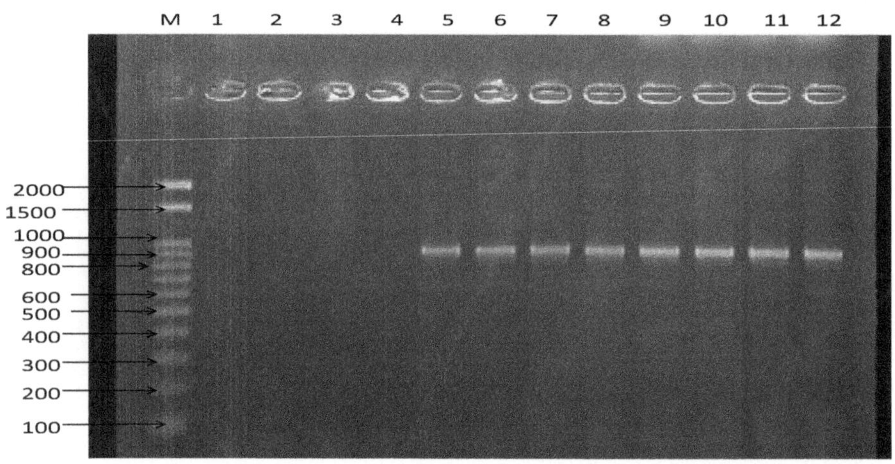

Plate 3. The antibiotic resistance determinant *tet A* genes for 12 isolates that were resistant to tetracycline. Lanes 3, 5 and 11 are for *Salmonella typhimurium* while lanes 1, 2, 4, 6, 7, 8, 9, 10 and 12 are for *E. coli*. M is the molecular marker, Hyper ladder (band sizes are in base pairs).

Conjugation outcomes

Conjugation experiments were done with 15 isolates that had plasmids that were resistant to at least ampicillin and 10 of them had successful transconjugants. Resistance to tetracycline and SXT was transferred in all cases. In 90% of all the successful transconjugants, the donor phenotype and recipient phenotype remained the same. Upon plasmid gel profiling of the transconjugants as shown in Plate 6, it was noted that there was plasmid transfer to all the transconjugants except one which did not have any plasmid at all after the mating experiments.

DISCUSSION

The prevalence of the three microorganisms identified varied as follows; *E. coli* (91.58%), *Salmonella* (4.21%) and *Shigella* (3.16%). *E. coli* therefore had the highest prevalence in the studied samples followed by *Salmonella* then *Shigella*. Data obtained indicated that foods (meat, vegetables and other dry foods), are potential reservoirs for many pathogenic organisms which were shown to be resistant to many antimicrobials, suggesting a potential public health hazard. These results emphasize the need to implement proactive measures,

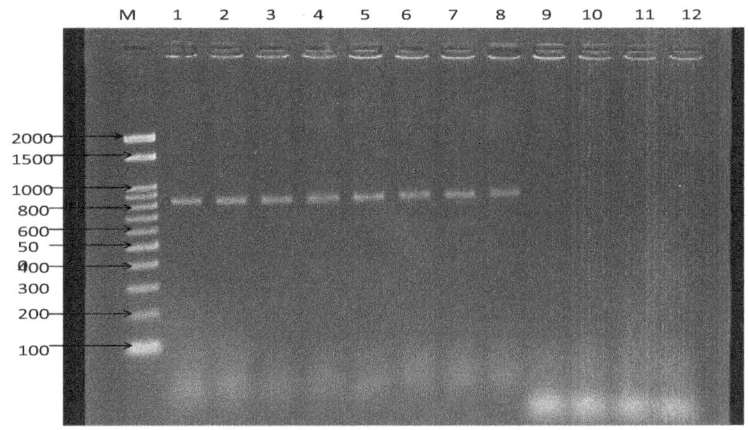

Plate 4. The resistance determinant *sul 1* genes for 12 isolates that were resistant to SXT. Lanes 3, 5 and 11 represent *Salmonella typhimurium*..

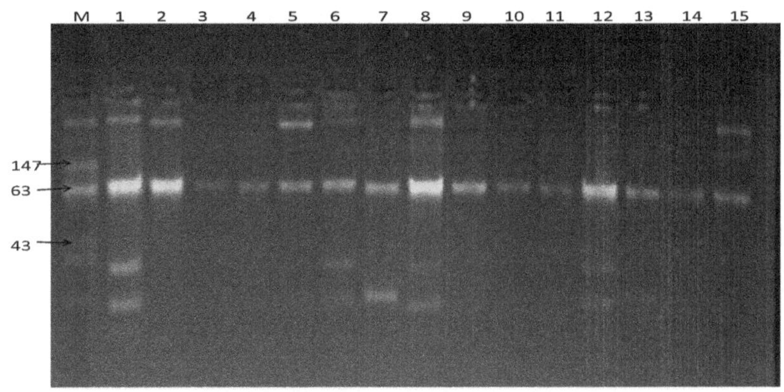

Plate 5. Plasmid DNA obtained from 15 isolates that were confirmed to have both *tet A* and *Sul 1* genes from PCR. M is the plasmid size marker *E. coli* strain 39R861 with band sizes 98, 42, 24 and 4.6 MDa. Lanes 5, 6, 9, 10 and 13 are for *Salmonella* ser Typhimurium while 1, 2, 3, 4, 7, 8, 11, 12, 14 and 15 are for *E. coli* isolates (band sizes are in base pairs).

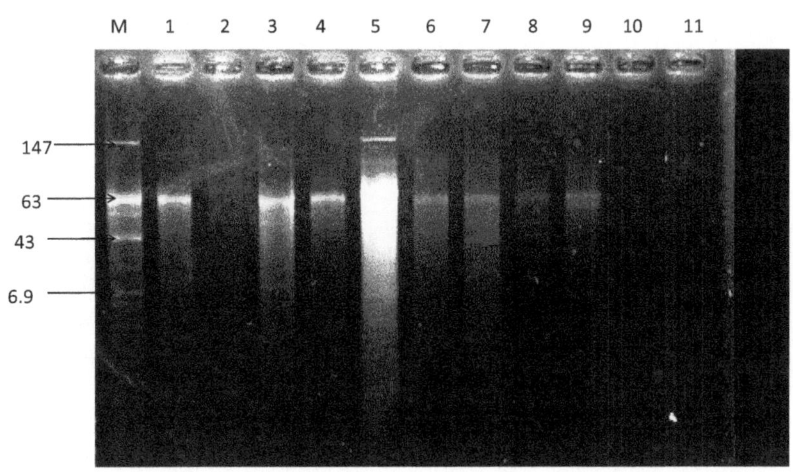

Plate 6. Plasmid DNA extracted from 10 successful transconjugants. M is the molecular marker *E. coli* R39, 7, 3 and 1 are *Salmonella* ser Typhymurium while 2, 4, 5,6,8,9 and 10 are *E. coli* (band sizes are in base pairs).

and more emphasis to be placed on the application of hygienic practices to reduce the risk of infection. Epidemiological reports suggest that meat products are a major agent for the spread of diarrheal illnesses (FAO/WHO, 2002). The isolates were tested for their susceptibility to nine antibiotics that are routinely prescribed for treatment and management of diarrheal infections that is SXT, ampicillin, tetracycline, chloramphenicol, gentamycin, nalidixic acid, ciprofloxacin, streptomycin and augmentin (USPHS/IDSA, 2003). 60% of the isolates were resistant to at least one antimicrobial while the rest (40%) were resistant to two or more antimicrobials (multiple antibiotic resistant). Drug susceptibility tests revealed that the isolates resistance to ampicillin (56.79%), sulfamethoxazole + trimethoprim (29.63%), augmentin (27.60%) and tetracycline (18.52%) were the highest just as was also concluded by Varga et al. (2008).

It also revealed that the most effective drugs to which the microorganisms were most sensitive were ciprofloxacin, gentamycin, chloramphenicol and nalidixic acid. These findings are in agreement with Mengo et al. (2010) where they concluded that the antibiotics that form the mainstay of therapy for typhoid patients in developing countries are ampicilin, cotrimoxazole and chloramphenicol but due to the increasing resistance to anti-bacterials used traditionally for therapy, the use of fluoroquinolones, such as ciprofloxacin and ofloxacin, for the treatment of typhoid has become more common. Gentamycin and ciprofloxacin were the most effective antibiotics in our study. 96.30% of the isolates tested were sensitive to gentamycin while 98.6% of all the isolates tested were sensitive to ciprofloxacin. Gentamycin is rarely abused because of its mode of administration which is mostly intramuscular.

This explains the near zero resistance to this antibiotic, as it cannot be easily abused. Gentamycin was introduced in 1969 as a broad spectrum amynoglycoside which was effective *in vitro* against most gram negative bacilli (Sierbert et al., 1979). Musoke and Revathi (2000), and Kariuki et al. (1997) also isolated gentamycin resistant enteric bacteria. Ciprofloxacin is one of the most expensive antibiotics usually prescribed for typhoid treatment. For all episodes of diarrhea with blood in the stool, the WHO currently recommends treatment with ciprofloxacin or one of the three second-line antibiotics, pivmecillinam, azithromycin and ceftriaxone (WHO, 2007). This has resulted in a cure rate of 99% (Traa et al., 2011). The frequency of resistant isolates varied widely among the different samples for different antibiotics. For instance meat had the highest number of resistant isolates against nalidixic acid, water had the highest number of resistant isolates against chloramphenicol while vegetables had the highest number of resistant isolates against ampicillin. The nature of sample thus did not affect the frequency of resistance.

The gel profiling of the PCR products revealed that the genes for resistance to sulphonamides and tetracycline are available in most of the phenotypically resistant isolates. The antibiotic resistance in those isolates which seem not to possess plasmids was associated with chromosome and/or transposons instead of being plasmid-mediated. This implies that there is no consistent relationship between antibiotic resistance pattern and the number of plasmid bands present. Such findings were also reported by Yah and Eghafona (2008). Sulphamethoxazole in this case was used in combination with trimethoprim but the gene for resistance to trimethoprim was absent in the resistant isolates. This phenomenon indicated the existence of a plasmid carrying multiple resistance to antibiotics. Plasmids were detected virtually in all the MDR isolates. Hence, plasmid mediated resistance is evident in our study. These findings are in agreement with Leversteine et al. (2002b). Plasmids are widespread among enteric bacteria of meat production animals and these emergent plasmids have flexibility in their acquisition of MDR-encoding modules, necessitating further study to understand the evolutionary mechanisms involved in their dissemination and stability in bacterial populations (Fernandez et al., 2011).

Conjugation studies showed the presence of self-transmissible plasmids that were responsible for the transfer of resistance phenotype to the recipient (*E. coli* K12). Resistance to sulphamethoxazole and tetracycline among other antibiotics was shown to be transferred to the recipient through conjugation. This shows that besides being chromosomally encoded, sulphamethoxazole and tetracycline resistance is also plasmid mediated. The acquisition of a new gene may occur by genetic transformation or through mobilization by conjugative transfer. The latter may occur at high frequency and efficiency, and several resistance genes can be acquired simultaneously (Carattoli, 2003). Leverstein et al. (2002a) reported the possibility of transfer of a complete resistance phenotype as is evidenced in 90% of isolates in this study. The high percentage of *E. coli* and *Salmonella* isolates that were MAR to tetracycline, ampicillin, and sulphamethoxazole + trimethoprim, suggested that there has been a misuse of these drugs, which has resulted in these environmental samples that is water, vegetables, meat, soil and dry foods posing a potential threat to humans in the area.

Of interest, ampicillin or augmentin, are not appropriate empiric therapy for common enteric bacteria such as *Shigella* spp. and *Salmonella* spp. because of the high frequency of resistance. The high antibiotic resistance also indicates a negative impact on therapy with these classes of antibiotics, which with time may be rendered completely unhelpful. According to the results of the conjugation experiments, resistance can be fully transmitted from one organism to the other, hence enteric pathogens can transfer resistance to susceptible organisms that share the same environment. Plasmid analysis of the transconjugants showed that the antibiotic sensitive *E. coli K12* isolates could acquire the R-plasmid from other enteric bacteria such as *E. coli* for a suitable adaptation for survival in the changing antibiotic environment.

Conclusion

The high percentage of *E. coli* and *Salmonella* isolates that were MAR to tetracycline, ampicillin, and sulphamethoxazole + trimethoprim suggested that there has been a misuse of these drugs, which has resulted in these environmental samples that is water, vegetables, meat, soil and dry foods posing a potential threat to humans in the area. According to the results of the conjugation experiments, resistance can be fully transmitted from one organism to the other, hence enteric pathogens can easily transfer resistance to susceptible organisms that share the same environment. Plasmid analysis of the transconjugants showed that the antibiotic sensitive *E. coli K12* isolates could acquire the R-plasmid from other enteric bacteria such as *E. coli* for a suitable adaptation for survival in the changing antibiotic environment.

AKNOWLEDGEMENTS

I greatly appreciate the contribution made by the Kenya Medical Research Institute in funding this project and the staff of the bacteriology lab, for assisting during the benchwork which has led to the success of this work.

REFERENCES

Bauer A, Kirby W, Sherris WM, Turck M (1966). Antibiotic susceptibility testing by a standardized single disk method. Am. J. Clin. Pathol. 45:493-496.

Birnboim HD, Doly J (1979). A rapid alkaline Extraction procedure for screening recombinant plasmid DNA. Nucleic Acids Res. 7:1513-1523.

Carattoli A (2003) Plasmid-Mediated Antimicrobial Resistance in *Salmonella enterica.* Curr. Issues Mol. Biol. 5:113-122.

Chemuliti JK, Gathura PB, Kyule MM, Njeruh M (2010). Bacteriological qualities of indoor and outdoor drinking water in Kibera sub-location of Nairobi, Kenya. East Afr. Med. J. 79:5.

Duijkeren EV, Wannet WJB, Houwers DJ, Pelt WV (2003). Antimicrobial susceptibility of *Salmonella* strains isolated from humans, cattle, pigs, and chickens in the Netherlands from 1984 to 2001. J. Clin. Microbiol. 41:(8):3574-3584.

FAO/WHO (2002). Global Forum of Food Safety Regulators. Improving Efficiency and Transparency in Food Safety Systems Sharing Experiences, Marrakesh, Morocco: www.fao.org/docrep/meeting/004/Y3680E/Y3680E02.htm

Fernandez AC, Singer RS, Johnson TJ (2011). Comparative Genomics of Multidrug Resistance-Encoding IncA/C Plasmids from Commensal and Pathogenic *Escherichia coli* from Multiple Animal Sources. PLoS ONE: 6:8.

Finlay BB, Falkow S (1988). Comparison of the invasion strategies used by *Salmonella cholerae-suis*, *Shigella flexneri* and *Yersinia enterocolitica* to enter cultured bacterial cells: Endosome acidification is not required for bacterial invasion or intracellular replication. J. Biochimie 70:1089-1099.

Fricke WF, Meredith SW, Lindell AH, Harkins DM, Craig B, Ravel J, Ramunas S (2008). Insights into the environmental resistance gene pool from the genome sequence of the multidrug-resistant environmental isolate *Escherichia coli* SMS-3-5. J. Bacteriol. 190:6779-6794.

Kariuki SM, Gilks CF, Kimari J, Muyodi J, Waiyaki P, Harts CA (1997). Plasmid diversity of multi-drug resistant *Escherichia coli* isolated from children with diarrhea in poultry farming area in Kenya. Ann. Trop. Med. Parasitol. 91:87-94.

Kinge W, Constance N, Njie AC, Kawadza DT (2010). Antibiotic resistance profiles of *Escherichia coli* isolated from different water sources in the Mmabatho locality, North-West Province, South Africa. South Afr. J. Sci. 106:1-2.

Leverstein-van HM, Wang MD, Schael F, Jiang PI, O'Brien TF (2002a). Multi-drug resistance among Enterobacteriaceae is strongly associated with the presence of integrons and is independent of species or origin. J. Infec. Dis. 187:251-259.

Leverstein-van HM, Adriene MB, Hetty B, Armand P, Jan V (2002b). Evidence of extensive interspecies transfer of integron mediated antimicrobial resistance genes among Enterobacteriaceae in clinical settings. J. Infec. Dis. 186:49-56.

Mengo MD, Kariuki SM, Muigai AW, Revathi NG (2010). Trends in *Salmonella enterica serovar Typhi in* Nairobi, Kenya from 2004 to 2006. J. Infect. Dev. Ctries. 4(6):393-396.

Murray PR, BaronEJ, Pfaller MA, Tenovar MC, Yolked RH (1999). Manual of Clinical Microbiology. ASM Press. Washington.

Musoke RN, Revathi BG (2000). Emergence of multidrug-resistant gram-negative organisms in a neonatal unit and the therapeutic implications. J. Trop. Paediatr. 46:86-91.

Sambrook J, Russell DW (2007). Molecular cloning: a laboratory manual, 3rd (Ed). Cold Spring Harbor Laboratory Press, Cold Spring Harbor, NY.

Sang WK, Kariuki SM, Boga HI, Waiyaki PG, Wamae CN, Schnabel D (2011). Antibiotic susceptibility patterns of enteric pathogens from the Maasai community, Narok and Kajiado districts, Kenya. Afr. J. Health Sci. 1:9.

Schierack P, Römer A, Jores J, Heike K, Sebastian G, Matthias F, Jürgen E, Lothar HW (2009). Isolation and characterization of intestinal *Escherichia coli* clones from wild boars in Germany. Appl. Environ. Microbiol. 75:695-702.

Sharma A, Kumar S, Divya B (2010). Phenotypic and genotypic characterization of *Shigella* spp. with reference to its virulence genes and antibiogram analysis from River Narmada. Natl. Res. Counc. Canada. 165:33-42.

Traa SB, Fischer LC, Munos M, Black ER (2011). Antibiotics for the treatment of dysentery in children. Int. J. Epidemiol. 39:70-74.

UN–Habitat (2006). State of the World's Cities 2006–2007, Earthscan, London, p. 202.

USPHS/IDSA (2003). Guidelines for prevention of opportunistic infectionsin persons infected with human immunodeficiency virus.

Varga C, Andrijana R, Margaret EM, Richard JR, Anne ED, David LP, Brent PA, Sylvia LC, Scott A (2008). Comparison of antimicrobial resistance in generic *Escherichia coli* and *Salmonella* spp. cultured from identical fecal samples in finishing swine. Can. J. Vet. Res. 72(2):181-187.

Yah SC, Eghafona NO (2008). Plasmids: A vehicle for rapid transfer of antibiotic resistance markers of *Salmonella* species in animals. Int. J. Integr. Biol. 2(1):55.

Molecular detection of *Chlamydia trachomatis* among gynecological patients attending Khartoum Teaching Hospital

Mohammed A. Mohammed[1]* and Al Fadhil A. Omer[2]

[1]Department of Medical Microbilogy, Faculty of Medicine, Medical Institute Center, Al Neelain University, Khartoum, Sudan.
[2]Department of Medical Microbilogy, Faculty of Medical Laboratory Science, Al Neelain University Khartoum, Sudan.

To perform moleular detection of *Chlamydia trachomatis* among gynecological patients attending Khartoum Teaching Hospital. 200 endocerical smears were collected randomly from out-patient infertile women attending the Department of Obsetrcs and Gyencology, Khartoum Teaching Hospital (Sudan). Detection of *C. trachomatis* was performed by the polymerase chain reaction (PCR) technique. About 191 out of 200 patients having vaginal discharge were investigated, 43 (22.5%) of them were found positive for *C. trachomatis* by PCR test. Most positive cases were in the age range 26-40 years, whereas negative cases were in the age range 61 years and above. The age range (20-40) years, is a risk factor that exposes sexually-active women to *C. trachomatis* infection. Also, *C. trachomatis* infection in pregnant women may be associated with abortion and pelvic inflammatory disease. *C. trachomatis* infection is becoming a public health problem in Khartoum due to its highest frequency among infertile Sudanese women.

Key words: *Chlamydia trachomatis*, polymerase chain reaction, Sudanese women.

INTRODUCTION

Chlamydia trachomatis is one of the most common sexually transmitted bacterial infection in the world and sexually active young persons are at highest risk. The inciedence of chlamydia infection in women with tuboperitareal sterility is approximatlely 80%. Clinically inflammotary disease of the pelvic organs in such an infection is characterised by primary chronic course and involvement of the cervix (Akush, 1993). *C. trachomatis* is the most common bacterial sexually transmitted infection (STI) in many developed countries. The highest prevalence rates are found among young adults who have frequent partner change rates (Christian et al., 2012).

In Sudan there is no recent published data about the prevalence of *C. trachomatis* infection among infertile and pregnant women, however, there is highly sensitive and specific techniques have been developed for diagnosos of *C. trachomatis*. *C. trachomatis* infection have been reported to cause silent infections (asymptomatic) in communities which becomes endomic and could remain unnoticed for very long time. In most parts of Sudan these organisms are not screened for, and hence relative infermation about frequencies of the organisims are sparse (Okoror et al., 2007). The aim of this study is to perform moleular detection of *C. trachomatis* among gynecological patients attending Khartoum Teaching Hospital.

MATERIALS AND METHODS

This is a facility based qualitative study to determine the prevalence

*Corresponding author. E-mail: msud22@hotmail.com.

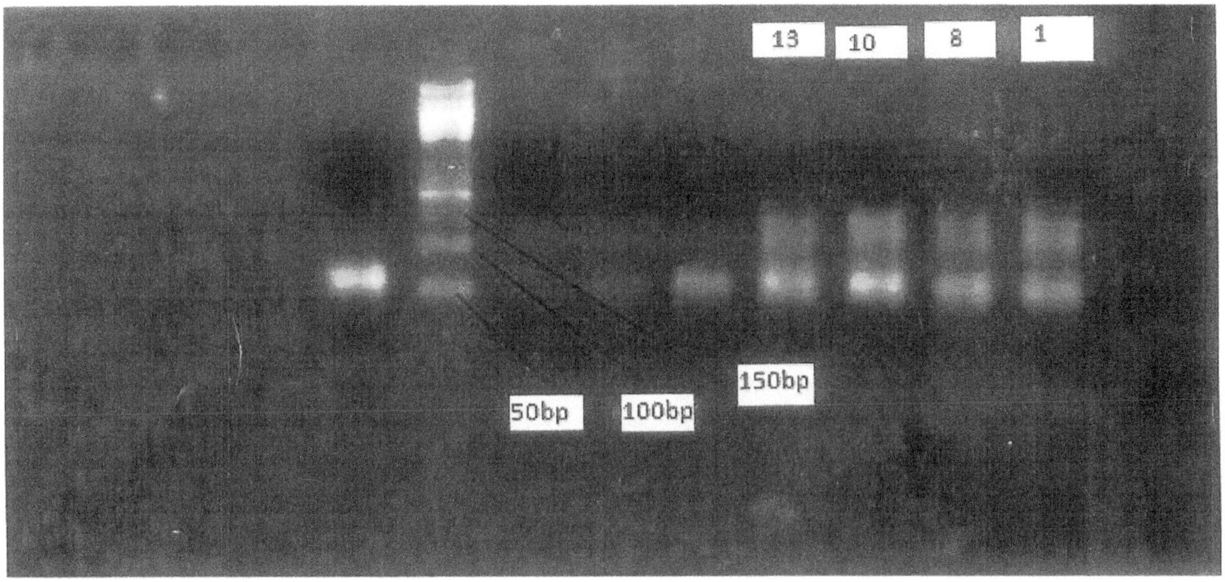

Figure 1. *C. trachomatis* PCR result.

of *Chlamydia* infection among Sudanese women. It is a descriptive cross-sectional study to determine the frequency of this infection. The study was carried out in two hundred women, attending Khartoum Maternity Hospital (Sudan).

DNA extraction

Endocevical samples were collected in 1 ml lysis buffer containing 5 µl proteinase K, 500 µl guanidine chloride, and 150 µl NH_4 acetate. All were mixed and incubated at 37°C overnight. The mixture was boiled, cooled to room temperature, and then 2 ml pre-chilled chloroform were added. It was then vortexed and centrifuged for 5 min at 3000 rpm. The upper layer was transferred to another tube and 6 ml of cold absolute ethanol were added, shaken, and kept at -20°C for at least 2 h. After that it was centrifuged at 3000 rpm for15-20 min, the supernatant was carefully drained, and the tube was inverted on a tissue paper for 5 min. Then the pellet was washed with 4 ml 70% ethanol. Another centrifugation at 3000 rpm for 15 min was made, and the supernatant was poured off and the pellet was allowed to dry for 10 min. Later the pellet was re-suspended in 200 µl deionized water, briefly vortexed and kept at 40°C overnight. Lastly the DNA extract was aliquoted as a stock solution and stored at –20°C (Fallah et al., 2005).

C. trachomatis primers such as CT1: (CCT/GTG/GGG/AAT/ GCT/GCT/GAA) and CT4: (GTC/GAA/AAC/AAA/GTCATCCAGTA/ GTA were drived from highly conserved regions of the published DNA sequences for *C. trachomatis* species were used in this study.

Polymerase chain reaction (PCR) analysis

Two microliter of DNA extracts were processed in a 30 µl reaction volume containing PCR buffer (10 Mm Tris , 50 Mm KCL , 0.01% gelatin 200 µM deoxynucleoside triphosphate, 2.5 mM $MgCl_2$, 0.5 µM of each primer, and 1 U of tag polymerase. The first cycle, consisting of 5 min denaturation at 94°C, followed by 35 cycles each of 30 s at 94°C, 45 s at 56°C and 1 min at 72°C with a final extension for 10 min at 72°C. The PCR products were visualized in 2% agarose gels containing 0.5 µg of ethidium bromide/ml. The edges of the product were sealed carefully, using sealing tape, and then 1.5% agarose gel was prepared by adding 54 ml DW, 6 ml 10x tris base boric acid EDTA, 90 µm tris borate, and 2 µm EDTA (pH 8.0) to 0.9 g agarose gel. The mixture was melted in a boiling water bath, then it was cooled to 60°C, and 3 µm of ethidium bromide (10 mg/ml) were added to the gel, and mixed thoroughly. The mixture was then poured into horizontal electrophoresis gel tank with suitable size combs, and the gel was left for 30-45 min to polymerize. Running buffer was added containing 250 ml distilled water and 15 ml buffer to cover the gel. 7 µl of PCR product was located into the comb wells, and 5 µl of DNA marker were also located. The run was performed at 100 Volt, and current range was 3-8 ml Ampere for 30 min. The gel was visualized over ultraviolet trans-illuminator and photographed using gel documentation system. Fragements size was estimated from the distance of migration velative to the positive control (Fallah et al., 2005). We used positive control of *C. trachomatis* with length of 50 bp to determined the size of the amplified DNA.

RESULTS

A total of 200 endocerical smears were collected randomly from out-patient infertile women attending the Department of Obstetrics and Gyencology (Khartoum Teaching Hospital). The endocervical smears were tested for *C. trachomatis* using PCR technique (Figure 1). From the 200 patients investigated, 191 were presenting with vaginal discharge, and 43 (51.2%) of them were found positive for *C. trachomatis* by PCR technique. Most positive cases were in the age range 26- 40 years, where as negative cases were aged 61 years and above (Table 1).

Nine patients without vaginal discharge were investigated for *C. trachomatis*, 5 of them (55. 5%) were found positive for *C. trachomatis* by PCR technique (Table 2).

Table 1. Detection of *Chlamydia trachomatis* among infertile vaginal discharge women according to age incidence.

Age range (years)	Positive PCR (%)	Negative PCR (%)	Total
15-25	9 (45)	11 (55)	20
26-40	30 (50.8)	29 (49.2)	59
41-60	4 (80)	1 (20)	5
61 and above	0 (0)	0 (0)	0
Total	43 (51.2)	41(48.8)	84

Table 2. Detection of *Chlamydia trachomatis* among infertile women without vaginal discharge according to age incidence.

Age range (years)	Positive PCR (%)	Negative PCR (%)	Total
15-25	1 (10)	9(90)	10
26-40	3 (50)	3(50)	6
41-60	1 (10)	9 (90)	10
61 and above	0 (0)	20 (100)	20
Total	5 (10.5)	41(89.5)	46

Table 3. Detection of *Chlamydia trachomatis* infection among infertile PID patients according to age incidence.

Age range (years)	Positive PCR (%)	Negative PCR (%)	Total
15-25	8 (53.3)	7(46.7)	15
26-40	23 (47.9)	25(52.1)	48
41-60	4 (40)	6(60)	10
61 and above	0 (0)	23(100)	23
Total	35 (36.4)	61(63.6)	96

Table 4. Detection of *Chlamydia trachomatis* among infertile women according to risk factors.

Risk factor	Positive PCR (%)	Negative PCR (%)	Total
Low education	19 (44.2)	24(55.8)	43
Smoking	15 (44.1)	19(55.9)	34
Promosecuity	26 (52)	24(48)	50
Total	60 (47.2)	67(52.8)	127

73 pelvic inflammatory disease patients (Known case which were already diagnosed by labroscopy), out of the 200 infertile patients were investigated for *C. trachomatis*. 35 (47.9%) of them were positive by PCR technique. The frequency of PID was higher in the age range 26-40 years (Table 3). 60 (47.2%) of PCR positive women were found exposed to *Chlamydia* infection risk factors, such as smoking, low education, and promosecuity (Table 4). We found that there are association with *C. trachomatis* and smoking, then those people have highly sexualty active behavior.

On the other hand, *C. trachomatis* was detected among 11 cases (36.0%) with PID, 13 cases (41%) with vaginal discharge, 10 cases (38%) with past history of abortion, 12 cases (41%) with PID and vaginal discharge, and 9 cases (56.2%) with PID, vaginal discharge, and abortion.

DISCUSSION

Screening of *C. trachomatis* is needed to define measures of prevention, modes of transmission to newborn, and ways to reduce sexual spread. *C. trachomatis* is most comon agents leading to congenital infection in both men and women. Worldwide the estimated annual incidence goes up to 50 million cases of *C. trachomatis* infection. *C. trachomatis* in women has a clinical course varying from asymptomatic infections to ascending infections leading to pelvic inflamatory disease associated with late ectopic pregnancy and tubal infertility (Land, 2010).

A total of 191 patiens with vaginal discharge were investigated, 43 (51.2%) of them were found positive for *C. trachomatis* by PCR tehnique. This indicates that the molecular detection of *C. trachomatis* by PCR is a useful technique. Most positive cases were in the age range of 26- 40 years, where as negative cases were aged 61 years and above (Table 1).

In this study 43 (51.2%) patients were positive by PCR technique. This result was higher than results of a previous study performed by Ortashi et al. (2004) who found a prevelance of 7.3% of *Chlamydia* infection among Sudanese women attending an obestetric and gynecology clinic in Kartoum. Also in the present context, the frequency rate of *Chlamydia* infection among women investigated was 51.2% which was higher than the frequency rate of *Chlamydia* infection among men (10.4%) that was reported by Omer et al. (1985).

In this study 17% of the fertile pregnant women were found positive for *C. trachomatis*. This figure was similar to that reported (16.4%) by Nafi (2006) study in Khartoum (Sudan), where he found a frequency of 16.4% *C. trachomatis* infection among pregnant fertile women. This author also reported a frequency of 43.4% among women with past history of abortion and 52% among PID patients. In this study *C. trachomatis* infection detected was (38%) among women with past history of abortion and (56.2%) among PID patient .

In the present context *C. trachomatis* was detected in 43 patients (21.5%) using PCR tchnique. This result is higher than that reported by Fallah et al. (2005) who detected *C. trachomatis* in 14 patients (14.9%).

Land (2010) in Netherlands studied asymtomatic cervical *C. trachomatis* infections in association with age incidence using PCR technique. They found the overall prevelance rates of *C. trachomatis* infection were 9.2 and 11.8% among patients youger than 30 years and above 30 years respectively. While in this study the frequency rates of *C. trachomatis* detected by PCR were 69.8% in the age range 26-40, 21% in the age range below 26 years, 9.3% in the age range above 41 years. This result reflects a significant association between the age range 26-40 years and *C. trachomatis* infection.

In conclusion, the age range (20-40) years (Table 3) is a risk factor that exposes sexually-active women to *C. trachomatis* infection. Also *C. trachomatis* infection in pregnant women may be associated with abortion and pelvic inflammatory disease. *C. trachomatis* infection is becoming a public health problem in Khartoum due to its highest frequency among fertile and non-fertile Sudanese women.

REFERENCES

Akush G (1993). The role of Chlamydial infection in the genesis of tub peritoneal sterility in women. ACP Med. 5:36-39.

Christian LA, Katherine MET, Boris VS, Janneke CMH, Mirjam K, Nicola L (2012). Transmission of *Chlamydia trachomatis* through sexual partnerships: a comparison between three individual-based models and empirical data. J. Royal Soc. interface 9(66):136-146.

Fallah F, Kazemi B, Goudarzi H, Badami N, Doostdar F, Ehteda A, Naseri M, Pourakbari B, Ghazi M (2005). Detection of *Chlamydia trachomatis* from urine speciments by PCR women with cerrivitis, Iran. J. Pub. Health 34(2):20-26.

Land JA (2010). Epidemiology of *Chlamydia trachomatis* infection in women and the cost-effectiveness of screening. Hum. Reprod. Update 16(2):189-204.

Nafi M (2006). MSc study, detection of *Chlamydia trachomatis* in urine specimens by polymerase chain reaction in pregnant women with vaginal discharge in Omdurman, Alneelain University.

Okoror LE, Agbolahor DE, Esumeh FI, Umolu PI (2007). Prevalence of *Chlamydia trachomatis* in patients attending gynecological clinics in South eastern Nigeria. Afr. Health Sci. 7(1):18-24.

Omer EE, Forsey T, Darougar S, Ali MH, El Naeem HA (1985). Seroepidemological survey of *Chlamydia* genital tract infections in Khartoum, Sudan. Genitourin Med. 61(4):261-263.

Ortashi OM, Elkhidir I, Herieka E (2004). Prevalence of HIV, syphilis, *Chlamydia trachomatis*, *Neisseria gonorrhoea*, *Trichomonas vaginalis* and candidiasis among pregnant women attending an antenatal clinic in Khartoum, Sudan. J. Obstet. Gynecol. 24(5):313-315.

Antibiotic resistance pattern in bacterial isolates obtained from frozen food samples of animal origin in Sanandaj and Ahvaz

Kalantar Enayat[1,2], Amin Mansour[3]*, Bahmani Nasrin[4], Tamimi Mohammad[5], Hashemzadeh Mohammad[3] and Narenji Hanar[4]

[1]School of Health, Environmental Health Research Center, Kurdistan University of Medical Sciences, Sanandaj, Iran.
[2]Department of Pathobiology, Alborz University of Medical Sciences, Karaj, Iran.
[3]Department of Microbiology, School of Medicine, Infectious and Tropical Diseases Research Center, Ahvaz Jundishapur University of Medical Sciences, Ahvaz, Iran.
[4]Department of Microbiology, School of Medicine, Kurdistan University of Medical Sciences, Sanandaj, Iran.
[5]Food and Drug Deputy, Kurdistan University of Medical Sciences, Sanandaj, Iran.

One of the major public health challenges nowadays is the rapid emergence of antibiotic resistant pathogens. This study investigated the antibiotic resistance pattern in bacterial isolates obtained from frozen food samples of animal origin in Sanandaj and Ahvaz Iran. A total of 250 packages of food of animal origin were purchased from different separate grocery stores in Sanandaj and Ahvaz during one year. Bacteria were isolated and identified based on national procedures. Antibiogram was done according CLSI guideline. From a total of 250 samples collected from Sanandaj and Ahvaz, 68 (54.4%) and 71 (56.8%) were positive for bacteria respectively. In Sanandaj Prevalence of *Escherichia coli, Citrobacter* Spp, *Enterobacter* Spp, *Staphylococcocus* Spp, *and Bacillus cereus* was 40, 18, 05, 03 and 2% respectively. The most contaminated food was chicken nuggets (80%). Antibiogram results clearly showed that resistance to Tetracycline and Ampicillin was 55 and 50% respectively. Similarly, in Ahvaz also prevalence of *E. coli, Salmonella* Spp, and *Proteus* Spp, were 52, 03.2 and 02.4% respectively. The above bacteria were 39 and 26% resistance to Ampicillin and Tetracycline respectively. The frozen food samples are perhaps the potential vehicles for transmitting food-borne diseases. Antimicrobial resistance among enteric organisms in food animals was generally low; however, the significance of this study lies in the detection overall resistance of bacterial isolates from the frozen food samples of animal origin in Ahvaz and Sanandaj against different antibiotics.

Key words: Frozen food, antimicrobial resistance, *Escherichia coli.*

INTRODUCTION

There is an increasing trend in production and consumption of meat, poultry meat, fish and their products. This, of course, requires adequate control and inspection both during poultry rearing and in slaughterhouses, processing plants and shops. Consumers are also a link in the chain of food-borne human diseases, because of the way they store and cook meat, poultry meat, fish and their products (Kozacinski et al., 2006; Lei et al., 2008). Therefore, microbial safety of food is a significant concern of consumers and industries today.

The rapid and accurate identification of food-borne pathogenic bacteria in food is important; both for quality assurance and to trace pathogens within the food supply (Kye et al., 2008; Hyun-Joong et al., 2008). Major food-borne pathogens include *Bacillus cereus*, *Listeria*

*Corresponding author. E mail: mnsamin@yahoo.com.

Table 1. Prevalence of bacterial contamination of frozen food samples in Ahvaz and Sanandaj.

Frozen food sample	City	Total number	Percent
	Ahvaz	25	12 (48)
Chicken	Sanandaj	25	36 (9)
	Ahvaz	25	0 (0.0)
Fish	Sanandaj	25	52 (13)
	Ahvaz	25	21 (84)
Kabab	Sanandaj	25	80 (20)
	Ahvaz	25	20 (80)
Minces	Sanandaj	25	60 (15)
	Ahvaz	25	18 (72)
Beef burger	Sanandaj	25	44 (11)

monocytogenes, *Staphylococcus aureus*, *Clostridium botulinum*, *Clostridium perfringens*, *Campylobacter jejuni*, *Vibrio parahaemolyticus*, *Yersinia enterocolitica*, *Escherichia coli* O157:H7 and *Salmonella enteric* serovar Typhimurium (Laboratório, 2006). Some of these foodborne pathogens can cause life-threatening diseases to humans and animals. Whilst all are well recognized, some are considered emerging because they have recently become more common.

In addition, food contamination with antibiotic resistant bacteria could be a major threat to public health as the antibiotic resistance determinants can be transferred to other pathogenic bacteria, causing compromise in the treatment of severe infections. The prevalence of antimicrobial resistance among food-borne pathogens has increased during recent decades (Boerlin et al., 2005; Choi et al., 2001). Therefore, this study aimed to determine the antibiotic resistance pattern in bacterial isolates obtained from frozen food samples of animal origin in Sanandaj and Ahvaz.

MATERIALS AND METHODS

A total of 250 packages of food of animal origin consisting of frozen chicken, chicken nuggets, frozen fish, mince, and beef burger were purchased from different separate grocery stores in Sanandaj and Ahvaz during one year. Field personnel collected the samples on a weekly basis. Retail samples was sealed in a plastic bag, labeled with a unique identifying number, and placed into a cooler with ice packs. Field personnel transport the specimens to food and drug laboratory, within 1 h of collection.

Isolation and identification of bacteria

Bacteria were isolated and identified based on Institute of Standards and Industrial Research of Iran. Antibiogram was done using the agar disc diffusion method on Muller-Hinton agar (Merck, Germany), according to CLSI (Clinical and Laboratory Standards Institute, 2010).

RESULTS

From a total of 125 samples collected from Sanandaj, 68 (54.4%) were positive for bacteria (Table 1). Prevalence of *E. coli, Citrobacter* Spp, *Enterobacter* Spp, *Staphylococcocus* Spp, and *Bacillus cereus* was 40, 18, 05, 03 and 2% respectively (Table 2). The most contaminated food was chicken nuggets (80%).

Antibiogram results clearly showed that resistance to Tetracycline and Ampicillin was 55 and 50% respectively. We also observed resistance to Chloramphenicol, and Erythromycin (Table 3). Similarly, Ahvaz results are as follows: out of 125 food samples, 71 (56.8%) were contaminated. A total of 76 bacteria were isolated and identified. Prevalence of *E. coli, Salmonella* Spp, and *Proteus* Spp, were 52, 03.2 and 02.4% respectively. The above bacteria were 39 and 26% resistance to Ampicillin and Tetracycline respectively.

DISCUSSION

A major goal for the food processing industry is to provide safe, wholesome and acceptable food to the consumer. Control of microorganisms is essential to meeting this goal. Many food pathogenic and spoilage bacteria are able to attach to food and remain viable even after cleaning and disinfection. This can seriously affect the quality and safety of the food processed and pose a potential risk to the consumer (Bagge-Ravn et al., 2003).

The present study demonstrated that three major

Table 2. Distribution of bacteria isolated from the frozen food samples of animal origin in Ahvaz and Sanandaj.

City		E.c	Cit	Ente	S. a	Co.NS	B.c	Sal	Prot	List
Ahvaz	Chicken	05	-	-	-	-	-	04	03	-
Sanandaj		08	-	-	01	-	-	-	-	-
Ahvaz	Fish	-	-	-	-	-	-	-	-	01
Sanandaj		04	09	-	-	-	-	-	-	-
Ahvaz	Kabab	20	-	-	-	-	-	-	-	01
Sanandaj		-	07	03	-	-	-	-	-	-
Ahvaz	Mince	20	-	-	-	-	-	-	-	-
Sanandaj		11	-	-	-	-	04	-	-	-
Ahvaz	Beef burger	20	-	-	-	02	-	-	-	-
Sanandaj		07	02	-	01	-	01	-	-	-

E. c= *E. coli*, Cit= *Citrobacter*, Ent= *Enterobacter*, S.a. *Staphylococcus aureus*, CoNS= Coagulase negative *Staphylococcus*, B.a= *Bacillus cereus*, Sal= *Samonella*, Prot= *Proteus*, Lis= *Listeria.*

Table 3. Overall resistance of bacterial isolates from the frozen food samples of animal origin in Ahvaz and Sanandaj against different antibiotics.

Antibiotic	Resistance to	
	Ahvaz	Sanandaj
Ampicillin	03	50
Ciprofloxacin	01	09
Clindamycin	0.0	08
Cephotoxim	0.0	05
Nalidixic Acid	0.0	09
Erythromycin	0.0	13
Gentamycin	0.0	0.0
Tetracycline	02	ND
Chloramphenicol	01	19
Trimetoprim-Sulfametoxazol	ND	10

enteric bacterial taxa were present in frozen food samples of animal origin obtained from supermarkets in the Sanandaj and Ahvaz. The findings presented herein suggest that there was not much a significant difference in the occurrence of bacteria among the evaluated food samples. The most prevalent bacteria in both the cities were *E. coli*. A study performed in the Washington, D.C reported *E. coli* prevalence of 38.7% in poultry (Cuiwei et al., 2001), which is much lesser to the results of the present survey. Several studies have shown that *E. coli* O157:H7 and other STEC are present in meat products, mostly beef products (Brooks et al., 2001; Heuvelink et al., 1996; Samadpour et al., 1994).

As regards antibiotic resistance, 55 and 50% of the bacterial strains which isolated from Sanandaj were resistant to Tetracycline and Ampicillin respectively. Many studies have reported almost same results concerning Tetracycline and Ampicillin resistant (Soltan et al., 2007; Bywater and Deluyker, 2004). Similarly, Tansuphasiri et al. (2006), evaluated 239 isolates of enterococci (113 from frozen foods and 126 from environmental water) for their resistance to 8 antibiotics by agar disk diffusion method. Most isolates from both sources were resistant to tetracycline (64.1% food strains; 46.8% water strains) and ciprofloxacin (53.4% food strains; 48.4% water strains). A relatively high prevalence of chloramphenicol, trimethoprim-sulfamethoxazole and vancomycin resistance was present, ranging from 9.7 to 27.2% for food strains and 10.3 to 15.9% for water strains.

A recent study reported that the prevalence of antimicrobial resistance in commensal microflora is very useful in monitoring and understanding the process of antimicrobial-mediated selection in a population (Caprioli et al., 2000). The previous studies on resistance from developing countries have concentrated on pathogenic bacteria (Byarugaba et al., 2005; Kassa et al., 2007), but our study lead to isolation of one important resistance indicator; *E. coli.*

ACKNOWLEDGEMENTS

This study is financially supported by research affairs Ahvaz Jundishapur University of Medical Sciences and Kurdistan University of Medical Sciences with grant number of U-88182.

REFERENCES

Bagge-Ravn D, Ng Y, Hjelm M (2003). The microbial ecology of processing equipment in different fish industries-analysis of the micro flora during processing and following cleaning and disinfection. Intl. J. Food Microbiol., 87: 239-250.

Boerlin P, Travis R, Gyles C (2005). Antimicrobial resistance and virulence genes of *Escherichia coli* isolates from swine in Ontario. Appl. Environ. Microbiol., 71:6753-6761.

Brooks HJ, Mollison BD, Bettelheim KA, Matejka K, Paterson KA, Ward VK (2001). Occurrence and virulence factors of non-O157 Shiga toxin-producing *Escherichia coli* in retail meat in Dunedin, New Zealand. Lett. Appl. Microbiol., 32:118-122.

Bywater R, Deluyker H (2004). A European survey of antimicrobial susceptibility among zoonotic and commensal bacteria isolated from food-producing animals. J. Antimicrob. Chemother., 54, 744–754.

Caprioli A, Busani L, Martel JL (2000). Monitoring of antibiotic resistance in bacteria of animal origin: epidemiological and mmicrobiological methodologies. Int. J. Antimicrob. Agents, 14: 295-301.

Choi C, Kwon D, Chae C (2001). Prevalence of the enteroaggregative *Escherichia coli* heat-stable enterotoxin gene and its relationship with fimbrial and enterotoxin genes in *E. coli* isolated from diarrheic piglets. J. Vet. Diagn. Invest., 13:26-29.

Clinical and Laboratory Standards Institute (2010): Performance standards for antimicrobial susceptibility testing; 20th informational supplement. CLSI/NCCLS M100-S20: Clinical and Laboratory Standards Institute, Wayne, pp. 3 (1).

Cuiwei Z, Beilei G, Juan DV (2001). Prevalence of *Campylobacter* spp., *Escherichia coli*, and *Salmonella* Serovars in Retail Chicken, Turkey, Pork, and Beef from the Greater Washington, D.C., Area. Applied and Environmental Microbiology. Dec, pp. 5431–5436.

Heuvelink AE, Wernars K , Boer DE (1996). Occurrence of *Escherichia coli* O157 and other verocytotoxin-producing *E. coli* in retail raw meats in the Netherlands. J. Food Prot., 59:1267–1272.

Hyun-Joong K, Si-Hong P, Tae-Ho L (2008). Microarray detection of food-borne pathogens using specific probes prepared by comparative genomics. Biosensors Bioelectron., 24: 238–246.

Kozacinski L, Hadziosmanovic M, Zdolec N (2006). Microbiological quality of poultry meat on the Croatian market. *VETERINARSKI ARHIV. 76* (4): 305-313.

Kye MC, Devaiah MK, Seong WK (2008). Simultaneous Detection of Food-borne Pathogenic Bacteria in Readyto- eat Kimbab Using Multiplex PCR Method. Food Sci. Biotechnol., 17(6): 1240 -1245.

Laboratório SP (2006). Microbiologia - Nº. de Análises Realizadas por Tipo de Amostras e Parâmetros. In Boletim Estatístico, Serviços de Saúde, Macau SAR; Annu. Rep: 89.

Lei JF, Roffey P, BlanchardC (2008). Development of a Multiplex PCR Method for the Detection of Six Common Foodborne Pathogens. Journal of Food Drug Analysis. 16(4): 37-43.

Samadpour M, Ongerth JE, Liston J (1994). Occurrence of Shiga-like toxin-producing *Escherichia coli* in retail fresh seafood, beef, lamb, pork, and poultry from grocery stores in Seattle, Washington. Appl. Environ. Microbiol., 60:1038– 1040.

Soltan Dallal MM, Vahedi S, Zeraati H (2007). Evaluating the effects of cooking on the decrease microbial contamination of kebabs and hamburgers supplied for selling in southern areas of Tehran. Payavard Salamat., (1): 24-31.

Tansuphasiri U, Khaminthakul D, Pandii W (2006). Antibiotic resistance of enterococci isolated from frozen foods and environmental water. Southeast Asian J. Trop. Med. Public Health, 37(1):162-170.

Isolation and identification of a pathogen, *Providencia rettgeri*, in *Bombyx mori*

Jiao Zhang[1], Zhong-yuan Shen[1,2,3]*, Xu-dong Tang[1,2], Li Xu[1,2] and Feng Zhu[1]

[1]Jiangsu University of Science and Technology, Zhenjiang 212018, Jiangsu Province, China.
[2]The Sericultural Research Institute, Chinese Academy of Agricultural Sciences, Zhenjiang 212018, Jiangsu Province, China.
[3]The Key Laboratory of Genetic Improvement of Silkworm and Mulberry of Agricultural Ministry, Zhenjiang 212018, Jiangsu Province, China.

A Gram-negative, rod-shaped bacterium, SY-1 strain, was isolated from a naturally infected silkworm *Bombyx mori* that was suffering from the common infection septicemia. The colony was round, slightly convex, smooth, moist, translucent, regular-edged and milky white on nutrient agar medium. The SY-1 was identified as a putative strain of *Providencia rettgeri* through its physiological and biochemical properties, and phylogenetic analyses performed using the 16S rRNA gene sequence. The SY-1 was identified as a putative strain of *P. rettgeri*. This is the first report demonstrating that *P. rettgeri* is pathogenic to *B. mori*. The identification of this strain adds to known kinds of pathogens that cause bacterial septicemia in the silkworm. This study aids the understanding of the prevention and control of the bacterial septicemia pathogens of *B. mori*.

Key words: *Bombyx mori*, bacterial septicemia, pathogen, Providencia.

INTRODUCTION

The silkworm, *Bombyx mori*, is an insect with high economic value which has been domesticated for silk production for more than 5000 years. It also occupied an important position in the human economic life and culture history. In China, sericulture is an important part of agriculture. However, bacterial diseases often cause severe economical loss in sericulture. Although, some kinds of bacterial diseases have been researched and described already, many bacteria which are pathogenic to the silkworm are unknown before now. So, the identification of new pathogenic bacteria for silkworm is very important.

Bacterial infection of silkworm, also known as flacherie, causes the silkworm cadavers to lose elasticity, become soft and rapidly rot. Silkworm bacterial flacherie are classified into three groups, namely, septicemia, bacterial toxicosis and bacterial gastroenteric disease. Most bacterial septicemia pathogens of silkworm induce acute attacks. The time from pathogen infection to death may be more than 10 h at high temperature (~28°C), and approximately 24 h at low temperature (Jing, 2000). A diverse range of pathogenic bacteria can cause septicemia in the silkworm, including those belonging to the following genera: *Bacillus*, *Enterobacter*, *Serratia*, *Aeromonas*, *Streptococcus*, *Pseudomonas* and *Staphylococcus*. The virulence of these bacteria to the silkworm differ, and therefore, the symptoms of septicemia and course of disease are not uniform (Chen et al., 2008).

In our study, a previously unknown pathogenic bacterial strain SY-1 was isolated from a naturally infected silkworm. The cultured colony was translucent and milky white when grown on nutrient agar medium. The strain was identified as *Providencia rettgeri* according to its morphological, physiological and biochemical features, and by 16S rRNA sequence analysis. In addition, the isolated strain was pathogenic to healthy silkworm when

*Corresponding author. E-mail: szysri@yahoo.com.cn.

injected into the healthy fifth-instar larvae in the laboratory. *P. rettgeri*, which is widespread in nature and weakly virulent, is a member of the family Enterobacteriaceae, *P. rettgeri* has been reported to infect insects of Heterorhabditis spp. (Jackson et al., 1995).

It has also been reported that this bacteria can cause disease in animals, especially aquatic animals (Fan et al., 2001). Indeed, Fan and colleagues (2001) isolated *P. rettgeri* from caseous necrotic organs of a sick turtle. They then injected the bacteria into a healthy turtle, which induced disease and subsequent death. These results confirmed that this bacterium is pathogenic to turtle. Zhan Wen-bing et al. (1997) also isolated the bacteria from shrimp, and artificially infected shrimp in the laboratory, thereby demonstrating that this bacterium is pathogenic to marine aquaculture animals. John et al. (1979) also isolated this bacteria from polluted water and frog. In addition, *P. rettgeri* can cause human diseases, such as urinary tract infections and others external to the intestinal tract as well as many outbreaks in hospitals (Zhang, 2003). Yoh et al. (2005) demonstrated that *Providencia* spp., especially *P. rettgeri*, is a causative agent of traveler's diarrhoea, and that these species are important pathogens. Besides, *P. rettgeri* is also often found in bacteria as it is easy for food to get contaminated with this bacterium during processing, transportation and sales (Zhang, 2005). In addition, *P. rettgeri* has been reported to infect insects of *Heterorhabditis* spp.

The silkworm infected with the SY-1 exhibited symptom similar to those of bacterial septicemia. First, the infected silkworm stopped eating mulberry leaves and their movements slowed. Then, their bodies curled up, and they subsequently began vomiting and excreting soft waste. The corpses were stiff shortly after death. Furthermore, their bodies were soft and discolored, and the body wall easily ruptured, finally, foul-smelling liquid was released. To the best of our knowledge, this is the first report that *P. rettgeri* is pathogenic for the silkworm *B. mori*.

MATERIALS AND METHODS

Test silkworm, strain and medium

Experimental silkworm was provided by the pathology laboratory of the Sericultural Research Institute of Chinese Academy of Agricultural Sciences. Healthy newly molted fifth-instar larvae, whose variety are qiufeng × baiyu were used for the experiment. A naturally infected silkworm was collected in a farm in Sheyang City, Jiangsu Province. The corpse was disinfected with 70% (v/v) ethanol for 1 to 2 min and washed twice with sterile distilled water after it had just began to soften. Then, we used sterilized insect needles to pierce the body wall and withdraw body fluid with an inoculation loop. The body fluid was then inoculated onto nutrient agar and cultured at 30°C for 48 h (Tao et al., 2011). Dominant single clones were purified by streaking and re-streaking the same agar plates. The purified strain, SY-1, was injected into prepared test silk-worms for the purpose of confirming the bacteria's pathogenicity. The clinical symptoms presented by the infected silkworm were observed and Koch's laws were used to investigate the pathogenicity of the strain (Liu et al., 1995).

Observation of colony characteristics and bacterial morphology

The strain was purified and incubated on a nutrient agar plate at 30°C for 24 h. During this period, the colony characteristics, including size, color, transparency, humidity, shape and edge were observed.

The bacterial colony that had been cultured for 24 h was subjected to Gram's staining and observed using an oil immersion objective under a microscope. The cell morphology was examined under a JEM-1230 (Japan) scanning electron microscope at 40,000× magnification (Zhao et al., 2000).

Phenotypic and biochemical characterization of SY-1

Biochemical and physiological analyses were performed as described in Bergey's Manual (Holt et al., 1994) and the Manual for the Identification of Medical Bacteria (Barrow et al., 2004) together with relevant references of Brenner et al. (1978) and Penner et al. (1975).

Preparation of genomic DNA of SY-1

The phenol-chloroform method for extracting DNA was used as previously described (Xia et al., 2005).

PCR amplification and analysis of the 16S rRNA sequence

Universal primers for amplification of the bacterial 16S rRNA gene were used as previously reported (Weisburg et al., 1991). The forward primer was FT1 (5'-AGAGTTTGATCATGGCTCAG -3') and the reverse primer was RT5 (5'-ACGGTTACCTTGTTACGACTT-3'). The PCR reaction were performed as follows: 95°C for 5 min; 30 cycles of 95°C for 20 s, 50°C for 20 s and 72°C for 3 min; and a final extension at 72°C for 10 min. The PCR product was stored at 4°C. Following the completion of the reaction, 5 μL of the reaction solution and the stored PCR product (45 μL) were subjected to 1% agarose gel electrophoresis. A DNA fragment (~1.5 kb) was purified from the 1% agarose gel by using the commercial product for DNA purification PureLink quick extraction kit (Beijing Dingsheng Biotechnology Company). The PCR product was sequenced in the automated sequencing facilities at Shanghai Sangon Biotechnologies Co. Ltd. (China). The homology comparison of the almost complete 16S rRNA gene sequence of SY-1 was performed using the NCBI BLAST database (http://www. ncbi.nlm.nih.gov/). Subsequently, the sequence was analyzed on the Ribosomal Database Project (RDP) website (http://rdp.cme.msu.edu) (Cole et al., 2009) in order to obtain the 16S rRNA gene sequence (Holt et al., 1994). Sequence alignment was performed using ClustalX 1.83 software and the MEGA 4.0 software package. The phylogenetic tree was constructed using the neighbor-joining (NJ) method and the genetic distances were calculated using the Kimura 2-parameter method and 1,000 bootstrap replications to assess the stability of the relationships (Tamura et al., 2007).

Pathogenicity to silkworm

The cultured SY-1 bacteria was washed with 3.0 to 5.0 mL sterile

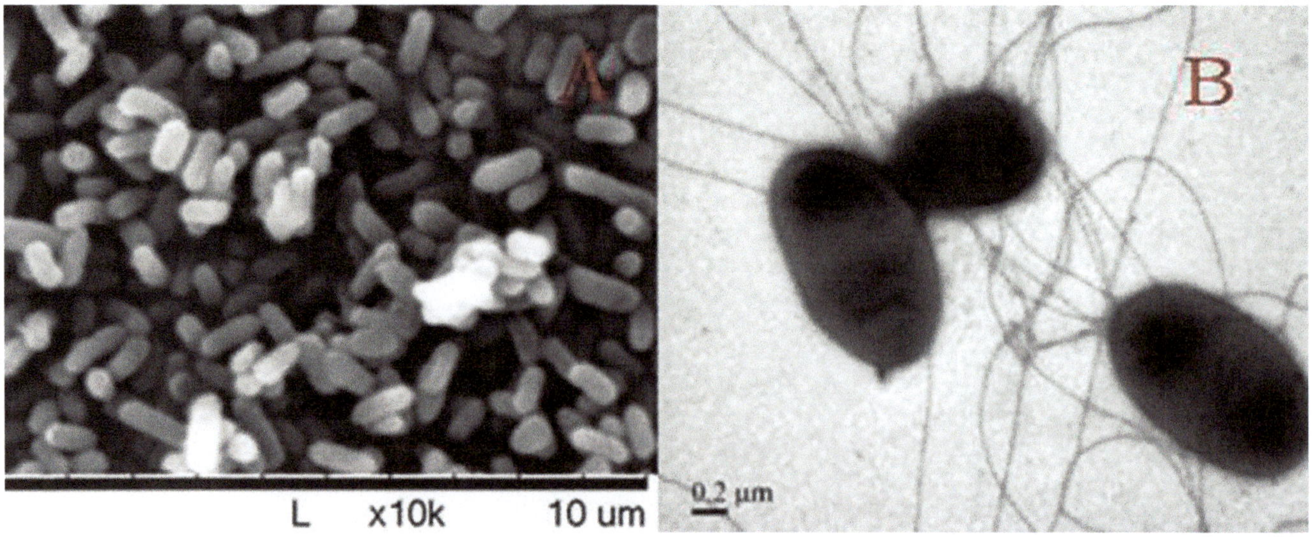

Figure 1. Morphological features of SY-1 using a scanning electron microscope. A: Stereoscan photograph, 10,000×. B: negatively stained electron microscope image, 40,000×.

phosphate-buffered saline (PBS) (2 mol/L; pH 7.4) by repeated blowing and suction. Then, the bacterial suspension of SY-1 was transferred to a new sterile test tube and centrifuged at 4,000 rpm for 15 min. The supernatant was removed and resuspended in 5 mL of PBS by oscillation to ensure the bacteria were uniformly distributed. The plate colony-counting method was used to count the number of bacteria (Luo, 2007).

The bacterial suspension of SY-1 was diluted into 7 solutions (10^{-1}, 10^{-2}, 10^{-3}, 10^{-4}, 10^{-5}, 10^{-6} and 10^{-7}) with sterile phosphate-buffered saline (PBS). Thirty healthy newly molted fifth-instar larvae in one experimental group were treated with the bacterial suspension by using a 10-μL microinjector and 2 μL of bacterial suspension was injected into each larva. The same concentration of bacterial suspension was tested for three replicates under the same conditions. Concurrently, the control was set with 2 μL of PBS only. Then, the infected larvae were reared at 25.0 ± 1.0℃. After 20 h, the mortality of silkworm was recorded. Statistical analysis was completed using SPSS 16.0 (Jia, 2006), and the median lethal concentration (LC50) and regression equations were obtained.

RESULTS

Description of colony and morphology of SY-1

After routine culture on nutrient agar for 48 h, the colony of the SY-1 strain was slightly convex, smooth, moist, translucent, regular-edged and milky white. The bacterial cell was Gram-negative, rod-shaped. Furthermore, electron microscopy results revealed that the cells were straight short rods measuring 1.3 ± 0.2 × 0.7 ± 0.1 μm with rounded ends and flagella (Figure 1).

Physiological and biochemical characteristics of SY-1

The results of the phenotypic and biochemical tests for SY-1 are presented in Table 1. The bacterium is found to be a facultative anaerobe which can utilize and produce acid by fermenting d-glucose, mannitol, rhamnose, d-sorbitol, mannose, d-galactose, d-fructose and 10% lactose. It produced peroxidase and was tested to be positive for the deoxidization of nitrate. In addition, indole, urease, amylolysis, phenylalarine deaminase, utilization of malonate and methyl red tests were all positive. SY-1 was negative for produced acid from sucrose, maltose and l-arabinose. Voges–Proskauer reaction, gelatin liquefaction, tryptophan deaminase, ornithine decarboxylase, phenylalarine deaminase and arginine decarboxylase were also negative.

16S rRNA gene sequence and phylogenetic analysis

The 16S rRNA sequence of SY-1 was 1504 bp (GenBank accession number: JQ954968), as expected. Alignment of the aforementioned sequence by using the RDP clearly demonstrated the taxonomic status of the SY-1 strain as follows: Bacteria (domain), Proteobacteria (phylum), Gammaproteobacteria (class), Enterobacteriales (order), Enterobacteriaceae (family), Providencia (genus). In the SeqMatch, 20 *Providencia* strains that had highly similar 16S rRNA gene sequences to SY-1 were retrieved from the RDP database. The S_ab scores obtained from the aforementioned alignment are shown in Table 2. Furthermore, of the 20 strains, 8 were identified as *P. rettgeri*, and one each as *Providencia stuartii*, *Providencia alcalifaciens* and *Providencia vermicola*. Another 8 strains were identified in only the genus level (Providencia). The remaining sequence was an Enterobacteriaceae bacterium.

We also compared the 16S rRNA gene sequences by using NCBI BLAST http://www.ncbi.nlm.nih.gov/).

Table 1. Physiological and biochemical characterization of the SY-1 strain.

Characteristics	Response	Characteristics	Response
Peroxidase	+	Argnine dihydrolase	-
Growth temperature/4 °C	+	Arginine decarboxylase	-
Growth temperature/41 °C	+	Utilization of malonate	+
Facultative anaerobic	+	Lipase (Tween 80)	-
Motility	+	Deoxidization of nitrate	+
Oxidase	-	d-Glucose (acid production)	+
ONPG production	-	d-Glucose (gas production)	-
Indole test	+	d-fructose	+
Trehalose	-	Sucrose	-
Gelatin liquefaction	-	rhamnose	+
Growth in 6.5% NaCl	+	Maltose	–
Methylredtest	+	Lactose	–
VP reaction	-	10% lactose	+
Amylolysis	+	d-galactose	+
Urease test	+	d-mannitol	+
Utilization of sodium citrate	+	d-mannose	+
Tryptophan deaminase	-	d-sorbitol	+
Ornithine decarboxylase	-	d-cyclohexanhexol	+
Phenylalarine deaminase	+	L-arabinose	–
Lysine decarboxylase	-	Ethanol oxidation	+

Table 2. Similarity between the 16S rRNA of the SY-1 strain and 20 other *Providencia* strains.

Short ID orientation	Strains	S_ab score with the 16S rRNA of SY-1
S000445102	*Providencia stuartii*; PGS13; AY803746	0.977
S000544654	Providencia rettgeri (T); type strain: DSM 4542; AM040492	0.968
S000550620	Providencia sp. UTDM314; AY870456	0.984
S000728634	Providencia rettgeri; NCTC 11801; DQ885263	0.985
S000769862	Enterobacteriaceae bacterium ACU; EF151985	0.985
S000968901	Providencia sp. SBS1; EU195872	0.970
S001044813	Providencia rettgeri; CM7; EU660316	0.971
S001044848	Providencia rettgeri; CT23; EU660351	0.979
S001095422	Providencia sp. B282Ydz-ds; EU070350	0.969
S001095844	Providencia sp. YRL09; EU373416	0.980
S001572543	Providencia alcalifaciens; Acj 108; AB480755	0.963
S001613563	Providencia rettgeri; KR-3; GQ923882	0.969
S001794733	Providencia rettgeri; IITRP2; GU193984	0.972
S002034338	Providencia rettgeri; VITJCSTT3; GU457413	0.968
S002167916	Providencia vermicola; KSVDL-09-71600; HM071065	0.965
S002228783	Providencia sp. DGC-5; HM347050	0.985
S002233507	Providencia rettgeri; KPE62307H; HQ009881	0.989
S002233997	Providencia sp. R4-2A; HQ154555	0.970
S002234011	Providencia sp. R8-1A; HQ154569	0.967
S002951763	Providencia sp. BIHB 1402; JF766697	0.963

Sequence similarities (>99%) were identified between the 16S rRNA gene of SY-1 and many *Providencia* strains. Subsequently, 13 sequences that demonstrated high sequence homology with SY-1 or the other strains of

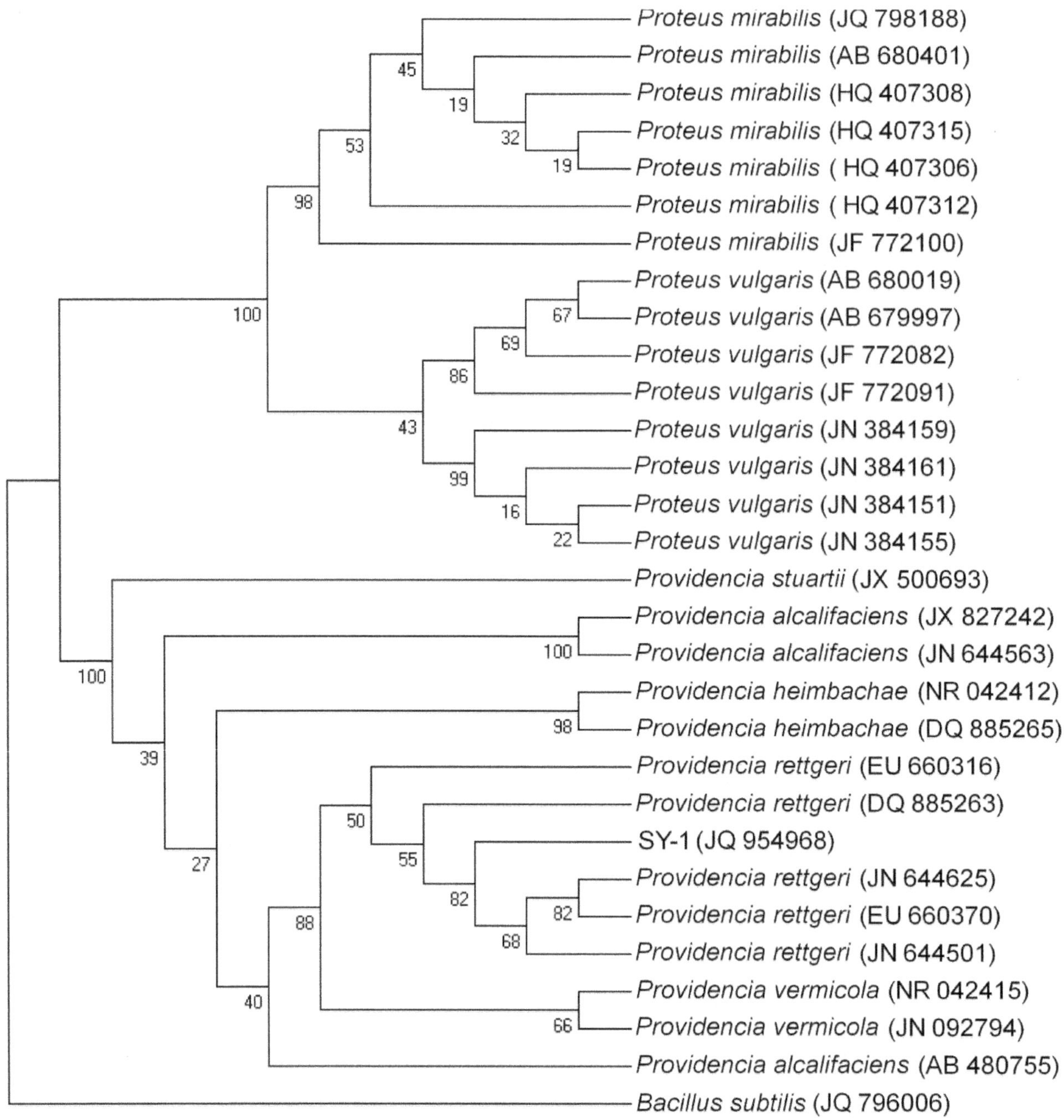

Figure 2. Phylogenetic tree of the 16S rRNA sequences of the SY-1 strain. The phylogenetic tree was constructed by the NJ method. *B. subtilis* was used as the outgroup. The NJ consensus tree used 1,000 bootstrap replicates, and the number represents the percentage of bootstrap values.

species of *Providencia* besides *P. rettgeri* were selected. Previous DNA relatedness tests revealed that *Providencia* spp. are more closely related to *Proteus vulgaris* and *Proteus mirabilis* than to other members of the family Enterobacteriaceae (Holt et al., 1994). So, we selected 15 sequences of different *P. vulgaris* and *P. mirabilis* strains and compared them with the 16S rRNA gene sequence of SY-1. Furthermore, we used *Bacillus subtilis* (GenBank accession number: JQ796006), which is distantly related to members of the genus *Providencia*, as the outgroup. Thirty sequence alignments were performed using Clustal software and the MEGA 4.0 software package to generate the phylogenetic tree (Tamura et al., 2007) and assess the stability of the relationships (Figure 2).

The phylogenetic tree showed that several *P. rettgeri* strains had high sequence homology with SY-1. Therefore, according to its morphological characteristics

Figure 3. Dead silkworm infected by SY-1 bacteria.

Table 3. Pathogenicity of SY-1 strains against the silkworm.

Strain	Concentration of bacterium ($\times 10^7$ cfu/mL)	Corrected mortality (%)	Linear regression equation	LC_{50}	95% confidence limits	Pearson correlation
SY-1	0.0084	20	y=-0.7870+0.2350x	1.4411×10^7	0.6561 - 2.9525	0.944
	0.084	33.33				
	0.84	53.33				
	8.4	96.66				

and the physiological and biochemical properties, and the phylogenetic relation, the SY-1 was a strain of *P. rettgeri.*

Pathogenicity to silkworm

Symptom in the infected silkworm

The silkworms infected with SY-1 exhibited symptoms similar to those of bacterial septicemia. First, the sick silkworms stopped eating mulberry leaves and their movements became slow. Afterwards, some silkworm vomited, excreted soft waste and shrank, which resulted to the bodies of the diseased silkworm been much smaller than the normal larvae. Finally, the diseased silkworm went into convulsion and died. The corpses were stiff shortly after death, and the middle portion of each sector swelled due to tightening. The initial color of dead silkworm's body did not significantly differ from that of the normal silkworm. The body wall gradually contracted and some curled up, resulting in a C-shape. In addition, their color changed. Finally, the whole body was soft, flat and deflated, and the internal organs were broken down and liquefied, resulting in a foul-smell. Only the chitin in the skin remained, and the body gradually darkened and released a black liquid as it came across slight vibration (Figure 3).

Pathogenicity test

To test the ability of the SY-1 isolate to cause mortality of silkworm, different concentrations of the bacterial suspension were injected into healthy newly molted fifth-instar silkworm larvae. The peak mortality of the silkworm occurred between 23 and 26 h after inoculation. The control setting with 2 μL of PBS only were all survived. And the corrected mortality for silkworm injected with 2 μL of $8.4 \times 10^4 \sim 8.4 \times 10^7$ colony forming units (CFU)/mL bacterial suspension of SY-1was 20.0 to 96.66% at 26 h after injection. The linear regression relationship was $y = -0.7870 + 0.2350x$, and the medium lethal concentration was $LC_{50} = 1.4411 \times 10^7$ CFU/mL (Table 3).

DISCUSSION

In our study, on the basis of traditional morphology, the

strain is Gram-negative and the cells of SY-1 bacteria were straight short rods measuring 1.3 ± 0.2 μm × 0.7 ± 0.1 μm with rounded ends and flagella. The results of the physiological and biochemical characteristics showed it produced peroxidase and was tested to be positive for the deoxidization of nitrate. In addition, indole, urease, amylolysis, phenylalarine deaminase, utilization of malonate and methyl red tests were all positive. SY-1 tested negative for produced acid from sucrose, maltose and l-arabinose. Voges–Proskauer reaction, gelatin liquefaction, tryptophan deaminase, ornithine decarboxylase, phenylalarine deaminase and arginine decarboxylase were also negative. The bacterium is found to be a facultative anaerobe which can utilize and produce acid by fermenting d-glucose, mannitol, rhamnose, d-sorbitol, mannose, d-galactose, d-fructose and 10% lactose. Especially, the characteristic of producing acid by fermenting mannitol is different from other species of the genus Providencia and 16S rRNA gene sequence analysis. The phylogenetic tree showed that several *P. rettgeri* strains had high sequence homology with SY-1. It was finally identify as *P. rettgeri*. SY-1 has a higher pathogenicity to silkworm (*B. mori*). The linear regression relationship was $y = -0.7870 + 0.2350x$, and the medium lethal concentration was $LC_{50} = 1.4411 \times 10^7$CFU/mL. *P. rettgeri* belongs to pathogenic strain, which cannot only infect insects such as *Heterorhabditis* spp., having all kinds of symptoms, but also aquatic animals, such as turtle, shrimp, etc. Additonally, it can also lead to human diseases like urinary tract infections. Therefore, research on it has important public health implications.

To date, a wide variety of bacteria have been identified as pathogenic to silkworms, including *Bacillus*, *Enterobacter*, *Serratia*, *Aeromonas*, *Streptococcus*, *Pseudomonas* and *Staphylococcus*. However, in sericulture production, septicemia is mainly caused by the *Bacillus* spp., *Serratia marcescens Bizio* and *Aeromonas* spp. This is the first study to report that *P. rettgeri* is pathogenic to *B. mori* and that it causes typical septicemia symptoms. These results broaden our knowledge regarding the types of pathogens capable of inducing silkworm bacterial septicemia, as well as the requirements for the prevention and treatment of silkworm bacterial diseases. Furthermore, this study has enhanced our understanding of hosts for *Providencia* spp. These findings on silkworm disease in conjunction with the widespread of *Providencia* spp. in nature, and its influence on human and other animals, highlights the need for further research focused on the prevention and control of epidemic disease caused by this bacterium.

ACKNOWLEDGEMENTS

This work was supported by the earmarked fund for the China Agriculture Research System. We are grateful to all who provided the means for us to access free software, which we used and cited in this article. We thank all partners and laboratory members for their kind help and criticism.

REFERENCES

Barrow GI, Feltham RKA (2004). Cowan and Steel's manual for the identification of medical bacteria . UK:Cambridge University Press.

Brenner, DJ, Farmer JJ, Fanning GR, Steigerwalt AG, Klykken P, Wathen HG, Hickman FW, and Ewing WH(1978). Deoxyribonucleic acid relatedness of Proteus and Providencia species. Int. J. Syst. Bacteriol. 28:269-282.

Chen YM, Huang JR (2008).Occurrence and control of the silkworm bacterial septicemia. Yun Nan Agricultural Science and Technology, z3.

Cole JR, Wang Q, Cardenas E, Fish J, Chai B, Farris RJ, Kulam-Syed – Mohideen AS, McGarrell DM, Marsh T, Garrity GM, Tiedje JM (2009).The Ribosomal Database Project: improved alignments and new tools for rRNA analysis. J. Nucleic Acids Res. 37(Database issue), pp. 141–145.

Fan SH, Huang GQ, Guo AZ, Lu CP (2001).Organ abscesses of Chinese turtle caused by *Providencia rettgeri*. J. Nanjing Agric. University 24(4):71-74.

Holt JG, Krieg NR, Sneath PHA (1994). Bergey's manual of determinative bacteriology, 9th edn. Baltimore:Williams and Wilkins.

Jackson TJ, Wang HY, Miriam JN, Christine TG, Ann MB, Barbara CAD (1995). Isolation of insect pathogenic bacteria,*Providencia rettgeri*, from *Heterorhabditis* spp. J. Appl. Microbiol. 78(3):237-244.

Jia CS (2006). Calculating the LC50 of insecticides with software SPSS. Chin. Bull. Entomol. 43(3):414-417.

Jing W (2000). Silkworm Pathology . Beijing : China Agriculture Press. pp. 122-130 .

John LP, Joan NH (1979). Application of O-Serotyping in a Study of *Providencia rettgeri* (*Proteus rettgeri*) Isolated from Human and Nonhuman Sources. J. Clin. Microbiol. 10(6):834-840.

Liu S, Zhu W, Yang ZR, Ge SR, He XY (1995). Isolation and identification of a pathogen of grasshoppers. Acta Microbiol. Sin. 35:86-89.

Luo TL, Shen LL, Bai YL, Xue Y (2007) Comparative study of Mycobacterium counting methods. J. Fourth Military Med. Univ. 28(4):353-357.

Penner JL, Hinton NA, Hennessy J(1975). Biotypes of *Proteus rettgeri*. J. Clin. Microbiol. 1:36-142.

Standardization Administration of the People's Republic of China.GB/T 4789.2-2008 Microbio-logical examination of food hygiene-Aerobic plate count. China Standard Press, Beijing.

Tamura K, Dudley J, Nei M, Kumar S (2007).MEGA4:Molecular evolutionary genetics analysis (MEGA) software version 4.0. Mol. Biol. Evol. 24(8):1596–1599.

Tao HP, Shen ZY,Zhu F,Xu XF,Tang XD, Xu L (2011). Isolation and Identification of a Pathogen of Silkworm *Bombyx mori*. Curr. Microbiol. 62(3):876-883.

Weisburg WG, Barns SM, Pelletier DA, Lane DJ (1991). 16S ribosomal DNA amplification for phylogenetic study. J. Bacteriol. 173(2):697-703.

Xia H, Fu WL, Chen M, Huang Q, Zhao M, Zhao YH, Wang F , Luo Y (2005). Research about rapid extraction of bacterial DNA method. Mod. Prev. Med. 32:5.

Yoh M, Matsuyama K, Ohnishi M, Takagi K, Miyagi H, Mori K, Sam Park K, Ono K, Honda T (2005). Importance of *Providencia* species as a major cause of travellers' diarrhoea. J Med. Microbiol. 54(11):1077-1082.

Zhan WB, Zhou L, Chen ZQ,Yu KK, Meng QX (1997). A new shrimp pathogen—*Providencia rettgeri*. Fish. Sci. China 4(1):38-44.

Zhang GH (2005). The investigation of food poisoning caused by *Providencia rettgeri*. Shang Hai Prevention Medicine, 17:3.

Zhang ZR (2003). Clinical Microbiology and Microbiological testing.3rd ed. Beijing : People Hygiene Press, pp.136-140.

Zhao BG, Guo DS, Gao R (2000).Observation of the Site of Pine Wood Nematode where Bacteria are Carried with SEM and TEM. J. Nanjing For. Univ. 24(4):69-71.

Spatial and temporal variation of *Escherichia coli* in water and soil with relation to water parameters at the estuary of Karnafuly River, Bangladesh

Md. Wahidul Alam* and Mohammad Zafar

Institute of Marine Sciences and Fisheries, University of Chittagong, Chittagong-4331, Bangladesh.

This study examined the concentrations of *Escherichia coli* in soil and water samples, and some physicochemical properties of water collected from the Karnafuly River estuary, Bangladesh during March-June 2012. Among the water parameters, temperature was in the range 26.5 to 32°C, salinity varied between 0 and 5 ppt, pH values ranged from 7.5 to 7.9 and dissolved oxygen (DO) was recorded at concentrations of 2.28 to 2.91 mg/l. The average total *E. coli* were 26 cells/ml in water, but the sediment contained 96.42 cells/g of *E. coli.* The population of *E. coli* was always higher in sediment than those that occurred in water among all sampling stations. The growth of *E. coli* were positively correlated ($P < 0.01$) with water temperature and negatively correlated with salinity, pH and DO. The data on water quality parameters and occurrence of *E. coli* indicated that water of the Karnafuly estuary is polluted and unhygienic for drinking, even impossible to use for washing purposes without any treatment.

Key words: Karnafuly, *Escherichia coli,* physicochemical, estuary.

INTRODUCTION

Microbiological impairment of drinking, bathing, irrigation, or recreational waters is commonly monitored using concentrations of fecal indicator bacteria (FIB). *Escherichia coli* are the FIB commonly used to evaluate the microbiological water quality. The concentration levels of *E. coli* are significantly influenced by various nonpoint sources such as surface runoff, bank soils, recreational activity, and animal excreta (Kim et al., 2007). Bottom sediments have been recognized as a major reservoir of *E. coli* in freshwater environments (Geldreich, 1970). Many studies indicate that sediments can harbor much higher populations of both fecal coliforms and *E. coli* than the overlying water column (Rehmann and Soupir, 2009).

The Karnafuly River estuary is an important estuary in Bangladesh and hydro-biologically it is the place where freshwater from upstream is continuously mixing with saltwater from the Bay of Bengal. Combinations of diverse fluctuating parameters are responsible for occurrences and distribution of different microorganisms in estuarine environment. It is also blessed with estuarine water, sediment, marine resources and varieties of fish species. Besides a lot of industry, fishing boats, vessels or trawlers and container ships are operating along the river hampering the status of water, sediment and fish species therein. Different researchers reported the estuarine environment polluted owing to the continuous discharge of waste material from the industries and sewage of the town. Certainly, pollution of water courses associated with industrial discharge and refuse from human settlements is a global problem (Joy et al., 1990). Everything carried by rivers ultimately goes into oceans through estuaries. The pollutants get dispersed by turbulence, ocean currents and tidal action firstly in the estuary, then concentrated in the food chain components through microbial action or deposited in the bottom sediments (Islam, 1998). People living in the Chittagong city are mostly depending on the Karnafuly River for drinking or other house-hold purposes. Besides, lots of people including a large number of fishermen live on both

*Corresponding author. E-mail: wahid.imsf.cu@gmail.com.

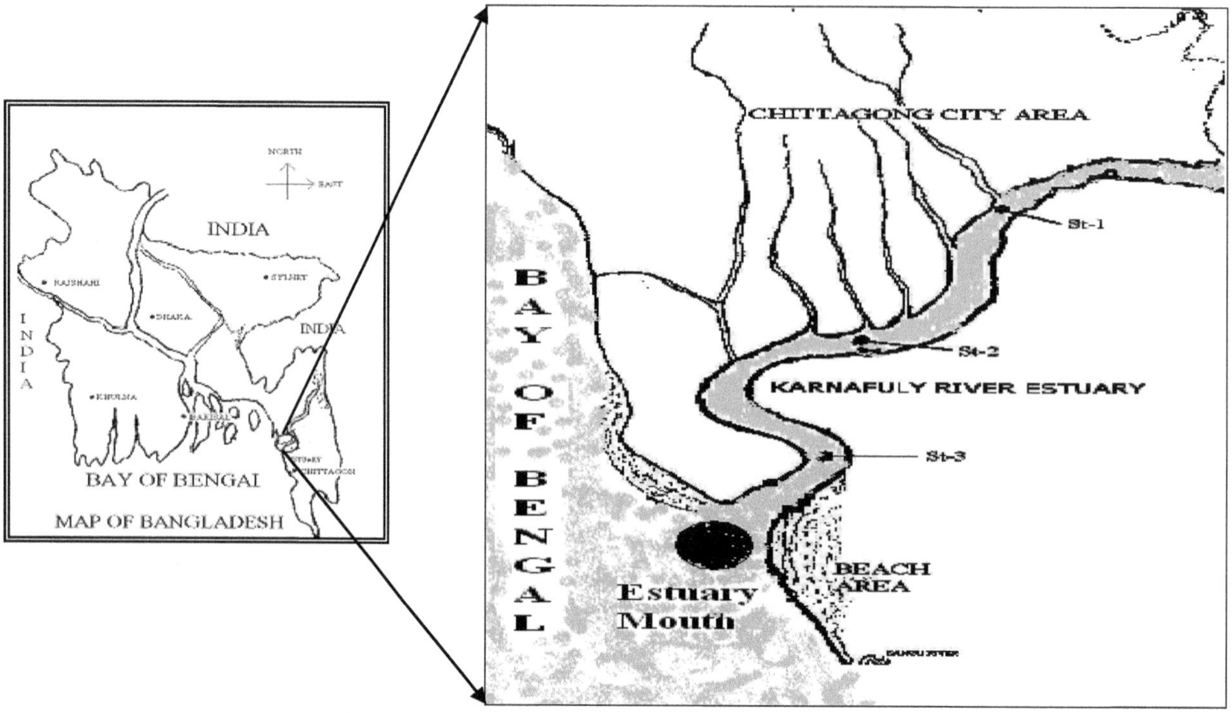

Figure 1. Geographical location of the study area showing different sampling stations.

sides of the river. They use the water for bathing and drinking. They also used sediment and soil for household purpose. Many countries in the world have developed drinking water criteria and standards. Bangladesh developed the first Water Quality Standards in 1976 based on International Drinking Water Standards (WHO, 1984). The Bangladesh standard specification for drinking water (BDS 1240: 1989) was prepared and published by the Bangladesh Standard and Testing Institution (BSTI) for the control of drinking water including bacterial load.

According to World Health Organization (WHO, 1993) and Environment Quality Standard for Bangladesh (EQSB, 1991), the standard value of total load of *E. coli* for drinking water is nil/ml of water. Different researchers identified different pathogenic bacteria among contaminated fish of the Karnafuly River, but seldom *E. coli* was isolated from water and sediment samples of this area. Considering the importance of water and sediment in everyday life, the present research work aimed to determine the concentrations of *E. coli* in water and soil of Karnafuly River estuary, including their relationship with some water parameters.

MATERIALS AND METHODS

Description of the study area

The Karnafuly is the principal river of Chittagong District of Bangladesh. It originates in the Lushai Hills of Mizoram (India), flows through Rangamati and the port city of Chittagong and discharges into the Bay of Bengal near Patenga. The river is flashy and its length is about 131 km. Geographically, this estuary is located between latitude 22°53' N and longitude 92°27' E and enters the districts of Chittagong from the north eastern side. It travels about 121 miles in zigzag ways to the west and south west side and finally falls on the Bay of Bengal at latitude 22°12' N and longitude 91°47' E at near Patenga, Chittagong. The average channel depth of the basin at the river estuary is 8 m, although the depth varies greatly in its upstream portion for strong current and siltation.

Geologically, the entire catchment consists of a substratum of tertiary rocks covered with alluvial deposits. The overlying deposits show that it consists of successive layers of mud and sand (Rizbi, 1971). Hydrological parameters that is temperature, salinity, pH, D.O etc. fluctuate seasonally. Three sampling stations (Figure 1) were selected for the present investigation with the basis of different types of pollution. The first sampling station (St-1) located near Chaktai canal (most polluted canal of Bangladesh), receives heavy discharges from different domestic and industrial sources to the estuary. The second sampling station (St-2) is located at the middle portion of the River, where small city canal falls into the river and carry industrial and city wastes. The third sampling station (St-3) is located at the mouth of the estuary and receives discharges composed of different kinds of city wastes through a canal which ultimately falls into the estuary.

Data collection and analysis

This work was carried out from March to June 2012. Water and soil samples were collected from three stations of the Karnafuly River estuary during this study period. *In situ* measurement of water temperature was measured with a centigrade thermometer while water pH and salinity were measured by pen pH meter and

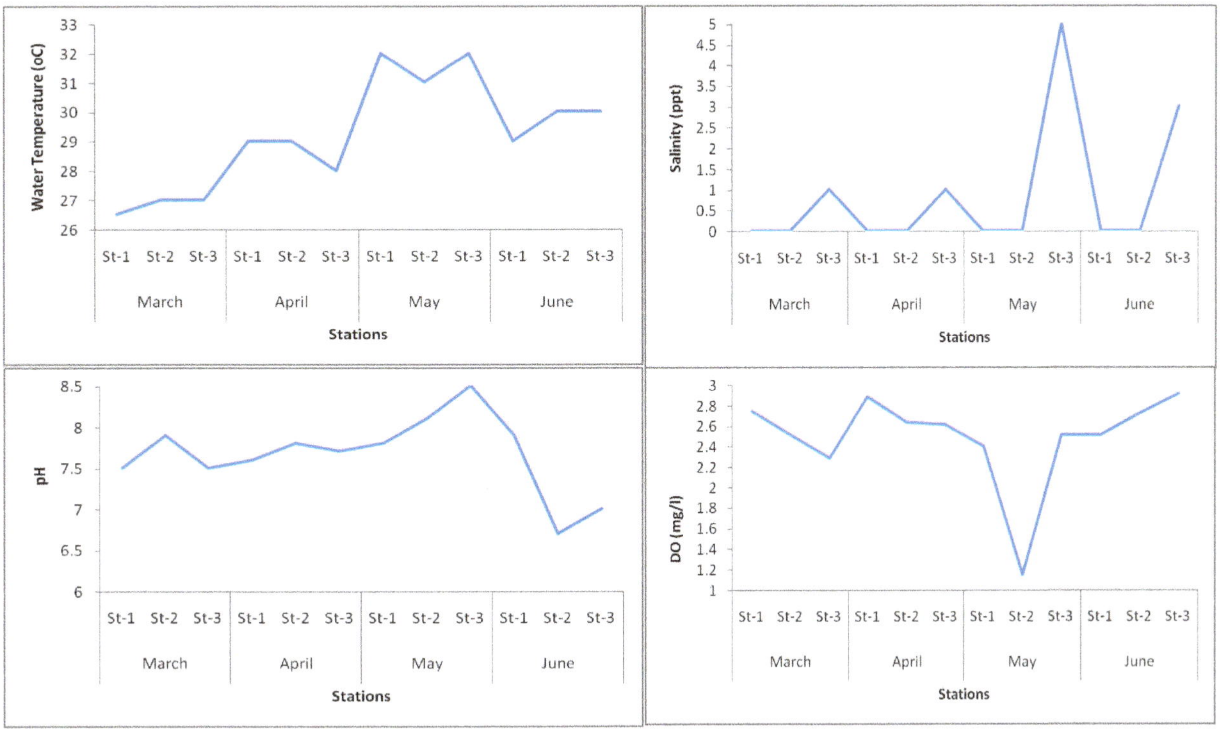

Figure 2. Physicochemical parameters of Karnafuly River at different stations in different months.

refractometer respectively. Dissolved Oxygen (DO) concentration was measured in the laboratory by following standard method (APHA, 1976). Collected water and soil samples were preserved and microbial analysis was conducted in the laboratory to identify *E. coli* in water and soil of the Karnafuly Estuary. *E. coli* of water and soil were identified by standard plate count (SPC) techniques (APHA, 1976). SPC was conducted by making serial dilution, plating and total plate count by pour plate techniques. In case of water, 1 ml of water sample was transferred with a sterile pipette to a sterile screw capped vial containing 9 ml of distilled water and the vial was shaken thoroughly. The sample was made to give 1:10 or 10^{-1}. In case of soil, 1 g of soil sample was weighted by weight meter and then transferred to a sterile screw capped vial containing 9 ml of distilled water and then the vial was shaken thoroughly.

The sample was made to give 1:10 or 10^{-1}. Each dilution was plated by sterile pipette with 1 ml of dilution into a Petri dish and the respective sterile melted media were poured into the plates. The plates were rotated by hand, a minimum of 5 times clockwise and 5 times in the anticlockwise direction and several times crosswise for equal distribution of the media. After solidifying the media the plates were inverted. Then the plates were incubated at room temperature. Levin's Eosin Methylene Blue (EMB) agar media were used for identification of the quantity of total *E. coli* population. Triplicate plates of each dilution were prepared for total *E. coli* load and the average of the colonies formed in the three plates was recorded. Total plate count was done by multiplying the colonies on the plate by the dilution factor. The average number of the total plate counts of each dilution was considered as total counts. In order to enumerate the *E. coli,* Levin's EMB media were used for each sample and incubated at room temperature for 48 h. As the *E. coli* was very rare in water and soil, so the author did the work without serial dilution. Density of *E. coli* was estimated by using pour plate techniques. Colonies having a green metallic sheen with dark centre were confirmed as *E. coli*. The colonies were picked and streaked on the nutrient agar slants and the slants were incubated at room temperature for 48 h. Then the cultures of the slant were subjected to Gram staining, Biochemical test and Fermentation test.

For the isolation of single colony, procedures of Sharp et al. (1969) were followed. The selected colonies were marked and their cultural and morphological characteristics were recorded. Simple correlation have been calculated to depict the relations using the formula:

Coefficient $r_{xy} = (n \sum x\, y - \sum x \sum y) / [n\sum x^2 - (\sum x)^2]\, [n\sum y^2 - (\sum y)^2]$

Where, x = i th sample values of one variable (i= 1, 2, 3......n); y = i th sample value of other variable; n = total number of samples.

Correlation coefficient has been tested by using the t-test for testing significance or non significance. The formula used is given as follows:

For t-test: $t_{(n-2)} = (r\sqrt{n-2})/\sqrt{(1-r^2)}$

RESULTS

Hydrographic conditions

Results from the water samples showed distinct hydrographic conditions (Figure 2). Water temperature varies between 26.5 to 32.0°C. Maximum water temperature was recorded to be 32.0°C in June at St-1 and St-3 while minimum temperature was recorded in March at St-1. Monthly maximum average temperature

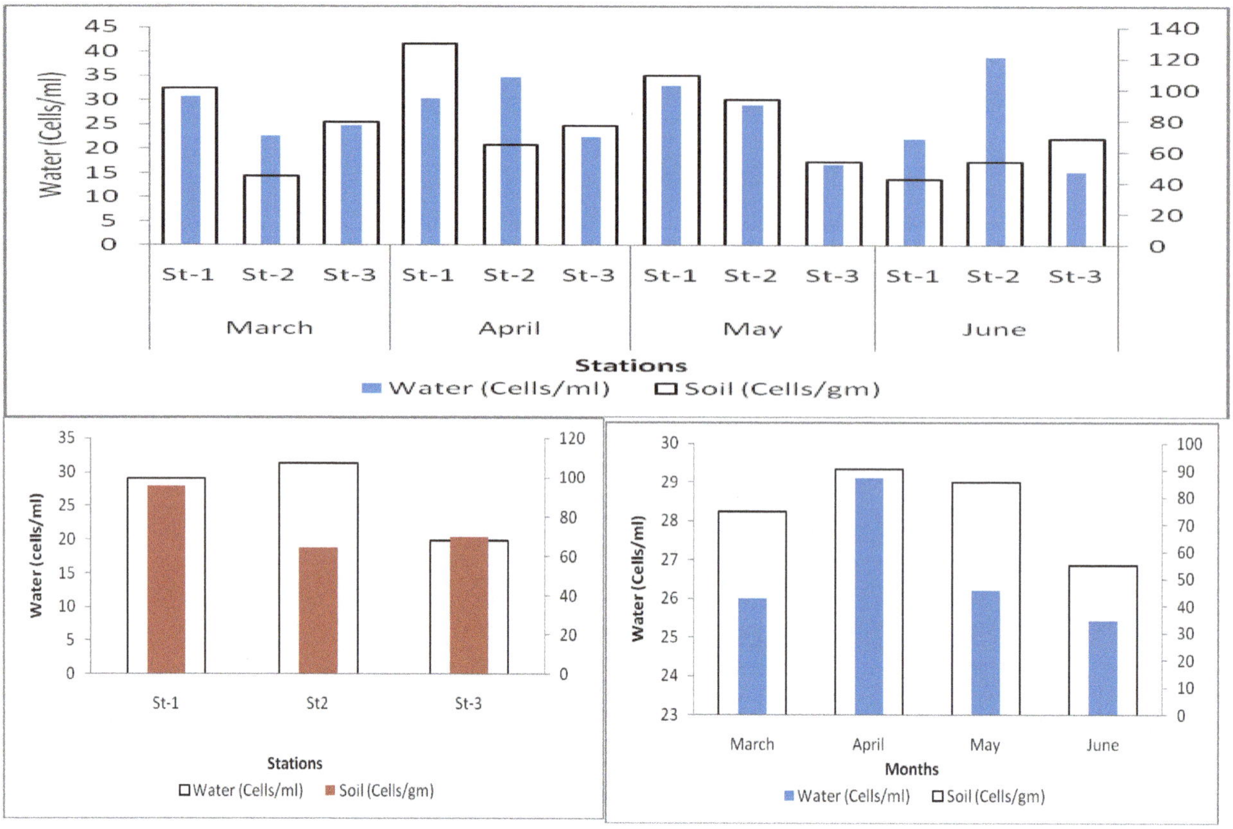

Figure 3. Variation in *E. coli* concentration in water and soil sample at different stations during different months.

was found to be 31.67°C during May while minimum was 26.83°C during March. On the other hand, average maximum water temperature was recorded 29.25°C at St-2 and St-3 and a minimum of 29.13°C at St-1. No significant difference was observed in water temperature among the stations but significant differences were found among the months (F=23, P<0.01). Salinity of the present study was found between 01 ppt to 05 ppt. The maximum was recorded (1.6 ppt) in June at St-12 and minimum 0.3 ppt at St-14 during March and April. Monthly maximum average salinity was found as 1.67 ppt during May where minimum salinity was recorded 0.33 ppt during March and April. Mean maximum salinity was observed at 2.3 ppt at St-3 and no salinity was found at St-1 and 2. No significant difference was observed among the stations but there was significant difference among the months (F=26, P<0.001) for salinity value. Dissolved Oxygen (DO) concentration in water sample varies between 2.40 to 2.88 mg/l. The maximum DO content was recorded as 2.91 in July at St-3 and the minimum was recorded in June at St-2. Monthly maximum mean DO was found as 2.71 mg/l during June and a minimum of 2.02 mg/l during May. On the other hand, maximum mean DO 2.63 mg/l was recorded at St-1 while a minimum of 2.25 mg/l was recorded at St-2. No significant difference was found in DO among the stations though significant differences were observed among the months (F=25, P<0.01). Water pH varies between 7.5 to 8.5 and maximum pH value was recorded as 8.5 at St-3, during June and minimum 7.5 during March at St-3. Monthly maximum mean pH was recorded as 8.13 during May and a minimum of 7.2 during June. Average pH at maximum level occurred at St-1(7.7) where the minimum was found at St-2 (7.6). No significant difference was found in pH values among the stations but significant difference was found among the months (F=31, P<0.01).

E. coli concentration in water and soil samples

Variation in *E. coli* concentration in water and soil samples at different stations during different months is shown in Figure 3. *E. coli* counted in water sample was found to contain the highest concentration, 35 cells/ml at St-2 during April while minimum concentration was found to be 22 cells/ml at St-1 during June. Average maximum concentration of *E. coli* was found to be 29 cells/ml during April while minimum concentration 25 cells/ml was found during June. During March and May, concentration of *E. coli* was found to be 26 cells/l. Mean concentration in

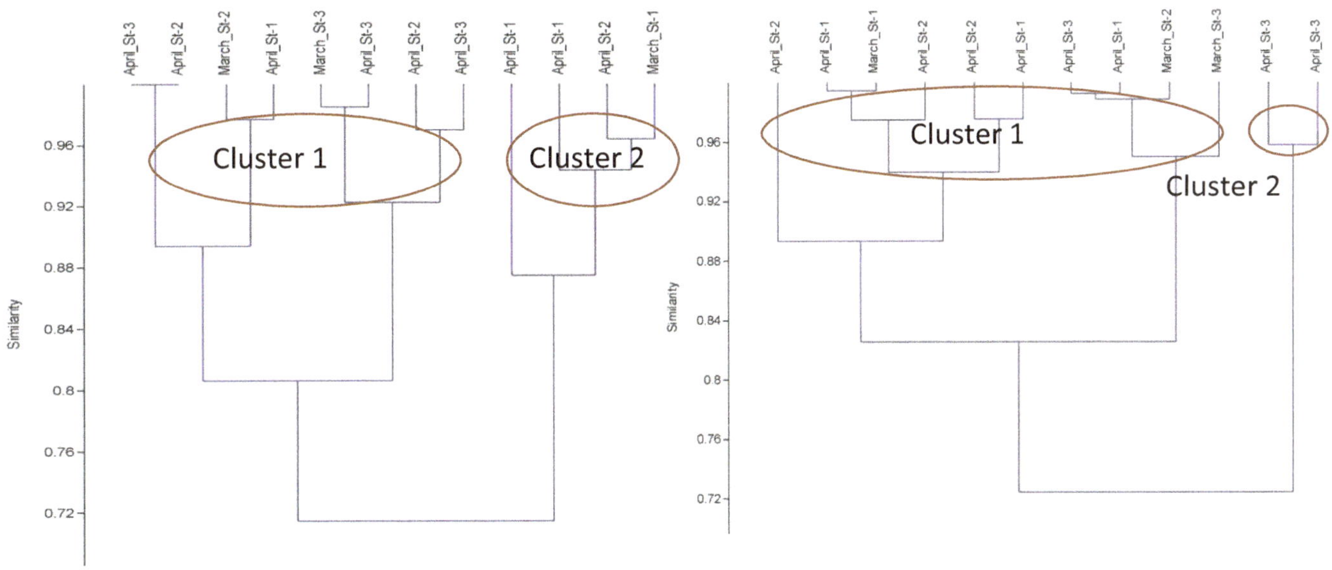

Figure 4. Dendrogram showing the cluster of *E. coli* occurrence in soil and water sample.

maximum level found 31cells/l at St-2 and minimum concentration of 20 cells/l at St-3 where 29 cells/l was recorded at St-1. *E. coli* counted in soil sample was found to have the highest concentration, 109 cells/g at St-1 during May while minimum concentration was found to be 54 cells/g at St-3 during May and St-2 during June. Average maximum concentration of *E. coli* was found to be 90 cells/g during April while minimum concentration was found to be 55cells/g in June. During March and May concentration of *E. coli* was found to be 75 and 85 cells/g respectively. Mean concentration in maximum level was found as 95 cells/g at St-1 and a minimum of 64 cells/g at St-2 where 70 cells/g was recorded at St-3.

Spatial and temporal similarity of *E. coli* occurrence

Cluster analysis of *E. coli* concentration in water and soil samples is based on different sampling stations and months as shown in Figure 4. Regarding temporal and spatial assemblage in soil sample two clusters were formed at 72% similarity level. Cluster 1 consists of April St-2, April St-1, March St-1, April St-3, March St-2 and March St-3 which showed similarity to April St-3. Similar to soil sample, two clusters were also found for soil samples regarding temporal and spatial assemblage of *E. coli* occurrence at 72% similarity level. Cluster 1 consists of April St-2, April St-3, March St-3 April St-1 which showed similarity with cluster 2 and consists of April St-1, April St-2 and March St-1.

Effects of water parameters on *E. coli* concentration

Regression line (Figure 5) has been drawn to see the influence of water parameters on *E. coli* concentration in water sample at Karnafuly River estuary. With the increase of water temperature *E. coli* concentration has increased a little in water sample (y=0.0237x+26.003) whereas in case of salinity, a decreased trend (y= -3.2751x+29.424) in occurrence of *E. coli* was found with increased salinity level. On the other hand similar to salinity, a decreased trend was found in *E. coli* concentration in water sample at Karnafuly River estuary with the increase of water pH (y= -5.0745x+65.599) and Dissolve Oxygen (y= -1.091x+29.408).

Similar to water sample, regression line (Figure 6) has also been drawn to observe the influence of water parameters on *E. coli* concentration in soil sample. In case of soil sample, similar trend of water sample *E. coli* concentration was found for all parameters. *E. coli* concentration has increased in soil sample of Karnafuly River estuary with the increase of water temperature (y=0.8078x+52.823).

On the other hand decreased trend was observed for *E. coli* concentration in soil sample with increase of salinity (y= -4.7896x+80.41), pH (y= -1.9996x+91.749) and Dissolve oxygen (y= -4.9654x+88.766) concentration in water.

DISCUSSION

From the present physicochemical study of the water quality of the Karnafuly, it can be concluded that the condition of the Karnafuly River is critical. Decreasing trend of DO of the Karnafuly River water was observed from the present and previous study (Majid et al., 1977). Minimum DO value found for the Karnafuly River was as

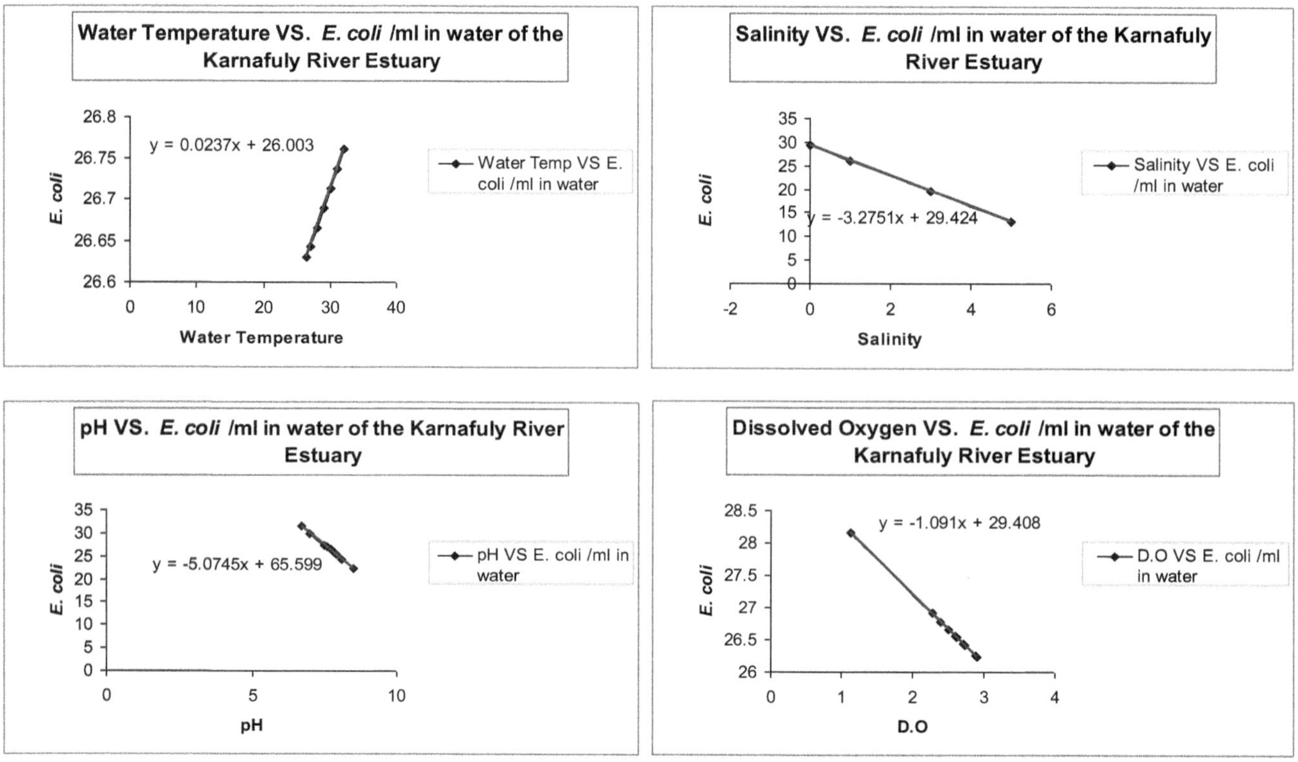

Figure 5. Regression analyses showing the influence of water parameters on *E. coli* concentration in water sample.

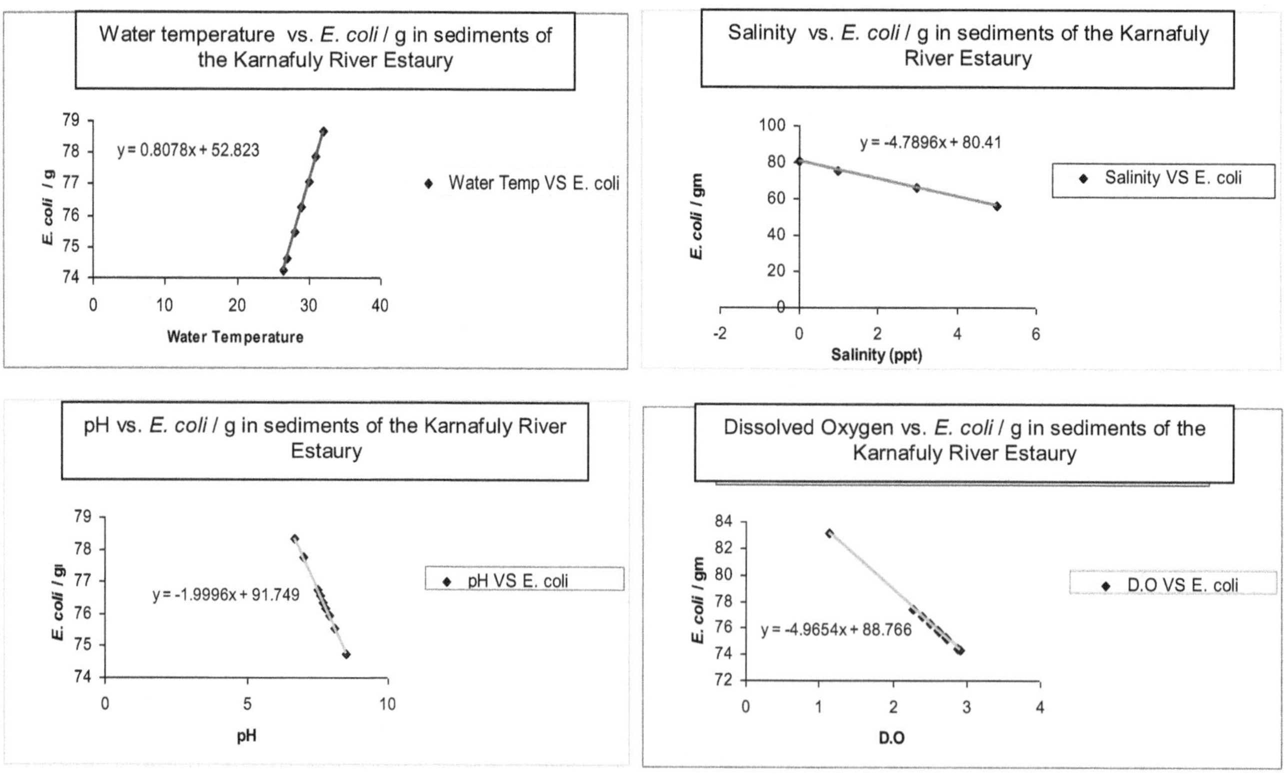

Figure 6. Regression analyses showing the influence of water parameters on *E. coli* concentration in soil sample.

low as 2.40 mg/l. This indicates the critical condition of this River. Similar situation was observed by Gasim (2007). Alam et al. (1996) reported DO values in the range (2.74 to 5.12 mg/l) and (3.95 to 5.97 mg/l) for river water samples and (3.73 to 5.01 mg/l) and (5.04 to 5.49 mg/l) for lake water samples in the dry and rainy seasons, respectively. Mean pH value of the Naf River water was found in the alkaline range. Similar pH values 7.5 to 8.5 were reported by Chowdhury et al. (2009) for the Naf River, which is almost similar to the present findings. McLusky and Elliot (2004) stated that salinity of an estuary ranged between 0.50 to 35 ppt and Ahammdd (2004) showed the salinity ranged between 14.43 to 25.92 ppt; the salinity of the present work was found at a similar value. Alam (1993) also reported variation of water temperature for seasonal change. Kamal (1992) and Zaman (1994) observed that seasonal variation of water temperature ranged between 7.54 and 10°C. Belaluzzaman (1995) also found 10°C water temperature variations at Bangladesh coast. During the present study monthly variation of water temperature was observed.

Several non-point and point sources can contribute to the presence of fecal indicator bacteria (*E. coli* and enterococci) in aquatic systems that can be detrimental to public health such as, humans, agriculture, water run-off, tidal actions, animal traffic, sustained winds, boats, dredging and polluted groundwater and environmental sources such as soil. Toxic metals pollution is predominant in the Karnafuly River water as the toxic metals were included from drainage of domestic waste. A large number of chemical and fertilizer industries have been established since the independence on both the banks of the Karnafuly. Effluents from these industries are reportedly being directly discharged into this river. Huge amount of solid wastes and effluents are discharged through Chaktai, Sundari, Noakhal, Mazirghat, Gupta, Mohesh, Shikalbhaha and Ferighat canals into the Karnafuly River; as a result, pollution of this river is increasing day by day. Leaking and leaching of oil from ships and boats are also polluting the water of the Karnafuly River.

Due to the climate change during flood, soil erosion and land slide increases the suspended solids, metallic and other pollutants. Polluted water and soil create a favorable condition for *E. coli* multiplication. The previous study revealed high concentrations of *E. coli* and total coliforms bacteria in water, soil and mussels from Guion Creek (Yeung-Cheung et al., 2009). 18000 coliforms/100 ml is recorded in the water sample of Karnafuly River estuary near sewage disposal areas (Hossain, 1980), which is far higher than safer levels (0/100 ml) for drinking water and greater than 200 cells/100 ml for bathing recommended by WHO (1982) which is similar to the present study. A study of the seasonal occurrence and distribution of microbial indicators and pathogens were done in the Rhode river of Chesapeake Bay where they found the highest frequency of occurrence of total *E. coli* load in muddy creek, a marsh area receiving pasture land runoff.

The second highest frequency of occurrence of those bacteria was in Coddle creek a populated area and the lowest measurements of those microbial parameters were obtained at stations in the central portion of the river (Carney et al., 1975). A study was done about the presence of pathogenic organisms in the River Ganga (Sinha and Vajpayee, 1972). Soil appear to serve as a hospitable environment for bacterial survival due to the availability of soluble organic matter and nutrients (Jamieson et al., 2005), protection from predators such as protozoa (Decamp and Warren, 2000), and shielding from exposure to UV sunlight (Koirala et al., 2008). The present study found high concentration of *E. coli* in soil. The concentration of *E. coli* from both water and soil increased during rainfall, but the bacterial concentration quickly decreased to the pre-rain levels (Whitman et al., 2006).

The present study found increased concentration of *E. coli* during the starting of the rainy season. Studies about the physiological states of *E. coli* can be affected by various environmental conditions, e.g. temperature, pH, salinity etc (Miller and Litsky, 1976.) An observation was found that the *E. coli* can survive longer in saline medium than in distilled water (Winslow and Falk, 1923) which is similar to present findings. This result is also consistent with previous results which showed that *E. coli* in distilled water, die most rapidly of elevated temperature and high pH values (Cohen, 1992).

Conclusion

Water of Karnafuly River estuary used for any purpose, may affect people by several health hazard diseases like diarrhea, dysentery, cholera, typhoid etc. With the relation of some relative physio-chemical parameters against *E. coli* bacteria both in water and soil, it was observed that against temperature, *E. coli* was insignificantly (1% confidence level) positively correlated whereas negatively correlated against salinity, pH and DO. In the present work, with comparison between the water and soil, it was found that soil sample carries higher total *E. coli* load than water sample. It may be 2-3 times more than water in all stations of the Karnafuly River estuary.

In Karnafuly River area, people suffer from various health hazard diseases due to use of water for drinking, bathing, washing and soil for various household purposes. Water and soil resources and fish species of Karnafuly estuary are contaminated by *E. coli*. After the study, it is concluded that *E. coli* bacteria from three stations of the Karnafuly, exceed the standard level both in water and soil, so this water is polluted and unhygienic for drinking and even washing without treatment, which directly affects human health.

ACKNOWLEDGEMENTS

The authors are grateful for the technical lab and instrument support of the Institute of Marine Sciences and Fisheries, University of Chittagong, Bangladesh to complete the research work successfully.

REFERENCES

Ahmed F (2004) Catch Composition of Estuarine Set Bag Net (ESBN) in the Moheshkhali Channel of the Bay of Bengal, Bangladesh. MSc Thesis, Institute of Marine Sciences, University of Chittagong, Bangladesh.

APHA (1976). Standard methods for examination of water and waste water (14th ed.). American Public Health Association Inc. Washington D.C.

Carney JF, Carty CE, Colwell RR (1975). Seasonal occurrence and distribution of microbial indicators and pathogens in Rhode River of Chesapeake bay. Appl. Microbiol. 30:771-780.

Cohen GF (1992). The effect of ecological factors on bacteria in water. Vern. Inst. Verein. Angeu. Limnol. 17:731-743.

Decamp O, Warren A (2000). Investigation of *E. coli* removal in various designs of subsurface flow wetlands used for wastewater treatment. Ecol. Eng. 14:293-299.

EQSB (1991). Environmental Quality Standard for Bangladesh. Department of Environment, Government of the People's Republic of Bangladesh.

Geldreich EE (1970). Applying bacteriological parameters to recreational water quality. J. Am. Water Works Assoc. 62:113-120.

Hossain MM (1980). Pollution as revealed by macro benthic organisms in the Karnafuly river estuary. M. Sc. Thesis. Institute of Marine Sciences, University of Chittagong, Chittagong, Bangladesh.

Islam MN (1998). Study on the pollution indicating bacteria in the brackish water environment with reference to Karnafuly river estuary. M. Sc. Thesis. Institute of Marine Sciences, University of Chittagong, Chittagong, Bangladesh.

Jamieson RC, Joy DM, Lee H, Kostaschuk R, Gordon RJ (2005). Resuspension of soil-associated *Escherichia coli* in a natural stream. J. Environ. Qual. 34:581-589.

Joy CM, Bakkrishnan KP, Joseph A (1990). Physicochemical aspects of a tropical river receiving industrial effluents. New Delhi, India. pp. 220-233.

Kim GH, Yur JH, Kim JK (2007). Diffuse pollution loading from urban storm water runoff in Daejeon city, Korea J. Environ. Manag. 85:9-16.

Koirala SR, Gentry RW, Perfect E, Schwartz JS, Sayler GS (2008). Temporal variation and persistence of bacteria in streams. J. Environ. Qual. 37:1559-1566.

McLusky DS, Elliott M (2004). The Estuarine Ecosystem. Oxford. Oxford University Press. p. 214.

Miller SK, Litsky BS (1976). Ecological effect of sewage pollution in Biscayne Bay. Florida. In: Soil and micro organisms. Bull. Mar. Sci. 11:364.

Rehmann CR, Soupir MR, (2009). Importance of interactions between the water column and the soil for microbial concentrations in streams. Water Res. 43:4579-4589.

Rizbi SNH (1971). Bangladesh district gazetteers, Chittagong. Govt. of Bangladesh, Service and general administration department. Bangladesh Government Press. Dacca. p. 64.

Sinha AK, Vajpayee SL (1972). Pathogenic organisms in River Ganga between Shuklagonj and Kalkankar. Ashish Publishing House. New Delhi. pp. 207-216.

Whitman RL, Nevers MB, Byappanahalli MN (2006). Examination of the watershed-wide distribution of *E. coli* along southern Lake Michigan: an integrated approach. Appl. Environ. Microbiol. 72(11):7301-7310.

WHO (1984). Guidelines for drinking water quality. World Health Organization.

WHO (1993). Microbial specifications for drinking water. World Health Organization. Ecol. Microbiol. p. 173.

Winslow CK, Falk TS (1923). Estuarine oil pollution effect on the micro benthos. Fishing news (Books) Ltd. 23, Rosemount Avenue, West Byfleet Surrey, England. pp. 243-258.

Yeung-Cheung AK, Benevento NM, Pavel DM (2009). An evaluation of the effectiveness of the Gunderboom in protecting waters from bacteria. J. Environ. Health 71(8):42-47.

Community-acquired methicillin-resistant *Staphylococcus aureus* carrying *mecA* and Panton-Valentine leukocidin (PVL) genes isolated from the holy shrine in Najaf, Iraq

Ali M. Al-Mohana[1], Alaa H. Al-Charrakh[2]*, Fadhil H. Nasir[1], and Miaad K. Al-Kudhairy[3]

[1]Department of Microbiology, College of Medicine, Kufa University, Iraq.
[2]Department of Microbiology, College of Medicine, Babylon University. Babylon province, Iraq.
[3]Department of Community Health, Kufa Technical Institute, Iraq.

Two hundred and eighty six *Staphylococcus aureus* isolates were collected from separate places of the holy shrine in Najaf city, Iraq. Phenotypic and genotypic examination for community associated methicillin resistant *S. aureus* (CA-MRSA) isolates was carried out. Antibiotic and plasmid profiles of these isolates were also done. The CA-MRSA isolates were examined using polymerase chain reaction (PCR) primers for Panton Valentine leukocidin (PVL) gene. 54 (18.8%) of all of the *S. aureus* examined were identified as community associated methicillin resistant *S. aureus*, of which 11 isolates were CA-MRSA. CA-MRSA isolates were examined and subdivided into two groups according to their antibiotic profiles. Eight of the 11 MDR CA-MRSA isolates were alike in their plasmid profiles. Results of PCR revealed that 3 (27.2%) of CA-MRSA isolates carried PVL genes and 9 (72.8%) carried none. The study also revealed that CA-MRSA isolates were resistant to all ß-lactam and many of the non ß-lactam antibiotics and the frequency of resistance was higher among CA-MRSA isolates than methicillin sensitive *S. aureus* (MSSA), with low ratio of carrying PVL gene among CA-MRSA isolates. Surveillance and researches on CA-MRSA that carry PVL gene should continue to provide a significant insight into the prevalence and epidemiology of these important resistant pathogens.

Key words: Methicillin resistant *Staphylococcus aureus*, community associated methicillin resistant *S. aureus* (CA-MRSA), Panton Valentine leukocidin (PVL) genes, Iraq.

INTRODUCTION

Methicillin-resistant *Staphylococcus aureus* (MRSA) has become the most prevalent pathogen causing hospital and community infection throughout the world and MRSA incidence is still increasing in many countries (Crisostomo et al., 2001). MRSA is resistant to practically all ß-lactam antibiotics, a class of antibiotics represented by penicillins and cephalosporins (Lee, 2003). The ß-lactam resistance of MRSA is caused by the production of a novel penicillin binding protein (PBP) designated PBP 2' (PBP 2a), which has remarkably reduced binding affinities to ß-lactam antibiotics. Despite the presence of otherwise inhibitory concentrations of ß-lactam antibiotics, MRSA can continue cell wall synthesis solely depending upon the uninhibited activity of PBP 2' which is encoded by a *mecA* gene located on the chromosome of MRSA (Chambers, 2003).

Panton Valentine leukocidin (PVL) is a cytotoxin, one of the β-pore forming toxins (Panton and Valentine, 1932), and its presence is associated with increased virulence of certain strains of *S. aureus*. It is present in the majority of community associated methicillin resistant *S. aureus* (CA-MRSA) and is the cause of necrotic flesh eating lesions, an aggressive condition that often kills patients within

*Corresponding author. E-mail: aalcharrakh@yahoo.com, ahani67@gmail.com.

72h (Voyich et al., 2007). PVL creates pores in the membranes of infected cells. PVL is produced from the genetic material of a bacteriophage which infects *S. aureus*, making it more virulent. The aim of the present study was to monitor and limit the spreading factors of CA-MRSA and possibility of carrying PVL gene in the holy shrine in Najaf city, Iraq, especially in pilgrimage season, and religious visits of foreign visitors to Iraq from other countries who visit the holy shrine, who perhaps take share in transmission of virulent multidrug CA-MRSA.

MATERIALS AND METHODS

Bacterial isolates

A total of 286 *S. aureus* isolates were collected and isolated as swabs from different places of the holy shrine of Imam Ali in Najaf city, Iraq, from January 1, 2009, to December 31, 2010. All isolates of *S. aureus* were stored in 15% glycerol nutrient broth at -20°C.

Bacterial identification

Isolates were identified as *S. aureus* by testing their ability to ferment mannitol salt agar and DNase production on agar plates (India, Himedia). Gram's stain was performed and isolates were identified by their ability to produce catalase and coagulase enzymes (Barrow and Feltham, 2003) then, the identification was confirmed using API systems strips as recommended by Biomérieux (France).

Antimicrobial susceptibility testing

The susceptibility of the bacterial isolates to antimicrobial agents was determined by using disk diffusion method (NCCLS, 2003) on Muller-Hinton agar (India, Himedia) and interpreted according to Clinical and Laboratory Standards Institute documents (CLSI, 2010). The following antimicrobial agents were obtained from Oxoid, U.K, as standard reference disks as known potency for laboratory use: amoxyclav, ampicillin-cloxacillin, azithromycin, cefixime, ceftazidime, ceftriaxone, chloramphenicol, ciprofloxacin, clarithromycin, cloxacillin, erythromycin, gentamicin, penicillin G, rifampicin, trimethoprim, and vancomycin (Turkey, Bioanalyse). All antibiotic plates without NaCl supplementation were incubated at 35°C for 16 to 20 h. Multiresistance was defined as resistance to more than two antimicrobial agents (Merlino et al., 2002).

Detection of MRSA isolates

Methicillin resistance among isolates of *S. aureus* was determined by disk diffusion susceptibility testing using a 30 µg cefoxitin disk (Turkey, Bioanalyse) as recommended by NCCLS (2003). Using breakpoints nearest ≤ 21 mm for *mecA* positive and nearest ≥ 22 mm for *mecA* negative isolates of *S. aureus*, the sensitivity at 18 h was 98 to 100% while specificity was 100% (CLSI, 2010; Cauwelier et al., 2004; Skov et al., 2003; Swenson and Tenover, 2005).

DNA extraction

Genomic DNA used as target for plasmid profile and PCR assays was extracted by using a standard phenol chloroform procedure (Sambrook and Rusell, 2001).

PCR analysis

Oligonucleotide primers were designed for *mecA* gene detection by using oligo MRS1F (5'-TAGAAATGACTGACGTCCG-3') and oligo MRS2R (5'-TTGCGATCAATGTTACCGTAG-3') (Monterial, Alpha) (Santos et al., 1999).

Amplification of PVL genes was accomplished by using LUK-PVL-1 (5'-ATCATTAGGTAAAATGTCTGGACATGATCCA-3') and the LUK-PVL-2 (5'-GCATCAASTGTATTGGATAGCAAAAGC-3') primers (Monterial, Alpha) (Scott et al., 2005). Oligonucleotide primers for standard PCR amplifications utilized PCR Master Mix (USA, Pomega) and crude cell lysates on high purity genomic DNA templates (Sambrook and Rusell, 2001).

DNA agarose gel electrophoreses

All genomic DNA and all amplified PCR products were visualized following electrophoresis in 1% agarose gels run at 70 V with ethidium bromide staining (USA, Sigma), and comparison to standard positive control (Monterial, Alpha). *mecA* gene and PVL gene positive strains yielded an amplification product of shining band with standard positive control(Santos et al., 1999; Scott et al., 2005).

RESULTS AND DISCUSSION

Antibiotic profiles of *S. aureus*

The antibiotic profiles of all 56 *S. aureus* isolates were examined (Figure 1). All *S. aureus* isolates were susceptible to cefixime, gentamicin, chloramphenicol and vancomycin. Majority (45, 83.8%) of the isolates were highly resistant to penicillin G and resistant to ceftazidime (42, 77.7%), ampicillin-cloxacillin (29, 53.7%), amoxyclav (26, 48.1), azithromycin, erythromycin (20, 37%), clarithromycin (21, 38.8%), cloxacillin (17, 32.2%), ceftriaxone and trimethoprim (15, 27.7%), ciprofloxacin (9, 16.6%), and were resistant to rifampicin (4, 7.4%).

Resistance to penicillins in this study was found commonly among *S. aureus* isolates tested and is closely related to many studies (Katayama et al., 2000; Lee, 2003; Merlino et al., 2002). These results are in agreement with those results being reported by other researchers who found that production of β-Lactamase is common, under plasmid control, and makes the organisms resistant to many penicillins (penicillin G, ampicillin, ticarcillin, piperacillin, and similar drugs) (Kuehnert et al., 2006).

The resistance to penicillins produces an additional penicillin-binding protein, PBP2a or PBP2' (Kuehnert et al., 2006), which confers resistance to all currently available ß-lactam agents. This resistance is under chromosome control (Merlino et al., 2002).

Four variants of *S. aureus* β-lactamase (A, B, C, D) can be distinguished. Although the genes encoding types A, C, and D staphylococcal β-lactamases are usually located

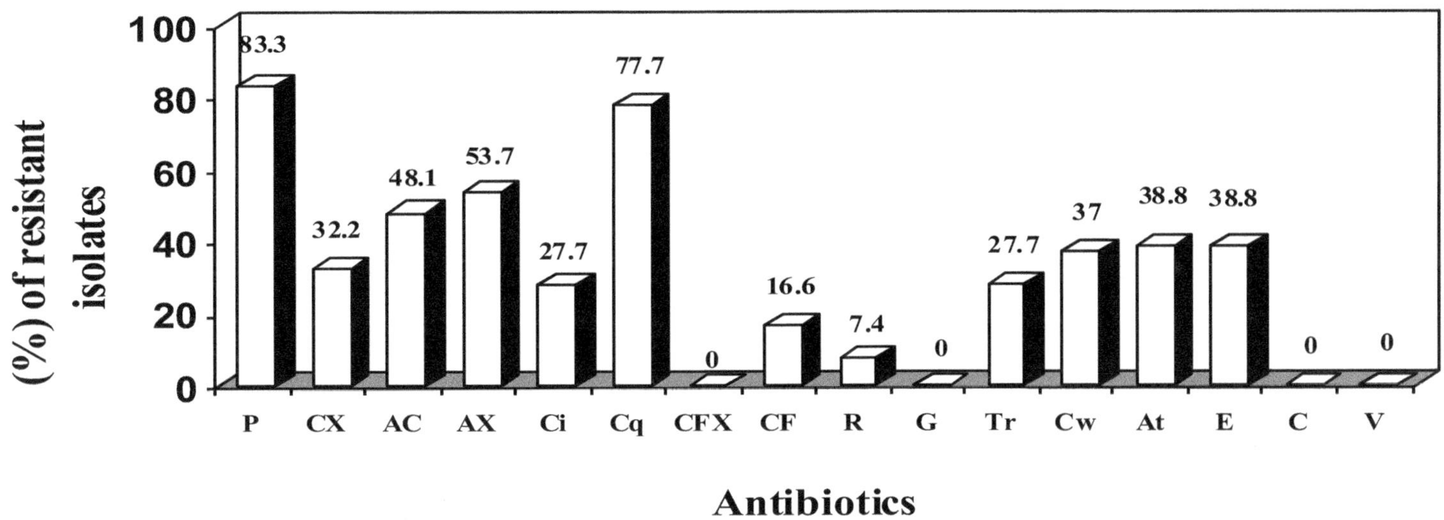

Figure 1. Antibiotic resistance patterns of *S. aureus* isolates.

Table 1. Numbers and percentages of CA-MRSA isolates detected by using a 30 µg cefoxitin disk and PCR analysis.

CA-MRSA isolates carrying *mecA* gene	No. of isolates	No. of isolates (%)
mecA-positive isolates (CA-MRSA)	11	20.4
mecA-negative isolates (MSSA)	43	79.6
Total	54	100

on a plasmid, the gene for type B β-lactamase is believed to reside on the chromosome of phage group II isolates. For these reasons, multiple methods should be used for detection of the different types of resistance to β-lactam antibiotics (Olsen et al., 2006).

On the other side, there are many reasons to the variations in a result for resistance to tetracyclines, erythromycins, aminoglycosides, and other drugs because the frequent plasmids are mediated in staphylococci, and these plasmids can be transferred from one strain to another by conjugation or transduction. Therefore, resistance carried by plasmids can spread rapidly amongst many different strains and genera (Black, 2002). Despite these, there are many studies describing the resistance to tetracyclines, erythromycins, aminoglycosides, carrying on the *mecA* of chromosome MRSA strains (Cauwelier et al., 2004; Skov et al., 2003).

The mechanisms responsible for resistance to erythromycin in staphylococci are target site modification and active drug efflux (Black, 2002; Berg et al., 1998). Target site modification is mediated by the presence of *erm* genes [*erm* (A), (B) and (C)] conferring resistance to Macrolide, lincosamide and streptogramin type B (MLS_B) antibiotics. Phenotypic expression of MLS_B resistance can be inducible or constitutive (Schmitz et al., 2000).

All isolates of *S. aureus* were susceptible to vancomycin, indicating the absence of *vanA* gene. This susceptibility to vancomycin encourages using it as the last step for treatment of hard infection with multidrug resistant *S. aureus* (Moise-Broder et al., 2004; CLSI, 2010).

Antibiotic resistance among isolates of *S. aureus* in the present study was comparable to reports from other parts of the world, which also revealed multiple drug resistance among staphylococci (Merlino et al., 2002; Santos et al., 1999; Tiemersma et al., 2004).

CA-MRSA detection

A total of 54 *S. aureus* isolates were investigated in this study using phenotype [cefoxitin disk diffusion (DD)], and genotype methods PCR analysis. The phenotypic and genotypic characteristics of the *S. aureus* are shown in Table 1. PCR of the *mecA* gene confirmed the identification of all (43, 79.6%) methicillin sensitive *S. aureus* isolates and (11, 20.4%) methicillin resistant *S. aureus* isolates as previously indicated by cefoxitin disk. The results of PCR are shown in Figure 2.

Resistance to methicillin is independent of β-lactamase production. The *mecA* gene for methicillin resistance resides on the chromosome. Accurate detection of *mecA*

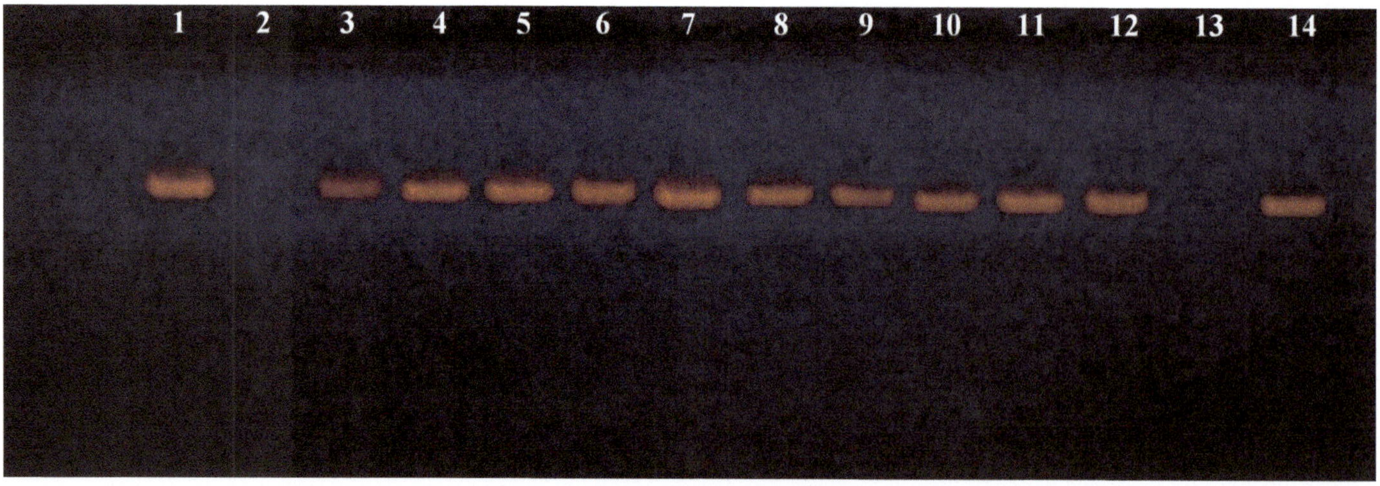

Figure 2. Agarose gel electrophoresis demonstrative the results of PCR for *mecA* gene to CA-MRSA. Lane 1: positive *mecA* gene control; lane 2: negative *mecA* gene control; lanes 3, 4, 5, 6, 7, 8, 9, 10, 11, 12, 14, *S. aureus* carrying *mecA* gene; lane 13: *S. aureus* without *mecA* gene.

mediated resistance to methicillin and other penicillinase stable penicillins (PSPs) such as, oxacillin, nafcillin, cloxacillin, dicloxacillin, flucloxacillin, and mecillinam, is necessary to ensure appropriate antimicrobial chemotherapy of staphylococcal infections, particularly those from community associated infections. Oxacillin has been the agent recommended by CLSI (2010) for phenotypic tests to predict resistance to PBPs because of its stability and superior sensitivity over other PBPs for susceptibility tests (Pinho et al., 2001). However, antimicrobial susceptibility tests using oxacillin are often difficult to read despite changes in techniques to improve the discrimination between oxacillin susceptible and resistant results (Skov et al., 2003).

The recent development of cefoxitin disc diffusion tests is likely to alter the recommendations for these methods as studies all suggest that tests with cefoxitin are more reliable than those with oxacillin (CLSI, 2010; Skov et al., 2003).

It is suggested that no special medium or incubation temperature is required with cefoxitin (Cauwelier et al., 2004).

In disc diffusion tests, hyper producers of penicillinase may show small methicillin or oxacillin zones of inhibition, whereas most true methicillin resistant isolates give no zone. Resistance mediated by *mecA* may be confirmed by PCR analysis. Some hyper producers of penicillinase give no zone, particularly with oxacillin, and will therefore be falsely reported as MRSA. Tests with cefoxitin do not appear to be affected to the same extent as oxacillin by hyper production of penicillinase. Expression of methicillin resistance in *S. aureus* depends on environmental factors such as temperature and osmolarity (Swenson and Tenover, 2005). As previously reported, better phenotypic expression of oxacillin resistance was observed at 30°C or after addition of NaCl (4%) to the culture medium (CLSI, 2010).

The cefoxitin DD test and PCR can be used to predict the presence of *mecA* gene in *S. aureus* with a high degree of sensitivity and specificity when compared to *mecA* detection using oxacillin disk. Those using disk diffusion as their routine susceptibility testing method for staphylococci should replace their 1 µg oxacillin disks with 30 µg cefoxitin disks for routine testing because the oxacillin is affected by high production of β-lactamase enzymes (Swenson and Tenover, 2005).

Antibiotic profiles of CA-MRSA

Eleven CA-MRSA of the *S. aureus* isolates underwent antibiotic susceptibility testing, and the results are shown in Figure 3. All the CA-MRSA were resistant to penicillin, cloxacillin, amoxyclav, ampicillin-cloxacillin, ceftriaxone, ceftazidime, cefixime, 8 (72.7%) were highly resistant to azithromycin erythromycin, and clarithromycin, 5 (45.4%) were resistant to ciprofloxacin, 4 (36.3%) to trimethoprim, and 15 (27.7%) to gentamicin. In addition all of the CA-MRSA isolates were susceptible to rifampicin, chloramphenicol, and vancomycin.

MRSA is practically resistant to all β-lactam antibiotics, a class of antibiotics represented by penicillins and cephalosporins (Katayama et al., 2000; CLSI, 2010). Resistance due to β-lactamases and PBPs mediated *mecA* gene is an emerging global problem that needs to be monitored. Resistance organisms are increasing in number and causing more severe infections because of their continuous mutation. The pharmaceutical industry makes little effort in developing new antibiotic agents because of the emergence of resistance worldwide. It takes at least 10 years to establish new agents, and after 2 years in clinical use, resistance probably emerges.

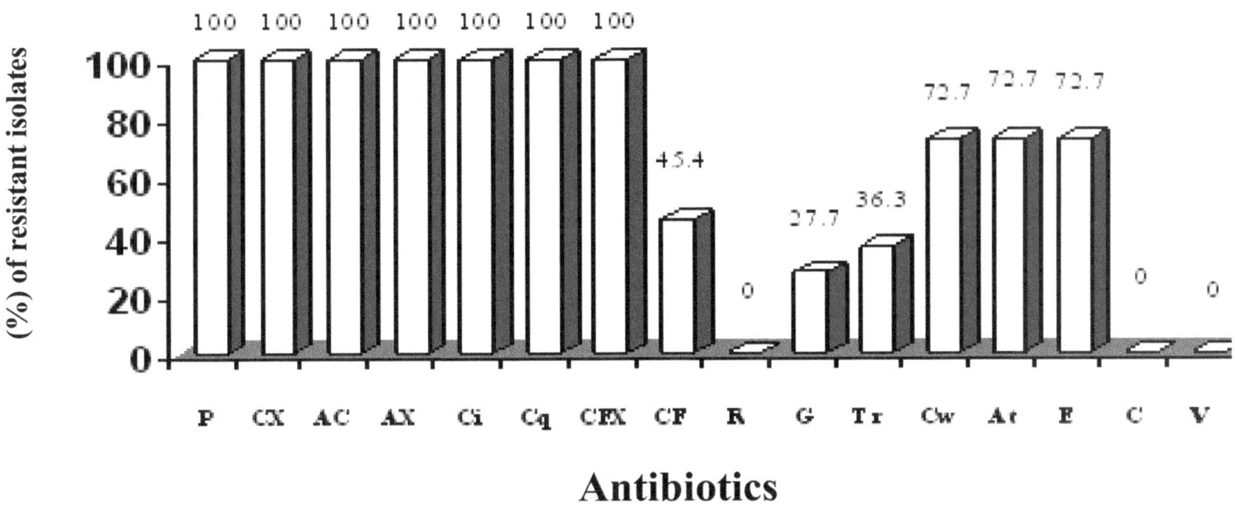

Figure 3. Antibiotic resistance pattern of CA-MRSA isolates. Note: P: penicillin G, CX: cloxacillin, AC: amoxyclav, AX: ampicillin-cloxacillin, Ci: ceftriaxone, Cq: ceftazidime, CFX: cefixime, CF: ciprofloxacin, R: rifampicin, G: gentamicin, Tr: trimethoprim, Cw: clarithromycin, At: azithromycin, E: erythromycin, C: chloramphenicol, and V: vancomycin.

Therefore, reliable, rapid and accurate detection methods have to be developed (Pinho et al., 2001).

It was observed that the High resistance of azithromycin and erythromycin in this study is in line with many studies which reported that transposon Tn*554* (Ito et al., 1999), encoding resistance to macrolides (erythromycin and spectinomycin), was located in the upstream of the *mecl-mecR1-mecA* gene complex that directly carries MRSA strains, while the Plasmid pUB110, encoding resistance for aminoglycosides (gentamicin, kanamycin, tobramycin and bleomycin), was inserted between two insertion sequences of IS431 (or IS257) at the left of *mecA* gene (Katayama et al., 2000).

The result of vancomycin resistance is expected, because the resistance is very rarely reported (CLSI, 2010; Moise-Broder et al., 2004). This pattern is most commonly found among MRSA isolates tested at the department of health public laboratories in USA, UK and most European countries, and it is closely related to the other pattern of the primary types causing community acquired infections nationwide (Schmitz et al., 2000; Tiemersma et al., 2004; Gemmell et al., 2006; Wannet, 2002).

Antibiotic profiles of MDR CA-MRSA, NMDR CA-MRSA, and MSSA

The antibiotic profiles of all 54 *S. aureus* isolates were examined. NMDR CA-MRSA had variable antibiotic profiles. The results are shown in Figure 4. The 54 CA-MRSA isolates examined were subdivided into two groups according to their antibiotic profiles and comprised 3 (27.3%) non multidrug resistant NMDR CA-MRSA isolates, resistant to less than two non ß-lactam antibiotics, and 8 (72.7%) multidrug resistant MDR CA-MRSA isolates, resistant to three or more non ß-lactam antibiotics.

All NMDR CA-MRSA were susceptible to azithromycin, erythromycin, clarithromycin, ciprofloxacin, chloramphenicol, rifampicin, trimethoprim, whereas all MDR CA-MRSA were resistant to azithromycin, erythromycin, clarithromycin, 5 (45.3%) of MDR CA-MRSA were resistant to ciprofloxacin, and 4 (36.3%) were resistant to trimethoprim, but they were susceptible to rifampicin, and chloramphenicol. All 54 staphylococcal isolates (NMDR CA-MRSA, MDR CA-MRSA and MSSA) were susceptible to vancomycin.

NMDR CA-MRSA isolation has been reported with increased frequency worldwide, despite universal guidelines published by the NCCLS for the susceptibility testing of *S. aureus to* methicillin (CLSI, 2010; Merlino et al., 2002).

Laboratories without genotypic or monoclonal based assays may have problems in detecting or confirming low level *mecA* phenotypic heterogeneous expression of methicillin resistance in NMDR CA-MRSA (Wannet, 2002). The degree of heterogeneity is variable in NMDR CA-MRSA strains and modification of growth conditions can increase or decrease the proportion of cells that express resistance to various degrees (Merlino et al., 2002). In this study, it was found that using PCR for the detection of *mecA* gene is an excellent assay, and confirms that methicillin resistance tests for these NMDR CA-MRSA shows heterogeneous resistances. The direct implementation of such assays will allow more rapid establishment of effective antibiotic therapy and should help control the spread of MRSA in hospitals and the

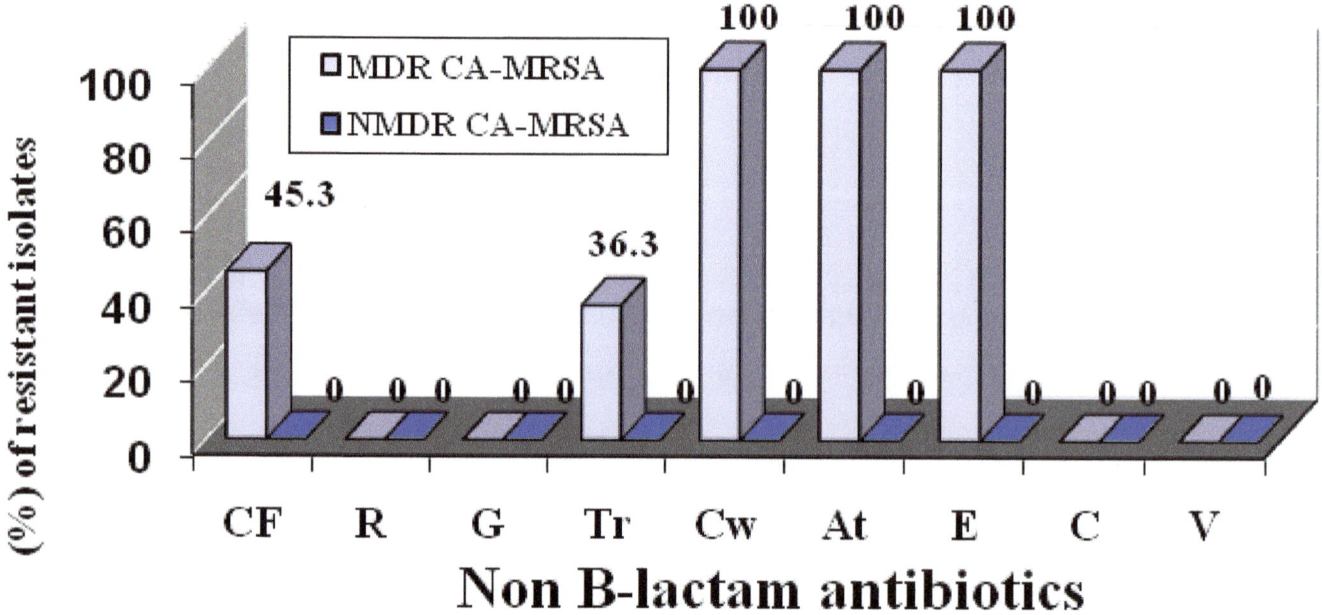

Figure 4. Antibiotic resistance pattern of MDR CA-MRSA and NMDR CA-MRSA isolates.

community (Merlino et al., 2002; Wannet, 2002). Further investigative research is in progress to elucidate factors involved in heterogeneous expression of methicillin resistance in NMDR CA-MRSA and the effects of different non ß-lactam antibiotics on these strains as an alternative to vancomycin use (Merlino et al., 2002).

Because there are epidemiological and microbiological differences between community associated methicillin resistant *S. aureus*, these modifications could include more frequent culturing and susceptibility testing of *S. aureus* isolates of clinical infections, and careful selection of empirical antimicrobials when such treatment is indicated for suspected staphylococcal infections (that is, clinicians should be aware that MRSA organisms are non susceptible to ß-lactam antimicrobials) (CLSI, 2010). Because most community associated MRSA isolates were susceptible to several already available antimicrobial agents, and because most patients had noninvasive infections, the treatment of community associated MRSA infections should not routinely require the use of vancomycin. However, patients with MRSA infections should be evaluated and treated appropriately and receive follow-up evaluation to ensure resolution of their infection (Merlino et al., 2002).

To improve prevention strategies, more work is needed to characterize specific risk factors for CA-MRSA infections. However, at least one study has implicated prior exposure to antimicrobial agents as an independent risk factor for CA-MRSA infection (CDC, 2007). This suggests that the judicious use of antimicrobials, particularly in outpatient settings, could help control the emergence of CA-MRSA strains and limit the acquisition of additional antimicrobial resistance genes in existing strains (CDC, 2007; Andenesch, 2003; Okuma et al., 2002). Currently, there is no Arabic or Iraqi data to suggest that decolonization protocols for MRSA patients or their families are necessary or have long term effectiveness (CDC, 2007). Resistance to antimicrobial agents used for decolonization has evolved rapidly in settings in which such strategies have been attempted (Boyle-Vavra and Daum, 2007; CDC, 2007; Skov et al., 2003).

Plasmid profiles pattern of MDR CA-MRSA

All eight MDR CA-MRSA were selected as ß-lactam resistance, and multidrug resistant to non ß-lactam antibiotics. They are use to determine their plasmid profiles patterns. Isolates in Figure 5 were alike in their plasmid profiles and there were two bands of different plasmids that appeared identical in shape and position when electrophoresis was performed. These bands must be divided depending on their molecular weight, by using restriction enzymes and ladder as marker, but in this study, plasmid profiles were used just to ensure that isolates have DNA of plasmid that encodes antibiotic resistance and many virulence characteristics. Only one study was in agreement with our results (Santos et al., 1999).

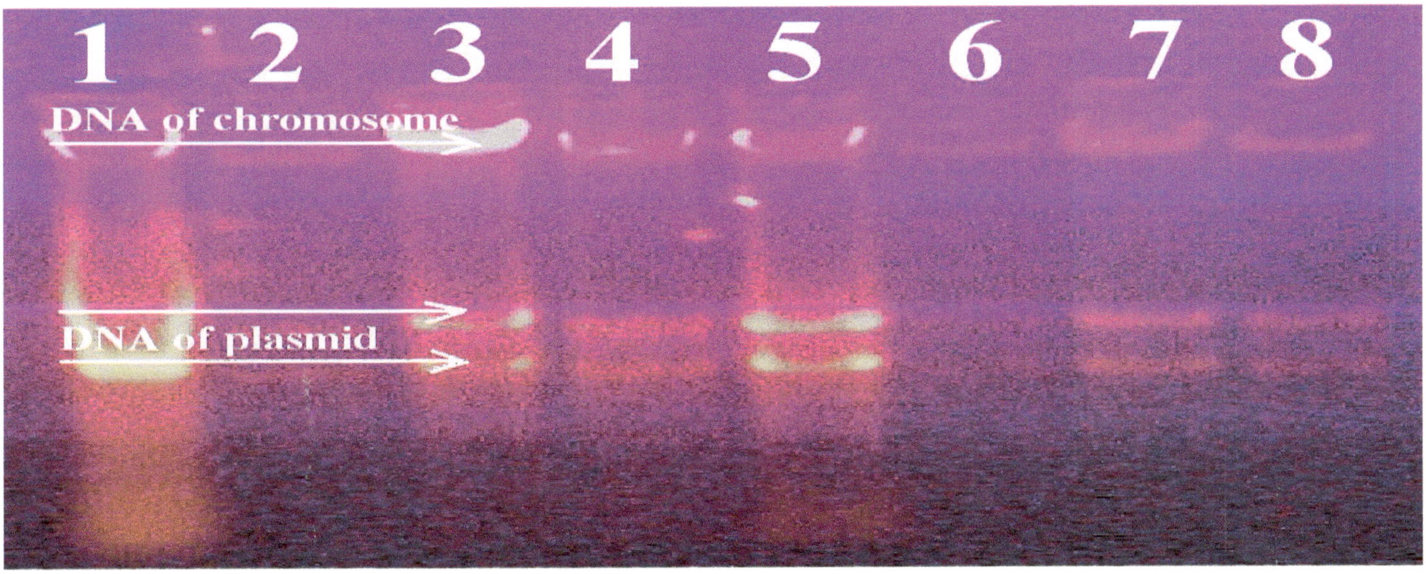

Figure 5. Agarose gel electrophoresis demonstrative the results of DNA content for MDR CA-MRSA.

Table 2. Numbers and percentages of CA-MRSA isolates carrying PVL gene detected by using PCR analysis.

CA-MRSA isolates carrying PVL gene	No. of isolates	No. of isolates (%)
PVL-positive isolates	3	27.2
PVL-negative isolates	9	72.8
Total	11	100

Plasmid analysis has been applied to determine the evolution and spread of antibiotic resistance among isolates with different profiles or among different species of organisms. Plasmids are not generally helpful in the differentiation between strains, because plasmids are often mobile extrachromosomal DNA fragments that can be acquired and deleted (Feil et al., 2003). Many plasmids carry antibiotic resistance determinants that are contained within mobile genetic elements (transposons) that can move in or out of plasmids and the chromosome, allowing for the DNA composition of a plasmid to potentially change rapidly (Singh et al., 2006). The analysis of plasmid content is limited to investigations in which a plasmid epidemic is responsible for the spread of a resistance trait. The selective pressure for organisms to express antibiotic resistance may cause such plasmids to spread rapidly among strains and among different species (Feil et al., 2003; Singh et al., 2006).

Several of the antibiotic resistances in *S. aureus* determinants are carried by pSK41-like plasmids (Berg et al., 1998; Firth et al., 2000). The sequence of pSK41, a prototypical multiresistance plasmid from *S. aureus*, has been determined, representing the first completely sequenced conjugative plasmid from a Gram-positive organism.

The pSK41-like conjugative plasmids usually carry a Tn*4001*-like element encoding the bifunctional aminoglycoside resistance gene *aacA-aphD* and often contain cointegrated remnants of small plasmids encoding genes such as *aadD*, mediating resistance to the aminoglycosides neomycin and kanamycin; *smr*, conferring antiseptic and disinfectant resistance; and *dfrA*, encoding resistance to trimethoprim (Singh et al., 2006; Firth et al., 2000). In addition, members of the pSK1 plasmid family typically confers resistance to antiseptics and disinfectants, encoded by *qacA*, and may also carry Tn*4001*, a Tn*552*-like transposon, and/or a cointegrated remnant of a *dfrA*-carrying plasmid previously designated Tn*4003*. Analysis of the sequence has enabled the identification of the probable replication, maintenance, and transfer functions of the plasmid (Firth et al., 2000; Simpson et al., 2003).

Detection of CA-MRSA carrying PVL gene

Using polymerase chain reaction for the determination of the presence of PVL gene, Scott et al. (2005) found that CA-MRSA isolates with PVL genes were detected in 3 (27.2%). PVL genes were not detected in 9 (72.8%) of CA-MRSA isolates. The results are found in Table 2 and

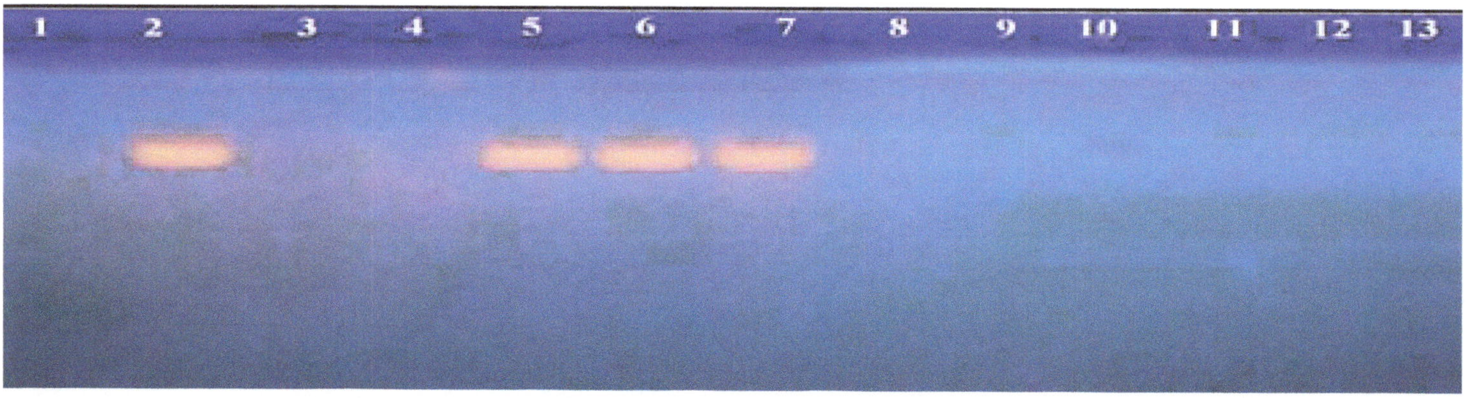

Figure 6. Agarose gel electrophoresis demonstrative the results of PCR for PVL gene to CA-MRSA. Lane 1: negative PVL gene control; lane 2: positive PVL gene control; lanes 3, 4, 8, 9, 10, 11, 12, 13: CA-MRSA without PVL gene; lanes 5, 6, 7: *S.* carrying PVL gene.

Figure 6. For comparison, there is no study in Iraq about the PVL gene with CA-MRSA, and there is no professional center for studying or researching about it. In recent years, the increased prevalence of CA-MRSA has become a major public health concern. In contrast to HA-MRSA, CA-MRSA strains are commonly susceptible to many antibiotics.

Also, CA-MRSA appears to have a distinct exotoxin PVL, which has been associated with severe infections (Okuma et al., 2002; Kazakova et al., 2005).

PVL is one of the many toxins associated with *S. aureus* infection. Because it can be found in virtually all CA-MRSA strains that cause soft-tissue infections, several research groups previously proposed that PVL is the key virulence factor. But new evidence strongly suggests that this is not the case (Andenesch et al., 2003; Dufour et al., 2002; Lina et al., 1999; Melles et al., 2006; Panton and Valentine, 1932).

The *S. aureus* bacterium is a growing global public health menace because of its rapid spread from hospital settings into communities of healthy people. This is an evolving pathogen that in recent decades has developed resistance to common medical treatments and is now finding new mechanisms to spread and cause severe illness. About 75% of CA-MRSA infections are localized to the skin and soft tissues and can usually be treated effectively (Boubaker, 2007; Lina et al., 1999; Voyich et al., 2007). CA-MRSA strains have enhanced virulence, meaning they can infect otherwise healthy people. One of the biggest problems with CA-MRSA skin infections is that they spread rapidly and have the potential to cause illness much more severe than traditional hospital associated MRSA infections, where PVL is less common (Boubaker, 2007; Boyle-Vavra and Daum, 2007; Lina et al., 1999, ,and Valentine 1932, Scott et al., 2005).

Scientists had recognized a connection between MRSA strains that contained PVL and the increased occurrence and severity of CA-MRSA disease, though no one had directly tested the role of PVL in CA-MRSA virulence (Lina et al., 1999). In striving to learn more about CA-MRSA, the RML scientists, with their colleagues at the International Center for Public Health (ICPH) in Newark, and the Université Claude Bernard in Lyon, France, decided to test the PVL virulence theory, thinking that if they could understand the role of this toxin in disease, they could more quickly diagnose serious cases and develop effective treatments (Voyich et al., 2007; Lina et al., 1999). The strong association between PVL and CA-MRSA makes the toxin an excellent marker in tracking community strains, but the assumption that it is the major virulence determinant driving this epidemic is simply not true (Voyich et al., 2007).

Conclusion

In summary, PVL positive CA-MRSA appears to be an increasingly prevalent pathogen worldwide. Clonal expansion of this virulent pathogen appears to be increasing in North American, Australia, Asia, and Europe (Scott et al., 2005; Wannet, 2007). Surveillance and researches on CA-MRSA that carry PVL gene should continue to provide a significant insight into the prevalence and epidemiology of these important resistant pathogens.

REFERENCES

Andenesch F, Naimi T, Enright M, Lina G (2003). Community-acquired methicillin resistant *Staphylococcus aureus* carrying Panton Valentine leukocidin genes worldwide emergence, Emerg. Infect. Dis., 9(8): 978-984.

Barrow GI, Feltham RKA (2003). Cowan and Steels' manual for identification of medical bacteria. 3rd edition. Cambridge Univ. press, U.K.

Berg T, Firth N, Apisiridej S, Hettiaratchi A, Leelaporn A, Skurray AR (1998). Complete nucleotide sequence of pSK41: evolution of staphylococcal conjugative multiresistance plasmids, J. Bacteriol., 180: 4350-4359.

Black JG (2002). Antimicrobial therapy the resistance of microorganisms In Microbiology principles and explorations. 5th edition. Wiley and Sons, NY, pp. 342-343.

Boubaker K (2007). Panton-Valentine leukocidin and staphyloccoccal skin infections in schoolchildren, Emerg. Infect. Dis., 13: 22-24.

Boyle-Vavra S, Daum RS (2007). Community acquired methicillin resistant *Staphylococcus aureus* the role of PVL. Lab. Invest., 87:3-9.

Cauwelier B, Gordts B, Descheemaecker P (2004). Evaluation of a disk diffusion method with cefoxitin (30μg) for detection of methicillin-resistant *Staphylococcus aureus*, Eur. J. Clin. Microbiol. Infect. Dis., 23: 389-392.

Centers for disease control and prevention (CDC) (2007). CA-MRSA community associated methicillin resistant *Staphylococcus aureus* (CA-MRSA).

Chambers HF (2003). Solving staphylococcal resistance to beta-lactams, Trends Microbiol., 11(4): 145-148.

Clinical and Laboratory Standards Institute (CLSI) (2010). Performance standards for antimicrobial susceptibility testing. Approved standard M100-S20. Vol. 30, No. 1. National Committee for Clinical Laboratory Standards, Wayne, PA. USA.

Crisostomo MI, Westh H, Tomasz A (2001). The evolution of methicillin resistance in *Staphylococcus aureus* similarity of genetic backgrounds in historically early methicillin-susceptible and methicillin-resistant isolates and contemporary epidemic clones, Proc. Nat. Acad. Sci., 98: 9865-9870.

Dufour P, Gillet Y, Bes M, Lina G, Vandenesch F, Floret D, Etienne J, Richet H (2002). Community-acquired methicillin-resistant *Staphylococcus aureus* infections in France emergence of a single clone that produces Panton Valentine leukocidin, Clin. Infect. Dis., 35:819-824.

Feil EJ, Cooper JE, Grundmann H, Robinson DA (2003). How clonal is *Staphylococcus aureus*, J. Bacteriol., 185:3307-3316.

Firth N, Apisiridej S, Berg T, O'Rourke BA, Curnock S, Dyke KGH, Skurray RA (2000). Replication of staphylococcal multiresistance plasmids, J. Bacteriol., 182:2170-2178.

Gemmell CG, Edwards DI, Fraise AP, Gould FK (2006). Guidelines for the prophylaxis and treatment of methicillin-resistant *Staphylococcus aureus* infections in the UK, J. Antimicrob. Chemother., 57(4):589-608.

Ito T, Katayama Y, Hiramatsu K (1999). Cloning and nucleotide sequence determination of the entire *mec* DNA of pre-methicillin-resistant *Staphylococcus aureus* N315. Antimicrob. Agents Chemother., 43:1449-1458.

Katayama Y, Ito T, Hiramatsu K (2000). A new class of genetic element, staphylococcal cassette chromosome *mec*, encodes methicillin resistance in *Staphylococcus aureus*. Antimicrob. Agents Chemother., 44:1549-1555.

Kazakova SV, Hageman JC, Matava M, Srinivasan A, Phelan L, Garfinkel B, Boo T, McAllister S, Anderson J (2005). A clone of methicillin-resistant *Staphylococcus aureus* among professional football players, N. Engl. J. Med., 3:468-475.

Kuehnert MJ, Hill HA, Kupronis BA (2006). Methicillin-resistant *Staphylococcus aureus* hospitalizations. United States Rep., 121:17-35.

Lee JH (2003). Methicillin (oxacillin)-resistant *Staphylococcus aureus* strains isolated from major food animals and their potential transmission to humans, Appl. Environ. Microbiol., 69:6489-6494.

Lina G, Piémont Y, Godail-Gamot F, Bes M, Peter M, Gauduchon V, Vandenesch F, Etienne J (1999). Involvement of Panton-Valentine leukocidin producing *Staphylococcus aureus* in primary skin infections and pneumonia, Clin. Infect. Dis., 29(5):28-32.

Melles DC, van Leeuwen WB, Boelens HAM, Peeters JK, Verbrugh HA, van Belkum A (2006). Panton-Valentine leukocidin genes in *Staphylococcus aureus*, Emerg. Infect. Dis., 32-38.

Merlino J, Watson J, Rose B (2002). Detection and expression of methicillin/oxacillin resistance in multidrug resistant and non multidrug resistant *Staphylococcus aureus* in central Sydney, Australia, J. Antimicrob. Chemother., 49:793-801.

Moise-Broder PA, Sakoulas G, Eliopoulos GM (2004). Accessory gene regulator group II polymorphism in methicillin-resistant *Staphylococcus aureus* is predictive of failure of vancomycin therapy, Clin. Infect. Dis., 38:1700-5.

National Committee for Clinical Laboratory Standards (NCCLS) (2003). Performance standards for disc susceptibility tests, 8th ed. Approved standard M2-A8. National Committee for Clinical Laboratory Standards, Wayne, Pa.

Okuma K, Iwakawa K, Turnidge JD, Grubb WB, Bell JM, O'Brien FG (2002). Dissemination of new methicillin-resistant *Staphylococcus aureus* clones in the community, J. Clin. Microbiol., 40:4289-4294.

Olsen JE, Christensen H, Aarestrup FM (2006). Diversity and evolution of *blaZ* from *Staphylococcus aureus* and coagulase-negative staphylococci. J. Antimicrob. Chemother., 57:450-460.

Panton PN, Valentine FC, (1932). Staphylococcal toxin. Lancet, 1:506-508.

Pinho MG, Filipe SR., De Lencastre H, Tomasz A (2001). Complementation of the essential peptidoglycan transpeptidase function of penicillin-binding protein 2 (PBP2) by the drug resistance protein PBP2A in *Staphylococcus aureus*, J. Bacteriol., 183:6525-6531.

Sambrook J, Rusell DW (2001). Molecular cloning. A laboratory manual. Third ed. Cold Spring Harbor, NY.

Santos KRN, Teixera LM, Leal GS, Fonseca LS, Filho PPG (1999). DNA typing of methicillin-resistant *Staphylococcus aureus*: isolates and factors associated with nosocomial acquisition in two Brazilian university hospitals. Med. Microbiol., 48:17-23.

Schmitz FJ, Petridou J, Fluit AC (2000). Distribution of macrolide resistance genes in *Staphylococcus aureus* blood culture isolates from fifteen German university hospitals, Euro. J. Clinic. Microbiol. Infect. Dis., 19:385-387.

Scott T, Micek PD, Dunne PD, Marin H, Kollef MD (2005). Pleuropulmonary complications of Panton-Valentine leukocidin-positive community-acquired methicillin-resistant *Stapylococcus aureus,* Chest, 128:2732-2738.

Simpson AE, Skurray RA (2003). Firth N. A single gene on the staphylococcal multiresistance plasmid pSK1 encodes a novel partitioning system, J. Bacteriol., 185:2143-2152.

Singh A, Goering RV, Simjee S, Foley SL, Zervos MJ (2006). Application of molecular techniques to the study of hospital infection, Clin. Microbiol. Rev., 19:512-530.

Skov R, Smyth R, Clausen M (2003). Evaluation of a cefoxitin 30 μg disc on Iso-Sensitest agar for detection of methicillin-resistant *Staphylococcus aureus*, J. Antimicrobiol. Chemother., 52:204-207.

Swenson JM, Tenover FC (2005). Results of disk diffusion testing with cefoxitin correlate with presence of *mecA* in *Staphylococcus* spp. J. Clin. Microbiol., 43: 3818-3823.

Tiemersma EW, Bronzwaer SL, Lyytikainen O, Degener JE (2004). Methicillin-resistant *Staphylococcus aureus* in Europe, 1999-2002. European antimicrobial resistance surveillance system. Emerg. Infect. Dis., 10:1627-1634.

Voyich JM, Otto M, Mathema B (2007). Is Panton-Valentine leukocidin the major virulence determinant in community-associated methicillin-resistant *Stapylococcus aureus* disease, J. Infect. Dis., 194(12):1761-1770.

Wannet W (2002). Spread of an methicillin-resistant *Staphylococcus aureus* clone with heteroresistance to oxacillin in the Netherlands, Euro. Surv., 7:73-74.

DNA microarray-based detection of antibiotic resistance genes of human isolates of *Escherichia coli* in Nigeria

Chijioke A. Nsofor[1,2] , Christian U. Iroegbu[1], Douglas R. Call[2] and Margaret A. Davies[2]

[1]Department of Microbiology, University of Nigeria Nsukka, Enugu State, Nigeria.
[2]Department of Veterinary Microbiology and Pathology, Washington State University, Pullman, USA.

Molecular epidemiology of human ecovariants of *Escherichia coli* from different regions of Nigeria were studied using their antibiotic susceptibility patterns, plasmid profile and DNA microarray techniques. *E. coli* was isolated using Eosin Methylene Blue agar (EMB) and identified by conventional microbiological technique. The isolates were tested against 14 antibiotics using the disc diffusion method. Genotyping was done using DNA microarray. Overall, 42 different antibiotics resistance clusters were observed, with each isolate showing resistance to at least four or more drugs tested. Of the 60 isolates genotyped with DNA microarray, 57 were identified as having, at least, one antimicrobial resistance gene. Among the 90 antibiotic resistance genes detected, *bla-CMY-2* was the most prevalent occurring in 38 (63.3%) of the isolates. Other highly prevalent genes occurring in the human isolates, include *strA* 28(70%) *aadE* 28(70%); *TEM1* 11(27.5%); *Sul2* 14(35%); and *TetA* 21(52.5%). The microarray genotyping corresponded with the phenotype of the strains. Presence of drug-resistance genes/plasmids in commensal strains isolated from apparently healthy individuals is of great public health importance.

Key words: DNA microarray, *E. coli,* Nigeria.

INTRODUCTION

The large-scale genome sequencing effort and the ability to immobilize thousands of DNA fragments on a surface, such as coated glass slide or membrane, have led to the development of DNA microarray technology (Cassone et al., 2006). An entire microbial genome can be easily represented in a single array, making it feasible to perform genome-wide analysis (Ye et al., 2006) The two common applications of DNA microarray technology in molecular biology are the exploration of genome-wide transcriptional profiles and the measurement of the similarities or differences in genetic contents among different microbes (Peterson et al., 2010). DNA microarray technology is being used to study many bacterial species ranging from standard laboratory strains and pathogens to environmental isolates (Murakami et al., 2002).

DNA microarrays are basically a miniaturized form of dot blot, but in a high-throughput format. There are two major types of DNA microarrays; one is the oligonucleotide-based array and the other is the PCR product-based array (Panicker et al., 2004). A DNA microarray experiment consists of array fabrication, probe preparation, hybridization and data analysis (Call et al., 2001). Although the basic array technology is the same, there are fundamental differences in its application to prokaryotes and eukaryotes.

For example, total RNA is usually labeled for a bacterial

array experiment, while poly(A) RNA is often used for eukaryotic arrays.

Detection of single bacterial genes (antibiotic resistance genes or species-specific genes) in diagnostics and in epidemiological studies is typically carried out by PCR, whereas DNA microarrays have been developed to perform a large number of different hybridization experiments simultaneously on a single membrane or glass substrate. They are well-suited to comprehensively investigate and quantitatively compare the expression levels of a large number of genes, but they can also be easily used in qualitative studies to detect selected DNA sequences (Call et al., 2003a; Call et al., 2003b; Perreten et al., 2005). To better understand the emergence and dissemination of resistance phenotypes from clinical, agricultural, and environmental settings, it is therefore necessary to perform molecular epidemiological analysis of resistant isolates at different levels, comparing whole genomes, single plasmids, and individual resistance gene cassettes. In this study, a method based on DNA microarrays was used to comprehensively assess the presence of antibiotic resistance genes in Nigerian human *E. coli* isolates.

MATERIALS AND METHODS

Study population

The study population included humans (who were either ill or presumptively healthy) residing in the five geopolitical zones of Nigeria viz: South-East, South-West, South-South, North-Central and North-North. In the South-South and South-East, clinical specimens were collected at the University of Port Harcourt Teaching Hospital, Port Harcourt, Rivers State and the Abia State University Teaching Hospital, Aba, Abia State respectively. The Lagos State University Teaching Hospital, Ikeja, Lagos was the site of specimen collection for the South-West, while the National Hospital, Abuja and Military Reference Hospital, Kaduna State were the sources of specimens from the North-Central and North-North respectively. All samples from these hospitals were clinical specimens from patients who were having gastroenteritis or similar illness. Apparently healthy undergraduates' students of Madonna University Elele were included in the study for the isolation of human commensal *E. coli.* These individuals reported no exposure to antibiotics for six months prior to sampling and each person received an explanation of the study objectives and consent form for inclusion in the study. All sampling procedures were in accordance with guidelines of the National Health Research Ethics Committee, Nigeria (www.nhrec.net).

Specimen collection, cultivation and identification of *E. coli*

Sample collection, cultivation, identification of *E. coli* and antibiotics susceptibility testing was based on our previous published work Nsofor and Iroegbu (2013). Briefly, human fecal specimens were streaked directly on EMB agar. No antibiotic was included in the EMB agar plates used for the cultivation. The inoculated plates were incubated overnight at 37°C. A single colony on EMB with green metallic sheen taken to be *E. coli* was selected from an individual fecal sample for further characterization. *E. coli* was fully identified using conventional microbiological tests-Indole positive, Methyl red positive and Citrate negative (Cheesbrough, 2000).

Antibiotics susceptibility testing

The antibiotics susceptibility pattern of the isolates was determined using the disk diffusion method (Cheesbrough, 2000), on Mueller-Hinton agar (Oxoid, England). Inhibition zone diameter values were interpreted using standard recommendations of the Clinical Laboratory Standard Institute (CLSI, 2006). Susceptibility was tested against ampicillin (10 µg), amoxycillin/ clavulanic acid (20/10 µg), tetracycline (30 µg), gentamicin (10 µg), cefpodoxime (10 µg), Cefoxitin (30 µg), cefpirome (30 µg), streptomycin (10 µg), chloramphenicol (30 µg), nalidixic acid (30 µg), sulfamethoxazole-trimethoprim (10 µg), cephalothin (30 µg), nitrofurantoin, ceftriaxone (30 µg), and cefotaxine (30 µg) (Oxoid, England). *E. coli* ATCC 25922 was included as a reference strain.

DNA microarray

The isolates were genotyped using DNA microarray technique (Call et al., 2001). This is based on the principle that nick translated genomic DNA is hybridized to oligonucleotide gene probes printed on a Teflon masked glass slide. The hybridization is detected with florescence marker like the tyramide signal amplification (TSA) system.

Preparation of microarray slides

Multiple DNA microarrays were printed on glass slides so that independent arrays were contained within ten individual wells defined by Teflon masking slides (Erie Scientific, Portsmouth, N.H. USA); the hydrophobic nature of the masking permitted independent samples to be hybridized within each well. Slides were derivatized with epoxysilane (3-glycidoxypropyltrimethoxysilane; (Sigma-Aldrich, Milwaukee, WS, USA) as described by Call *et al* (2001). Prior to printing, the slides were soaked in 2.5% Contrad 70 detergent (Fisher Scientific, Pittsburgh, PA, USA.) for 2 min, rinsed three times with distilled water, and dried using compressed air. Slides were then soaked for 1 h in 3 N HCl, rinsed three times with deionized water, and dried with compressed air.

Construction of DNA microarray

Oligonucleotide probes of known antibiotics resistance genes were reconstituted in TE buffer diluted to 60 µm in print buffer (0.1 M Na_2HPO_4, 0.2 M NaCl, 0.01% sodium dodecyl sulfate) with a pH of 11 and transferred to 384-microwell plates for printing. Arbitrary biotinylated oligonucleotides (70-mer; 5 µM) were included with every array. These biotin pseudoprobes served as positive controls for the detection chemistry and to orient the array for image processing. All probes were deposited as four replicates at a fixed location within each masked well using a Robotic Microgrid II arrayer (Bio-Robotics, Woburn, Mass.USA) with humidity held at 45%. Printing parameters included washing the pins in a recirculating bath (four pins washed twice for 4 s each time), followed by 0.5 s of flushing and 6 s of drying. This washing procedure was repeated twice between probes to minimize possible probe carryover. Printed slides were baked under vacuum (22 Hg/mm) for 1 h (130°C) and stored away from light at room temperature until used.

Genomic DNA extraction

The bacterial total DNA was extracted using the Qiagen DNeasy silica-gel adsorption method (Qiagen, Valencia, CA USA).

A 1.0-ml volume of overnight broth culture of the test isolate was pelleted in a 1.5 ml microcentrifuge at 10000 rpm for 10 min and resuspended in180 µl of buffer ATL from the Qiagen DNeasy kit. Then 20 µl of Qiagen proteinase K solution was added, mixed by vortexing and the cell was incubated for 3 h in a 55°C shaker water bath for lysis. After the lysis, 20 µl of RNase A (100m g/mL) (Qiagen, Valencia, CA USA) was added to each tube (to degrade RNA) and the tubes were incubated at room temperature for 2 mim. This was followed by the addition 200 µl of buffer AL, vortexing, and incubation at 70°C for 10 min. Then, the genomic DNA (gDNA) was concentrated by the addition of 200 µl of 100% ethanol. To separate the DNA from other cellular contaminants, the treated DNA lysate was pipetted into a DNeasy column in a collection tube, and centrifuged for 1 minat 10,000 x *g*. The remaining contaminants were washed out by using 500 µl each of buffer AW1 and AW2 in a new collection tube at each time. The purified gDNA was eluted in a fresh 1.5 ml micro-centrifuge tube by using 200 µl AE buffer and centrifugation for 1 minat 10,000 x *g*. Finally, the Nanodrop spectrophotometer was used to quantify the DNA. DNA was quantified to properly scale the subsequent nick translation and any sample that failed to reach the value of A260/A280 ratio of 1.7 to 2 or below 25ng/µl was re-extracted. All the buffers, enzymes and columns used in this extraction came from the Qiagen DNeasy kit (Qiagen, Valencia, CA USA; Cat. No. 69504).

Nick translation: biotinylation and fragmentation of DNA

This reaction is designed to generate small (50 - 100 base) biotin-labeled DNA probes by nick translation which are important for successful in situ hybridization.

Approximately 1.0 µg (up to 40 ul) of the quantified g DNA, 5 µl of 10X dNTP mix [(0.2 mM each of dCTP, dGTP, dTTP; 0.1 mM of dATP; 0.1 mM of biotin-14-dATP; 500 mM of Tris-HCl, pH 7.8; 100 mM of β-mercaptoethanol and 100 µg/ml of nuclease-free BSA) (Invitrogen, USA)] and 5 µl of 10X enzyme mix [0.5 U/µl of DNA polymerase 1, 0.007 U/µl of DNase 1.50 mM of Tris-HCl pH7.5, 5 mM of magnesium chloride, 0.1 mM of phenylmethylsulfonyl fluoride, 5% (v/v) of glycerol and 100 µg/ml of nuclease-free BSA) (Invitrogen, USA)] were combined in 0.2 ml PCR tubes on ice. The total volume was brought to 50 µl with PCR water. The mixture was incubated at 16°C in a thermal cycler for 2 hand then held at 4°C for nick translation of DNA. To precipitate the nick translated DNA, the samples were transferred to 1.5 ml micro-centrifuge tubes followed by the addition of 5 ul of 3 M sodium acetate, (pH 5.2), 110 µl of 100% ethanol and incubation at -80 for 30 min. After the incubation, the DNA was pelleted by centrifugation at 14000 rpm for 30 minat 4°C. Then, the pellets were resuspended with 400 µl of 70% ethanol. For more purification, the above steps were repeated once and the pellets were dried with a vacuum centrifuge for 10 min. Finally, the purified nick-translated DNA was resuspended with 100 µl 1x hybridization buffer.

Microarray slide pre-hybridization preparation

Microarray slides were prepared by immersing them in 50 ml of 1% BSA blocking solution in a Coplin staining jar followed by incubation at room temperature for 10 min, with shaking at 80 rpm to eliminate bubbles on the slide surface. The slides were rinsed 20 times in double de-ionized after which their back and edges were wiped with a Kimwipe and spin dried with slide centrifuge for 15 seconds.

Sample application/hybridization

The nick translated gDNA was boiled for 3 min, chilled on ice and briefly vortexed for 15 s. Then, the microarray slides were placed on a humidified chamber (200 µl tip box and lid with de-ionized water covering the bottom of the box) and 45 µl of the gDNA sample was placed in each well (2 wells per nick translated gDNA sample) on the microarray slide. The droplets were carefully spread to fully cover the well without touching the slide surface with the pipette. Carefully, the slide was sealed (face-up and frosted end toward the cap) in a hybridization chamber (50 ml conical tube with filter paper moistened with 1x hybridization buffer). The slide was placed on top of the filter paper in the hybridization chamber without allowing the damp filter paper to touch the wells. The hybridization chamber was placed in a rack and lead weight on top of the rack, then the rack was submerged in the 55°C water bath. Finally, the sample DNA was allowed to hybridize with the probes on the array for 16 h.

Post-hybridization stringency washes

After hybridization, the slides were removed from the hybridization chamber with forceps and excess hybridization solution was aspirated off the slides. Then, the slides were completely immersed (frosted end up) in a 55°C pre-warmed low stringency array wash solution (1X SSC, 0.2% SDS) contained in a Coplin jar. The above procedure was repeated in medium stringency (0.1XSSC, 0.2% SDS) and high stringency (0.1XSSC) array wash solutions respectively. At each time, the slides were washed for 4 min at room temperature on an Orbital shaker at 80 rpm. After the stringency washes, the slides were transferred to a horizontal staining jar that contains enough TNT buffer to cover the slide and were shaken for 1 min at 80 rpm at room temperature to remove the stringency wash buffers. This TNT buffer washing was repeated three times.

Microarray development

For the following applications, 45 µl of each solution was added directly to each well. The slides were gently tapped to distribute the reagent over the full well surface without allowing the reagents to cross over to other wells. The slides were spin-dried for 5 s using a slide centrifuge followed by incubation with 1:100 Streptvadin-Horseredish peroxidase (SA-HRP) in TNB for 30 min. After the incubation, the slides were washed 3 times for 1 min each in horizontal staining jars at 80 rpm shaking. The above procedure was repeated with 10% FES, 2XSSC; 1:50 BioT, 1xAmp Dil; and 1:500 SA-Alexa 555, 1XSSC, 5X Den. This last incubation was done for 1 h in the dark. All incubation were done at room temperature in a humidified chamber (made from a covered tip box with ~10 ml PCR water in the bottom). At the end of these development reactions, the slides were spin-dried for 15 s using the slide centrifuge and were stored in the dark prior to scanning.

Scanning/imaging of slides

After hybridization and development, slides were scanned or imaged by standard DNA microarray slide scanners or imagers. The florescence marker used in this experiment (Alexa555) has an optimal excitation wavelength of 555 nm and emission wavelength of 565 nm. The scanner/imager we used (Applied Precision array WoRx scanner) had a white light source and an emission filter for Cy3 that functions well for Alex555. We used an excitation wave-length of 540 nm (25 nm bandwidth) and an emission wave-length of 595 nm (50 nm bandwidth).

There were five pairs of Teflon-masked wells on each slide, with each well containing a full array and our normal protocol calls for two wells to be hybridized to the same sample. Within each well, there were two spots per probe so in effect there were four individual probe-target hybridizations (2 wells total). Each full array has dimensions of 22 horizontal and 20 vertical spots. The distance between spots is approximately 250 µm.

RESULTS

Antimicrobial resistance genes for microarray construction

Ninety (90) antimicrobial resistance genes oligonucleotide probes were employed in the microarray; they include 21 aminoglycoside resistance genes, *aac(3)-Id, aac(3)-III, aac(3)-Iva, aac(6')-Ib, aac(6')-IIa, aacC2, aacCA5, aadA1, aadA2, aadA21, aadA5, aadA7, aadB, aadE, aph(3)-Ia, aph(3)-IIa, aphA7, aphD, AphE, strA* and *strB;* 21 beta-lactam resistance genes, *blaACC-01, bla-CMY-2, blaCTX-M-1, blaCTX-M-12, blaCTX-M-15, blaCTX-M-2, blaCTX-M-8, blaDHA-1, blaFOX-2, blaIMP-2, blaKPC-3, blaMIR, blaOXA-1, blaOXA-2, blaOXA-7, blaOXY-K1, blaPSE-1, blaPSE-4, blaROB-1, blaSHV-37,* and *TEM1;* 10 chloramphenicol resistance genes, *cat4, catB2, catB3, catB8, catI, catII, catP, cmlA, cmlB,* and *floR;* 2 integrase genes, *intI1,* and *intI2,* 4 qinolone resistance genes, *qac delta E, qnrA1, qnrB,* and *qnrS;* 11 trimethoprim resistance genes, *dfrA1, dfrA14, dfrA16, dfrA21, dhfrII, dhfrV, dhfrVI, dhfrVII, dhfrXII, dhfrXIII,* and *dhfrXV;* 3 sulfonamide resistance genes, *Sul1, Sul2,* and *sul3;* and 18 tetracycline resistance genes, *tet(C), tet(Y), Tet30, TetA, tetB, TetD, TetE, tetG, TetH, TetJ, TetK, TetM, TetQ, TetS, TetT, TetV, TetW* and *TetX.*

The sequence identity of each gene was compared with GenBank sequences, therefore, all the 90 genes were used to construct the DNA microarray; biotin was used as the positive control and 16s rDNA as orientation.

To determine the specificity of microarray hybridization, all of the labeled genes probes were hybridized to the microarray. In most cases, there was a one-to-one correspondence for hybridization signal to respective target, orientation gene, and positive control gene spots. There was minor cross-hybridization between some genes and they were marked as abnormal during analysis, thus these genes are not included in the net results shown here.

Detection of antimicrobial resistance gene with microarray

The antimicrobial resistance genes of 60 *E. coli* isolates were tested with the microarray. Considering resistance according to class of antibiotics, the following resistance genes were most commonly present in the isolates: for aminoglycoside, strA was most prevalent (65.0%); the isolates from the south-east harbored this gene most (88.2%), among the human isolates (Table 1). In beta-lactam, *bla-CMY-2* was most frequent in the isolates, 38(63.3%) and 28(70%) (Table 2).

However, analysis of isolates from different geopolitical zones shows that *TEM1* occurred highest among the individual zones with 14(82.4%) in south-east. The analysis of chloramphenicol resistance genes encountered in this study indicated that 26 isolates, (43.3%) harbored this gene. Integrase gene, int1 was most prevalent in the sample sources with 35 isolates (58.3%) harboring the gene in human specimens. The most detected qinolone resistance gene was *qnrA1;* 25 isolate (41.7%) (Table 3).

Table 4 shows the prevalent of trimethoprim and sulfonamide resistance genes detected in this study. *dfrA16* was most frequent in the specimens; 17 isolates (28.3%). For sulfonamide resistance genes, 35 (58.3%) of the isolates, harbored *Sul2* at highest rate. Tetra-cycline resistance gene profile observed in this study indicates that *TetA* was most frequent in the sample sources with 36 isolates (60%) of human specimens bearing this gene (Table 5). The *TetM* (one isolate from north-north), *TetQ* (10 isolates 16.7%) and *TetW* (one isolate from north-central) were detected in this study. Figure 1 is a representative of the microphotograph of microarrays hybridized with genomic DNAs of *E.coli* isolates.

DISCUSSION

The microarray was designed with oligonucleotides of 50-100 bases. This enabled us to find consensus sequences within a family of genes sharing high DNA identities (Barlow et al., 2008). The consensus sequences allow for identification of the few different bases which distin-guish these genes and also indicate which family they belong.

Two different oligonucleotides probes were chosen for each resistance gene; the use of two different oligonucleotides for the detection of each resistance gene has the advantage of increased specificity and sensitivity of the method. Lack of sensitivity was found in three out of 60 isolates genotyped by microarray: (PH18, south-south, JA10, north-north and LA13, south-west). This demonstrated that, the oligonucleotide was effectively spotted on the microarray and indicated that the detection of the probe depend on the labeling procedure. The hybridization analysis using genomic DNAs of our isolates enabled verification of the sensitivity of the oligonucleotide targets and identification of 94 resistance genes.

The resistance gene array allowed us to quickly characterize a collection of resistance genes in *E. coli* of diverse origin

For example, the high tetracycline resistance observed in this study could be explained by the presence of variety of tetracycline resistance genes (*tet(C), tet(Y), Tet30, TetA, tetB, TetD, TetE, tetG, TetH, TetJ, TetK, TetM, TetQ, TetS, TetT, TetV, TetW* and *TetX),* known to be involved in resistance to tetracycline antibiotic (Guay, 2008). This is, to our knowledge, the first report of the detection of aph*(3')-III,* and *ant(6)-Ia* genes in a *E. coli* strain, suggesting the presence of a Tn*5405*-like structure. Transposon Tn*5405* carries an *ant(6')-Ia-sat4-aph(3')-III* cluster which is widespread among Salmonella and Shigella and might have been transferred from one

Table 1. The prevalence of aminoglycosides resistance genes in human *E. coli* isolates from Nigeria.

Gene	Source of samples					
	South – East (N=17)	South South (N=13)	South –West (N=13)	North – North (N=13)	North-Cnetral (N=4)	Total (N=60)
aac(3)-Id	01(5.9)	00	00	00	03(75)	04(4.7)
aac(3)-III	03(17.6)	06(46.2)	07(53.8)	00	00	16(26.7)
aac(3)-IVa	00	01(7.7)	00	00	00	01(1.7)
aac(6')-Ib	00	03(23.1)	03(23.1)	00	01(25)	07(11.7)
aac(6')-IIa	01(5.9)	01(7.7)	00	01(7.7)	00	03(5.0)
aacC2	01(5.9)	03(23.1)	03(23.1)	00	00	07(11.7)
aacCA5	07(41.2)	00	01(7.7)	01(7.7)	00	09(15.0)
aadA1	13(76.5)	02(15.4)	05(38.5)	02(15.4)	01(25)	23(38.3)
aadA2	13(76.5)	01(7.7)	06(46.2)	03(23.1)	01(25)	24(40.0)
aadA21	13(76.5)	01(7.7)	05(38.5)	03(23.1)	01(25)	23(38.3)
aadA5	04(23.5)	00	04(30.8)	00	03(75)	11(18.3)
aadA7	00	01(7.7)	00	00	00	01(1.7)
aadB	00	00	00	00	01(25)	01(1.7)
aadE	08(47.1)	06(46.2)	07(53.9)	08(61.5)	04(100)	33(55.0)
aph(3)-Ia	00	00	00	00	00	00
aph(3)-IIa	01(5.9)	01(7.7)	04(30.8)	01(7.7)	02(50)	09(15.0)
aphA7	00	01(7.7)	01(7.7)	01(7.7)	01(25)	04(6.7)
aphD	06(35.3)	04(30.8)	02(15.4)	02(15.4)	03(75)	17(28.3)
AphE	00	00	02(15.3)	01(7.7)	01(25)	04(4.70
strA	15(88.2)	07(53.8)	07(53.8)	06(46.2)	04(100)	39(65.0)
strB	15(88.2)	03(23.1)	06(46.2)	03(23.1)	02(25)	19(48.3)

N=Number of isolates hybridized.

Table 2. The prevalence of Beta-lactam resistance genes in human *E. coli* isolates from Nigeria.

Gene	Source of samples					
	South – East (N=17)	South South (N=13)	South –West (N=13)	North – North (N=13)	North-Central (N=4)	Total (N=60)
blaACC-01	01(5.9)	00	00	02(15.4)	00	03(5.0)
bla-CMY-2	12(70.6)	07(53.8)	06(46.2)	09(69.2)	04(100)	38(63.3)
blaCTX-M-1	05(29.4)	00	02(15.4)	00	04(100)	11(18.3)
blaCTX-M-12	00	00	00	00	00	00
blaCTX-M-15	05(29.4)	01(7.7)	00	01(7.7)	02(50)	09(15)
blaCTX-M-2	00	00	00	01(7.7)	00	01(1.7)
blaCTX-M-8	04(23.5)	00	00	00	01(25)	06(8.3)
blaDHA-1	00	00	00	00	00	00
blaFOX-2	00	01(7.7)	00	00	00	01(1.7)
blaIMP-2	00	00	00	00	00	00
blaKPC-3	12(70.6)	03(20.1)	06(46.2)	01(7.7)	03(75)	25(41.7)
blaMIR	00	01(7.7)	00	01(7.7)	01(25)	03(5.0)
blaOXA-1	04(23.5)	03(20.1)	02(15.4)	02(15.4)	01(25)	12(20)
blaOXA-2	01(5.9)	00	00	00	00	01(1.7)
blaOXA-7	00	01(7.7)	00	00	00	01(1.7)
blaOXY-K1	00	01(7.7)	00	00	00	01(1.7)
blaPSE-1	01(5.9)	00	00	00	00	01(1.7)
blaPSE-4	10(58.8)	03(20.1)	07(53.8)	02(15.4)	04(100)	26(43.3)
blaROB-1	00	01(7.7)	01(7.7)	00	00	02(3.3)
blaSHV-37	02(11.8)	00	00	01(7.7)	01(25)	04(6.7)
TEM1	14(82.4)	06(46.2)	08(61.5)	06(46.2)	02(50)	36(60)

N=Number of isolates hybridized.

Table 3. The prevalence of chloramphenicol and qinolone resistance genes in human *E. coli* Isolates from Nigeria.

Gene	Source of samples					
	South – East (N=17)	South South (N=13)	South –West (N=13)	North – North (N=13)	North-Central (N=4)	Total (N=60)
cat4	13(76.5)	01(7.7)	02(15.4)	00	01(25)	17(28.3)
catB2	00	02(15.4)	01(7.7)	00	00	03(5.0)
catB3	00	03(23.1)	02(15.4)	00	00	05(8.3)
catB8	02(11.8)	03(23.1)	03(23.1)	00	01(25)	09(15.0)
catI	12(70.6)	01(7.7)	02(15.4)	00	01(25)	
catII	00	00	01(7.7)	00	00	01(1.7)
catP	00	00	00	00	01(25)	01(1.7)
cmlA	00	00	01(7.7)	04(30.8)	00	05(8.3)
cmlB	05(29.4)	00	02(15.4)	03(23.1)	00	10(16.7)
floR	07(41.2)	03(23.1)	07(53.8)	05(38.5)	04(100)	26(43.3)
intI1	12(70.6)	09(69.2)	07(53.8)	06(46.2)	01(25)	35(58.3)
intI2	00	00	02(15.4)	00	02(50)	04(6.7)
qac delta E	12(70.6)	08(61.5)	07(53.4)	04(30.8)	02(20)	33(55)
qnrA1	08(47.1)	03(23.1)	07(53.4)	03(23.1)	04(100)	25(41.7)
qnrB	00	01(7.7)	017.7)	00	00	02(3.3)
qnrS	00	01(7.7)	00	01(7.7)	01(25)	03(5.0)

N=Number of isolates hybridized.

Table 4. The prevalence of trimethoprim and sulfonamide resistance genes in human E. coli isolates from Nigeria.

Gene	Source of sample					
	South – East (N=17)	South – East (N=17)	South – East (N=17)	South – East (N=17)	South – East (N=17)	South – East (N=17)
dfrA1	10(58.8)	00	03(23.1)	00	01(25)	14(23.3)
dfrA14	02(11.8)	04(30.8)	03(23.1)	04(30.8)	02(50)	15(25)
dfrA16	13(76.5)	00	01(7.7)	02(15.4)	01(25)	17(28.3)
dfrA21	00	00	00	02(15.4)	00	02(3.3)
dhfrII	00	00	00	00	00	00
dhfrV	01(5.9)	03(23.1)	02(15.4)	04(30.8)	01(25)	11(18.3)
dhfrVI	03(17.6)	00	01(7.7)	01(7.7)	00	05(8.3)
dhfrVII	03(17.6)	04(30.8)	01(7.7)	01(7.7)	01(25)	10(16.7)
dhfrXII	00	00	01(7.7)	04(30.8)	00	05(8.3)
dhfrXIII	00	00	01(7.7)	04(30.8)	00	05(8.3)
dhfrXV	00	00	0197.7)	00	00	1(1.7)
Sul1	13(76.5)	03(23.1)	05(38.5)	01(7.7)	00	22(36.7)
Sul2	14(82.4)	07(53.8)	07(53.8)	05(38.5)	02(50)	35(58.3)
sul3	00	00	02(15.4)	04(30.8)	00	06(10)

N=Number of isolates hybridized

of these species to *E.coli*. This demonstrate the efficiency of this technology to rapidly characterize antibiotic resistance genes in strains whose resistance genotype was completely unknown. Furthermore, automation of the hybridization procedures is conceivable, since all the hybridization steps are performed in the same tube. The microarray technology will then facilitate and speed the analysis of antibiotic resistance genes.

The microarrays have the particular advantage of detecting the presence of antibiotic resistance genes that are not phenotypically expressed *in vitro*. We have compared this microarray assay for the detection of antibiotic resistance genes with traditional phenotypic methods for the determination of antibiotic susceptibility. Overall, we found that, the discrepancies between the microarray results and the disk diffusion results are not significant (3 out of 60 isolates) and were mainly due to the phenoltypically susceptible, but resistance gene-positive isola-

Table 5. The prevalence of tetracycline resistance genes in human *E. coli* Isolates from Nigeria.

Gene	Sources of sample					
	South – East (N=17)	South – East (N=17)	South – East (N=17)	South – East (N=17)	South – East (N=17)	South – East (N=17)
tet(C)	05(29.4)	02(15.4)	03(23.1)	02(15.4)	03(75)	15(25)
tet(Y)	05(29.4)	03(23.1)	02(15.4)	02(15.4)	01(25)	13(21.7)
Tet30	05(29.4)	03(23.1)	02(15.4)	01(7.7)	02(50)	13(21.7)
TetA	12(70.6)	07(58.3)	06(46.2)	07(58.3)	04(100)	36(60)
tetB	14(82.4)	07(58.3)	03(23.1)	01(7.7)	01(25)	26(43.3)
TetD	13(76.5)	03(23.1)	03(23.1)	01(7.7)	04(100)	24(40)
TetE	12(70.6)	00	02(15.4)	00	01(25)	15(25)
tetG	06(35.3)	02(23.1)	02(15.4)	01(7.7)	00	11(18.3)
TetH	05(29.4)	05(38.5)	01(7.7)	02(15.4)	01(25)	14(23.3)
TetJ	05(29.4)	02(23.1)	03(23.1)	06(46.2)	03(75)	19(31.7)
TetK	05(29.4)	05(38.5)	02(15.4)	02(15.4)	01(25)	15(25)
TetM	00	00	00	01(7.7)	00	01(1.7)
TetQ	02(11.8)	05(38.5)	00	02(15.4)	01(25)	10(16.7)
TetS	05(29.4)	00	01(7.7)	01(7.7)	00	07(11.7)
TetT	00	00	00	00	01(25)	01(1.7)
TetV	00	00	00	00	00	00
TetW	00	00	00	00	01(25)	01(1.7)
TetX	00	00	00	01(7.7)	01(25)	02(3.3)

N=Number of isolates hybridized.

isolates. For example, we encountered cephalosporin – sus-ceptible, but *bla-CMY-2*-positive strains (*n* = 3), presumably because of the known heterogeneous expression of *bla-CMY-2* gene in *enterobacteriaceae in vitro* (Call et al., 2003). More examples are seen in the following isolates where microarray detected genes that were not phenotypically expressed by the isolates. PH 11-South-south (*aadE, bla-CMY-2)*; KD 4, JA3 –North-north *(floR, strB)*; LA 4- South-west (*aadE, bla-CMY-2, blaPSE-4, floR, qac delta E, qnrA1, TetJ-)*; For more frequently encountered gentamicin-susceptible but *aac*(6')-Ie-*aph* (2''')-positive isolates, the discrepancy in genotype and phenoltype might be attributed to the so-called silent antibiotic resistance gene that might become activated to express the resistance (Martineau et al., 2000a), and their presence may also facilitate the spread to other bacteria. From a clinical perspective, a susceptible strain harboring but not expressing an antibiotic resistance gene should be regarded as potentially resistant to that antibiotic (Martineau et al., 2000b). Thus, detection of the discrepancies between the presence of the resistance gene and the phenotypic susceptibility (false-positive) is important for physicians to guide prescription of appropriate functional antibiotic therapy so as to control the spread of the resistance due to antibiotic selection.

Furthermore, there are also discrepancies between the absence of the antibiotic gene test on the microarray and the phenotypic resistance (false negative). This was observed in isolates PH 18-South-south (Am-C-Sxt-T); LA 13-South-west (Am-C-Sxt-S-T-Amc-Nal); LA 15-South-west (Am-Sxt-S-T); Resistance was phenotypically observed against the antibiotics written against each of the isolates but the genes were not detected by the microarray. This could be attributed by the non inclusion of the oligo-nucleotide probes encoding theses genes in the construction of the microarray or the genes encoding the resistance are novel. However, more research is needed in this area before conclusion can be established.

The most common mechanism of cephalosporin resistance is through production of beta-lactamases. Broad-spectrum cephalosporins, like cephalothin, have increased ring stability against some beta-lactamases but are susceptible to cephamycinases (Call et al., 2003). Cephamycinase (*bla-CMY-2*) share extensive homology to chromosomal *ampC* beta-lactamases. The *bla-CMY-2* gene is closely related to the chromosomal *ampC* of *Citrobacter freundii*. Cephalothin resistance in *E. coli* is typically due to the presence of the *bla-CMY-2* gene (Call et al., 2003). Our findings support this, as the *bla-CMY-2* gene was found in 63.3% of human isolates in this study. Antibiotic-resistant bacteria today are present in a large variety of ecological niches such as hospitals, the environment, and food.

In conclusion, the DNA microarray presented in this study has been shown to be an efficient prototype that allows for rapid screening of resistance genes in *E. coli.* This technology should rapidly find application in surveillance programs of antibiotic resistance genes, industry, and research in order to limit the emergence and spread of antibiotic resistance genes and extend the therapeutic

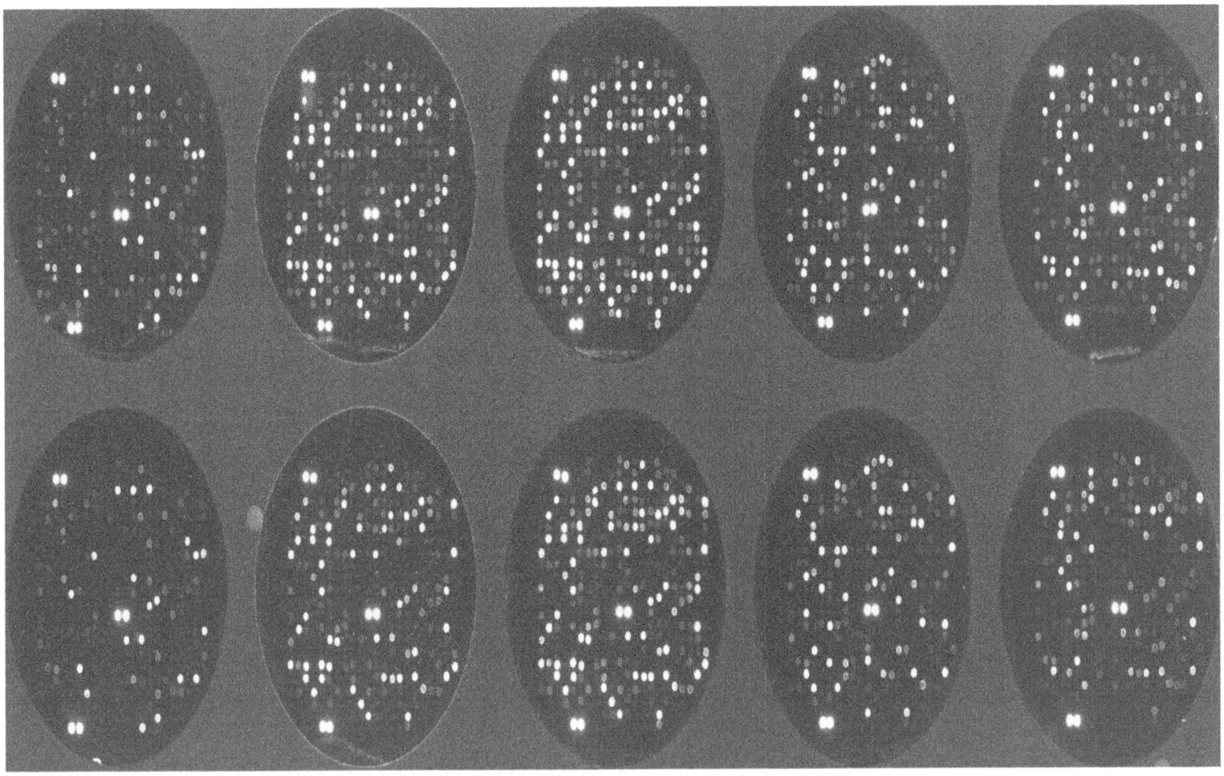

Figure 1. Microphotograph of microarrays hybridized with genomic DNAs of *E.coli* Isolates from South-West Nigeria.

action of existing drugs.

REFERENCES

Barlow M, Reik RA, Jacobs SD, Medina M, and Meyer MP. (2008). High rate of mobilization for bla_{CTX-M}s. Emerg. Infect. Dis. 14:423-428.

Call DR, Chandler DP, Brockman F (2001). Fabrication of DNA microarrays using unmodified oligonucleotide probes. BioTech. 30:368-379.

Call DR, Bakko MK, Krug MJ, Roberts MC (2003a). Identifying antimicrobial resistance genes with DNA microarrays. Antimicrob. Agents Chemother. 47:3290-3295.

Call DR, Borucki MK, Loge FJ (2003b). Detection of bacterial pathogens in environmental samples using DNA microarrays. J. Microbiol. Methods 53:235–243.

Cheesbrough M (2000). *District Laboratory Practice in Tropical Countries*, Part 2. Cambridge University Press, Cambridge, UK; pp.434.

Cassone M, Marco MD, Francesco I, Marco RO, Gian MR, Gianni P (2006). DNA Microarray for Detection of Macrolides Resistance Genes. Antimicrob. Agents Chemother. pp. 2038-2041.

Clinical Laboratory Standards Institute (2006). Performance standards for Antimicrobial susceptibility testing. National committee for clinical laboratory standards, Wayne pa.

Martineau F, Picard FJ, Grenier L, Roy PH, Ouellette M, Bergeron MG (2000a). Multiplex PCR assays for the detection of clinically relevant antibiotic resistance genes in Staphylococci isolated from patients infected after cardiac surgery. J. Antimicrob. Chemother. 46:527-533.

Martineau F, Picard FJ, Grenier L, Roy PH, Ouellette M, Bergeron MG (2000b). Correlation between the resistance genotype determined by multiplex PCR assays and the antibiotic susceptibility patterns of *Staphylococcus aureus* and *Staphylococcus epidermidis*. Antimicrob. Agents Chemother. 44:231-238.

Murakami S, Nakashima R, Yamashita E, Yamaguchi A (2002).Crystal structure of bacterial multidrug efflux transporter AcrB. Nature 419:587-593.

Nsofor CA, Iroegbu CU (2013) Antibiotic Resistance Profile of *Escherichia coli* Isolated from Five Major Geopolitical Zones of Nigeria. J. Bacteriol. Res. 5(3):29-34.

Panicker G, Call DR, Krug MJ, Bej AK (2004). Detection of Pathogenic *Vibrio* spp. in Shellfish by Using Multiplex PCR and DNA Microarrays. Appl. Environ. Microbiol. pp.7436-7444.

Perreten VL, Vorlet-Fawer P, Slickers R, Ehricht PK, Frey J (2005). Microarray-based detection of 90 antibiotic resistance genes of gram-positive bacteria. J. Clin. Microbiol. 43:2291-2302.

Peterson G, Jianfa B, Nagaraja TG, Sanjeev N (2010). Diagnostic microarray for human and animal bacterial diseases and their virulence and antimicrobial resistance genes. J. Microbiol. Methods 80:223-230.

Ye Y, He X, Szewczyk P, Nguyen T, Chang G (2006). Structure of the multidrug transporter EmrD from *Escherichia coli*. Sci. 312:741–744.

29

Antibiotic resistance patterns of strains of *Staphylococcus aureus* isolated from patients in three hospitals in Kumasi, Ghana

Sixtus Bieranye Bayaa Martin Saana, Francis Adu, Christian Agyare , Stephen Yao Gbedema, Vivian Etsiapa Boamah and Duredoh Freeman George

Department of Pharmaceutics, Faculty of Pharmacy and Pharmaceutical Sciences, Kwame Nkrumah University of Science and Technology, Kumasi, Ghana.

***Staphylococcus aureus* continues to offer challenges to medical science in the area of resistance to chemotherapeutic agents leading to treatment failures using common antibiotics. In this study, *S. aureus* isolated from patients from three hospitals in Kumasi, Ghana were tested for their sensitivity to some reference antibiotics using the Kirby-Bauer agar disc diffusion method. A total of 109 *S. aureus* isolates were obtained from wound and nose swabs of 300 patients. *S. aureus* was isolated from 44% female and 56.0% male patients. Majority of the *S. aureus* isolates were identified in patients aged between 20 and 29 years, while those within the age group of 60 to 69 years recorded the least number of *S. aureus* isolates. Vancomycin had the highest susceptibility of 74.1% followed by ceftriaxone with 67.6%, erythromycin with 49.0%, ampicillin with 47.0% and gentamicin with 44.4%. Out of the 109 *S. aureus* isolated from the three hospitals, 32.1% exhibited multiple drug resistance.**

Key words: Antibiotics, bacterial resistance, *Staphylococcus aureus*, sensitivity.

INTRODUCTION

In Ghana, *Staphylococcus aureus* is a common cause of human infections and is recognized as a pathogen of high public health significance (Newman et al., 2006). *S. aureus* is a normal inhabitant of the human skin and the anterior nares. In a healthy person, *S. aureus* is usually not a health concern, but an injury or poor hygiene can cause *S. aureus* infections (Brenda and Lee, 2008). It is also a serious opportunistic pathogen responsible for a number of infections in immuno-compromised individuals (Burnett et al., 1996). *S. aureus* has therefore emerged as one of the main important human pathogens, and has over the past decades, been a leading cause of hospital and community-acquired infections (Shittu and Johnson, 2006).

S. aureus is the most frequently occurring bacterial pathogen among clinical isolates from hospital inpatients and is the second most prevalent bacterial pathogen among clinical isolates from outpatients (Styers et al., 2006). According to Biedenbach et al. (2004), *S. aureus* is the most common cause of nosocomial bacteraemia in North America, Latin America and Europe. Tagoe et al. (2011) also found that 57.6% of all the pathogenic isolates causing hospital acquired infections in south eastern part of Ghana were *S. aureus*. Most cases of nosocomial infections are acquired through exposure to hands of health care workers after they have been transiently colonized with bacterial agents from their own reservoir or from contact with infected patients (Sheretz et al., 1996).

Attempts to control diseases caused by *S. aureus* through the use of antibiotics have resulted in increased prevalence of resistant strains of the organism (Levy, 2001; Crowder et al., 2006). Therefore, in order to effectively treat infections caused by *S. aureus*, culture

Table 1. Distribution of *S. aureus* isolates among males and females in relation to specimen types.

Samples	Hospitals	Males	Females	Total no. of *S . aureus*
Blood (No = 1 50)	Tafo (No = 50)	0	0	0
	North Suntreso (No = 50)	0	0	
	Kumasi South (No = 50)	0	0	
Wound swabs (No = 150)	Tafo (No = 50)	11	5	53
	North Suntreso (No = 50)	12	7	
	Kumasi South (No = 50)	5	13	
Nose swabs (No = 150)	Tafo (No = 50)	8	6	56
	North Suntreso (No = 50)	12	6	
	Kumasi South (No = 50)	11	13	

and antibiotic sensitivity tests must first be determined. Once culture and sensitivity results confirm the type of bacterial infection and sensitivity pattern, treatment may be modified (Paterson, 2000). However, most health facilities in Ghana lack adequately resourced laboratories to culture and test the antibiotic sensitivity of bacteria causing infections. Even, where laboratory facilities are available, culture and sensitivity test results take 48 to 72 h to be ready (Newman et al., 2006). In many cases, sensitivity test may not be carried out at all due to lack of microbiologists and the extra cost it constitutes for the patients (Ohene, 1997). This study is intended to isolate *S. aureus* from patients in Kumasi South, North Suntreso and Tafo hospitals in Kumasi, Ghana and determine their susceptibility patterns to some common antibiotics used for the treatment of staphylococcal infections.

MATERIALS AND METHODS

The protocols for the study were approved by the individual Hospital's Ethics Committees and the patients' consent for the study was done. Clinical samples including wound swabs, nose swabs and blood samples were collected from 300 patients suspected to have bacterial infection at Kumasi South, North Suntreso and Tafo hospitals and cultured to isolate *S. aureus*. All the clinical samples collected at the hospitals were inoculated into a transport medium (cooked meat broth, Oxoid Basingstoke, UK) and incubated at 37°C for 72 h (Ajello et al., 2003).

The samples were then inoculated on plates of blood and Mannitol salt agar (Oxoid Basingstoke, UK). Suspected *S. aureus* colonies were isolated, Gram stained and identified through biochemical tests as described by Betty and colleagues (Forbes et al., 2007).

In vitro sensitivity tests of the *S. aureus* isolates against some reference antibiotics (ampicillin 10 µg, ceftriaxone 30 µg, gentamicin 10 µg, erythromycin 15 µg and vancomycin 30 µg (Oxoid, Basingstoke, UK) was carried out using the Kirby-Bauer agar disc diffusion method as recommended by the Clinical and Laboratory Standards Institute (2010). A standardized suspension of the isolated *S. aureus* was prepared by inoculating a colony into 10 ml peptone water and incubated at 37°C for 24 h. It was then diluted to 0.5 MacFarland turbidity standards. A sterile swab was dipped into the standardized inoculum and used to inoculate evenly the surface of already prepared Mueller-Hinton agar (Oxoid Basingstoke, UK). The agar was left for 15 min for the surface moisture to dry (Ajello et al., 2003). The antibiotic discs were applied to the surface of the inoculated agar plate using a disc dispenser (Oxoid 6-place, 90 mm) and incubated for 18 h at 37°C (Lalitha, 2004). The zones of inhibition for each antibiotic was measured in millimeters and compared with values provided by the Clinical and Laboratory Standards Institute (CLSI, 2010). *S. aureus* (ATCC 25923) was used as positive control.

The mean diameter of zone of growth-inhibition observed was measured and compared with the reference data provided by the Clinical and Laboratory Standards Institute (CLSI, 2010).

Statistical analysis

Data obtained in this study were statistically analyzed using Graph Pad prism version 5.0 (GraphPad Software, San Diego, CA, USA) and Microsoft excel 2010 edition. Two-way ANOVA was used to compare the means. $P<0.05$ was considered significant.

RESULTS

Results from the study showed 48.6% *S. aureus* from wound swabs and 51.4% were from nose swabs. The occurrence of *S. aureus* in the samples according to gender as shown in Table 1. The *S. aureus* isolates were

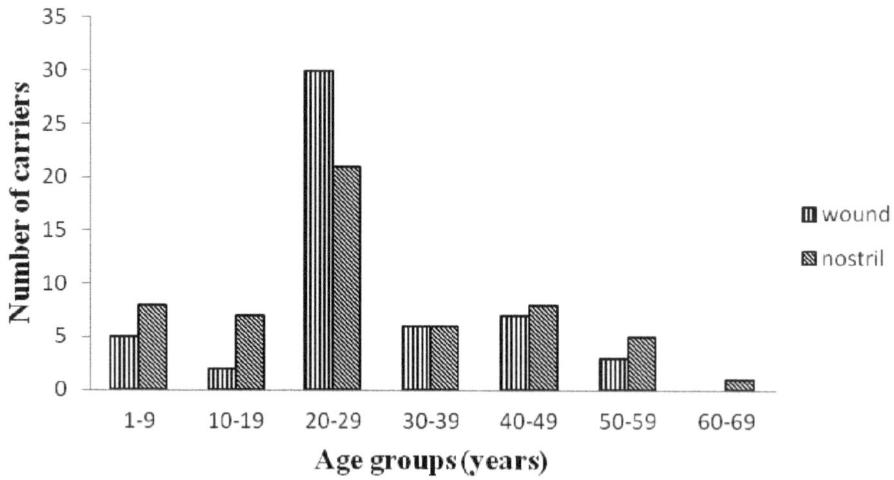

Figure 1. The number of *S. aureus* isolated in relation to age (years) of patients.

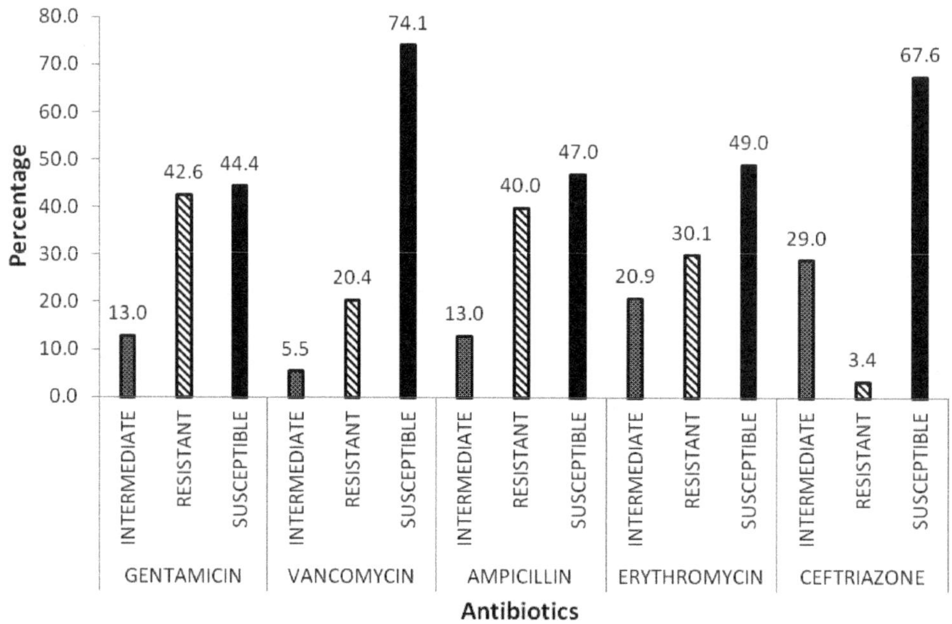

Figure 2. Antibiotic sensitivity patterns of *S. aureus* isolates to reference antibiotics determined by Kirby-Bauer agar disc diffusion method and the mean zones of inhibition when compared with the values provided by the Clinical and Laboratory Standards Institute (CLSI, 2010).

from 44% of the female and 56.0% male samples. *S. aureus* occurred more frequently between the ages of 20 and 29 years, whilst 60 to 69 years age group registered the least number of *S. aureus* isolates (Figure 1).

Antibiotic sensitivity patterns of *S. aureus* isolates

S. aureus isolates showed variable susceptibility to the antibiotics tested. Vancomycin recorded the highest susceptibility of 74.1% followed by ceftriaxone with 67.6%, erythromycin with 49.0%, ampicillin with 47.0% and gentamicin with 44.4%. Resistance to vancomycin, ceftriaxone erythromycin, ampicillin and gentamicin were 20.4, 3.4, 30.1, 40.0 and 42.6.0%, respectively (Figure 2). The *S. aureus* isolates exhibited intermediate resistance of 29.0% to ceftriaxone, 13.0% to ampicillin, 20.9% to erythromycin, 13.0% gentamicin and 5.5% to vancomycin

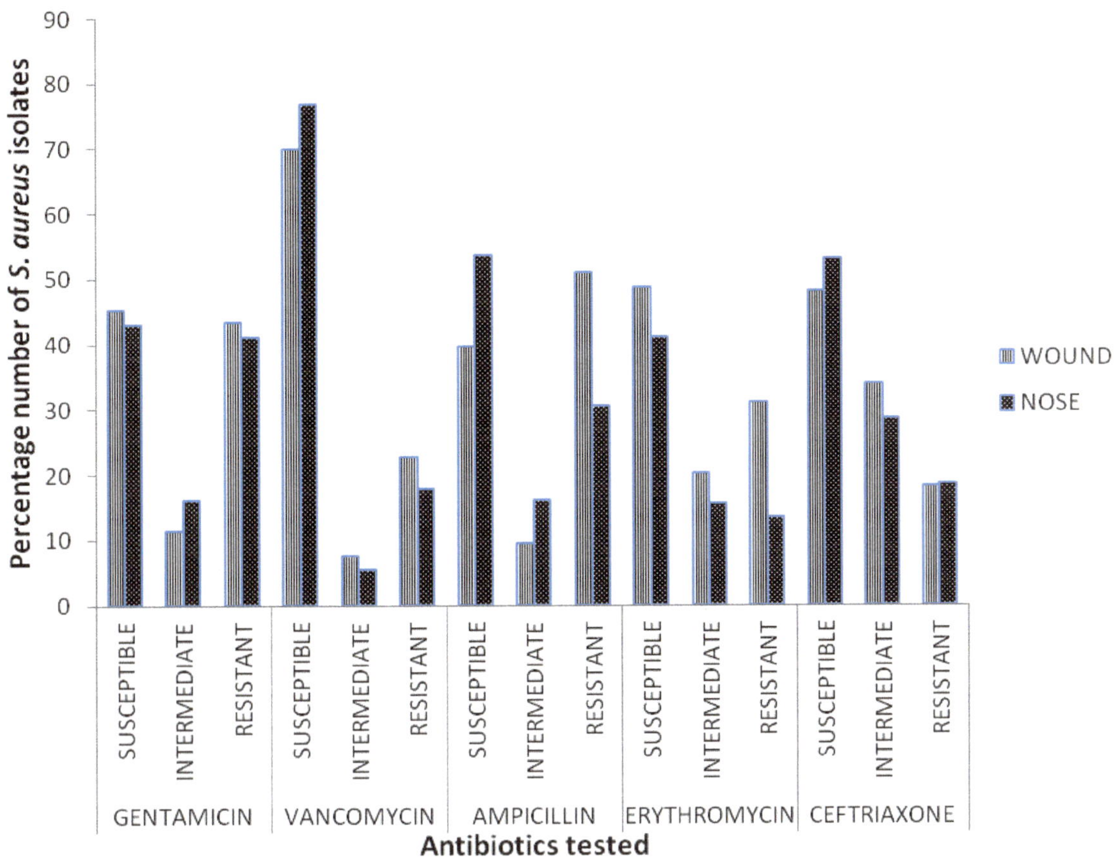

Figure 3. Antibiotic susceptibility patterns of *S. aureus* isolates from wound and nose swabs determined by Kirby-Bauer agar disc diffusion method and the mean zones of inhibition when compared with the values provided by the Clinical and Laboratory Standards Institute (CLSI, 2010).

based on comparison with the values provided by the Clinical and Laboratory Standards Institute (CLSI, 2010).

Antibiotic susceptibility patterns of wound and nose isolates of *S. aureus*

The wound and nose isolates of *S. aureus* were sensitive to vancomycin with susceptibilities of 69.9 and 76.8%, respectively.

Generally, the number of the *S. aureus* isolates from both wound and nose swabs that showed resistance to gentamicin, ampicillin, erythromycin and ceftriaxone were slightly high (Figure 3).

Out of the 109 *S. aureus* isolated from the three hospitals, 35 (32.1%) exhibited resistance to at least three different classes of antibiotics, that is, showing multiple drug resistance (Table 2).

DISCUSSION

S. aureus was isolated in nose (37.3%) and wound swabs (35.3%) but not in blood samples. The high yield of *S. aureus* from nose swabs confirms the nostrils as a habitat for the organism (Prescott, 2002) and could therefore be one of the major sources of *S. aureus* infection. There was no significant difference between the number of nose swabs and wound swabs that yielded *S. aureus*.

From the results, no patient had *S. aureus* in the blood, an indication that none of the patients recruited for the study had bacteraemia. Blood is sterile in healthy persons and the presence of *S. aureus* is normally an indication of bacteraemia and possible sepsis (Mylonakis and Calderwood, 2001). Some of the risk factors for bacteraemia include old age, immunosuppression, chemotherapy and invasive procedures (Newman et al., 2006).

More males had *S. aureus* than females even though there was no significant difference. This observation is corroborated by Manal et al. (2006) who also reported a higher percentage of *S. aureus* isolates (64.4%) in male patients from a Saudi Arabian hospital. There was a significant difference of *S. aureus* isolates among the various age groups with the 20 to 29 years group

Table 2. Multiple-drug resistant (MDR)* pattern of *S. aureus* isolates determined by Kirby-Bauer agar disc diffusion method and mean zones of inhibition when compared with the values provided by the Clinical and Laboratory Standards Institute (CLSI, 2010).

Hospital	No. of MDR *S. aureus*		Total no. of MDR *S. aureus*	Total no. of *S. aureus* isolates	Percentage of MDR
	Wound	Nose			
Kumasi South	5	3	8	42	19.0
North Suntreso	11	3	14	37	37.8
Tafo	8	5	13	30	43.3
Total	24	11	35	109	32.1

*Multiple drug resistance (MDR) is defined as resistance of organism (bacteria) to at least three different antibiotics.

recording the highest carriers (51 patients). The highnumber of isolates of *S. aureus* in this age group could be due to the level of contamination arising from the habitat of the patients, low personal hygiene and poor health education which still persists in many African countries (Agwu et al., 2006).

S. aureus exhibits remarkable versatility in its susceptibility towards antibiotics and the capability of this bacterium to cause human diseases has not diminished even with the availability of antibiotics (Obiazi et al., 2007). In this study, the susceptibility of the *S. aureus* isolates to gentamicin, vancomycin, ampicillin, erythromycin and ceftriaxone were 44.4, 74.1, 47.0, 49.0 and 67.6% respectively. This trend of susceptibility is similar to a report by Reshma et al. (2007) but contrary to another report by Obiazi et al. (2007) who observed higher susceptibility of *S. aureus* to the antibiotics mentioned above. All the antibiotics used in this study except vancomycin and ceftriazone were considered to be ineffective against *S. aureus*. Vancomycin is not recommended for the management of *S. aureus* infections in public hospitals by the Ghana National Drugs Programme (GNDP, 2004). The use and subsequent abuse of this antibiotic by patients is therefore limited and could be the reason why vancomycin was active against the *S. aureus* isolates.

The isolates from both wound and nose exhibited similar intermediate resistance to gentamicin, ampicillin and erythromycin. However, many of the isolates showed intermediate resistance to ceftriaxone from both sites (Figure 3).

The number of resistant isolates from wounds were 31.1% to erythromycin, 50.9% to ampicillin, 43.4% to gentamicin and 18.2% to ceftriaxone, whereas 18.5% of those isolated from the nose were resistant to ceftriaxone, 41.0% to gentamicin, 13.3% to erythromycin and 30.4% showed resistance to ampicillin (Figure 3). Considering the results above, *S. aureus* has developed resistance to most of the older, less expensive antibiotics used in the treatment of infections. These inexpensive antibiotics are widely available to patients in pharmacies without prescription from authorized health personnel (Newman et al., 2006) and this leads to indiscriminate use which promotes selective pressure favouring the emergence of resistant bacteria (Levy, 2001). Not only are these resistant bacterial strains potential causes of recurrent infections but they are also reservoirs of resistance genes that could be transferred to other pathogens. For this reason, the antibiotic susceptibility trends seen with the *S. aureus* isolates may also occur with other bacterial pathogens.

Multiple-drug resistance among *S. aureus* isolates did not vary much from one hospital to another (Table 2). Kumasi South hospital recorded a multiple-drug resistance of 19.0% among the *S. aureus* isolates. North Suntreso hospital had 37.8% of the *S. aureus* being multiple-drug resistant, while Tafo hospital had 43.3% multiple-drug resistant *S. aureus* isolates. The overall multiple-drug resistance among the 109 *S. aureus* isolates was 32.1%. The multiple-drug resistance values in this study are lower than the findings of Newman et al. (2006) who reported that the multiple-drug resistance of *S. aureus* isolated in nine regions of Ghana was 42.3%.

The presence of multiple-drug resistant strains of *S. aureus* among the isolates may be attributed to antibiotic misuse arising from self-medication in suspected bacterial infections (Newman et al., 2006). Self-medication prevents early reporting of patients to hospitals at the onset of disease symptoms, except where complications had occurred, also, some other factors such as unnecessary prescriptions and substandard antibiotics. And hence there is a need to monitor the resistance and susceptibility patterns of pathogenic bacteria in patients suffering from bacterial infections from other parts of the country. Also, antibiotic sensitivity tests should be conducted on isolates from patients suffering from bacterial infections before antibiotics are prescribed.

Conclusion

S. aureus was isolated from nose and wound swabs but not from blood samples. The age group of 20 to 29 years recorded the highest number of *S. aureus* carriers. Over 30% of the *S. aureus* isolated were resistant to ampicillin and gentamicin and 32.1% of the isolates exhibited multidrug resistance to at least three of the antibiotics used.

ACKNOWLEDGEMENTS

The authors would like to thank the management and laboratory staff of Kumasi South, North Suntreso and Tafo Hospitals, Kumasi, Ghana for their assistance ad cooperation during the study. We also thank the Ghana Education Trust Fund (GETFUND) for their financial support to S. B. M. Saana.

REFERENCES

Agwu E, Ohihion AA, Agba MI, Okogun GRA, Okodua M, Tatfeng YM, Nwobu GO (2006). Incidence of *Streptococcus pneumoniae* infections among patients attending tuberculosis clinics in Ekpoma, Nigeria. SEMJ 7(1).

Ajello G, Bopp C, Elliott J, Facklam R, Knapp J S, Popovic T, Wells J, Donell SF (2003). Manual for the Laboratory Identification and Antimicrobial Susceptibility Testing of Bacterial Pathogens of Public Health Importance in the Developing World. Center for Disease Control and Prevention, Atlanta, Georgia, USA and World Health Organization: Department of Communicable Disease Surveillance and Response. pp. 103-118.

Biedenbach DJ, Moet GJ, Jones RN (2004). Occurrence and antimicrobial resistance pattern comparisons among bloodstream infection isolates from the sentry antimicrobial surveillance program (1997–2002). Diagn. Microbiol. Infect. Dis. 50:59-69.

Brenda WL, Lee KL, (2008). Infectious Diseases in Context, Farmington Hills, Mich: Thomson Gale. pp. 173-210.

Burnett GW, Henry WS, Schuster SG, (1996). *Staphylococcus* and Staphylococcal infections. In Oral Microbiology and Infectious Disease., Williams and Wilkins, pp. 405-416.

Clinical and Laboratory Standards Institute (2010). Performance standards for antimicrobial susceptibility testing; Seventeenth informational supplement.M100-S17.CLSI, Wayne, PA.

Crowder MW, Spencer J, Vila AJ (2006). Metallo-β-lactamases: Novel Weaponry for Antibiotic Resistance in Bacteria. Acc. Chem. Res. 39(10):721-728

Forbes BA, Sahm DF, Weissfeld AS (2007). Bailey & Scott's Diagnostic Microbiology.12th edition. Elsevier Inc, New York, pp. 322-325.

Ghana National Drugs Programme (Ministry of Health), (2004). Standard treatment guidelines, 5th edition, Yemens Press Ltd, Accra, 6:172-181.

Lalitha MK (2004). Manual on Antimicrobial Susceptibility Testing. 7th edition. Indian Association of Medical Microbiologists. Delhi, pp. 10-47.

Levy S (2001). Antibiotic resistance: Consequences of inaction. Clin. Infect. Dis. 33(3):S124-S129.

Manal MB, Manal MA, Amal JF (2006). Trends in antibiotic susceptibility patterns and epidemiology of MRSA isolates from several hospitals in Riyadh, Saudi Arabia. Ann. Clin. Microbiol. Antimicrob. 5(30):176-179.

Mylonakis E, Calderwood SB (2001). Infective Endocarditis in adults. N. Engl. J. Med. 345(18):1318-1330.

Newman MJ, Frimpong E, Asamoah-Adu A, Sampane-Donkor E (2006). Resistance to antimicrobial drugs in Ghana. The Ghanaian – Dutch collaboration for Health Research and Development, pp. 1-6.

Obiazi HAK, Nmorsi OPG, Ekundayo AO, Ukwandu NCD (2007). Prevalence and antibiotic susceptibility pattern of *Staphylococcus aureus* from clinical isolates grown at 37 and 44°C from Irrua, Nigeria. Afr. J. Microbiol. Res. 1(5):57-60.

Ohene A (1997). Bacterial pathogens and their antimicrobial susceptibility in Kumasi, Ghana. East Afr. Med. J. 74(45):450-455.

Paterson DL (2000). Recommendation for treatment of severe infections caused by Enterobacteriaceae producing extended-spectrum beta-lactamases (ESBLs). Clin. Microbiol. Infect. 6(9):460-463.

Prescott LM, (2002). Introduction to Microbiology, 5th edition, McGraw Hill Companies. pp. 919-920.

Reshma F, Asma B, Talat Y, Mujahid R, Nayar J (2007). Antibiotic resistance profile: Isolation and characterization of clinical isolates of *Staphylococcus aureus* from patients with community acquired infections. Pak. J. Pharm. Sci. 20(4):295-299.

Sheretz RJ, Reagan DR, Hampton KD (1996). A cloud adult: The *Staphylococcus aureus*-virus interaction revisited. Ann. Int. Med. 1(24):539-547.

Shittu, AO, Johnson L, (2006). Antimicrobial susceptibility patterns and characterization of clinical isolates of *Staphylococcus aureus.* BMC Infect. Dis. 12(5):2334-2336.

Styers D, Sheehan DJ, Hogan P, Sahm DF (2006). Laboratory-based surveillance of current antimicrobial resistance patterns and trends among *Staphylococcus aureus:* 2005 status in the United States. Ann. Clin. Microbiol. Antimicrob. 5(2):1-9.

Tagoe DNA, Baidoo SE, Dadzie I, Tengey D, Agede C (2011). Potential sources of transmission of hospital acquired infections in the Volta Regional Hospital in Ghana. Ghana Med. J. 45(1):22-25.

30

Antibiotic resistance profile of *Escherichia coli* isolated from five major geopolitical zones of Nigeria

Chijioke A. Nsofor and Christian U. Iroegbu

Department of Microbiology, University of Nigeria, Nsukka, Enugu State, Nigeria.

Information on the resistance profiles of clinical and non clinical human bacteria isolates in the developing countries can serve as important means of understanding the human pathogens drug resistance interactions in the zone. *Escherichia coli* isolated from five geopolitical zones of Nigeria were screened for anti-microbial resistance profile against 14 antibiotics using the disc diffusion method. Overall, the 89 isolates tested were 94.4% that showed resistance to Ampicillin; 85.5% to Cotrimoxazole, 92.1% to Cephalothin; 78.7% to Streptomycin, 70.8% to Nitrofurantoin; 79.8 to Tetracycline; 67.4% to Chloramphenicol; 74.2% to Amoxicillin clavulanic acid; 61.8 to Cefpirome; 52.8% to Cefpodoxime; 46.1% to Cefotaxine; 46.1% to Ceftriaxone; 31.5% to Cefoxitin; 38.2% to Nalidixic acid and 24.7% to Gentamycin. These isolates showed a total of 42 different antibiotics resistance profiles, with all the isolates showing resistance to at least four or more of the drugs tested. Statistical analysis showed there was no statistical difference in the proportion of isolates that were resistant to ≥1 antibiotic for human clinical (84.6%) and non-clinical (91.7%) isolates. This result indicates that antimicrobial use in humans has driven the emergence of multi-drug resistant clones in developing countries such as Nigeria that has resulted in an increasingly high prevalence of multiple resistance.

Key words: *Escherichia coli*, resistance profiles, antibiotics, isolates.

INTRODUCTION

The increasing occurrence of drug-resistant *Escherichia coli* isolates of human origin is a global public health problem. Drug resistant *E. coli* infections are presently associated with higher morbidity, mortality and invariably higher expenditure in treatment when compared with infections by strains susceptible to the drugs. Drug resistant commensal *E. coli*, which may not directly cause disease, remains significant in public health as a reservoir of drug resistance genes. These genes can be transferred to human pathogens, such as *E. coli*, *Salmonella*, or to other Gram-negative bacteria in the gut.

Emergence of drug resistance in *E. coli* may arise either because an erstwhile susceptible strain acquires resistance *de novo* or already drug-resistant clones are selected by drug administration and elimination of the susceptible strains. Examples of the latter phenomenon include the outbreak in South London, United Kingdom, involving trimethoprim-sulfamethoxazole-resistant (TS-R) strains of *E. coli* O15:K52:H1 (1986 to 1987); disseminated urinary tract infections (UTIs) in the United States due to TS-R strains of *E. coli* clonal group A (CGA) (1990s); and widespread infections in Canada, Europe, and Asia due to strains of *E. coli* sequence type 131 (ST131; O25:H4) that exhibit extended-spectrum cephalosporin (ESC) resistance through the production of CTX-M-15, an extended-spectrum beta-lactamase (2000s)

(Mshana et al., 2009).

There is emerging evidence that anti-microbial use in humans may be the major selective force for multi-drug resistant clones in enterobacteriaceae in many developing countries that has resulted in an increasingly high prevalence of multidrug resistance in these countries (Kariuki et al., 1996; Okoli et al., 2002; Ling et al., 1987). This is contrary to the information from the developed countries where overwhelming evidence seems to suggest that antimicrobial use in agriculture is the major driving force in the selection and dissemination of bacterial resistance (Khachatourians, 1998; Apley et al., 1998; Witte, 1998).

For *E. coli* and other classes of enterobacteriaceae in which asymptomatic colonization of the intestine usually precedes infection, acquisition of one or several new genes, rather than point mutation in existing genes has been shown to be the common anti-microbial resistance mediation route (Lipsitch and Samore, 2002). For example, segments of new genes could be replaced with alleles having entirely different sequences from the new type at multiple positions. These new mechanisms thus arise and spread in the bacterial population under conditions of anti-microbial selective pressure (Lipsitch and Samore, 2002).

Again, it has been shown that exposure of most *E. coli* strains to anti-microbial agents usually occurs during treatment directed at infections caused by other unrelated organisms. Anti-microbial treatment that alters the population of *E. coli* in a given host or environment, will usually affect bacterial contacts of the host or environment (Halloran and Struchiner, 1991). Thus, use of a particular antibiotic in for example human hosts, in an environment may increase the risk of colonization by or infection with resistant organisms in other humans or even animals that have not received that set of antibiotics but are sharing common environment with the humans (Lipsitch and Samore, 2002). Anti-microbial use may also increase the density of resistant organisms within a host that already harbor such organisms at a lower density resulting in enhanced shedding of these organisms and increased risk to other hosts (Donskey et al., 2000).

Problems associated with inappropriate use of antibiotics continues to pose very serious hindrance to adequate and qualitative therapeutic management particularly in a developing country such as Nigeria. It is particularly worrisome that despite numerous studies and identification of specific problems, the problem of antibiotics resistance still persists like a canker worm at all levels of our healthcare delivery system. The consequences of aberrant drug utilization (overuse, under use, etc.) cannot be overemphasized, so also the utility of prescription database as a monitoring tool; it is therefore imperative that we continuously perform these studies. In this study, we evaluated the prevalence of antibiotic resistant *E. coli* collected from the major tertiary hospitals located in the five geopolitical zones of Nigeria. We also studied the susceptibility pattern of commensal isolates, we got from apparently healthy individuals.

MATERIALS AND METHODS

Study population

The study population included humans (who were either ill or presumptively healthy) from five geopolitical zones of Nigeria viz: South-East, South-West, South-South, North-Central and North-North. In the South-South and South-East, clinical specimens were collected at the University of Port Harcourt Teaching Hospital, Port Harcourt, Rivers State and the Abia State University Teaching Hospital, Aba, Abia State, respectively. The Lagos State University Teaching Hospital, Ikeja, Lagos was the site of specimen collection for the South-West, while the National Hospital, Abuja and Military Reference Hospital, Kaduna State were the sources of specimens from the North-Central and North-North, respectively. All samples from these hospitals were clinical specimens from patients who were having gastroenteritis or similar illness. Apparently, healthy undergraduate students of Madonna University Elele, Rivers State were included in the study for the isolation of human commensal *E. coli*. These individuals reported no exposure to antibiotics for six months prior to sampling and each person received an explanation of the study objectives and consent form for inclusion in the study. All sampling procedures were in accordance with guidelines of the National Health Research Ethics Committee, Nigeria (www.nhrec.net).

Specimen collection, cultivation and identification of *E. coli*

Human fecal specimens were streaked directly on Eosin Methylene Blue agar (EMB) (Oxoid, England). No antibiotic was included in the EMB agar plates used for the cultivation. The inoculated plates were incubated overnight at 37°C. A single colony on EMB with green metallic sheen taken to be *E. coli* was selected from an individual fecal sample for further characterization. *E. coli* was fully identified using conventional microbiological tests- Indole positive, Methyl red positive and Citrate negative (Cheesbrough, 2000).

Antibiotics susceptibility testing

The antibiotics susceptibility pattern of the isolates was determined using the disk diffusion method (Cheesbrough, 2000), on Mueller-Hinton agar (Oxoid, England). Inhibition zone diameter values were interpreted using standard recommendations of the Clinical Laboratory Standard Institute (CLSI, 2006). Susceptibility was tested against ampicillin (10 µg), amoxycillin/ clavulanic acid (20/10 µg), cefpodoxime (10 µg), cefoxitin (30 µg), cefpirome (30 µg), ceftriaxone (30 µg), cefotaxine (30 µg), cephalothin (30 µg), chloramphenicol (30 µg), gentamicin (10 µg), nalidixic acid (30 µg), nitrofurantoin (30 µg) and streptomycin (10 µg), sulfamethoxazole-trimethoprim (10 µg), tetracycline (30 µg), (Oxoid, England). *E. coli* ATCC 25922 was included as a reference strain.

RESULTS

The antibiotic resistance pattern of the *E. coli* isolates is shown in Table 1. Overall, 42 different antibiotics resistance profiles were observed, with all the isolates showing resistance to at least four or more of the drugs

Table 1. Antibiotics resistance profiles of *E. coli* isolated from human specimens.

Antibiotics resistance profile	Percentage
AMP, SXT, NA, TE, S, C, KF, F, AMC, CPO	18.0
AMP, SXT, GN, NA, S, C, KF, F, AMC, CPO	2.2
AMP, SXT, NA, TE, S, C, KF, F, AMC	3.4
AMP, SXT, GN, NA, TE, S, C, KF, CPO	3.4
AMP, SXT, TE, S, C, KF, F, AMC, CPO	6.7
AMP, SXT, GN, TE, S, C, KF, F, AMC	3.4
AMP, SXT, GN, TE, S, C, KF, F, CPO	9.0
AMP, SXT, NA, TE, C, KF, F, CPO	3.4
AMP, SXT, NA, TE, S, C, KF, CPO	1.1
AMP, GN, TE, C, KF, F, AMC, CPO	1.1
AMP, SXT, TE, C, KF, F, AMC, CPO	2.2
AMP, SXT, TE, S, C, KF, AMC, CPO	1.1
AMP, SXT, TE, S, C, KF, AMC, CPO	10.1
AMP, SXT, TE, S, C, KF, F, AMC	00
SXT, TE, S, C, KF, F, AMC, CPO	1.1
AMP, SXT, NA, S, KF, AMC, CPO	00
AMP, SXT, S, KF, F, AMC, CPO	3.4
AMP, SXT, TE, C, KF, AMC, CPO	1.1
AMP, NA, S, KF, F, AMC, CPO	1.1
AMP, NA, TE, S, C, AMC, CPO	00
AMP, TE, C, KF, F, AMC, CPO	1.1
AMP, SXT, TE, S, KF, AMC, CPO	7.9
AMP, SXT, TE, S, KF, F, CPO	1.1
AMP, GN, NA, TE, S, KF	1.1
AMP, GN, TE, S, C, F, CPO	00
AMP, SXT, KF, F, AMC, CPO	00
AMP, SXT, S, C, KF, AMC	00
AMP, SXT, TE, S, KF, AMC	7.9
AMP, C, KF, F, AMC, CPO	3.4
AMP, NA, TE, S, C, KF	1.1
AMP, KF, F, AMC, CPO	1.1
AMP, SXT, C, KF, CPO	00
AMP, SXT, GN, KF, F	00
AMP, SXT, GN, TE, CPO	00
AMP, SXT, KF, AMC, CPO	1.1
SXT, NA, C, KF, CPO	1.1
SXT, TE, KF, F, CPO	00
SXT, NA, TE, KF	1.1
SXT, S, AMC, CPO	00
AMP, TE, AMC, CPO	00
AMP, C, KF, CPO	00
SXT, TE, S, F	1.1

AMP = Ampicillin, SXT= cotrimoxazole, GN = gentamycin, NA = nalidixic acid, TE = tetracycline, S = streptomycin, C = chloramphenicol, KF = cephalothin, F = nitrofurantoin, AMC = amoxycillin, CPO = cefpirome.

tested. Among the 89 isolates tested, 94.4% showed resistance to ampicillin; 85.5% to cotrimoxazole, 92.1% to cephalothin; 78.7% to streptomycin, 70.8% to nitrofurantoin; 79.8 to tetracycline; 67.4% to chloramphenicol; 74.2% to amoxicillin clavulanic acid; 61.8 to cefpirome; 52.8% to cefpodoxime; 46.1% to cefotaxine; 46.1% to ceftriaxone; 31.5% to cefoxitin; 38.2% to nalidixic acid and 24.7% to gentamycin.

Comparing samples from clinical and non clinical human specimens, the results showed that, the isolates were resistant to ampicillin (93.9% and 100%), cotrimoxazole (87.9 and 50%), cephalothin (92.8 and 83.3%), streptomycin (78.3 and 83.3%), nitrofurantoin (71.1 and 66.7%), tetracycline (81.9 and 50%), chloramphenicol (67.5 and 66.7%), amoxicillin clavulanic acid (74.7 and 66.7%), cefpirome (60.2 and 50%), cefpodoxime (51.8 and 50%), cefotaxine (45.8 and 33.3%), ceftriaxone (44.6 and 66.7%), cefoxitin (28.9 and 66.7%), nalidixic acid (38.6 and 33.3%) and gentamycin (22.9 and 33.3%), respectively (Figure 1).

Taking the susceptibility pattern source after source; the isolates from the south east zone of Nigeria showed a unique pattern of resistance, they were resistant to ampicillin and cotrimoxazole (91.3% each); gentamycin and nalidixic acid (17.4% each); tetracycline and amoxicillin clavulanic acid (87% each); chloramphenicol and ceftriaxone (47.8% each); streptomycin, cefpodoxime and cefotaxine (65.2% each) (Table 2). A similar pattern was observed in the isolates from the south west zone; 93.1% of the isolates were resistance to cotrimoxazole, tetracycline and cephalothin respectively. Also, common resistance pattern of 66.7% to cefpodoxime and nitrofurantoin; 33.3% to ceftriaxone and gentamycin; and 53.3% to cefoxitin and nalidixic acid, respectively, was observed among these isolates (Table 2). The isolates from the south-south zone showed a unique resistance to cephalosporins; 41.2% to cefoxitin, ceftriaxone and cefotaxine, respectively and a very high resistance of 76.5% to cefpirome- a third generation cephalosporin (Table 2). Statistical analysis showed there was no statistical difference in the proportion of isolates that were resistant to ≥1 antibiotic for human clinical (84.6%) and non-clinical (91.7%) isolates (Fisher's exact test, $P = 1.0$).

DISCUSSION

The introduction of antibiotics is one of the most important medical interventions with regards to reducing human morbidity and mortality. However, the intensive use of antibiotics (which was estimated in 2002 to be 100,000 to 200,000 tonnes per annum (Andersson and Hughes, 2010), worldwide has dramatically increased the frequency of resistance among clinical and commensal isolates and threatens a loss of therapeutic options. Resistance dramatically reduces the possibility of treating infections effectively and increases the risk of complications and of a fatal outcome.

In this study, we examined antimicrobial resistance in

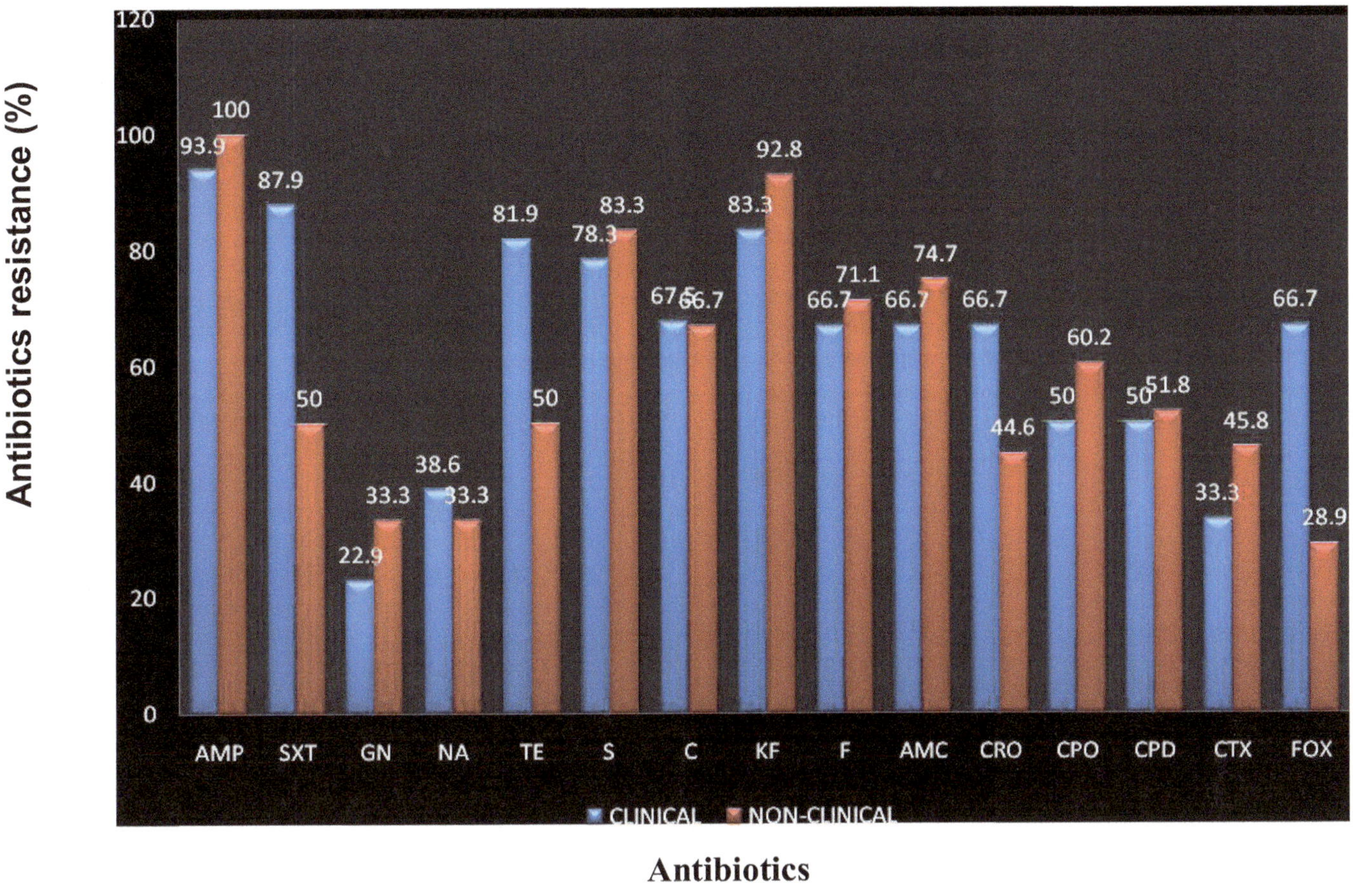

Figure 1. Comparison of percentage antibiotics resistance of *E.coli* Isolates from clinical and non-clinical human specimens. AMP = Ampicillin, SXT = cotrimoxazole, GN = gentamycin, NA = nalidixic acid, TE = tetracycline, S = streptomycin, C = chloramphenicol, KF = cephalothin, F = nitrofurantoin, AMC = amoxycillin clavulanic acid, CRO = ceftriaxone, CPO = cefpirome, CPD = cefpodoxime, CTX = cefotaxine, FOX = cefoxitin.

Table 2. Antibiotics resistance profiles of *E. coli* isolated from human specimens in different regions of Nigeria.

Sample	Antibiotics													
source	AMP	SXT	GN	NA	TE	S	C	KF	F	AMC	CRO	CPO	CPX	CTX	FOX
SE	91.3	91.3	17.4	17.4	87.0	65.2	47.8	95.7	70.0	87.0	47.8	74.0	65.2	65.2	39.0
SW	86.7	93.3	33.3	53.3	93.3	100	73.3	93.3	66.7	60.0	33.3	46.7	66.7	40.0	53.3
SS	100	76.5	29.4	23.5	58.8	88.2	76.5	88.2	58.8	82.4	41.2	76.5	47.1	41.2	41.2
NC	100	100	20.0	100	80.0	80.0	80.0	100	100	80.0	40.0	60.0	60.0	40.0	20.0
NN	96.6	79.3	27.6	44.8	75.8	72.4	70.0	90.1	72.4	65.5	52.2	55.2	41.4	41.4	10.3

AMP = Ampicillin, SXT = cotrimoxazole, GN = gentamycin, NA = nalidixic acid, TE = tetracycline, S = streptomycin, C = chloramphenicol, KF = cephalothin, F = nitrofurantoin, AMC = amoxycillin, CPO = cefpirome, SE = South-East, SW = South-West, SS = South-South, NC = North-Central, NN = North-North.

clinical *E. coli* isolates from human clinical specimens collected from five major tertiary hospitals located in the five geopolitical zone of Nigeria. Because of the geographic sampling of the isolates, this surveillance provides a representative sample of the resistance trends in Nigeria. Generally, when compared with the other antimicrobial agents used in this study, the isolates were susceptible to gentamicin, nalidixic acid and cefoxitin. However, increasing resistance frequencies were noted for ampicillin, cotrimoxazole, cephalothin, streptomycin, tetracycline and nitrofurantoin.

High antimicrobial resistance to tetracycline was noted

among the isolates (79.8%). The presence and frequency of tetracycline resistance agree with findings of other studies on antibiotic resistance in *E. coli* (van den Bogaard et al., 2001; Saya et al., 2005). The patterns of resistance to tetracycline have been attributed in part to widespread and lengthy use of tetracycline in human medicine (Piddock, 2006; Vu and Nikaido, 1985; van den Bogaard et al., 2001). Since tetracycline is a naturally derived compound, bacteria can be exposed to these agents in nature and outside any human use for disease treatment, for prophylaxis, or for livestock growth promotion. Tetracycline is a commonly used first line antibiotic for humans and is often used before the antibiotic resistance profile of a pathogen has been determined. Resistance to tetracycline is plasmid mediated, with a wide variety of genetic determinants (van den Bogaard et al., 2001). This makes it more possible for a susceptible bacterium to acquire resistance factors by conjugation, or other genetic exchange mechanisms.

The high resistance to cefpodoxime (50% in clinical and 51.8% in non-clinical specimens) and cefpirome (50% in clinical and 60.2% in non-clinical specimens), observed in this study is of public health concern. This is probably the first documented evidence of such high resistance figure against any third or fourth generation cephalosporins in *E. coli* isolates from Nigeria. In fact, Okoli et al. (2005) reported the highest resistance rate of 31.5% to cefuroxime (second generation cephalosporins). It is probable that even though these drugs belong to the third and fourth generation cephalosporins, they may have been introduced into clinical practice in the country. Current information indicates that, third and fourth generation cephalosporins resistance can also be plasmid mediated, involving the *blaCMY-2* plasmids, which is quite distinct from the known cephalosporins resistance genes. The gene has been isolated from *E. coli* and *Salmonella* spp. and has been identified in USA (Daniels, 2009). It is probable that this plasmid-mediated gene may be contributing to the wide distribution of high bacterial resistance to cephalosporins observed in this study. This assumption was later confirmed by DNA microarray, as a remarkable number of this gene was detected in this study (data not shown)

Multi-drug resistance (resistance to at least three antimicrobials) was found in *E. coli* from both clinical and commensal sources, but was higher in frequency and proportion in clinical isolates. However, when these multi-drug resistant strains were compared, it is clear that they do not have common sources of resistant bacteria. Most clinical multi-drug resistant isolates exhibited resistance to a combination of antimicrobials that included streptomycin, tetracycline and cephalothin, while most commensal isolates were resistant to ampicillin and tetracycline or nalidixic acid and gentamicin. This suggests that clinical isolates that were Tet-R are more likely to become resistant to additional antimicrobial agents. Resistance to tetracycline may be conserved in bacterial populations over time, regardless of selection pressure, which might result in an overall increase in resistance over time. In addition, it is likely that clinical sources could act as reservoirs for Tet-R for environmental contamination and human colonization. Furthermore, the difference in resistance between the two groups of isolates can be explained in terms of the interactions of the organisms (associated with the host) and potential horizontal gene transfer in their respective environments.

The very low resistance recorded against gentamicin here is in agreement with results of similar studies in Nigeria (Chah et al., 2003, Okoli et al., 2002). They are however at variance with the results reported by Uwaezuoke et al. (2000). It is possible that these organisms originated as direct human contaminants of the feed ingredients through handling. Gentamicin is marketed mostly as a 2 ml injectable solution in a glass vial. This probably makes the drug unpopular and therefore safe from common abuse in Nigeria.

In conclusion, 42 resistance profiles were observed in the *E. coli* isolates, with predominant patterns being distributed widely across both sample sources, indicating a striking diversity of resistance patterns in Nigeria. Multi-drug resistance was a common feature in these isolates, highlighting the fact that the resistance genes for these drugs are linked to plasmids (Schroeder et al., 2002). Moreover, the wide spread resistance to cotrimoxazole may be implying the presence of class 1 integrons, which are also important in conferring resistance to multiple antimicrobials (Jonhson et al., 2007). This is strong evidence that antimicrobial use in humans has driven the emergence of multi-drug resistant clones in developing countries such as Nigeria which has resulted in an increasingly high prevalence of multiple resistance.

REFERENCES

Andersson AD, Hughes EF (2010). Patterns of antimicrobial resistance observed in *Escherichia coli* isolates obtained from domestic- and wild-animal fecal samples, human septage, and surface water. Appl. Environ. Microbiol. 71:1394-1404.

Apley MD, Brown SA, Fedorka-Cray PJ (1998). Role of veterinary therapeutics in bacterial resistance development: animal and public health perspectives. J. Am. Vet. Med. Assoc. 212:1209-1213.

Chah KF, Okafor SC, Oboegbulem SI (2003). Anti-microbial resistance of none clinical *E. coli* strains from chicken in Nsukka, Southeastern Nigeria. Nig. J. Anim. Prod. 30(1):101-106.

Cheesbrough M (2000). District Laboratory Practice in Tropical Countries, Part 2. Cambridge University Press, Cambridge, UK; p. 434.

Clinical Laboratory Standards Institute (2006). Performance standards for antimicrobial susceptibility testing. National committee for clinical laboratory standards, Wayne pa.

Daniels JB, Call DR, Hancock DD, Sischo WM, Baker K, Besser TE (2009). The role of ceftiofur in the selection and dissemination of bla_{CMY-2} – mediated cephalosporin resistance in *Salmonella enterica* and commensal *Escherichia coli* isolated from cattle. Appl. Environ. Microbiol. 75:3648-3655.

Donskey CJ, Chowdhry TK, Hecker MT (2000). Effect of antibiotic therapy on the density of vancomycin-resistant enteroccoci in the stool of colonized patients. N. Engl. J. Med. 343:1925-1932.

Halloran ME, Struchiner CJ (1991). Study designs for dependent happenings. Epidemiology 2:331-338.

Johnson TJ, Wannemuehler YM, Johnson SJ, Logue CM, White DG, Doetkott C, Nolan LK (2007). Plasmid replicon typing of commensal and pathogenic *Escherichia coli* isolates. Appl. Environ. Microbiol. 73:1976-1983.

Kariuki S, Gilks C, Corkill J (1996). Multi-drug resistant non-typhoid salmonellae in Kenya. J. Antimicrob. Chemother. 38:425-434.

Khachatourians GG (1998). Agricultural use of antibiotics and the evolution and transfer of antibiotic resistant bacteria. CMAJ 159:1129-1136.

Ling J, Chau PY, Rowe B (1987). *Salmonella* serotypes and incidence of multiple resistant salmonellae isolated from diarrhoeal patients in Hong Kong from 1973-1982. Epidemiol. Infect. 99:295-306.

Lipsitch M, Samore MH (2002). Anti-microbial use and anti-microbial resistance: A population perspective. Emerg. Infect. Dis. 8:347-354.

Mshana SE, Can I, Hamid H, Torsten H, Eugen D, Chakraborty T (2009). Using Conjugative IncFI plasmids carrying CTX-M-15 among *Escherichia coli* ESBL producing isolates at a University hospital in Germany. BMC Infect. Dis. 9(97):1471-1498.

Okoli CI, Chah KF, Ozoh PTE, Udedibie ABI (2005). Antimicrobial resistance profile of *E. coli* isolates from tropical free range chickens. Online J. Health Allied Sci. 3:3.

Okoli IC, Okeudo NJ, Onwuchekwa CI (2002). New trends in antibiotic resistance among *E. coli* isolates from southern Nigeria. In book of abstracts for the 39th annual national congress, Nigerian Veterinary Medical Association, 2002 October 27- 31; Sokoto, Nigeria. p. 16.

Piddock LJV (2006). Multidrug-resistance efflux pumps—not just for resistance. Nat. Rev. Microbiol. 4:629-636.

Sayah RS, Kaneene JB, Johnson Y, Miller R (2005). Patterns of antimicrobial resistance observed in *Escherichia coli* isolates obtained from domestic- and wild-animal fecal samples, human septage, and surface water. Appl. Environ. Microbiol. 71:1394-1404.

Schroeder CM, Zhao C, DebRoy C, Torcolini J, Zhao J, White DG (2002). Antimicrobial resistance of *Escherichia coli* O157: H7 isolated from humans, cattle, swine and food. Appl. Environ. Microbiol. 68:576-581.

Uwaezuoke JC, Ogbulie JN, Njoku JN, Obiajuru IOC, Njoku AJ (2000). Antibiotic sensitivity patterns of bacterial isolates from poultry feed. Int. J. Environ. Health Hum. Dev. 1:23-28.

van den Bogaard AE, London N, Driessen C, Stobberingh EE (2001). Antibiotic resistance of faecal *Escherichia coli* in poultry, poultry farmers and poultry slaughterers. J. Antimicrob. Chemother. 47:763-771. doi: 10.1093/jac/47.6.763.

Vu H, Nikaido H (1985). Role of β-lactam hydrolysis in the mechanism of resistance of a β-lactamase-constitutive *Enterobacter cloacae* strain to expanded-spectrum β-lactams. Antimicrob. Agents Chemother. 27:393-398.

Witte W (1998). Medical consequences of antibiotic use in agriculture. Science 279(5353):996-997.

The presence of cagA genome in blood of infected patient with *Helicobacter pylori* as new marker

Hatami saeideh[1], Esmaeili Davoud[2] , Fallah Jali I.[2] and Rezaee Jafar[2]

[1]Science and Research Branch, Islamic Azad University, Department of Biology Tehran, Iran.
[2]Applied Microbiology Research center and Department of Microbiology, Faculty of Medical science, Baqyiatallah University, Tehran, Iran.

Helicobacter pylori **is a motile and gram negative bacterium, that has been associated with gastritis, peptic ulcer, and gastric adenocarcinoma. The strains of** ***Helicobacter pylori*** **contain CagA associated with severe infections. To investigate the presence of cagA gene in** ***Helicobacter pylori*** **strains among patients' sera, about 100 serum specimens that had positive antibody against** ***H. pylori*** **were collected. Serum samples were studied using designed primers. The scale of triglyceride and cholestrol serum samples was measured. PCR method was done on sera of** ***H. pylori*** **samples that had antibody titers against** ***H. pylori*** **.The results indicated that 18% from all the serum samples contain cagA DNA. This study indicates that 12 serum samples contain triglyceride and cholesterol (TG & Chol) titers higher than normal scale. According to the findings, it is suggested that the presence of CagA DNA in strains of** ***Helicobacter pylori*** **may play an important role in aggravating disorders. In this investigation we were able to detect** ***H. pylori*** **cagA DNA in patients' sera.** ***H. pylori*** **probably attaches to red blood cell and causes bacteremia, but is also able to secrete CagA protein and cagA DNA to host cells by T4 secretory system. Thus, there is risk of integrating cagA DNA in host cell chromosome and inducting antibodies against it. Between the presence of cagA DNA and increase of triglyceride and cholesterol, there exists rational relation.**

Key words: Detection, cagA genome, blood, marker.

INTRODUCTION

Helicobacter pylori is one of the most important bacteria causing gastric infection which infects more than half of the people in whole world, causing chronic gastritis, ulcer peptic and adenocarcinoma (Bruce and Maaroos, 2008l; Choung and Talley, 2008; Costa et al., 2009; Herrera and Parsonnet, 2009; Maeda and Mentis, 2007). *H. pylori* strains have been divided into types I and II. Type I strains express CagA and VacA but type II strains do not show these antigens (Umeda et al., 2009; Sicinschi et al., 2009). Patients have been told that that to be infected with type I strain means to be exposed to adenocarcinoma ulcer peptic, atherosclerosis and gastritis (Jafarzadeh et al., 2009; Zeaiter et al., 2008). The *cag* pathogenicity island, a 40-kb DNA region secretes cagA genome and CagA protein to blood that may be associated with secretion of virulence factors. Several investigations have reported an increased prevalence of CagA positive *H. pylori* in gastric cancer (Hatakeyama, 2006; Handa et al., 2007). Once *H. pylori* CagA is injected into gastric epithelial cells, it can change the host cells structure, cytokines release, cell cycle, and gene expression (Talebkhan etal., 2008, Suriani et al., 2008). The CagA protein is common in East-Asian and associated with increased pathogenicity. The active CagA induces antibodies. About 60 to 80% of

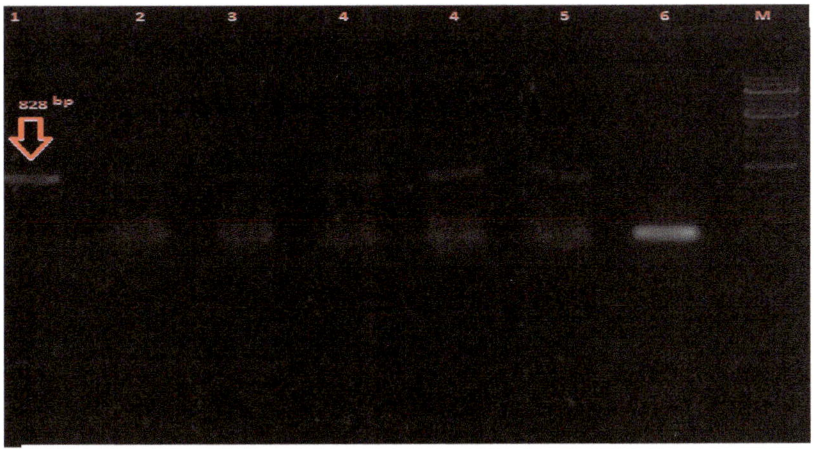

Figure 1. 10 g/L agarose gel electrophoresis of cagA DNA fragment amplified by PCR from *H pylori*. Lane M, Marker 1 kb, Lanes 1, 2, 3 and 4, CagA amplified. Lane 5, Positive control; Lane 6, negative control.

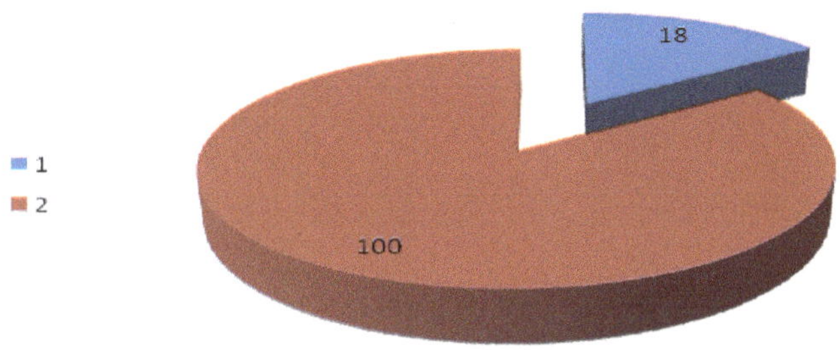

Figure 2. The specimens of positive for cagA genome of *H. pylori* in sera. Number 1 is total specimens and number 2 is positive specimen for *cagA* gene. This figure indicates that from 100 serum specimen, 18 specimens had cagA genome.

H. pylori strains in the world expressed CagA protein with molecular weight of 120 to 128-kDa (Hatakeyama, 2006). Serologic detection of CagA protein by enzyme linked immunosorbent assay (ELISA) is a risk specific indicator for *H. pylori* infection (Cooke et al., 2005). Furthermore, the presence of antibodies in CagA protein in either serum or mucosal secretions is a high risk (Mardh et al., 2002). Several studies have shown that infection with *H. pylori* CagA positive strains induces epithelial gastric cells to secrete increased amounts of interleukin-8 and cytokines that induce inflammatory response of mononuclear and neutrophil cells (Ye et al., 2005; Choi et al., 2007; Esmaeili et al., 2009; Zhang et al., 2006). The aim of this study was to investigate the presence of cagA genome as risk marker in sera of infected patients with *H. pylori.*

MATERIALS AND METHODS

The 100 samples positive with high antibody level against *H. pylori* were collected from Baqyiatallah and Booali hospitals in Tehranand; then were stored at −20°C until assayed. There is ethics regarding the collection and use of the sera in this study. All of the samples were collected after fasting.

IgG and IgA antibodies against *H. pylori* infection were tested by enzyme-linked immunosorbent assay (ELISA).we study sera that had positive seum titers against *H. pylori*. Single primer pair was used to amplify *H. pylori* cagA gene target fragment based on Gene Bank. Primers in early 828 bp of cagA fragment were candidate and were designed with the use of Gene Runner program; considering conserved fragment. These primers were designed by us and their sequences are as follows: Primer forward: ATGACTAACGAAACTATTGATC and Primer reverse: TATCGCCAAGAGTGAATTTAG. PCR was performed in a standard enzyme Taq DNA polymerase concentration of 1 U/50ul reaction, Mg^{2+} 1.5 mM, dNTP 200 uM each; primer, 20pmol/5 0ul; template DNA (5 ul serum) or 50 ng. The mixtures were incubated for 5 min at 94°C for primary denaturation, 1 min at 94°C for secondary denaturation of the target DNA and then, annealing at 55°C for 1 min, and extension at 72°C for 1 min and final extension for 10 min where 35 cycles were performed. The amplified products were analyzed by electrophoresis on 1% agarose gel (cinnagen) containing 0.1 g of ethidium bromide per ml in Tris-borate-EDTA (TBE) buffer. The PCR product was visualized under UV light and photographed. The controls used in the PCR include: negative control being *E. coli* DNA and positive control, cagA DNA from *H. pylori* ss1.

RESULTS

Serum analysis

The serum samples were studied for antibody titers against *H. pylori*. The measurement of serum titer was done according to ELISA kit Mouse Monoclonal Antibody Isotyping. 100 samples that contain high titers antibody against *H. pylori* were candidate for this study.

PCR amplification of *H. pylori* cagA gene

H. pylori with cagA fragment was amplified by PCR from the above primers and the PCR product was electrophoresized and visualized by 10 g/L agarose gel (Figure 1). It revealed that the size of cagA DNA fragment amplified by PCR was 828 bp.The 18% from 100 specimens were positive for cagA genome (Figure 2).

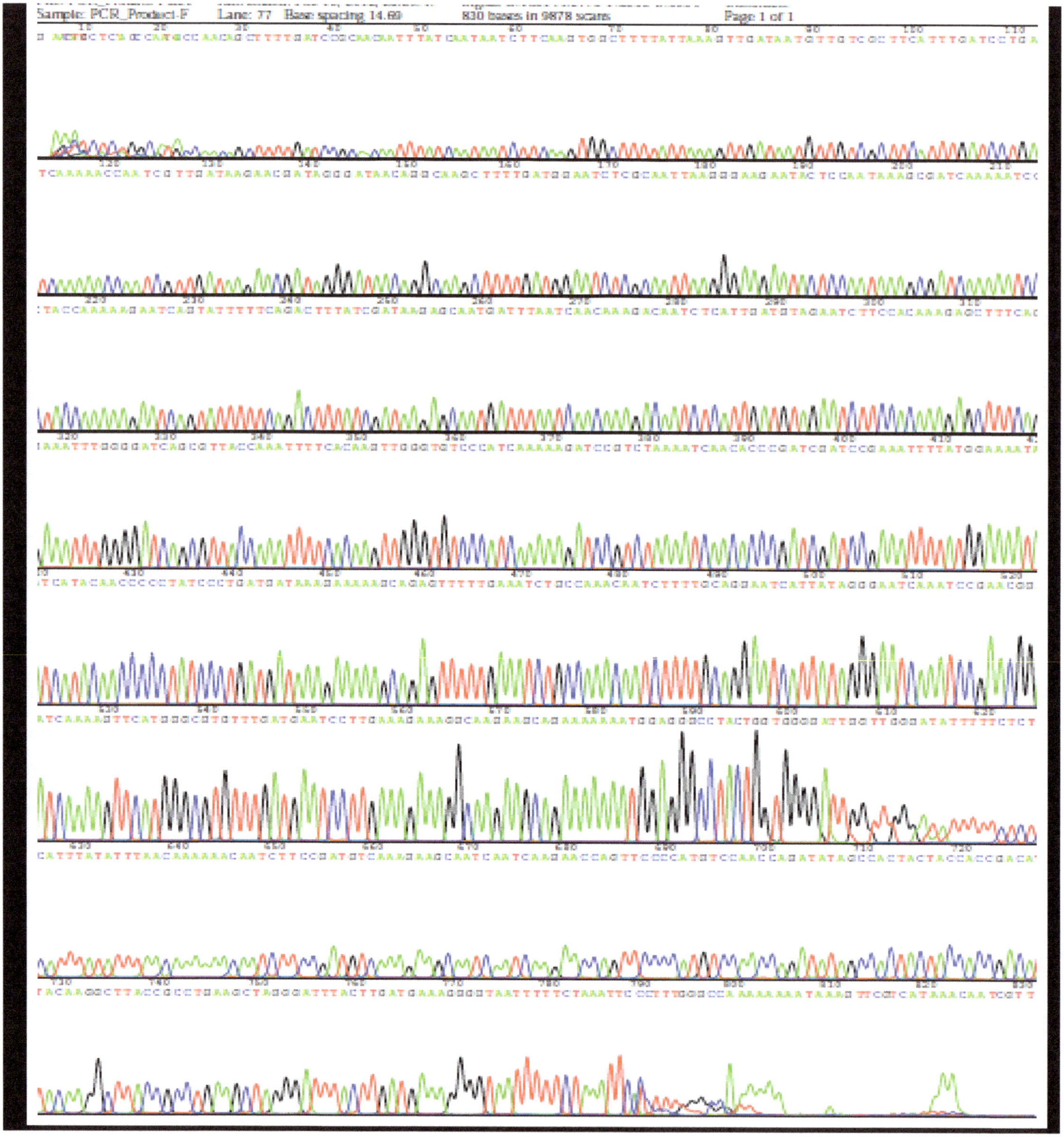

Figure 3. *H. pylori* Sequence of amplified cagA DNA.

Sequence analysis of cagA gene of *H. pylori*

H. pylori Sequence of cagA DNA was analyzed with Forward Primer and primer reverse using automatic sequence analyzer by Sanger dideoxy chain termination method by gene fanavaran. The result of analysis showed that the size of amplified DNA was about 828 bp (Figure 3).

Table 1. The rate of TG&Chol in sera of contains cagA genome in specimens.

Total serum specimen	%
cagA DNA	18%
High TG&Chol titer in total sera	30%
High TG&Chol titer in sra posive with cagA DNA	12%

This table illustrate that from 18 specimen (100%) contain cagA genome there was 12 specimen (66.6%) with high cholesterol and triglyceride titers

Relationship between cagA genome and triglyceride and cholesterol (TG&Chol) titers

This study indicates that 12 from 18 samples with cagA DNA in sera contain TG &Chol titers higher than normal scale (Table 1).

DISCUSSION

H .pylori is a bacterium that exists in the whole world and expresses CagA surface protein with a MW of 120-128 KD. A manifest characteristic of this protein is high antigenicity and immunogenicity (Jafarzadeh et al., 2009). Since the 5'-end of cagA gene contains sequence with conserved regions we used PCR to amplify the 5'-end fragment of cagA with length of 828 bp (Zeaiter et al., 2008). These selective primers are designed for the first time for detection of cagA DNA in serum and biopsy samples.

Several studies have shown that anti-CagA antibodies can be detected in patients infected with *H. pylori* that contain *CagA protein* (Esmaeili et al., 2009). The presence of anti-CagA antibodies correlating with aggravating disorders, high TG & Chol, aterosclerosis and patients infected with a *cagA* positive strain were shown to be more prone for the development of clinically significant *H. pylori*-related disease (Mardh et al., 2002).

Approximately, 60% of *H. pylori* isolates in the world possess the cytotoxin-associated gene A(cagA) and their presence is accompanied with risk factor for adenocarcinoma and atherosclerosis; but in this research we demonstrate 18% from serum samples receptacle cagA DNA. Importantly, infection with *cagA* positive strains is highly associated with peptic ulcer disease, lichen plan, cardiac syndrome, and the risk of developing intestinal metapelasia, atrophic gastritis and adenocarcinoma of the stomach (Mardh et al., 2002). This primer enables the detection of 5'cagA that contains conserved sequences. This primer is specific for conserved regions in all strains. In order to research for the presence of cagA DNA in serum, we designed the specific primers with length of 828 bp based on cagA DNA sequence reported in GenBank, and successfully amplified the cagA DNA of *H. pylori* by PCR. The secretory system type IV secretes CagA protein and cagA DNA into host cells. The number of studies has shown the presence of antibodies against CagA protein but so far does not report detection of the presence of cagA DNA in sera of infected patients with *H. pylori*. In this study, for the first time in the world, we were able to detect cagA DNA in patients' serum infected with H. pylori. Due to the presence of cagA DNA in infected sera many antibodies were secreted against CagA protein and also many probably were integrated into the host cell chromosomes. *H. pylori* does not create septicemia but enables secretion of CagA protein and DNA into host cells and blood. If this bacterium creates bacteremia, it must be able to culture it in synthetic medium. Thus, there may exist risk of integration of cagA DNA into host cell chromosome and induction of antibodies against it that were correlated to several disorders. This research indicates that from 18 samples of positive CagA about 12 samples had titers of high TG& Chol. This study demonstrates that from 100 samples of sera only 18% contain cagA DNA; thus there may be many cagA DNA without CagA protein cleavage in blood by nuclease.

In this study for the first time we can detect cagA DNA in sera of patients with high titers antibody against *H. pylori*. Furthermore, we demonstrate that the presence of cagA DNA is correlated with increase of TG&Chol titers. By in situ hybridization, *H. pylori* bacteria were observed in close association with erythrocytes in capillaries and post-capillary venules of the lamina propria of gastric mucosa in both infected humans and Rhesus monkeys (Aspholm et al., 2006).

Cases of *Helicobacter* bacteremia have been reported from time to time. *Helicobacter pylori* are the most important representative of *Helicobacterium*, but whether it can result in bacteremia has rarely been studied (Huang et al., 2006). In this study, we examined *H. pylori* cagA DNA in peripheral blood sera and gastric mucosa of patients with peptic ulcer and gastritis by polymerase chain reaction (PCR). The investigations indicate that 60-70% from gastric biopsy specimens patients were all positive for *H. pylori* cagA DNA. Our findings suggest that *H. pylori* exists not only in gastric mucosa but also can secrete DNA in peripheral blood, and it is possible that *H. pylori* can result in bacteremia or inject DNA to blood by secretory system type IV.

REFERENCES

Aspholm M, Olfat FO, Nordén J, Sondén B, Lundberg C (2006). SabA Is the *H. pylori* Hemagglutinin and Is Polymorphic in Binding to Sialylated Glycans. PLoS Pathog. 2(10): e110. doi:10.1371

Bruce MG, Maaroos HI (2008). Epidemiology of *Helicobacter pylori* infection. Helicobacter 13(1):1-6

Choi KD, Kim N, Lee DH (2007). Analysis of the 3' variable region of the cagA gene of *Helicobacter pylori* isolated in Koreans. Dig. Dis. Sci. 52:960-966

Choung RS, Talley NJ (2008). Epidemiology and clinical presentation of stress-related peptic damage and chronic peptic ulcer. Curr. Mol. Med. 8:253-257

Cooke CL, Huff JL, Solnick JV (2005). The role of genome diversity and immune evasion in persistent infection with *Helicobacter pylori*. FEMS Immunol. Med. Microbiol. 45:11-23

Costa AC, Figure ueiredo C, Touati E (2009). Pathogenesis of *Helicobacter pylori* infection. Helicobacter 14 (1):15-20

Esmaeili D, Mohabati Mobarez A, Hatef Salmanian A, Zavaran H.A, Hosseini doust R (2009). Immune Responses of Conserved 32 KD fragment from N-terminal of *H. pylori* cagA gene.J. Pure Appl. Microbiol. 4(1):117-123

Handa O, Naito Y, Yoshikawa T (2007). CagA protein of *Helicobacter pylori*: a hijacker of gastric epithelial cell signaling. Biochem. Pharmacol. 73:1697-1702

Hatakeyama M (2006). The role of *Helicobacter pylori* CagA in gastric carcinogenesis. Int. J. Hematol. 84:301-308

Hatakeyama M (2006). The role of *Helicobacter pylori* CagA in gastric carcinogenesis. Int. J. Hematol. 84:301-308

Herrera V, Parsonnet J (2009). *Helicobacter pylori* and gastric adenocarcinoma. Clin. Microbiol. Infect.15:971-976

Huang Y, Fan X, Tang Z (2006). Detection of Helicobacter pylori DNA in peripheral blood from patients with peptic ulcer or gastritis. APMIS. 114:851-6.

Jafarzadeh A, Mirzaee V, Ahmad-Beygi H, Nemati M, Rezayati MT (2009). Association of the CagA status of *Helicobacter pylori* and serum levels of interleukin (IL)-17 and IL-23 in duodenal ulcer patients. J. Dig. Dis. 10:107-112

Maeda S, Mentis AF (2007). Pathogenesis of Helicobacter pylori infection. Helicobacter 12 Suppl 1:10-14

Mardh E, Mardh S, Mardh B, Borch K (2002). Diagnosis of gastritis by means of a combination of serological analyses. Clin Chim. Acta 320:17-27

Sicinschi LA, Correa P, Peek RM (2009). CagA C-terminal variations in *Helicobacter pylori* strains from Colombian patients with gastric precancerous lesions. Clin. Microbiol. Infect.

Suriani R, Colozza M, Cardesi E (2008). CagA and VacA *Helicobacter pylori* antibodies in gastric cancer. Can. J. Gastroenterol. 22:255-258

Talebkhan Y, Mohammadi M, Mohagheghi MA (2008). cagA gene and protein status among Iranian *Helicobacter pylori* strains. Dig. Dis. Sci. 53:925-932

Umeda M, Murata-Kamiya N, Saito Y (2009). *Helicobacter pylori* CagA causes mitotic impairment and induces chromosomal instability. J. Biol. Chem. 284:22166-22172

Ye W, Held M, Enroth H (2005). Histology and culture results among subjects with antibodies to CagA but no evidence of Helicobacter pylori infection with IgG ELISA. Scand J. Gastroentero.l 40:312-318

Zeaiter Z, Huynh HQ, Kanyo R, Stein M (2008). CagA of *Helicobacter pylori* alters the expression and cellular distribution of host proteins involved in cell signaling. FEMS Microbiol. Lett. 288:227-234

Zhang MJ, He LH, Wong BC, Zhou ZF, Zhang JZ (2006). [Distribution of cagA 3' region, iceA, vacA and HP0519 on *Helicobacter pylori* isolated from China]. Zhonghua Liu Xing Bing Xue Za Zhi 27:508-512

Impact of *Bacillus thuringiensis* β – exotoxin to some biochemical aspects of *Musca domestica* (Diptera: Muscidae)

Faten, F. Abuldahab, Najlaa, Y. Abozinadah* and Nawal, S. Al- Haiqi

Department of Biology, Faculty of Science, King Abdul Aziz University, Jeddah, Saudi Arabia.

The present investigation aimed to evaluate the efficiency of *Bacillus thuringiensis* β – exotoxin as a biological agent to control *Musca domestica* and to study its impact on some biochemical parameters of 2nd larval instar. Laboratory studies are carried by applying topically the bacterium *B. thuringiensis* β – exotoxin at LC_{30} under laboratory conditions (27 ± 2°C and 60±5% RH). The total protein contents of the total body, the total lipids, total carbohydrate, enzymatic activities of carbohydrate hydrolyzing enzymes, alkaline and acid phosphatases and non-specific esterases were estimated. Our data revealed that, the total soluble protein contents in supernatant of the homogenated larvae post-treatment was generally decreased, as affected by the tested material at different time intervals as compared to control. As well as a significant reduction in the total lipid content was observed, this may be due to production of enzymes that utilize lipids to remove the invading agent. The total carbohydrate contents was significantly reduced and the reduction was (-14.989, -43.33, 63.93 and -51.69%) with respect to control at the four different time intervals. While the activities of α – and β – esterases were decreased in larvae treated with LC_{30} of *B. thuringiensis israelensis* and the reduction of enzymatic activity was highly significant. Alkaline and acid phosphatase showed a significant reduction in its activities. Bacterial treatments induced inhibition in the enzyme activity of invertase where the values of inhibition were (-27.85 , -5.92 , -19.46 and -39.96%) at different time intervals. Also, a similar reduction of trehalase enzyme activity of treated larvae was observed at different time intervals (-32.81 , 12. 95 , -18.20 and -5.16%). It may be possible in this instance to control flies by the use of this bacterium which incorporate spores and crystals of the appropriate strain of *B. thuringiensis israelensis*, by affecting the biochemical systems of the target insect. So, it is quite clear from our results that *B. thuringiensis israelensis* at concentration of LC_{30} significantly decreased the activities of all biochemical parameters, we investigated. The general disturbance in metabolism could originate primarily from inhibition of chitin synthesis.

Key words: Housefly, bacteria, biocides, sublethal effects, proteins, carbohydrates, lipids, enzymes.

INTRODUCTION

The house fly, *Musca domestica has* long been considered as a potential agent for disease transmission ever since its existence. The general truth of this assertion remains undisputed till the present day, inspite of increasing awareness toward an improved sanitation hygiene (Nazhi et al., 2005). The widespread use and massive application of hazardous insecticides, against insects of either agricultural or medical importance, frequently leads to serious problems such as the development of resistance and environmental pollution. Accordingly, attention has been given recently to safer insecticides. Therefore, they isolated and identified various naturally bioactive compounds to be used as pesticides. Many previous studies proved that these chemicals

*Corresponding author. E-mail: mohed@live.com

have a toxic action against insect species and dipterous in particular (Deb-Kirtaniya et al., 1980; Abbassy, 1981; Mesbah et al., 1985; Methela et al., 1989; El-Sherif et al., 1994; Soliman and El- Sherif, 1995; Wirth and Georghiou, 1996; Gheikh et al., 1998; Mulla et al., 2003; Mittal et al., 2004; Paul et al., 2005; Tikar et al., 2008).

There have been few studies of the effect of *Bacillus thuringiensis* toxins toward house flies. Yamvrias and Anagnou (1989) observed high levels of mortality of *Bactrocera* (Dacus) *oleae* (Gmelin) larvae, after application of an aqueous suspension of *B. thuringiensis;* although it is not known if β–endotoxin was present in their experimental preparation. In a different study, 55 strains of *B. thuringiensis* were tested for activity against larvae of *Anastrepha luden* with resulting mortality of 4 to 63%, although the β–exotoxin content of these preparations was not specified (Robacker et al., 1996; Roh et al., 2007). Cry toxins from *B. thuringiensis* are used for insect control.

Their primary action is to lyse mid-gut epithelial cells. In this review, we will summarize recent findings on the Cry toxin receptor interaction and the role of receptor recognition in their mode of action. Cry toxins interact sequentially with multiple receptors. These toxins have a synergistic effect and CytlAa overcomes Cry toxin resistance. Recently, it was proposed that CytlAa synergizes or suppresses resistance to Cry toxins by functioning as a membrane-bound receptor for Cry toxin (Gomez et al., 2007; Pigott et al., 2008; Jorge et al., 2010). The objective of the current study was to determine the potential of *B. thuringiensis* β – endotoxin as an agent for control of *M. domestica*. This was achieved by a bioassay of the direct effects of a pure β–exotoxin product on insect mortality and sublethal effects of biochemical aspects in survivors.

MATERIALS AND METHODS

Test insects

Sources of colony

Adult susceptible strain of house fly *M. domestica vicina* used in the present study were obtained from a well established colony originated from the Biology Department, Faculty of Science for girls, King Abdul Aziz University. All experiments were carried out from March to December 2009.

Rearing technique

Egg masses were used to maintain a colony in the laboratory under constant conditions of temperature and humidity (27 ± 2°C and 60 ±5% R.H.) Each egg mass was placed in a clean Petri dish (10 cm diameter), previously constant technique described by Lelwallen (1954). Full grown larvae were allowed to pupate in clean glass Petri dishes. Following emergence, the adults were provided with a piece of cotton soaked in 10% sugar, 2% milk solution as a source of food.

Source of the bacterial pathogen

The bacterium *B. thuringiernsis israelensis* was chosen as a pathogen for this study because of its wide use in biological control. The powder was obtained from Valent Biosciences, U.S.A.

Biochemical studies

Estimation of the total proteins

The total protein content of the total body was estimated according to the method described by Bradford (1976). This method is based on the observation that Coomassie Brilliant Blue G-250 exists in two different colour forms, red and blue. The red colour is converted to the blue colour upon binding of the dye to protein; this binding causes a shift in maximum absorption of the dye from 465 to 595 nm. The intensity of the colour was measured at 595 nm.

Preparation of samples for the assay: The starved (for about 12 h) larvae were homogenized in distilled water (5 larvae / 5 ml) using a chilled glass Teflon tissue grinder for 3 min. Homogenates were centrifuged at 3500 r.p.m for 10 min at 2°C in a refrigerated centrifuge. The supernatant can be used directly or stored at 5°C till use (Max. 2 weeks).

Preparation of protein reagent: Commassie Brilliant Blue (CBB) G-250 (100 mg) was dissolved in 50 ml of 95% ethanol. To this solution 100 ml of 85% (w/v) phosphoric acid were added. The resulting solution was diluted to a final volume of 1 L.

Procedure: Sample solution (5 µl) for preparation of standard curve 50 µl of serial concentrations containing 10 to 100 µg bovine serum albumin were pipette into test tubes. The volume in the test tube was adjusted to 0.1 ml with phosphate buffer (pH= 6.6). 5 ml of protein reagent were added to the test tube and the contents were mixed either by inversion or vortexing. The absorbance at 595 nm was measured after 2 min and before 1 h against blank prepared from 0.1 ml of phosphate buffer (pH 6.6) and 5 ml protein reagent. This procedure was repeated 3 times for each determination. The result is expressed as µg protein / larva.

Estimation of the total lipids

The total lipids were estimated by the method of Knight et al. (1972).

Preparation of samples of the assay: Homogenation was done as previously described in protein determination.

Phosphovanillin reagent: Pure vanillin (0.6 g) was dissolved in 10 ml ethyl alchohol and completed to 100 ml with distilled water. 400 ml of concentrated phosphoric acid was added, and the solution was stored in dark glass bottle at room temperature.

Procedure: 250 µl of samples solution was added to concentrated sulphuric acid (5 ml) in a test tube and heated in a boiling water bath for 10 min. After cooling to room temperature, the digest (500 µl) were added to phosphovanillin reagent (6.0 ml). After 45 min, the developed colour was measured at 525 nm against reagent blank prepared from 500 µl distilled water and 6.0 ml phosphovanillin reagent. The result is expressed as µg lipid / insect. For standard curve, serial concentrations of lipids (oleic and palmitic acids at a ratio of 7:3) from 1 to 5 µg were prepared in absolute ethanol (concentration of 1 to 5 µg/ml) were used treated in the same manner as the unknown. This procedure was repeated 3 times for each determination. The result is expressed as μg lipid/larva.

Estimation of total carbohydrate

The total carbohydrates were estimated by the phenol sulphuric acid reaction of Dubois et al. (1956).

Preparation of samples of the assay: The starved (for about 12 h) larvae (five to ten) were homogenized in 0.3 N $HClO_4$ (5 ml) at 0°C for 1 min in a glass homogenizer with a Teflon pestle. The homogenate was kept in ice for a further 10 min. Insoluble matter was removed by centrifugation for 3 min at 2000 r.p.m., and washed twice in ice–cold $HClO_4$ (5 ml) by re–dispersion and centrifugation.

Procedure: Our hundred microliters sample or 100 μl of serial concentrations of glucose solution containing between 10 and 70 μg glucose (for preparation of total carbohydrate standard curve) were added into a colorimetric tube to 0.5 ml of phenol (20% w/v). Then 5 ml of concentrated sulphuric acid was added rapidly with shaking. The tubes were allowed to stand 10 min, then they were shaken and placed for 10 to 20 min in water bath at 25 to 30°C before readings. Blanks were prepared by substituting distilled water for the sugar solution. The absorbance of characteristic yellow – orange color is measured at 490 nm against blank. This procedure was repeated 3 times for each determination. The result is expressed as μg carbohydrates/larva.

Estimation of enzymatic activities

Preparation of samples of the assay: Homogenation was done as previously described in protein and lipid determination.

Carbohydrates hydrolyzing enzymes: The principle based on the digestion of trehalose, sucrose and starch by trehalase, invertase and amylase, respectively according to the method described by Ishaaya and Swirisk (1976). The free aldehydic group of glucose formed after trehalose, sucrose and starch digestion were determined using 3.5 dinitrosalicylic acid reagent. Trehalase reaction mixture consisted of 0.2 ml of 3% trehalose (substrate), 0.18 ml acetate buffer (pH 5.40 and 20 μl of larval homogenate). Invertase reaction mixture consisted of 0.2 ml of 4% sucrose (substrate) and 20 μl of larval homogenate. Amylase reaction mixture consisted of 0.2 ml of 2% starch (substrate), 0.16 ml of phosphate buffer (pH 6), and 40 μl larval homogenate.

The dinitrosalicilic acid reagent was prepared by dissolving one gram of 3.5 diniotrosalicylic acid in 20 ml of 2 N NaOH and 50 ml of distilled water with the aid of a magnetic stirrer. Potassium sodium tartarte (30 g) was added, and magnetic stirring was continued until a clear solution was obtained. Distilled water was then added to bring the final volume to 100 ml. All test tubes were incubated at 37°C for 60 min, 0.8 ml of 3.5 dinitrosalicylic acid reagent were then added. The reaction mixture was heated for 5 min at 100°C in a boiling water bath followed by immediate cooling in an ice bath. The optical density (OD) of produced colour is measured at 550 nm using a spectrophotometer. The enzymatic activity was expressed as μg glucose/g body weight/min.

Preparation of standard curve of glucose: Serial concentrations of glucose solution containing 50, 100, 200, 300, 400 μg glucose / 0.4 ml distilled water were pippeted into test tubes, 0.8 ml of dinitrocalicylic acid solution was added to each tube .The mixture was heated for 5 min at 100°C in boiling water bath, and then cooled immediately in an ice bath. The resulting colour was measured spectrophotometrically at 550 nm. The optical densities was plotted against concentrations, thus a curve can be constructed.

Phosphatase enzymes (acid phosphatase and alkaline phosphatase): Acid phosphatase (Ac – pase) and alkaline phosphatase (Alk – pase) were determined according to the method described by Powell and Smith (1954). In this method, the phenol released by enzymatic hydrolysis of disodium phenylphosphate, reacts with 4 – aminoantipyrine, and by the addition of potassium ferricyanide, the characteristic brown colour is produced. The reaction mixture consists of 1 ml carbonate buffer (pH 10.4) for Alk – pase, or 1 ml of citric buffer (pH 4.9) for Ac – pase , 1 ml of 0.01 M disodium phenyl phosphate (substrate), and 0.1 ml larval homogenate. Mix gently and incubate for exactly 30 min at 37°C . At the end of incubation period, 0.8 ml of 0.5 N NaOH was added to stop the reaction. Then added 1.2 ml of 0.5 N $NaHCO_3$, followed by 1 ml of 4 – aminoantipyrine solution and 1 ml of potassium ferricyanide. The produced colour was measured, immediately, by spectrophotometer at 510 nm. The enzymatic activity is expressed as μg phenol released / g body weight / min.

Preparation of phenol standard curve: A stock of phenol was prepared by dissolving 1 g pure crystalline phenol in 1 L 0.1 N HCl, 10 ml of the stock solution (containing 10 mg) was diluted to 100 ml with distilled water. Aliquots of 0.05, 0.1, 0.2, 0.3 and 0.4 ml of the diluted phenol (equal to 5, 10, 20, 30, and 40 μg phenol) were pippeted into test tubes and the volume was completed to 1 ml with distilled water. 1.1 ml of buffer was added followed by 0.8 ml of NaOH, 1.2 ml of $NaHCOH_3$, 1 ml of aminoantipyrine and 1 ml of potassium fericyanide. Each tube was mixed well after each addition. The developed colour was measured at 510 nm.

Non–specific esterases: Alpha esterases (α – esterases) and beta esterases (β – esterases) were determined according to the method of Van Asperen (1962) using α–naphthyl acetate and β–naphthyl acetate as substrates, respectively. Naphthol produced as a result of hydrolysis of substrate can be prepared by the addition of diazoblue sodium laury sulphate solution which produce strong blue colour in case of α – naphthyl or strong red colour in the case of β–naphthyl, which can be measured spectrophotometrically. The reaction mixture consisted of 5 ml substrate solution ($3x10^{-4}$ M α or β–naphthyl acetate, 1% acetone and 0.04 M phosphate buffer pH 7) and 20 μl of larval homogenate. The mixture was incubated for exactly 15 min at 27°C then 1 ml. of diazoblue colour reagent (prepared by mixing 2 parts of 1% diazoblue and 5 part of 5% sodium lauryl sulphate) was added. The developed colour was read at 600 or 555 nm for α – and β – naphthol, respectively.

Preparation of standard curves of α – and β – naphthol: Stock solutions were prepared by dissolving 20 mg α – or β–naphthol in 100 ml of 0.04 M phosphate buffer (pH 7). Ten milliliter of stock solution was diluted up to 100 ml by the phosphate buffer. Aliquots of 0.1, 0.2, 0.4, 0.8 and 1.6 ml of diluted solution (equal to 2, 4, 8, 16 and 32 μg naphthol) were pippeted into test tubes and completed to 5 ml by phosphate buffer. One milliliter of diazoblue reagent was added and the developed colour was measured as stated before.

Statistical analysis

Data were expressed as mean ± standard error. The statistical significance of differences between individual means was determined by student "t" test paired observations. The level of significance of each experiment was stated to be non-significant ($p>0.05$), significant ($p<0.05$) and highly significant ($p<0.05$).

RESULTS

The present investigation aimed to evaluate the efficiency

Table 1. Total body protein contents (µg/larva) of homogenate 2nd instar larvae of M. domestica determined at different time intervals post – treatment with LC_{30} of *B. thuringiensis*.

Hours post-treatment	Total body protein content (µg/larva) (Mean ± SE)		Decrease (%)
	Control	Treated	
6	257.33±0.048	183.56±0.052**	-28.67
12	285.33±0.041	196.89±0.052**	- 31.00
24	649.33±0.087	256.44±0.019**	- 60.51
48	979.11±0.183	208.00±0.110**	- 99.78

n = 3 replicates per test. ** Highly significant (P < 0.01).

Table 2. Total body lipid contents (µg/larva) of homogenate 2nd instar larvae of *M. domestica* determined at different time intervals post–treatment with LC_{30} of *B. thuringiensis israelensis.*

Hours post-treatment	Total body protein content (µg/larva) Mean ± SE		Decrease (%)
	Control	Treated	
6	60.44±0.032	40.89±0.018**	-32.35
12	144.89±0.062	44.44±0.032**	- 69.33
24	286.22±0.085	90.67±0.071**	- 68.32
48	264.00±0.055	128.00±0.041**	- 51.52

n = 3 replicates per test. **Highly significant (P < 0.01).

Table 3. Total body carbohydrate contents (µg/larva) of homogenate 2nd instar larvae of *M. domestica* determined at different time intervals post – treatment with LC_{30} of *Bacillus thuringiensis israelensis.*

Hours post-treatment	Total body protein content (µg/larva) Mean ± SE		Decrease (%)
	Control	Treated	
6	125.88±0.020	107.03±0.012**	-14.98
12	200.47±0.029	113.60±0.011**	- 43.33
24	300.59±0.027	108.42±0.010**	- 63.93
48	299.84±0.029	1 144.85±0.038**	- 51.69

n = 3 replicates per test. ** Highly significant (P < 0.01).

of *B. thuringiensis israelensis* as a biological control agent on the 2nd instar larvae of *M. domestica.* Data in Table 1, indicate the changes in the level of total soluble protein in supernatant of the homogenated larvae post treatment which were generally decreased as affected by the tested material at different time intervals as compared to control, it recorded -28.67, -31.0, -60.51 and -99.78%, after 6, 12, 24, and 48 h post – treatment and this highly significant reduction was ($p< 0.01$). The total lipid content in the homogenate of 2nd instar larvae was significantly reduced when treatment was carried out with LC_{30} of the tested material at different time intervals post – treatment. The decrease in the haemolymph lipids may be due to produced enzymes that utilize lipids to remove the invading agent. The average of the decreasing percentage in the total lipid contents was -32.35, -69.33, -68.32 and – 51.52% at different time intervals (Table 2).

The total carbohydrate contents was significantly reduced and the reduction was (-14.9890, -43.33, -63.93 and -51.69%) with respect to control at the four different time intervals (Table 3). The activities of α and β – esterases were decreased in larvae treated with LC_{30} of Dipel 2X, and the reduction of enzymatic activity was highly significant (Figures 1 and 2). Alkaline phosphatase activity was highly significantly reduced ($p < 0.01$) at 6, 12 and 24 h post – treatment, whereas the reduction of enzymatic activities at these periods were (-18.04, -50.33, -51.77 and –20.07%), respectively (Figure 3). The same significance in the activity of acid phosphatase was also observed in Figure 4. Bacterial treatment induced inhibition in the enzyme activity of invertase where the values of inhibition were (-27.85, -5.92, -16.46 and -39.69%) at different time intervals of 6, 12, 24, and 48 h post – treatment, respectively (Figure 5).The percentages of reduction of trehalase enzyme activity of treated larvae was (-32.81, -12.59, -18.20, and -5.16%) at all inspected

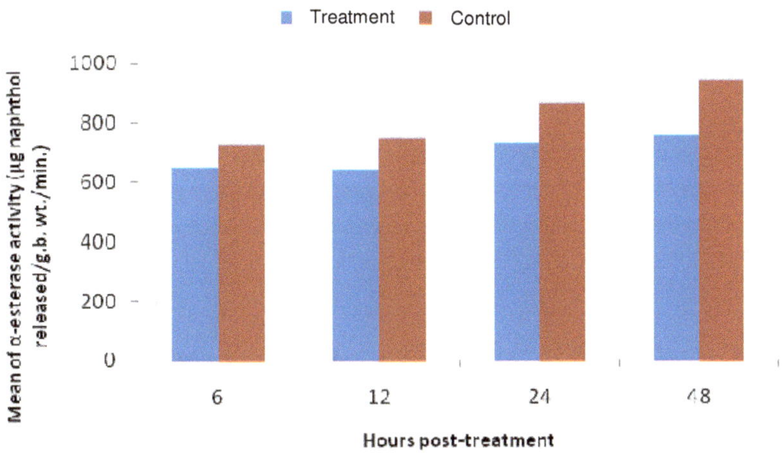

Figure 1. α-Esterase activity (µg naphthol released/g.b.wt./min) of homogenate of 2nd instar larvae of *M. domestica* determined at different time intervals post – treatment with LC_{30} of *B. thuringiensis israelensis*.

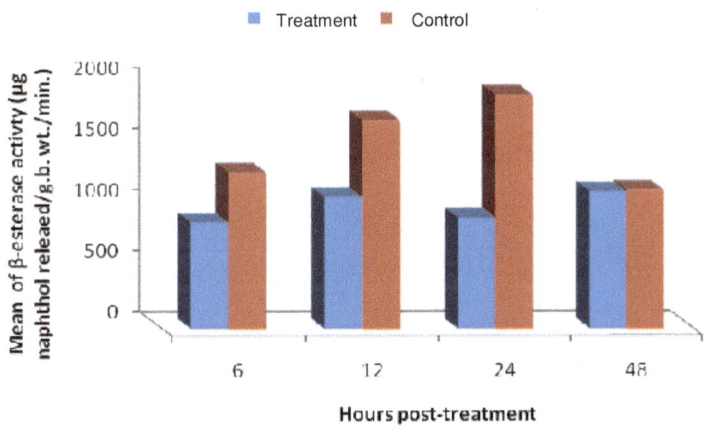

Figure 2. β-Esterase activity (µg naphthol released/g.b.wt./min.) of homogenate of 2nd instar larvae of *M. domestica*. determined at different time intervals post – treatment with LC_{30} of *B. thuringiensis israelensis*.

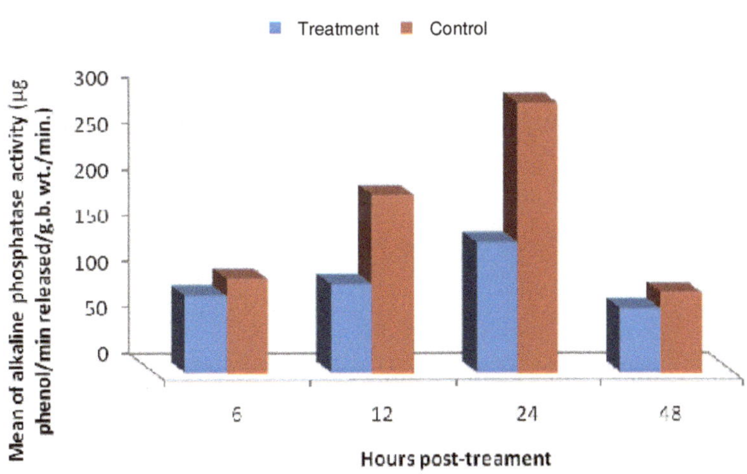

Figure 3. Alkaline phosphatase activity (µg phenol released/g.b.wt./min) of homogenate of 2nd instar larvae of *M. domestica* determined at different time intervals post – treatment with LC_{30} of *B. thuringiensis israelensis*.

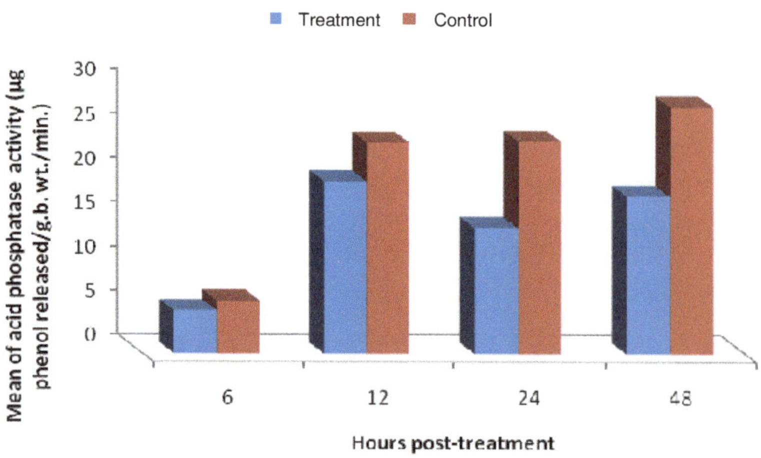

Figure 4. Acid phosphatase activity (µg phenol released/g. b. wt./min) of homogenate of 2nd instar larvae of *M. domestica* determined at different time intervals post – treatment with LC_{30} of *Bacillus thuringiensis israelensis*.

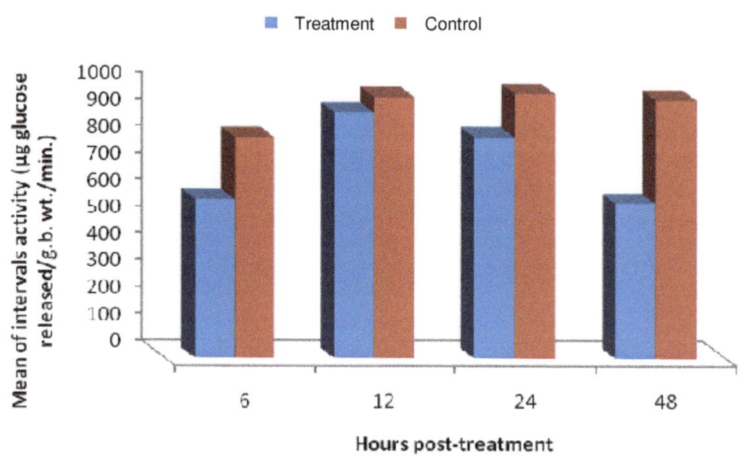

Figure 5. Invertase activity (µg glucose released/g.b.wt./min) of homogenate of 2nd instar larvae of *M. domestica* determined at different time intervals post – treatment with LC_{30} of *Bacillus thuringiensis israelensis*.

times (6, 12, 24 and 48 h), respectively (Figure 6). The corresponding figures of amylase were (-7.19, -42.63, -51.04 and -30.64%) (Figure 7)

DISCUSSION

The biochemical response of *M. domestica* larvae was assessed after treatment of the 2nd larval instar with LC_{30} of *B. thuringiensis israelensis* the changes in activities of total soluble protein ,total lipids , total carbohydrates, non specific esterases (α– and β–esterases), alkaline and acid phosphatases and carbohydrate hydrolyzing enzymes (invertase, trehalase and amaylase). It is clear from the present study that the LC_{30} of the tested material induced highly significant reduction in the level of total soluble protein at different time intervals post – treatment as compared to control. However, this reduction in the protein content may be due to inhibition of DNA and RNA synthesis and this may reflect the decrease in enzymatic activities. Our results agree with that obtained by many authors (Baker et al., 1991; Zeenath and Nair, 1994; Latha et al., 1996; Abd El-Aziz, 2000; Tawfik et al., 2002). The changes in total lipid contents were conducted as a result of bacterial treatment, in general, the total lipid contents in the homogenated larvae was significantly reduced as compared to control at different time intervals. The reason for decrease in the total lipid contents may be due to its conversion to proteins to substitute the reduction in protein content or produce supplementary

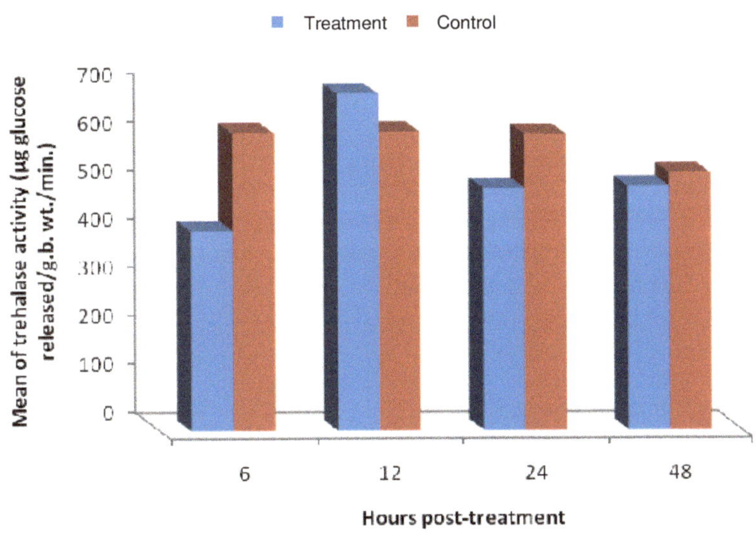

Figure 6. Trehalase activity (µg glucose released/g.b.wt./min.) of homogenate of 2nd instar larvae of *M. domestica* determined at different time intervals post – treatment with LC_{30} of *Bacillus thuringiensis israelensis*.

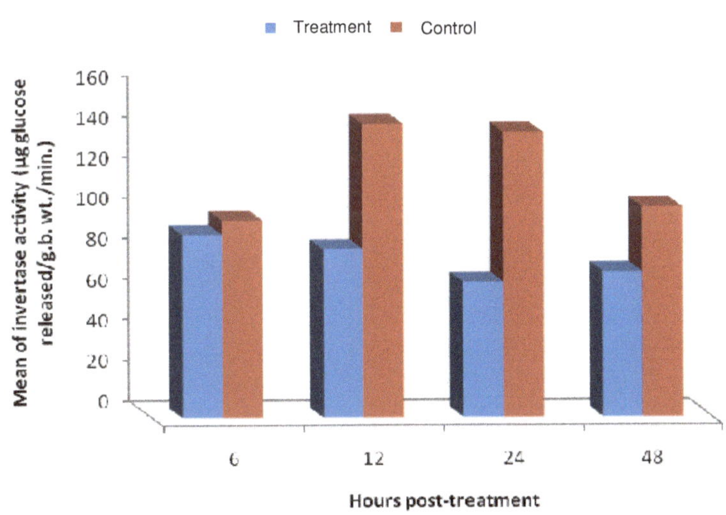

Figure 7. Amylase activity (µg glucose released/g.b.wt./min) of homogenate of 2nd instar larvae of *M. domestica* determined at different time intervals post – treatment with LC_{30} of *Bacillus thuringiensis israelensis*.

energy . Our results agree with those obtained by many investigators (Abo-Ela et al., 1998; Abde El-Aziz, 2000; Omar et al., 2005b). Nonspecific esterases in different insect species have been implicated in reproductive behavior, phormone and hormone metabolism digestion, neurotransmission and resistance to insecticides particularly organo phosphorus compounds (Georghiou and Pasteur, 1980). Esterases may contribute to resistance by hydrolyzing the pesticide (Devonshire and Moores, 1982) or by temporary binding to the pesticide when the catalytic activity is relatively low (Devonshire, 1989). It is quite clear from our results that the LC_{30} applied of *B. thuringiensis israelensis* induced decrease in the activity of both α– and β–esterases in the supernatant of the homogenate larnae of *M. domestica*. These results are in accordance with that demonstrated by El- Ghar et al. (1995b), Salem et al. (1995) and Hanafy et al. (2005).

Acid and alkaline phosphatases have been shown to be associated with insect development, especially in relation to nutrition, egg maturation (Tsumuki and Kanehisa, 1984). Our results clearly indicated that the alkaline and acid phosphatases activities were highly significantly reduced at all time interval post – treatment ($p < 0.01$). These results are in conformity with the findings of Abdel Hafez et al. (1988), Ayyangar and Rao (1990) and Sokar (1995). It is quite clear from our results that *B. thuringiensis israelensis* at concentration of LC_{30}, significantly decreased the activities of the carbohydrate hydrolyzing enzymes (invertase, trehalase and amaylase) as compared to control groups. The general disturbance in carbohydrate metabolism as expressed by reduction of carbohydrate hydrolyzing enzyme activities could be as a result from chain effect originating primarily from inhibition of chitin synthesis (Salem et al., 1995; Auda and Hedaya, 1997; Shakoori et al., 1998; Khedr, 2002; El-Mageed and Elgohary, 2006). Also, Bravo et al. (2007) stated that *B. thuringiensis* Crystal (Cry) and Cytolitic (Cyt) protein families are a diverse group of proteins with activity against insects of different orders, Lepidoptera, Coleoptera, Diptera and also against other invertebrates such as nematodes.

Their primary action is to lyse midgut epithelial cells by inserting into the target membrane and forming pores. Tilquin et al. (2008) concluded that they raise the issue of the persistence, potential proliferation and environmental accumulation of human-spread *B. thuringiensis israelensis* in natural mosquito habitats. Such *B. thuringiensis israelensis* environmental persistence may lengthen the exposure time of insects to this bio-insecticide, thereby increasing the risk of resistance acquisition in target insects, and of a negative impact on non-target insects. Mwamburi et al. (2009) suggested that the control of house fly larvae and the efficacy of *B. thuringiensis israelensis* applied as a larvicide, may be improved with frequent spray applications to chicken manure. Lysyk et al. (2010) was in agreement with our results, also Wirth et al. (2010) who suggest that recombinant bacterial strains have improved mosquito and vector management properties compared with the wild type strains used in current commercial formulations, and should prove useful in controlling important human diseases such as malaria and filariasis on a long term basis, even when used intensively under field conditions. Many authors investigate the role of an alpha-amylase receptor for *B. thuringiensis israelensis* Cry4Ba and Cry11Aa toxins in the malaria vector mosquito *Anopheles albimanus,* (Fernandez et al., 2010).

Generally, our results in this investigation concluded that there was a clear significant effect on all biochemical parameters discussed, as a result of treatment with the bacterium *B.* thuringiensis at the dose of LC_{30}. The total protein, lipid, carbohydrate, non-specific esterases, α- and β- phosphatases and carbohydrate hydrolysing enzymes in the supernatant homogenated 2nd larval instar of *M. domestica* was highly significantly decreased as compared to control groups at all different time intervals, the reduction was continual during all the tested periods.

REFERENCES

Abbassy MA (1981). Insecticidal and synergistic volatile oils isolated from certain food and medicinal plants . Proc. 4th . Arb. Pest. Conf. Tanta University. IIIA: 409 – 41 Egypt.

Abd El-Aziz HS (2000). Psychopathological studies on bacterial infection of cotton leaf worm, *Spodoptera littoralis* .M.Sc.Thesis.Fac.Sci.Ain Shams Univ.,Egypt.p :144.

Abdel- Hafez MM, Shaaban MM, El-Malla MA, Farag M, Abdel- Kawy AM (1988). Effect of insect growth regulators on the activity of transmirases with reference to protein and amino acids in the Egyptian cotton leafworm *Spodoptera littoralis* (Boisd.). Minia. J. Agric. Res. Dev., 10: 1357 – 1362.

Abou El-Ela, R. G.; Younes, M. W. F. and Gamal El-Din, M. M. A. (1998): Biochemical studies on the effect of three different insecticides on the cotton leafworm *Spodoptera exigua* (HB.) (Lepidoptera: Noctuidae). J. Union. Arab. Biol, Cairo, Vol., 9 (A), Zool: 455 – 475.

Auda M, Hedayd, S (1997): Effect of diflubenzuron on growth, development and digestive enzymes of the beet armyworm *Spdoptera exigua* larvae. Bull. Entomol. Sco. Egypt. Econ. Ser., 24: 95 – 106.

Ayyangar GSG, Rao PJ(1990) Changes in haemolymph constituents of *Spodoptera litura* (Fabr.) under the influence of azadirachin. Indian. J. Entomol. 52 (1): 69 – 83.

Baker RFA , Abdel-Razak NA, Hamed MS, Guneidy, AM (1991). Physiological effects of some insect growth regulators on the respirometric movements, total protein and free amino acids of the house fly *Musca domestica* . Ain Shams. Sci. Bull.28 pp: 169-177.

Bradford MM (1976).A rapid and sensitive method for the quantitation of microgram quantities of protein utilizing the principle of protein-dye binding.Anal.Biochem.,v.72pp: 248-254.

Bravo A, Gill SS Soberon ,M. (2007). Mode of action of *Bacillus thuringiensis* Cry and Cyt toxins and their potential for insect control. Toxicon., 15 ; 49 (4): 423 -35 .

Gheikh HB, Ali.Haouas ZB, Marquine M, Pasteur N (1998). Resistance to organophosphorus and pyrethroid insecticides in *Culex pipiens* (Diptera:Culicidae) from Tunisia. J.Med.Entomol.35: 251 – 260.

Ded – Kritaniya S, Ghosh MR, Mitra SA, Adityachaudhury N, Chatterjee A (1980). Extract of garlic as possible source of insecticides . Indian J. Agric. Sci. 50(6): 507- 510.

Devonshire AL (1989). The role of electrophoresis in the biochemical detection of insecticides resistance. In H.D. Loxdal And J. D. Hollander (eds.) Electrophoretic studies on agricultural pests. Clarendon, Oxford., 363 – 374

Devonshire AL Moores, GD (1982). A carboxylesterase with broad substrate specificity causes organophosphorous, carbamate and pyrethroid resistance in peach potato aphids (*Myzus persicae*) Pestic. Biochem. Physiol., 18: 235 – 246.

Dubois M, Gilla KI, Hamiltion JK, Rebers PA, Smith F. (1956). Colorimetric method for determination of sugars and related substances. Analytical chemistry, v. 28(2)pp:350-356.

El- Ghar GE, Radwan HS, El- Bermawy ZA, Zidan LT (1995b). Inhibitory effect of *Bacillus* thuringiensis and abamectin on digestive enzymes and non-specific esterases of *Spodoptera littorals* (Boisd.) (Lep. Noctuidae) larvae. J. Appl. Entomol., 119 (5): 355 – 3590

El-Mageed AEA, Elgohary LRA (2006). Impact of spinosad on some enzymatic activities of the cotton leafworm. Pakistan. J. Biol. Sci., 9 (4): 713 – 716.

EL- Sherif LS, Hussein MA, Hafez SE, Hewady MA NM (1994). Efficiency of certain plant oil extracts against newly hatched larvae of *Earias insulana* (Boisd) and *Pectinophora gossypiella* (Saunders). J. Fac. Educ. Ain Shams University . Egypt. 19: 131 -139.

Fernandez-Luna MT, Lanz-Mendoza H, Gill SS, Bravo A, Soberon M, Miranda-Rios J (2010). An alpha-amylase is a novel receptor for Bacillus thuringiensis ssp. israelensis Cry4Ba and Cry11Aa toxins in the malaria vector mosquito Anopheles albimanus (Diptera: Culicidae). Environ Microbiol. 12(3):746-57.

Georghiou GP, Pasteur N (1980).Organophosphate resistance and esterase patterns in natural populations of the southern house mosquitoes from California. J. Econ. Entomol., 73: 489 – 492.

Gomez I, Pardo-Lopez L, Munoz-Garay C, Fernandez LE, Perez C, Sanchez J, Soberon M,Bravo A (2007). Role of receptor interaction in the mode of action insecticidal Cry and Cyt toxins produced by *Bacillus thuringiensis*. J. Peptides, 28: 169-173.

Hanafy HEM, Sakr HEA, El-Sayed W (2005).Toxicological and biochemical effects of bio-agent products on the cotton leafworm *Spodoptera littoralis* (Boisd.)(Lepidoptera: Noctudae). Arab Univ. J. Agric. Sci., 13(3): 939-950.

Jorge T, Pablo L, Trevor W, Jorge I (1999).Toxicity of *Bacillus thuringiensis* β – exotoxin to three species of fruie flies (Diptera : Tephritidae). J. Econ. Entomol., 92 (5): 1052 – 1056.

Ishaaya I,Swiriski E (1976). Trehalase, invertase and amylase activities in the black scale, *Saisstia oleae* and their relation to host adaptability. J. Insect Physiol. 22: 1025-1029.

Khedr MM (2002). Effect of some plant extracts and insect growth regulators applied to control cotton leaf worm on honeybee, *Apis mellifera*. L. M. Sc. Thesis. Faculty of Agriculture Zagazig University Egypt, P 204.

Knight JA, Anderson S, Rawle JM (1972). Chemical basis of the sulfophospho-vanillin reaction for estimating total serum lipid. Clin. Chem., 18: 199-202.

Latha ES, Krishnayya PV, Subbaratnam GV (1996). Effect of *Bacillus thuringiensis* var. *kurstaki* Berliner and its additive chemicals on haemolymph constituents of *Spodoptera littoralis* (Fab.). Pest Manag. Econ. Zool., 4: 1 – 2, 65 – 69.

Lelwallen LL (1954).Biological and toxicological studies of the little house fly. Econ. Entomol., 47:1137 – 1141.

Lysyk TJ, Kalischuk-Tymensen LD, Rochon K, Selinger LB (2010). Activity of Bacillus thuringiensis isolates against immature horn fly and stable fly (Diptera: Muscidae). J. Econ. Entomol., 103(3):1019-29.

Mesbah AH, Mourad AK. Ebieda AM (1985). Toxicological studies of natural phyto – compounds. Synergism and antagonism of pyrethroid or organophosphorus – volatile oils of ornamental and medical plants against the house fly larvae *Musca domestica*. L. Proc. 6th Arb. Pest. Conf. Tanta University. Egypt, 1: 423 – 434.

Methela CS,Shah GC, Melkani AB, Pant AK (1989). Terpenoids of *Mentha longifolia himalaiensis*. Fitoterapia, 60(4): 349- 350.

Mittal PK, Wijeyaratne P, Sabeena P (2004). Status of insecticide resistance of malaria Kala-azar and Japanese encephalitis. Vector in Bangladesh, Bhutan, India and Nepal (BBIN). Prepared under EPH Project 26568 E.X. ANE. MRCCOORE. Office of Health. Infectious Disease and Nutrition. Bureau for Global Health, U.S. Agency for International Development. Washington. DC. 20523.

Mulla MS, Thavara U, Tawatsin A, Chompoosri J, Zaim M, Su T (2003). Laboratory and field evaluation of novaluron, a new acylurea insect growth regulator, against *Aedes aegypti* (Dipetra: Culicidae). J. Vect. Ecol., 28: 241 – 254.

Mwamburi LA, Laing MD, Miller R (2010).Interaction between Beauveria bassiana and Bacillus thuringiensis var. israelensis for the control of house fly larvae and adults in poultry houses. Poult. Sci., 88(11):2307-14.

Nazhi WA, Seleena B, Lee HI, Jeffery JT, Rogayah TA, Sofian MA (2005). Bacreria fauna from the house fly, *Musca domestica* (L.).Tro. Biomed., 22 (2): 225 – 31.

Omar NAM, Mousa A, El-Husseini MM, El-Bishry MH (2005b). Changes in lipid contents due to infection with *Bacillus thuringiensis kurstaki* in larvae of the greater wax moth *Galleria mellonella* L., (Lepidoptera: Galleridae). Egypt. J. Biol. Pest Control, 15(1/2) pp: 41-44.

Paul A, Harrington LC, Zhang L, Scott JF (2005). Insecticide resistance in *Culex pipiens* from New York. J. Am. Mosquito Control Assoc., 21: 305 – 309.

Powell MEA, Smith MJH (1954). The determination of serum acid and alkaline phosphatase activity with 4-aminoantipyrine. J. Clin. Pathol., 7: 245-248.

Pigott CR, King MS, Ellar DJ (2008). Flies as a source of enteric pathogens in a rural village in Thailand. App. Environ. Microbiol., 64(1): 32-66.

Roh JY, Choi JY, Li MS, Jin BR, Je YH (2007). *Bacillus thuringiensis* as a specific, safe, and effective tool for insect pest control. J. Microbiol. Biotechnol., 17(4): 547-59.

Salem IE, El-Sheakh AA, Gomaa EA, Desuky WM, Raslan SA (1995). Esterases and carbohydrate hydrolyzing enzymes determination in *S.littoralis* larvae treated with some IGRs. Zagazig J. Agric. Res., 22 (3):90 – 906.

Shakoori AR, Salem MA, Mantle D (1998). Some macromolecular abnormalities induced by a sub-lethal dose of Cymbush 10 EC in adult beetle of *Tribolium castaneum*. Pak. J. Zool., 30(2): 83 – 90.

Sokar LA (1995). Possible alternatives to classical insecticides in management program of *Spodoptera littoralis* (Boisd.). Ph. D. Thesis, Faculty of Agriculture, Moshtohor, Zagazig University, Egypt, P 209.

Soliman BA, El-Sherif LS (1995). Larvicidal effect of some plant oils on mosquito *Culex pipiens* L. (Diptera: Culicidae). J. Egypt. Ger. Soc. Zool. 16(e): 161 – 169.

Tawfik SM, Farghali AA, Abd-El-Wahab IS (2002). Biochemical studies of Dipel 2X and abamectin on the 4th instar larvae of cotton leafworm, *Spodoptera littoralis* (Boisd .). Egypt. J. Appl. Sci., 17(3): 371 -386.

Tikar SN, Mendki MJ, Chandel K,Parashar BD, Shri P (2008). Susceptibility of immature stages of *Aedes (Stegomyia) aegypti* vector of dengue and Chikungunya to insecticides from India. Parasitol. Res., 102: 907 – 913.

Tilquin M, Paris M, Reynaud S, Despres L, Ravanel P, Geremia RA Gury J (2008).Long lasting persistence of *Bacillus thuringiensis* subsp. *israelensis* (*Bti*) in mosquito natural habitats. PLoS ONE, 3(10): e3432.

Tsumuki H, Kanehisa K (1984). Phosphatases in the rice stem borer, *Chilo supperssalis* Walker (Lepidoptera: Pyralidae): Some properties and changes of the activities during hibernation. Cryobiology, 21: 177 – 182.

Van Asperen R (1962). A study of housefly esterase by means of sensitive colourimetric method. J. Insect. Physiol., 8: 401- 416.

Wirth MC, Georghiou GP (1996).Organophosphate resistance in *Culex pipiens* from Cyprus. J. Am. Mosq. Control. Assoc. 12: 112 – 118.

Wirth MC, Walton WE, Federici BA (2010). Evolution of resistance to the Bacillus sphaericus Bin toxin is phenotypically masked by combination with the mosquitocidal proteins of Bacillus thuringiensis subspecies israelensis. Environ. Microbiol., 12 (5):1154-60.

Zeenath C, Nair VSK (1994). Development profile of haemolymph proteins and amino acids in last larval instar of *Spodoptera mauritia* (Boisd.) following treatment of a juvenile hormone analogue. Indian J. Exp. Biol., 32(4): 271 – 273.

33

Antibiogram of *Escherichia coli* strains isolated from food of bovine origin in selected Woredas of Tigray, Ethiopia

Abebe M.[1] , Hailelule A.[1], Abrha B.[1], Nigus A.[1], Birhanu M.[1], Adane H.[2], Genene T.[3], Daniel H. [4], Getachew G.[1], Merga G.[4] and Haftay A.[1]

[1]College of Veterinary Medicine, Mekelle University, Tigray, Ethiopia.
[2]Wachemo University, SNNP, Ethiopia.
[3]Institute of Biodiversity Conservation, Addis Ababa, Ethiopia.
[4]Asella School of Health Science, Adama University of Science and Technology, Asella, Ethiopa.

***Escherichia coli* is a food borne pathogen causing a major public health problems. The use of antimicrobials in food animals produces resistant bacteria. To determine antimicrobial resistance of *E. coli* species isolated from food of bovine origin, a total of 384 of milk samples (n=192) and meat samples (n=192) were collected from different sources in 1:1 ratio in selected Woredas of Tigray, Ethiopia. Samples were cultured on sheep blood agar and sub-cultured on Eosin Methylene and further sub-cultured on Biolog Universal Growth Agar (BUG media). Pure colonies were taken and suspension was made and inoculated into micro plates. The bacteria were identified by BiOLOG Identification system. Antimicrobial resistance of *E. coli* isolates was done by disk diffusion method using twenty antimicrobials and minimum inhibitory concentration was determined for resistant isolates. The study revealed that out of 384 samples of milk and meat, *E. coli* 0_{157}:H_7 (10.4%), *E. coli, Non 157 STEC* (2.6%) and *E. coli enterotoxigenic* (10.7%) were isolated. Antimicrobial resistance pattern of *E. coli* isolates (n=91) revealed high resistance against cephalothin (84.6%), chloroamphenicol (83.3%), tetracycline (88.9%), gentamicin (65.9%), but low resistance for sulphoxazole-trimethoprim (16.5%), neomycin (15.4%), streptomycin (29.7%), kanamycin (30.8%), ciprofloxacin (10%), nitrofurantoine (3.3%), norfloxon (3.3%) and ciftriaxone (9.9%). Multidrug resistance was observed in 82 (93.2%) of species. The high prevalence of *0_{157}:H_7* and enterotoxigenic and high rates of multiple drug resistance indicate there is a need for timely designing prevention and control strategies.**

Key words: Antimicrobial, *Escherichia coli,* meat, milk, resistance, zoonoses.

INTRODUCTION

Food safety, safety of products of animal origin in particular, is an increasingly important issue with regards to human health. With increasing consumption of products of animal origin, the risk of food borne diseases of humans also increases. One product that is commonly distributed in raw form is milk. Raw milk is a known vehicle and medium for pathogens like *Escherichia coli*. Milk can become contaminated in many ways. There are mammary gland infection (mastitis) or a systemic infection, and contamination through the faeces of the animals and the hand of the milker usually during hand milking procedure or by equipment used for milk collection and storage (Leedom, 2006).

Similarly, meat and its products are important reservoirs for many of the food-borne pathogens, including *E. coli* O157:H7. Foodborne diseases remain a major public health problem across the globe. The problem is severe in developing countries due to difficulties in securing optimal hygienic food handling practices. In developing countries, up to an estimated 70% of cases of diarrheal disease are associated with the consumption of contaminated food (WHO, 2000). Reliable statistics on food borne diseases are not available due to poor or non-existent reporting systems in most developing countries.

Besides its high prevalence, the rising antimicrobial resistance (AMR) is partly due to the overuse and misuse of antimicrobials (e.g. as growth promoters for food animals) in food animal production, becoming a major problem.

In some countries, up to 70% of antibiotics are used for animals raised in industrial farms that are not sick, to offset the effects of crowding and poor sanitation. This practice promotes the development of drug-resistant bacteria that can spread to humans. Thus, food borne diseases, when associated with resistant bacteria, are harder to treat, resulting in longer hospitalization, higher mortality and morbidity, decreased productivity, and increased costs (WHO, 2011). Likewise, antimicrobial resistance is constantly evolving challenge. Further transfer of antimicrobial resistant bacteria to humans via food chain has been reported (Angulo et al., 2004). A limited number of investigations have been studied regarding the presence of antimicrobial resistance in food animals in Ethiopia (Mekonnen et al., 2005; Hundera et al., 2005). The finding of the present study on antimicrobial resistance of food borne pathogens will provide useful information on the development of public health policy in food animal production. Thus, the study was carried out with the aim to isolate *E. coli* species and to determine its antimicrobial resistance from food of bovine origin.

MATERIALS AND METHODS

Study area

The study was conducted in three districts of Tigray, Mekelle; Alamata and Adigrat. These districts were selected mainly because of their difference in the altitudes that may help us to obtain reliable evidence on the magnitude and epidemiology of disease in the region (RSITBARD, 2009).

Study design

A cross-sectional study was conducted from November 2012 to June 2013 in the selected districts of Tigray, Ethiopia.

Sample size and sampling technique

A total of 384 samples were collected from bovine raw milk and meat in the selected Woredas of Tigray, Ethiopia. The sample size was determined according the formula given by Thrusfield (2005) by taking prevalence of 50% so that the maximum sample size could be achieved. Accordingly, the calculated value for sample size was 384. Then, equal number of milk (n1=192) and meat (n2=192) samples were included purposely. In sampling of milk and meat samples, simple random sampling technique was applied until sample size was achieved.

Sample collection, transport and handling

Milk samples

Milk samples were collected according to the National Mastitis Council Guideline (1990) by principal investigator. Milk samples were aseptically collected directly from teats of lactating cows (n=64) and from distribution sites (shop=64 and restaurant=64) using sterile sample bottles. Samples were transported using icebox to Microbiology Laboratory of College of Veterinary Medicine, Mekelle University. Milk samples were immediately cultured or stored at 4°C for a maximum of 24 h until the samples were cultured.

Meat samples

Raw meat from slaughter house (n=64) during slaughtering and non pre-packed meat samples from beef were purchased randomly from selected butcher shops (n=64) and restaurant (n=64). Sections of meat (10 × 10 × 3 cm) from neck of each carcass were aseptically removed and placed in separate sterile plastic bags to prevent spilling and cross contamination. It was immediately transported to Microbiology Laboratory of College of Veterinary Medicine, Mekelle University in a cooler with ice packs. After culture, the prepared samples were transported with icebox to Microbiology Laboratory of Institute of Biodiversity Conservation, Addis Ababa for further confirmatory identification.

Culture and identification

Milk sample

Bacteriological examination was done according to the National Mastitis Council Guideline (1990). A 0.1 ml of milk was spread on tryptose blood agar base (Oxoid, UK) enriched with 7% defibrinated sheep blood using spread plate after centrifugation and discarding the supernatant. Blood agar plates were incubated aerobically at 37°C for 24 - 48 h. Then Gram staining was done for all suspected cultures of *E. coli* and Gram negative bacillus were sub-cultured into Eosin Methylene blue agar. Then, pure colony was taken and sub-cultured on BUG(BiOLOG Universal Growth Media) at 37°C for 18-24 h as a primary and secondary culture. Well-isolated fresh colonies from BUG (Biolog, USA) media were inoculated into 18-20 inoculation fluid to have bacterial suspension with turbidity equivalent to 20% transmittance as measured by turbidity meter. This suspension was poured into micro plates with multi-channel pipettes. The micro plates were loaded into Omnilog tray to be incubated, analyzed and interpreted for 18-24 h as per BiOLOG Users Guideline (2008) and finally identified bacteria were printed out.

Meat sample

The microbiological examination of each meat sample, 25 g was homogenized with 1 g of the homogenate and added to 5 mL of buffered peptone water (BPW- HiMedia Laboratories, Mumbai, India) and incubated. Cultures spread on tryptose blood agar base (Oxoid, UK) enriched with 7% defibrinated sheep blood using the spread plate techniques and the plates were incubated overnight at 37°C. From each plate (one plate for each meat sample), 5 to 10 suspected bacterial colonies were selected and sub-cultured onto Eosin Methylene Blue agar. Then pure colony was further sub-cultured on BUG at 37°C for 18-24 h as primary and secondary culture. Well-isolated fresh colonies from BUG (Biolog, USA) media were inoculated into 18-20 inoculation fluid to have bacterial suspension with turbidity equivalent to 20% transmittance as measured by turbidity meter. This suspension was poured into micro plates with multi-channel pipettes. The micro plates were loaded into Omnilog tray and incubated, analyzed and interpreted for 18-24 h as per BiOLOG Users Guideline (2008) and finally identified bacteria were printed out.

Antimicrobial susceptibility test

Antimicrobial susceptibility test was performed for all isolates according to the criteria of the Clinical and Laboratory Standards Institute (2008). For susceptibility test, a pure culture of all identified *E. coli* was taken from BUG media and transferred to a tube containing 5 ml of sterile normal saline and mixed gently to make homogenous suspension which was adjusted to a turbidity equivalent to a 0.5 Mc Farland standard as measured by turbidity meter. The bacterial suspension was inoculated on to Muller-Hinton agar (Oxoid, UK) with the sterile swab to cover the whole surface of the agar. The inoculated plates were left at room temperature to dry. The plates were prepared as per the manufacturer's instructions and checked for sterility before inoculation by incubating the plates over night at 37°C. Before using the antimicrobial disks, they were kept at room temperature for one hour and then dispended on the surface of media. Following this, the plates were incubated aerobically at 37°C for 24 h.

For susceptibility test, antimicrobials which were used for treatment of bovine mastitis or considered as important antimicrobial agents for human was selected for antibiogram based on the criteria of Clinical and Laboratory Standards Institute (2008). Thus, antimicrobials used in this study were cephalothin (30 µg), sulphoxazole-trimethoprim (25 µg), neomycin (5 µg), streptomycin (10 µg), kanamycin (30 µg), chloroamphenicol (30 mg), tetracycline (30 µg) and gentamicin (10 µg) (Oxoid, UK). Antimicrobials not used for treatment of bovine mastitis but important for human were ciprofloxacin (5 µg), nitrofurantoine (300 µg), norfloxon(10 µg), ciftriaxone(30 µg) (Oxoid, UK). The diameters of the zone of inhibition around the disks were measured to the nearest millimeter using calibrated rulers, and the isolates were classified as susceptible, intermediate and resistant according to the interpretative standards of Clinical and Laboratory Standards Institute (2008). In addition, minimum inhibitory concentration (MIC) was determined using broth dilution method with an antimicrobial concentration ranging from 0.25-512 µg/µL, in accordance with the guidelines of the Clinical and Laboratory Standards Institute (CLSI, 2008). Those isolates with minimum inhibitory concentrations (MIC) higher than the breakpoint for the respective antimicrobial agents were regarded as resistant, while those with MIC equal to or lower than the breakpoint were regarded as susceptible. Moreover, isolates showing resistance to three or more antimicrobial subclass were considered as multidrug resistant.

Quality control

Confidence in the reliability of test results was increased by adequate quality assurance procedures, and the routine use of control strains. Thus, *E. coli* ATCC-25922 was taken as an important part of quality control for culture, BiOLOG identification and antimicrobial susceptibility through this study.

Variables

Independent variables such as types of samples were interpreted against dependent variable of species isolates and antimicrobial sensitivity pattern of each isolates.

Ethical issues

Verbal consent was obtained from dairy farms, abattoirs and butcher shop owners/managers.

Statistical analysis

The collected data was entered into EPI data version 3.1 and exported to SPSS version 16 computer soft ware then the data was analyzed. Accordingly, descriptive statistics such as percentages and frequency distribution were used to describe/present bacterial isolates and antimicrobial susceptibility which were expressed as percent of resistant and susceptible. In addition, the proportion of bacteria resistant to at least one of the antibiotics and resistant to two or more were calculated.

RESULTS

Prevalence of subspecies isolated from milk and meat samples of bovine origin

The total number of species isolated from milk and meat samples of bovine origin are indicated in Figure 1. *O_{157}:H_7* (10.4%), *Non 157 STEC* (2.6%) and *E. coli enterotoxigenic* (10.7%) were detected in all the samples tested.

Antimicrobial resistance profile of species isolated from milk and meat samples

Analysis of subspecies specific resistance rates indicated for isolates from milk and meat are shown in Table 1. All *E. coli* stain showed high percentage resistance to cephalothin, chloramphenicol, tetracycline and gentamicin. On the other hand, most *E. coli* isolates were susceptible to sulphoxazole-trimethoprim, neomycin, streptomycin, kanamycin, ciprofloxacin, nitrofurantoine, norfloxon and ciftriaxon

The overall multiple antimicrobial resistance rate was 93.2%. The resistances against two or more antimicrobial

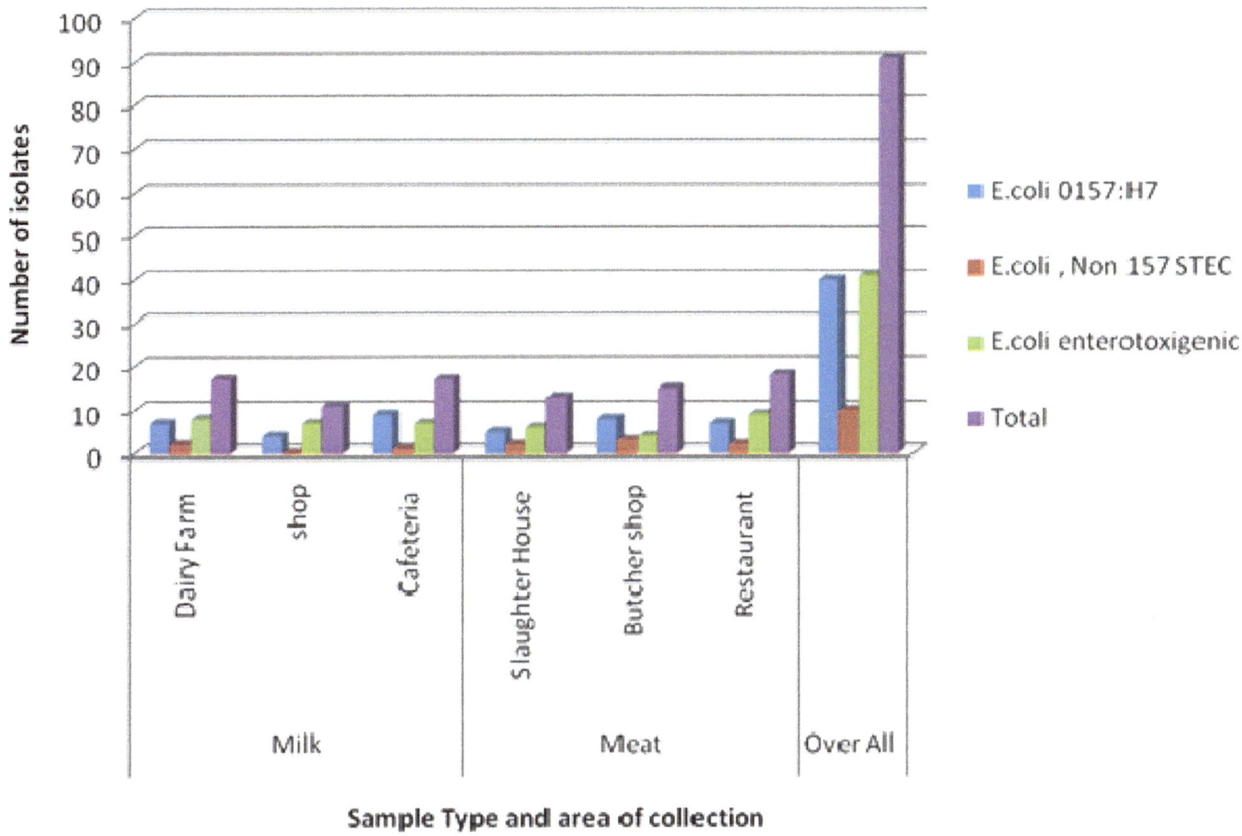

Figure 1. Prevalence of strains from milk and meat of bovine origin.

agents were observed in all 0_{157}:H_7 and non 157 STEC and 95% enterotoxigenic isolated from milk showed multiple drug resistance (Table 2). 89.5% 0_{157}:$H_{7,}$ 71.4% of Non 157 STEC and 94.7% enterotoxigenic isolated from meat samples showed multiple drug resistant.

DISCUSSION

In the present study, the presence of strains in food of bovine origin indicated that the bacteria originated from infected animals or unhygienic conditions during processing, handling and distribution. It did not only originate from infected animals but more likely as an indicator of poor hygiene and sanitary practices while handling food of animal origin.

The isolation rate of *E. coli* in the present study was 23.7% and it was mainly isolated from meat samples from restaurant (28.5%) and milk sample from cafeteria (26.6%). These findings are in conformity with reports by other researchers (Yismaw et al.., 2010; Al-Tawfiq, 2006; Gangoué et al., 2004). Higher prevalence was reported by Ali and Abdelgadir (2011) 63% and Lingathurai and Vellathurai (2010) 70%. In fact, if the methods of production, transportation, handling and sale of milk are entirely unhygienic there is high prevalence (Yismaw et al.., 2010).

Antibiotic resistance development among the bacteria poses a problem of concern. In all food samples of bovine origin in the present study, *E. coli* showed high resistance rates (greater than 80%) to cephalothin, chloramphenicol, tetracycline and (greater than 60%) to gentamicin. The results of this study are in line with the findings of other studies conducted in different parts of the world (Bharathi et al., 2008; Briscoe et al., 2005). However, antimicrobial resistance rates obtained in this study were higher as compared to susceptibility patterns reported from previous studies (Zhanel et al., 2006; Karlowsky et al., 2002; Barrett et al., 2000).

E. coli isolates were sensitive to sulphoxazole-trimethoprim, neomycin, streptomycin, kanamycin ciprofloxacin, nitrofurantoine, norfloxon and ciftriaxone. Similar studies conducted in Ethiopia by Tesfaye et al. (2009) and in Nigeria by Wariso and Ibe (2006) have reported comparable susceptibility rates. In this study, sulphoxazole-trimethoprim, neomycin, streptomycin, kanamycin ciprofloxacin, nitrofurantoine, norfloxon, and ciftriaxone were found to be the most effective antimicrobials against *E. coli* isolates. Furthermore in this study, a high rate of multiple antimicrobial resistance

Table 1. Antimicrobial susceptibility pattern of stains isolated from milk and meat sample of bovine origin.

Antimicrobial	0_{157}:H_7				Non 157 STEC				Enterotoxigenic				Overall (n=91)		MIC(µg/µl)	
	Milk (n=20)		Meat (n=20)		Milk (n=3)		Meat (n=7)		Milk (n=22)		Meat (n=19)				MIC 50	MIC 90
	S (%)	R (%)	S (%)	R (%)	S (%)	R (%)	S (%)	R (%)	S (%)	R (%)	S (%)	R (%)	S (%)	R (%)	128	512
cephalothin (30µg)	10	90	10	90	33.3	66.7	28.6	71.4	27.3	72.7	5.3	94.7	15.4	84.6	32	128
SXZ (25 µg)	80	20	85	15	33.3	66.7	85.7	14.3	95.5	4.5	78.9	21.1	83.5	16.5	64	512
Neomycin (5 µg)	80	20	95	5	100	0	71.4	28.6	181.8	18.2	84.2	15.8	84.6	15.4	128	512
streptomycin(10µg)	55	45	70	30	100	0	85.7	14.3	72.7	27.3	73.7	26.3	70.3	29.7	512	512
kanamycin (30 µg)	55	45	70	30	100	0	85.7	14.3	72.7	27.3	68.4	31.6	69.2	30.8	128	512
Chloramphenicol (30 mg)	10	90	25	75	0	100	28.6	71.4	22.7	77.3	5.6	94.4	16.7	83.3	128	512
Tetracycline (30 µg)	0	100	20	80	0	100	14.3	85.7	13.6	86.4	10.5	89.5	11.1	88.9	512	512
Gentamicin (10 µg)	25	75	30	70	0	100	57.1	42.9	40.9	59.1	36.8	63.2	34.1	65.9	64	128
Ciprofloxacin (5 µg)	94.7	5.3	85	15	100	0	85.7	14.3	95.5	4.5	84.2	15.8	90	10	16	64
Nitrofurantoine (300 µg)	95	5	95	5	100	0	100	0	100	0	94.7	5.3	96.7	3.3	4	128
Norfloxon (10 µg)	95	5	95	5	100	0	100	0	100	0	94.7	5.3	96.7	3.3	4	64
Ciftriaxone (30 µg)	90	10	95	5	100	0	71.4	28.6	100	0	89.5	10.5	90.1	9.9	2	32

S: Susceptible R: resistant MIC: minimum inhibitory concentration, SXZ: sulphoxazole-trimethoprim, n=number of positive isolate.

Table 2. Percentages of number of antimicrobials resistant strains isolated from milk and meat sample of bovine origin.

Number of antimicrobials	0_{157}:H		Non 157 STEC		Enterotoxigenic		Overall (n=91)
	Milk (n=20)	Meat (n=20)	Milk (n=3)	Meat (n=7)	Milk (n=22)	Meat (n=19)	
One (%)	0	10.5	0	28.6	5	5.3	6.8
MDR (%)	100	89.5	100	71.4	95	94.7	93.2

MDR: multi-drug resistance n=number of positive isolate.

(93.2%) was recorded, which is consistent with the reports of studies done elsewhere by other scholars (Orrett and Shurl, 2001; Kurutepe et al., 2005). Increases in rate of resistance to different antimicrobials have been reported from previous studies conducted in different parts of the world (Orrett and Shurl, 2001; Kurutepe et al., 2005).

Conclusions

Results clearly indicated that there is a possibility of potential public health threat of species originating from food of bovine origin. The high prevalence *0_{157}:H_7* (Shiga toxin producing) and enterotoxigenic, and high rates of multiple drug resistance indicates alarming situation for designing prevention and control methods.

Conflict of Interests

The author(s) have not declared any conflict of interests.

Acknowledgements

The authors are grateful to Mekelle University-NORAD Project III for giving financial support to the present investigation and also thankful to Institute of Biodiversity Conservation, Ethiopia for identification of *E. coli* strains.

REFERENCES

Ali AA, Abdelgadir WS (2011). Incidence of *Escherichia coli* in raw cow's milk in Khartoum State. British J. Dairy Sci.

2:23-26.
Al-Tawfiq JA (2006). Increasing antibiotic resistance among isolates of *Escherichia coli* recovered from inpatients and outpatients in a Saudi Arabian Hospital. Infect. Contr. Hosp. Epidemiol. 27:748-753.
Angulo FJ, Nunnery JA, Bair HD (2004). Antimicrobial resistance in zoonotic enteric pathogens. Rev. Sci. Tech. Aug. 23:485-96.
Barrett SP, Savage MA, Rebec MP, Guyot A, Andrews N, Shrimpton SB (2000). Antibiotic sensitivity of bacteria associated with community-acquired urinary tract infection in Britain. J. Antimicrob. Chemother. 44:359-365.
BiOLOG User Guide (2008). OmniLog Data Collection Software, OmniLog Data Collection, Version 2.1. Identification System, User Guide. USA. pp. 4-111.
Bharathi MJ, Ramakrishnan R, Maneksha V, Shivakuma C, Mittal S (2008). Comparative bacteriology of acute and chronic dacryocystitis. Eye 22:953-960.
Briscoe D, Rubowitz A, Assia EI(2005). Changing bacterial isolates and antibiotic sensitivities of purulent dacryocystitis. Orbit 24:95-98.
CLSI (2008). Performance Standards for Antimicrobial Disk and Dilution Susceptibility Tests for Bacteria Isolated From Animals. Clinical and Laboratory Standards Institute. 28:M31-A3.
Gangoué JP, Koulla-Shirob S, Ngassama P, Adiogo D, Njine T, Ndumbe P(2004). Antimicrobial resistance of Gram-negative bacilli isolates from inpatients and outpatients at Yaounde Central Hospital, Cameroon. Inter. J. Infect. Dis. 8:147-154.
Hundera S, Ademe Z and Sintayehu A (2005). Dairy cattle mastitis in and around Sebeta, Ethiopia. J. Appl. Vet. Med. 3:1525-1530.
Karlowsky JA, Kelly LJ, Thornsberry C, Jones ME, Sahm DF (2002). Trends in antimicrobial resistance among urinary tract infection isolates of *Escherichia coli* from female outpatients in the United States. Antimicrob Agents Chemother. 6:2540-2545.
Kurutepe S, Surucuoglue S, Sezgin C, Gazi H, Gulay M, Ozbakkaloglu B(2005). Increasing antimicrobial resistance in *isolates* form community-acquired urinary tract infections during 1998 - 2003 in Minisa, Turkey. Japan J. Infect. Dis. 58:159-161.
Leedom JM (2006). Milk of nonhuman origin and infectious diseases in humans. Clin. Infect. Dis. 43:610-615.
Lingathurai S, Vellathurai P (2010). Bacteriological Quality and Safety of Raw Cow Milk in Madurai, South India. http://www.webmedcentral.com.
Mekonnen H, Workineh S, Bayleyegne M, Moges A and Tadele K (2005). Antimicrobial susceptibility profile of mastitis isolates from cows in three major Ethiopian dairies. Rev. Med Vet. 176:391-394.
Orrett FA, Shurl SM (2001). Prevalence of resistance to antimicrobial of *E. coli* isolates from clinical sources at a private hospital in Trinidad. Japan. J. Infect. Dis. 54:64-68.
Regional State of Tigray Bureau of Agriculture and Rural Development (RSTBARD), (2009). Estimated animal population of Tigray region.
Tesfaye G, Asrat D, Woldeamanuel Y, Gizaw M (2009). Microbiology of discharging ears in Ethiopia. Asian Pac. J. Trop. Med. 2(91):60-67.
Thrusfield M (2005). Veterinary Epidemiology, 3rd edn. Blackwell Science Ltd, UK. 233-250.
Wariso BA, Ibe SN (2006). Bacteriology of chronic discharging ears in Port Harcourt, Nigeria. West Afr. J. Med. 25:219-222.
WHO (2000). Foodborne Disease: A focus for Health Education
WHO (2011). An integrated approach for containment of foodborne antimicrobial resistance (www.who.int/foodborne_disease/resistance/en/)
Yismaw G, Abay S, Asrat D, Yifru S, Kassu A (2010). A Bacteriological profile and resistant patterns of clinical isolates from pediatric patients, Gondar University Teaching Hospital, Gondar NorthwestEthiopia. Ethiop. Med. J. 48(4):293-300.
Zhanel GG, Hisanaga TL, Laing NM, DeCorby MR, Nichol KA, Weshnoweski B, Johnson J, Noreddin A, Low DE, Karlowski JA, Hoban DJ (2006). Antibiotic resistance in *Escherichia coli* outpatient urinary isolates: final results from the North American Urinary Tract Infection Collaborative Alliance (NAUTICA). Int. J. Antimicrob. Agents 27:468-475.

Isolation and identification of *Mannhemia* and *Pasturella* species from pneumonic and apparently healthy cattle and their antibiogram susceptibility pattern in Bedelle District, Western Ethiopia

Dereje Abera[1], Tesfaye Sisay[2] and Tadesse Birhanu[1]

[1]School of Veterinary Medicine, College of Medical and Health Sciences, Wollega University, P.O. Box 395, Nekemte, Ethiopia.
[2]Unit of Health Biotechnology, Institute of Biotechnology, College of Natural and Computational Sciences, Addis Ababa University, P.O.Box 34, Debre Zeit, Ethiopia.

The study was conducted from October 2012 to April 2013 at Bedelle district, Western Ethiopia. The aim was to isolate *Mannheimia* and *Pasteurella* species, and to assess the associated risk factors from pneumonic and apparently healthy cattle, and antibiotic susceptibility profiles of the isolates. Out of 329 samples from animals (185 nasal swabs from clinic and 144 lung tissues abattoir) examined, 50.2% was found to be positive to the disease. The bacteriological examination revealed 28 (8.51%) overall isolates of *Mannheimia haemolytica* (46.4%), *Pasteurella multocida* (39.3%) and *Bibersteina trehalosi* (14.3%) were recovered respectively in which 17 (9.19%) and 11 (7.63%) bacterial isolates were obtained from nasal swabs and pneumonic lungs, respectively. The higher isolation rate of *M. haemolytica* indicated it as the major cause in the study area. Age was found to be the potential risk factor in which young animals were highly affected. The antimicrobial susceptibility profiles of the isolates were carried out using disc diffusion method. The isolates were susceptible to most of the antibiotic disks used: amoxicillin, chloramphenicol, cephalexin, polymyxin-B, kanamycin and florifenicol. However, moderate resistance was observed to Tetracycline, Erythromycin and Penicillin-G. Thus, an integrated application of overall management and vaccination should be implemented as prevention and control measures.

Key words: Bedelle, cattle, *Mannheimia, Pasteurella*, pneumonic pasteurellosis, antimicrobial susceptibility test.

INTRODUCTION

Ethiopia has diverse animal genetic resource and its relatively large livestock population (approximately 100 million) is well adapted to and distributed among diverse ecological condition and management system (Aiello and May, 1998). In Ethiopia, like many developing countries, livestock play multiple roles. Despite the huge number of cattle and economic importance, the productivity is low due to the constraints of disease, nutrition, poor management and poor performance of the indigenous breeds. These constraints result in poor reproductive performance of these animals (Lobago et al., 2006). On the basis of statistics acquired from different sources, livestock provides 16% of the total GDP (equivalent to 30% of agricultural GDP) and generates 14% of the country's foreign exchange earnings (Central Statistical Agency, 2009).

Disease constraints like respiratory diseases contribute to the great financial losses and the socio-economic development of poor farmers in the area. These diseases cause a huge mortality and morbidity (Zewdie, 2004). The disease occurs in food animals due to complex factors that often interact to produce disease. Various conditions such as climate, weather, weaning, transportation, poorly ventilated housing and nutritional deficiencies are known to play a predisposing role as the animal's immunity weakens. In such conditions, occurrence of the normal flora of the upper respiratory tract and subsequent infection of the lungs is well documented (Radostitis et al., 2000).

Respiratory tract infections are of common occurrence in various species of domestic animals. However, pneumonic pasteurellosis, also known as respiratory mannheimiosis, is most common example with a wide prevalence in ruminants. The disease in its typical clinical form, is highly infectious, often fatal and with very serious economic mortality in many animals in which the disease accounts for approximately 30% of the total cattle deaths worldwide (Baundreaux, 2004).

The term pasteurellosis was broadly used to designate a number of infections in domestic animals caused by Gram- negative non-motile facultative anaerobic rods or coccobacilli formerly grouped under the genus *Pasteurella* (after Louis Pasteur). For several decades, the genus *Pasteurella* was believed to be only one single genus with numerous species infection of farm animals, particularly in ruminants (Angen et al., 1999b).

However, with more recent advancements in molecular biology involving DNA hybridization studies and 16s rRNA sequencing, most of the formerly recognized species were found to share a number of common features and become the subject of intensive revision and reclassification. In this respect, *Pasteurella haemolytica*, biotype A was allocated to a new genus and renamed *Mannheimia*. This new genus contains several species including *Mannheimia heamolytica, Mannheimia granulomatis, Mannheimia glucosida, Mannheimia ruminals* and *Mannheimia varigena*. The name *Mannheimia* was given in tribute to the Germen Scientist Welter Mannheimia for his significant contributions in the recent taxonomy of the family Pasturellacae. On the other hand, *Pasteurella haemolytica* biotype T was first reclassified as *Pasteurella trehalosi.* In addition avian *Pasteurella* species including *Pasteurella gallinarum, Pasteurella paragallinarum* and *Pasteurella volantinium* were similarly removed to new separate genus named as Avibacterium (Blackall et al., 2005).

Before the establishment of this newly revised classification, *P. haemolytica* was known to comprise two bio-types A and T, based on fermentation of arabinose and trehalose respectively. Within these two biotypes, 17 serotypes were further identified on the basis of soluble or extractable surface antigen by passive heamaglutination test (Carter and Cole, 1990). Serotypes 1, 2, 5, 6, 7, 8, 9, 11, 12, 13, 14, 16 and 17 belong to biotype A which was reclassified as *M. haemolytica*. However, serotype 11 was later reclassified as *M. glucosida*. The rest of serotypes (3, 4, 10 and 15) which belong to biotype T finally moved to separate new genus as mentioned. It is worth mentioning that *M. haemolytica, P. multocida* and *P. trehalosi (Bibersteina)* constitute the most important members of the family *Pasturellacae* that pose serious hazards in livestock industry. These species are commensally resident in the animal body as normal constituents of the nasopharyngeal micro flora and are all capable of causing infection when the body defense mechanisms are impaired. Their presence is mainly confined to ruminants with most adequately characterized strains originating from cattle, sheep and goats (Biberstin and Hirsh, 1999).

P. multocida, is associated with hemorrhagic septicemia in cattle and buffaloes and enzootic pneumonia complex in young ruminants (Jones et al., 1997). Concurrent infections of the respiratory tract by viruses, bacteria and lung worms have been described and such disease conditions are commonly known as respiratory disease complex (RDC), indicating the difficulty to attribute to only one etiology (Biberstin and Hirsh, 1999; Quninn et al., 2002). In the cool central highlands of Ethiopia, respiratory disease complex has been identified as leading. Irregular and insufficient vaccination program for diseases such as pastuerellosis and Pest des Petites Ruminant (PPR), lack of strategic mass drenching against lung worms and occurrence of viral infections may play significant roles in the persistence of respiratory disease complex in Ethiopia.

Even though pneumonic pasteurellosis is one of the most economically important infectious diseases of cattle in Ethiopia (Mohamed and Abdelsalam, 2008), there was lack of information on this regards in Southern Ethiopia especially in study area. Hence, the study was designed to isolate and identify Mannheimia and Pasteurella species that invade the lower respiratory tract of cattle causing pneumonic pasteurellosis, to assess potential associated risk factors and determine their antibiogram susceptibility pattern from pneumonic and apparently healthy cattle and in Bedelle District, Western Ethiopia.

MATERIALS AND METHODS

Study area

The study was conducted from October 2013 to April 2014 in the selected, Illubabor Zone, in Bedelle District of Oromia regional state, Western Ethiopia. The area is located 480 km from the capital city, Addis Ababa, to the West, along the way to Metu. The area is geographically positioned at an altitude of 1900-2000 m above sea-level (m.a.s) and 8-9° N latitude and 36-37°E longitude. The mean annual rain fall in the area is 1500 mm. The mean seasonal temperature varies from 20-25°C from October to January and decline to a level of 15-25°C during the rest of the months (Central Statistical Agency, 2009).

The area is within an agro-ecological zone and is prone to high rainfall and multi-species animal is preferred and most practiced system of livestock production. The area is among the known coffee-growing in the country. Agriculture is the main livelihood in the area in which cattle and sheep are kept as the major livestock which are highly important for the livelihood of the local population. The rearing system of cattle in study sites depends on natural grass and crop residues that are kept in traditional management system.

Study animals

The study population constitutes indigenous zebu cattle of varies sex, age groups, body conditions scores and managed under smallholder mixed crop-livestock farming system which are kept under traditional extensive husbandry system with communal grazing and watering points.

This study included all cattle with respiratory signs presented to the Bedelle Veterinary Clinic and/or site of veterinary clinics found in the District. For comparative study, live animals without pneumonic signs and non-pneumonic lungs from slaughter house were also included in the study. The study was conducted on local breed cattle selected from the district and 329 study animals were sampled. The origin, sex, age and body condition scores of the animals were the potential variables used to associate with the prevalence rate.

Sample collection

Each study animal was individually identified and restrained by an assistant and kept fixed before the sampling procedures. A total of 185 nasal swabs were collected aseptically out of which clinically pneumonic cattle (n=96) from animals showed respiratory signs such as an irregular breathing pattern and grunting on expiration, coughing, a serous nasal discharge, apparently healthy ones (n=89) for comparison purpose. All samples were processed bacteriologically.

The methods followed were based on the procedures of Sisay and Zerihun (2003). Sterile cotton-tipped, 20-25 cm long nasal swabs, moistened in sterile tryptose soya broth, were directed via the ventral nasal meatus in to nasopharynx, that is, the swabs were carefully inserted into nostril; rolled gently; put back to the test tube containing broth; and the tubes were capped.

The swabs were then kept in an ice box. The specimens collected were transported to Bedelle Regional Laboratory for bacteriological analysis. At the time of sample collection aseptic procedures were implemented to avoid getting contamination from the external nares; thus, the external nares were decontaminated with cotton soaked with 70% ethanol. A total of 144 lung tissue samples from both non-pneumonic (n=75) and pneumonic (n=69) from slaughtered animals were collected immediately after slaughter. The samples were placed in separate sterile universal bottles, labeled and kept cooled in the ice box and transported to the laboratory. Sampling method used was purposive sampling and comparative.

In addition, data were gathered from owners by using structured questionnaire survey to assess the risk factors for pneumonia in the study area. Finally, data gathered were sorted and analyzed using standard methods.

Sample processing and culturing

Nasal swab samples were streaked on sheep blood agar plates and incubated at 37°C for 24 to 72 h. The same samples were parallely streaked on MacConkey agar for primary differentiation of the pathogen following standard procedures. Colonies were characterized and those giving Gram-negative coccobacilli or short rods with or without bipolar staining on smears were subcultured for identification. A 24 h pure *Pasteurella* suspected culture (isolate) was subjected to biochemical tests using standard procedures.

A section of lung samples were aseptically taken from the edge of the lesion in the case of pneumonic cases and the samples were processed before inoculation into appropriate media as follows: the surface of the lung tissue passed through the Bunsen burner several times to avoid surface contamination and transferred to Petri dish. Then the surface was heated with spatula and incised and minced with sterile forceps, scalpel blade. Then the tissue specimens were placed i to screw capped test tubes containing TSB and was incubated for 2 h at 37°C and then streaked on to both blood agar plate (BAP, 5% blood in blood agar base) and MacConkey agar plates. The plates were incubated at 37°C for 24 h. The grown colonies were characterized and subcultured on nutrient agar plate to get pure culture for further biochemical tests following standard protocol (Quninn et al., 2002).

Isolation and identification of bacterial isolates

To obtain pure cultures of a single bacterial colonies type from blood agar and MacConkey agar only, bacterial colonies characteristics of rod, coccobacilli, rough or smooth, Gram negative, with or with no growth on MacConkey, presence or absence heamolysis on blood agar were subjected to sub culturing. Then, each bacterial colony was characterized and further bacterial identification was made using series of primary and secondary biochemical tests following standard procedures (Quninn et al., 2002).

Bacterial cultures were firstly characterized based on primary identification characteristics such as the cellular morphology of the bacteria; Gram staining, oxidase, catalase and oxidation fermentation tests to determine the genus of the bacterial isolates. After, pure colonies were obtained, transferred to nutrient agar media.

Furthermore, species of the bacterial isolates were determined using a variety of secondary biochemical tests. Indole production and motility was tested in SIM medium, hydrogen supplied production was tested on triple iron sugar agar and SIM medium. Fermentation reactions for sugars (glucose, lactose, maltose, mannitol, trehalose and arabinose) were conducted according to the procedure described by Quninn et al. (2002). All activities from streaking primary media to secondary biochemical tests were conducted for final identification of bacterial isolates to the species level. *P. multocida* on blood agar had colonies non-heamolytic round, smooth or muccoid, all the isolates failed to grow on MacConkey on the basis of Gram staining, the isolates were found to be Gram negative, coccobacillary rods with or without bipolar

Table 1. The overall prevalence of *Pasteurellacae* in Bedelle District, Western Ethiopia.

Total animal examined	+ve animals	Prevalence
Pneumonic (nasal swab)	96	58.2%
Pneumonic (lung tissues)	69	41.8%
Total (329)	165	50.2%

staining on primary identification. *M. haemolytica* and *B. trehalosi* were forms round, smooth, translucent, greyish with β- distinct zone of heamolysis on blood agar, grows on MacConkey agar. Therefore, the isolates were identified to the species level based on the series of results obtained from the primary identification to secondary biochemical test.

In vitro antimicrobial sensitivity test

Antimicrobial susceptibility profiles of the isolates were determined using standard disc diffusion methods as described by Kirby -Bauer on Mueller-Hinton agar supplemented with defibrinated 5% sheep blood. The antibiotic susceptibility tests of *Pasteurella* species from pure culture was carried out by standard disk diffusion methods and interpreted according to the National Committee for Clinical Laboratory Standards (NCCLS) guidelines (NCCLS, 1999). The following common antimicrobials applied in the country were assayed: chloramphenicol (CAF 30 g), erythromycin, florifenicol (FFC-30Ng), penicillin- G (10 IU), tetracycline (TTC-30 Ng), polymixin-B (30 g), amoxicillin (Amox-20 Ng), kanamycin (30 g), and cephalexin (30 g).

About 4-5 similar colonies from the blood agar were suspended in a propagating medium of nutrient broth and incubated for 2-8 h duration of incubation then the bacterium culture was taken and resuspended into saline water then it was adjusted until similar to 0.5 McFarland standards according to the guide line (NCCLS, 1999). Using sterile cotton swab with applicator stick, the suspended bacteria from the broth was uniformly distributed on the surface of Muller Hinton agar plate, containing 5% of sheep blood, at three different planes.

The plates were dried for some minutes by putting upside down. Antibiotic discs were applied on the plate using sterile forceps ensuring complete contact with the agar surface and 4-5 discs regularly placed in one plate 3 cm apart and 1.5 cm from the edge antibiotic impregnated paper discs and the plates were incubated at 37°C for 8-12 h. Finally, the diameters of the zone of inhibition was measured using a transparent ruler to the nearest mm. Based on the zone of inhibition, the isolates were categorized descriptively as resistant, intermediate and susceptible.

Questionnaire survey

A structured questionnaire with the primary objective of eliciting the multifactorial predisposing factors of pneumonic pasteurellosis was conducted. The information was collected with those pertinent questions about: risks associated respiratory problems of cattle in the study area.

In this study, structured questionnaires was prepared and administered to the livestock owners and animal health practitioners. The questionnaire was pre-tested on some farmers to assess the response rate and consistency of replies as well as to monitor any source of bias in the course of interview. Data on animal transportation, management, feed type, weather condition and stress causing factors was collected.

Data management and analysis

The collected data was coded and entered into Microsoft Excel, 2007 spread sheet. The data were analyzed using SAS software (SAS Version 9.2). Descriptive statistics called person Chi-square test was used to determine the statistical significance for categorical data. The data were also summarized using comparison of frequency, percentages (EpInfo, 6.04), drawing graphs. In addition, other statistical associations were also to summarize generated data on the rate of isolation, identification and sensitivity patterns difference.

The effect of age, adverse environmental condition, vaccination, nutritional and immune status observed through the questionnaire survey were interpreted and summarized. For data analyzed statistically, confidence level was held at 95% and $P<0.05$ was set for significance.

RESULTS

Out of 329 animals examined, 50.2% were found to be positive to the disease during the study period. The overall isolates were 28 (8.51%) with *M. haemolytica* (46.4%), *P. multocida* (39.3%) and *B. trehalosi* (14.3%) recovered on bacteriological analysis respectively. The culture results of samples from nasopharyngeal swabs (185) of pneumonic and non-pneumonic cattle and lung tissue (144) were taken from apparently healthy slaughtered cattle (Table 1 and 2).

Bacteriological isolation

From 185 nasal swab samples examined from respiratory tracts of cattle, 8 (4.32%), 6 (3.24%) and 3 (1.62%) of *P. multocida, M. haemolytica and B. trehalosi* were isolated respectively. While 144 lung tissue samples were collected from Bedelle Municipal Abattoir, 3 (2.08%), 7 (4.86%) and 1 (0.69%) of *P. multocida, M. haemolytica and B. trehalosi* were isolated. The total culture results of isolates from both the nasal swabs and lung tissues were 8.51% that comprises three different bacterial species (Table 3).

Among 96 nasal swabs collected from cattle showing respiratory signs, 15 (15.6%) were positive for *Pasteurella* and *Mannheimia* species. Whereas out of 89 nasal swabs collected from apparently healthy cattle, only2 (2.25%) were positive. From the total of 17 (9.2%) bacterial species isolated, nasal swabs originated from both pneumonic and healthy animals.

Of 144 lung tissues sampled from apparently healthy slaughtered cattle, 75 of the lung tissues were non-pneumonic and neither *Pasteurella* nor *Mannheimia* species were isolated. Although, the remaining 69 lung tissues were pneumonic and yield 11 (15.9%) isolates was recorded. The recovery rate of the bacterial isolates from animals showing signs of respiratory tract and non-pneumonic cattle of nasal swabs and lung tissues originated samples showed variation with higher rate of isolates from nasal swab as shown in Table 4.

Table 2. Summary of total isolate with corresponding health status and total sample processed.

Health status	Type of samples		Total sample	Total Isolate
	Nasal swabs	Lung tissues		
Pneumonic	96 (51.9%)	69 (47.9%)	165(50.2%)	26 (92.9%)
Non-pneumonic	89 (48.1%)	75(52.1%)	164 (49.8%)	2 (7.14%)
Total	185	144	329	28

Table 3. Distribution of *Pasteurellacae* species isolated from nasal swabs and lung tissues of cattle in and around Bedelle.

Types of sample	*Pasteurellacae* species isolated			Total
	P. multocida	*M. heamolytica*	*B. trehalosi*	
Nasal Swab	8 (4.32%)	6 (3.24%)	3 (1.62%)	17 (9.19%)
Lung tissues	3 (2.08%)	7 (4.86%)	1 (0.69%)	11 (7.63%)
Total	11 (3.34%)	13 (3.95%)	4 (1.22%)	28 (8.51%)

Table 4. Comparison on the recovery rate of bacterial isolates with respiratory signs and apparently healthy cattle.

Health status of animal	Type of samples	No. of samples	No. of +ve	(%) of +ve
Pneumonic cattle	Nasal swabs	96	15	15.6
Non-pneumonic	Nasal swabs	89	2	2.25
Apparently healthy	Lung tissue	75	0	0.00
Pneumonic	Lung tissue	69	11	15.9
Total		329	28	8.51

Cultural characteristics of isolates

On blood agar, the colonies of *P. multocida* isolates were non-hemolytic, round, smooth or muccoid. All *P. multocida* isolates were Gram negative, coccobacillary and did not grow on MacConkey agar. Whereas, *B. trehalosi* and M. *haemolytica* were able to grow on MacConkey agar and showed β-heamolysis on blood agar also grow as pin point red colonies.

Biochemical characteristics of the isolates

The biochemical test carried out showed that all the isolates were positive for oxidase, catalase and able to ferment glucose, mannitol, sucrose mannose. All *Mannheima* species ferment mannitol, glucose, maltose, arabinose and sucrose except trehalose.

M. haemolytica isolates were positive for catalase and oxidase. *B. trehalosi* unlike others can ferment trehalose but not arabinose and lactose. It was found to be catalase; indole and urease negative (Table 5).

The comparison result of isolated bacterial species between pneumonic and non-pneumonic lung tissues showed 3 (2.08), 7 (4.86) and 1 (0.69) of *P. multocida, M. haemolytica and B. trehalosi*, respectively with none of them isolated from of non-pneumonic lung tissues. There were no bacteria isolates revealed from samples of non-pneumonic lung tissues. However, 11 (7.63%) of the isolates were characterized from pneumonic lung (Table 6).

Antimicrobial susceptibility profiles of the bacterial isolates

The antibiotic susceptibility tests of bacterial species isolated during the study period from nasal swabs and lung tissues as carried out by disc diffusion method showed that almost *M. heamolytica* (n=13) were found to be sensitive to most of the drugs tested such as: florifenicol, amoxicillin, polymyxin B, kanamycin, cephalexin and chloramphenicol. However, moderate resistance was noted against erythromycin, penicillin and tetracycline.

The antimicrobial susceptibility tests conducted for *P.*

Table 5. Results of biochemical characteristics of bacterial species isolated from cattle.

Characteristics (tests)	Species		
	M. haemolytica	*P. multocida*	*B. trehalosi*
Heamolysis	+	-	+
Growth on MacConkey	+	-	+
Catalase	+	+	+
Oxidase	+	+	+
Indole	-	+	-
Urease	-	-	-
Lactose	+	-	+
Sucrose	+	+	+
Glucose	+	+	+
Maltose	+	-	+
Mannose	+	+	+
Arabinose	+	-	-
Mannitol	+	+	+
Trehalose	-	-	+

Table 6. The distribution of bacterial isolates from the lung tissue samples.

Bacterial isolate	Pneumonic lungs		Non pneumonic lungs	
	No. of isolate	Isolate (%)	No. of isolate	Isolate (%)
P. multocida	3	2.08	0	0
M. haemolytica	7	4.86	0	0
B. trehalosi	1	0.69	0	0

multocida also indicated that florifenicol, chloramphenicol, kanamycin, cephalexin, polymyxin and amoxicillin were the most effective antibiotics. Simililarily, antibiogram profile of *B. trehalosi* revealed that all isolates were susceptible to most of the drugs tested except against penicillin-G, erythromycin and tetracycline.

The antimicrobial susceptibility test result indicated that the bacterial isolates were more susceptible to six antibiotic discs used even if the degrees of susceptibility vary (Table 7). In this study, there was statistical significant difference between age and the occurrence of the disease.

The isolated bacteria indicate young cattle, 13 (92.9%) were more susceptible to pneumonic pasteurellosis than adult cattle, 2 (66.7%) (Table 8).

Questionnaire survey

The questionnaire survey was conducted to determine various aspects of bovine pneumonic pasteurellosis in the study areas. The result shows respiratory disease as the most important disease problem. In addition, the associated risk factors with bovine pneumonic pasteurellosis were identified by the respondents.

Among the assessed risk factors, age was one of the most predisposing factors to pneumonic pasteurellosis in which young were found to be more susceptible than adults according to the respondents. The main stresses causing factors were indicated by the respondents such as climatic change (80.0%), previous sickness (64.3%), feed shortage (37.0%), drought (23.7%) and commonly, dry season (53.5%) (Table 9).

DISCUSSION

In the present study, the overall prevalence of bovine pneumonic pasteurellosis was found to be 50.2% in which 58.2 and 41.8% were recorded at clinic and abattoir, respectively. The finding was slightly lower than that of Aschalew (1998) and Tesfaye (1997) who reported 63.8 and 67.6%, respectively. However, the result was higher than that of Tilaye (2010) who reported 28.4%. This might be due to the different ways of taking sample from purely pneumonic cattle, improved health

Table 7. Summary of antimicrobial sensitivity test result descriptively.

Bacterial isolate	PG	E	K	TTC	CAF	CP	PB	F	A
P. multocida	6(S)	7 (S)	11(S)	6(S)	11(S)	11 (S)	11 (S)	11 (S)	11(S)
M. haemolytica	7(S)	9 (S)	13(S)	7(S)	13 (S)	13(S)	13 (S)	13(S)	13(S)
B .trehalosi	2(S)	2 (S)	4 (S)	3 (S)	4 (S)	4(S)	4(S)	4 (S)	4 (S)

Where (S) is = Number of bacterial isolate susceptible to the corresponding antibiotic discs used. CAF = chloramphenicol, TTc = tetracycline, E = erythromycin, K = kanamycin, A = amoxicillin, PG = penicillin, PB = polymyxin-B, F = florifenicol, CP = cephalexin.

Table 8. Summary of isolation rate with age group.

Age group	Health status	Types of Isolates (%)			Total
		M. haemolytica	*P. multocida*	*B.trehalosi*	
Young	Healthy	-	1 (7.14%)	-	1 (7.14)
	Pneumonic	5 (35.7)	5 (35.7%)	3 (21.5%)	13 (92.9)
Adult	Healthy	-	1 (33.3)	-	1 (33.3)
	Pneumonic	1 (33.3)	1 (33.3)	-	2 (66.7)

X^2=8.59 Where, X^2 = Chi- square; P-value =0.04.

facilities, laboratory facilities and predisposing factors.

The prevalence of *M. haemolytica* was almost similar both at abattoir (4.86%) and clinic (3.24%). *P. multiocida* was discovered from nasal swabs and lung tissues. Pneumonic pasteurellosis was only positively associated with the age of suspected bovine particularly *P. multiocida*.

This result is lower than the findings of Eshetu (1991), Nurhusein (2005), Mohammed (1999) and Radostitis et al. (1994) that were reported as 13, 8.7, 40.8 and 56% respectively in pneumonic lungs. This difference might be due to the type of sample taken from purely pneumonic lung tissues and improved health facilities in the current study area.

B. trehalosi was also isolated both from nasal swab (1.62%) and lung tissues (0.69%). This was in agreement with the report of Tesfaye (1997) and Tilaye (2010) in which the species contributes in the occurrence of RDC. The pasteurella and Mannhemia species were isolated from nasal swabs as well as lung tissues, however, the isolation rate varies.

In the present study, the overall isolates were 28 (8.51%) with *M. haemolytica* (46.4%), *P. multocida* (39.3%) and *B. trehalosi* (14.3%) were recovered on bacteriological analysis respectively. Comparing the three *Pasteurella* species, *M. haemolytica* (46.4%) was the major causative agent involved in bovine pneumonic pasteurellosis. This was consistent with previous reports (Tesfaye, 1997; Tilaye, 2010; Eshetu, 1991; Mohammed, 1999).

M. haemolytica, which is a normal flora of the upper respiratory tract, may play a secondary role after the primary initiating agent suppressed the host defense mechanism, and favors the multiplication of *Pasteurella* species leading to bronchopneumonia in purely pneumonic animal (Aiello and May, 1998). Although the percentage isolation of *B. trehalosi* was relatively low (14.3%), it plays great role in the etiology and pathogenesis of bovine pneumonia should not be under estimated.

According to studies carried out by Sisay and Zerihun (2003) the prevalence of *M. haemolytica* and *P. trehalosi*, 75% from nasal swabs of apparently sheep, in the high lands of Wollo from local abattoir 205 lungs were investigated (34%) which showed pneumonia, *Mannheima* and *Pasteurella* were isolated from 59% of these pneumonic lungs and 69% of the respective tonsils. However, the present study shows lower rate of isolation of these species of bacteria. This lower rate of isolation of *Mannheimia* species and *Pasteurella* species may be due to the difference in epidemiological area and as well the anatomical structure of study sampling in addition to species study variation. Also, this higher isolation rate by the previous studies might be attributed also to the type of sample taken which were purely pneumonic, variations in the animal species in different study areas.

Comparing the three bacterial species, *M. haemolytica* constituted higher percentage of the total isolates, this implies that *M. haemolytica* is the major causative agent involved in bovine pasturellosis. This finding was consistent with other previous reports (Aschalew, 1998; Tesfaye, 1997; Mohammed, 1999; Daniel et al., 2006) that indicated *Mannheima* and *Pasteurella* species are the causative agent of pneumonic pasteurellosis in most animal species of all climatic zones in which ruminants are usually asymptomatic carriers (Thomson, 1998). *M.*

Table 9. Summary of the results of questionnaire survey.

Variable		Frequency	Percentage (%)
Management system	Extensive	151	81.6
	Semi intensive	34	18.4
	Intensive	0	0.00
Previous History Sickness	Yes	119	64.3
	No	66	35.7
Predisposing factors	Weaning	24	13.0
	Shipping	29	15.6
	Shortage of feed	68	37.0
	Drought and Crowding	44	23.7
	Unknown	20	10.7
Drenching of drugs	Yes	72	38.9
	No	113	61.1
History of stress before became sick	Yes	127	68.7
	No	58	31.3
Major stress causing factor	Transportation	19	10.3
	Housing system	18	9.70
	Climatic change	70	80.0
	Unknown	78	42.2
Season	Dry season	99	53.5
	Rainy season	86	46.5
Frequency of out break	Frequently	71	38.4
	Not Frequently	114	61.6
Seriousness of the disease	Very serious	138	74.6
	Moderate	45	24.3
	Not as such serious	2	1.10
Control method applied	Antibiotic	64	34.6
	Vaccination	67	36.2
	Improving management	54	29.2

haemolytica and *P. multocida are* considered to be normal inhabitants of nasopharyngeal mucosa (Ames et al., 2002) but not of the lung and considered to be opportunistic pathogens. *M. haemolytica* is considered to be the major cause of BRD but it is rarely isolated on nasal swabs in healthy animals. Similarly, the present study result also revealed neither *Pasteurella* nor *Mannheima* species were isolated from the non-pneumonic lung but only isolates of *P. multocida* were obtained from nasal swabs of nonclinical cases. *M. haemolytica* has been known to be the main bacterial agent responsible for the lung infection and has been known to be higher in acute pneumonia (Daniel et al., 2006).

In this study, higher rate of infection was associated with young age groups as compared to adults ($p<0.05$). This might be due to the immune status of the animal being able to predispose to the bacterial infection and other predisposing etiological agents. Furthermore, antimicrobial sensitivity patterns of the isolates were determined in this study. All the isolates (*P. multocida, M. haemolytica* and *B. trehalosi*) were susceptible to most of the antibiotics

discs used. Bacterial resistance to antibiotic is becoming a subject of interest and attention
of these days. However, moderate resistance against penicillin-G, erythromycin and tetracycline was noted. About half of *M. haemolytica* and of *P. multocida* isolates showed resistance against penicillin-G. Also some of *M. haemolytica, P. multocida* and about half of *B. trehalosi* did not show susceptibility to Erythromycin. As the result obtained indicates, the therapeutic success could be achieved using most of antibiotics in the study area. The finding was in line with the previous report (Disassa et al., 2013).

The quality of questionnaire is an important tool in individual cases detection. In the present study, the predisposing factors in the initiation of pneumonic pasteurellosis (Jones et al., 1997) which might be due to reduced immune status of the animal was surveyed based on the questionnaire. Accordingly, the main stresses causing factors were indicated by the respondents such as climatic change (80.0 %), previous sickness (64.3%), feed shortage (37.0%), drought (23.7%) and common in dry season (53.5%). This was in agreement with the previous findings (Bruere et al., 2002).

In the present study, the seasonal variation of the isolates was not determined due to shortage of time. Moreover, even if identification of serotypes is so important, it was not conducted due to lack of laboratory facilities.

However, the study has the following strengths: the sample was collected from both abattoir and clinic which shows the reliability of the data, the isolates were obtained from pneumonic bovine that made us compare the clinical case with ethological agents, all laboratory works were followed standard procedures and quality control was exhibited in each step of the work.

Conclusion

Generally, pneumonic pasteurellosis was the major disease of cattle in the study area and *M. haemolytica* is the most common cause. Young animals were the risk factor for the disease. It also demonstrated that pneumonic pasteurollosis is a highly complex multifactorial disease particularly in cattle which could be associated with stress, compromised immunity, adverse environmental condition, previous illness (co-infection), misuse of traditional medicine and this made the control of pasteurellosis more difficult.

Furthermore, the antibiotic sensitivity tests of the isolates showed them to be susceptible to most of the drugs. However, the isolates showed resistance to some of the antibiotic disks. Moreover, there must be an integrated management system, vaccination, controlling of the predisposing factors and use of broad spectrum antimicrobials as a prophylactic, and treating of sick animals is suggested.

Conflict of Interests

The author(s) have not declared any conflict of interests.

ACKNOWLEDGEMENTS

The authors would like to thank Oromia Regional Veterinary Laboratory, Bedelle Veterinary Clinic, Bedelle Municipal Abattoir, animal owners and all individuals who rendered help during the study period.

REFERENCES

Aiello SE, May A (1998). The Merck Veterinary Mannual. 8th edition. New Jersey. USA: Merck and Co, Inc., pp.1053-1057.

Ames TR, Baker JC, Wikes SE (2002). The bronchopneumonia (Respiratory disease complex of cattle, sheep, and goats). Mosby Inc. St Louis, pp. 551-570.

Angen OM, Quine W, Donachie BM (1999b). Investigations on the species specificity of *Mannhemia (pasturella) heamolytica* serotyping. Vet. Microbiol. 65:283-290.

Aschalew Z (1998). A study on Pneumonic Pasteurellosis in North Shewa. Debre-Zeit: DVM thesis, FVM, Addis Ababa University, Ethiopia, 59.

Baundreaux CM (2004). A novel strategy of controlling bovine pneumonic pasturellosis transfecting the upper respiratory tract of cattle with a gene coding for the antimicrobial peptide cecropin. MSc thesis, Louisiana State University, USA.

Biberstin L, Hirsh C (1999). Pasteurella. In; Veterinary Microbiology. Black Science Inc. USA, pp.135-139.

Blackall PJ, Christensen CH, Buckingham L, Blackall L, Bisgaard M (2005). Reclassification of pasturella galhanaun (Haemophilus) paragallinarium, pasturella ovum and *Pasturella* volantium as Auibacter, comb.nov. Int. J. Syst. Evol. Microb. 55:253-362.

Bruere AN, West D, Rieller AL (2002). Epizootic pneumonia. In the sheep health disease and production written for veterinarians and former, veterinary continuing education No. 2. palmerston North, Anassey University, pp.100-108.

Central Statistical Agency (CSA) (2009). The Federal democratic republic of Ethiopia central statistical agency, agriculture in figure key findings of the 2008/09-2010/11 agricultural sample surveys for all sectors and seasons country summary, FDRECSA Addis Ababa, Ethiopia, pp.14-18.

Daniel JA, Held JE, Boake DG, Wulf DM, Epperson WB (2006). Evaluation of the prevalence and onset of lung lesion and their impact on growth of lambs. Am. J. Vet. Res. 67:890-894.

Disassa H, Takele T, Adem A (2013). Study on Ovine pneumonic pasteurellosis: Isolation and Identification of Pasteurellae and their antibiogram susceptibility pattern in Haramaya District, Eastern Hararghe, Ethiopia. BMC Vet. Res. Pp. 9-239.

Eshetu M (1991). Pneumonic Pasteurellosis in sheep slaughtered at Addis Ababa abattoir. DVM thesis, FVM, Debre-Zeit, Addis Ababa University, Ethiopia.

Jones TC, Hunt RD, King MW (1997). *Pasturella multocida* pneumonia seminar in Respiratory Infections, 12:54-56.

Lobago F, Bekana B, Gustafsson H, Kindahl H (2006). Reproductive performance of dairy cows in small holder production system in Selalle, central Ethiopia. Trop. Anim. Prod. Health 38:333-342.

Mohamed RA, Abdelsalam EB (2008). A review on pneumonic pasteurellosis (respiratory mannheimiosis) with emphasis on pathogenesis, virulence mechanisms and predisposing factors. Bulg. J. Vet. Med. 11(3):139-160.

Mohammed R (1999). Bacteriological and histological examination of pneumonic lungs of small ruminants slaughtered at Hashim Nure export abattoir. Debre-Zeit: DVM thesis, FVM, Addis Ababa University, Ethiopia, p. 21.

Nurhusein Y (2005). Aerobic bacterial Flora in the respiratory tract of apparently healthy sheep slaughtered in Dessie Abattror, North Eastern Ethiopia. Debre-Zeit: DVM thesis, FVM, Addis Ababa University, Ethiopia.

Quninn PJ, Markey BK, Carter ME, Donnelly WJ, Lonard FC (2002). Veterinary Microbiology and Microbial Disease, Blackwell, Science, pp.137-143.

Radostitis OM, Blood DC, Gay CC (1994). Diseases of the respiratory system .In text book of the disease cattle, sheep, pigs, goats and horses. 8thed. Bailliere Tindal, London, pp. 391-420.

Radostitis O, Gay C, Blood D, Hinchcliff K (2000). Veterinary Medicine: A Textbook of Diseases of Cattle, Sheep, Pigs, Goats and Horses, 9th edn, W.B. Saunders, London, pp. 1550-1600.

Sisay T, Zerihun A (2003). Diversity of *M. haemolytica and P.trehalosi* serotypes from apparently health seep and abattoir specimens in the high lands of Wollo, North East Ethiopia. Vet. Res. Commun. 27:3-14.

Tesfaye S (1997). Serological and bacteriological investigation of *M. haemolytica* serotypes in sheep in the high lands of Wollo, North Eastern Ethiopia. DVM Thesis, AAU, FVM, Debre-Zeit, Ethiopia, pp.1-65.

Thomson RC (1998). Respiratory system In. Special Veterinary pathology 2nd ed. B. C. Decker inc. Philadelphia USA, pp.117-150.

Tilaye D (2010). Pathological, bacteriological and serological study on pneumonic sheep and goats with special emphasis on *Mannheimia*, *Babersteinia*, *Pasteurella* and *Mycoplasma* species. MSc Thesis, AAU, FVM, Debre Zeit, Ethiopia, pp.1-97.

Zewdie S (2004). Current status of veterinary services in Ethiopia. In proceedings of a national workshop on managing animal health constraint to export marketing of meat and livestock.

Production of biosurfactants by bacteria isolated from a mine tailing zone in Southern Mexico and their resistance to heavy metals

Jeiry Toribio-Jiménez[1], Miguel Ángel Rodríguez-Barrera[1], Monserrat Valdez Lucena[1], Ashanti Barrera Flores[1], Daniel Segura[2], Víctor Wilson-Corral[3], Eugenia Flores Alfaro[4] and Yanet Romero[1]

[1]Laboratorio de Investigación en Biotecnología y Genética Microbiana, Unidad Académica de Ciencias Químico Biológicas, Universidad Autónoma de Guerrero, Chilpancingo, Guerrero, México.
[2]Departamento de Microbiología Molecular, Instituto de Biotecnología, Universidad Nacional Autónoma de México, Cuernavaca, Morelos, México.
[3]Unidad Académica de Ingeniería, Universidad Autónoma de Sinaloa, Culiacan, México.
[4]Laboratorio de Investigaciones de Epidemiología clínica y Molecular, Unidad Académica de Ciencias Químico Biológicas, Universidad Autónoma de Guerrero, Chilpancingo, Guerrero, México.

Tailings generated through mining processes often create leachates containing high concentrations of heavy metals such as As, Fe, Mn, Zn and Pb. These high concentrations of heavy metals result in environmental damage such as contamination of soil, groundwater and air, which represents a huge problem for individuals living near mining areas. An alternative for soil metal removal is microbiological processes including the production of biosurfactants, possibly a survival mechanism for adverse conditions of mine tailings and leachates. Moreover, mine tailings are materials that have attracted interest among researchers, because they can be exploited by innovative techniques like phytomining. In this study, we sampled the leachates of the "El fraile" mine tailings and identified 103 bacteria capable of growth on these leachates. We observed that 11 bacteria produce a high amount of biosurfactants and developed the multi-metal tolerance with higher concentration gradient of Pb, Cd, Cu, Fe, Zn and As. We showed that the bacteria tolerate 853 nM of As and up to 12 nM of Pb, 17 nM of Cd, 10.6 nM of Cu, 22 nM of Fe and 10.5 nM of Zn. We determined that the bacterial isolates clustered within five phylogenetic groups that were very close: *Enterobacter, Klebsiella, Artherobacter*, *Pantoea* and *Solibacillus* groups. A bank of strains resistant to heavy metals and producers of biosurfactants was obtained for future studies on the mechanism of absorption or assimilation of heavy metals and light was shed on the alternative use of these bacteria in bioremediation of metal pollution.

Key words: Bacteria, metals, biosurfactants, mine tailings.

INTRODUCTION

Mining is one of the most important economic activities of Mexico since the 15th Century and is currently active in many parts of the country. The Taxco mining district has been one of the richest producers of silver and gold, among other metals, since colonial times (1523-1810). In this district, there are tailing impoundments of various ages and sizes

(La Concha, Santa Rosa, Taxco and El Fraile) with high levels of metals (As, Pb, Cd, Zn, Mn) (Armienta et al., 2003; Talavera et al., 2005). During mining activities, large amounts of material or mine wastes are removed and deposited into surrounding areas. Leaching of heavy metals from these dumps is common during the rainy season, which contaminates the surface water as well as ground water bodies (Dhakate and Singh, 2008). In the El Fraile impoundments, different types of acidic (pH=2.4-2.5) or near-neutral (pH=6.3-7.7) leachates were identified. Some environmental monitoring studies on these leachates have shown high levels of Ca-SO_4, (Ca+Mg)-SO_4, Mg-SO_4 and Ca-(SO_4+HCO_3) (Talavera et al., 2006). These deposits of mine tailings and their leachates are major sources of heavy metal contamination in and around the communities, because the leachates are sometimes used as alternative sources of domestic water.

Heavy metals are stable and persistent contaminants of the environment, since they cannot be degraded or destroyed (Sevgi et al., 2009) so, the accumulation of metals in soil represents a health risk for people (Moreno et al., 2010), and other living things in terrestrial and aquatic environments near mining areas. The use of micro-organisms for decontamination of heavy metals in the soil and water around industrial plants has gained growing attention (Doni et al., 2013; Gadd, 2004; Kavamura and Esposito, 2010). Tailings have very low concentrations of macronutrients for microbial growth, especially nitrogen, and conversely can contain high concentrations of heavy metals, which can be particularly problematic in tailings where the pH is low. In such instances, soil fertility and microbial communities are drastically reduced (Seaker and Sopper, 1988; Pepper et al., 2012). However, multiple investigations have been carried out with isolated microbial strains from soil, mud and water samples from metallurgically polluted environments for bioremediation of toxic heavy metals (Umrania, 2006; Das et al., 2013). Recent work indicated that most of the transitions between metal speciation forms are controlled by microbial activity (Johnson and Hallberg, 2004; Hall et al., 2005; Hall and Puhlmann, 2004). The potential for natural attenuation of acid mine drainage by microbial populations, and the subsequent removal of key toxic metal, particularly the removal of copper and zinc and the interaction of the root-organism in the activation of such decontamination procedures, were clearly demonstrated (Doni et al., 2013; Whitehead and Prior, 2004; Mulligan et al., 1999). Therefore, microbial processes that facilitate the detoxification and metal mobility using resistant microbial strains can be exploited for bioremediation of heavy metals from waste water and effluents containing heavy metals.

The production of biosurfactants by bacteria plays a role in the adaptation of these organisms to different environments (Toribio et al., 2011); therefore, this could be a mechanism by which the bacteria can absorb heavy metals and become resistant to growth in soils polluted by tailings. Biosurfactant producing bacteria are very diverse and have been isolated from a wide variety of environments, including contaminated soil or water (Bodour et al., 2003), but there are no reports of biosurfactant producing bacteria isolated from mining soils.

The present study is aimed at isolating, characterizing and monitoring biosurfactant producing bacterial isolates and their ability to resist heavy metal leachates from the mine El Fraile tailings located in the Taxco mining district in southern Mexico for their possible use in bioremediation in metal contaminated environments.

MATERIALS AND METHODS

Site description and tailing sampling

The present study was conducted near a mine site (El Fraile) located at Taxco, Guerrero, in southern Mexico (Figure 1). Thirty-two samples were randomly collected from heavy metal-contaminated mine tailings and leachates of the El Fraile and were put into aseptic screw capped bottles. The pH of the mine tailing and leachate was determined using a pH meter (Orion Research, Beverly, MA).

Isolation of strains, media and growth conditions

A quantity of 1 mL of leachate or 1 g of soil of El fraile mine tailings from each of the collected samples was mixed in 9 mL of sterile saline solution ($NaCl_2$ 0.85%) and serial dilutions were made (1:1,000 and 1:10,000). 100 µl of each dilution was plated on Luria Bertani (LB) agar plates supplemented with 0.025 mM Pb $(NO_3)_2$ and 0.0125 mM de Cd $(NO_3)_2$. Plates were incubated at 30°C for 48 h and colonies differing in morphological characteristics were selected.

Detection of biosurfactant production

The isolated bacteria were screened for their biosurfactant-producing ability by blood agar plate assay and PSVW agar plate assay. The screening was done culturing the bacteria on PSVW (PPGAS medium supplemented with 2.5 mg/ml of methylene blue, 200 mg/ml of the surfactant cetyl-thrimethyl-ammonium-bromide (CTAB) and 1.5% agar) plates and blood agar (15 g/L enzymatic digest of casein, 4 g/L enzymatic digest of animal tissue, 2g/L yeast extract, 1g/L corn starch, 5g/L sodium chloride and 14g/L agar) plates (Will et al., 1997). Plates were incubated at 30°C for 72 h. Bacteria producing a blue halo around the colonies and hemolysis were cultured on PPGAS liquid medium (0.02 M NH_4Cl2, 0.02 M KCl, 0.12 M Tris-HCl, 0.0016 M $MgSO_4$, 0.5% glucose, and 1% peptone, pH 7.2; Zhang and Miller, 2002) and were incubated at
30°C for 7 or 15 days shaken at 225 rpm. Foam-forming activity, drop collapse assay and emulsification ability (Sen, 2010; Toribio et al., 2011) were also evaluated.

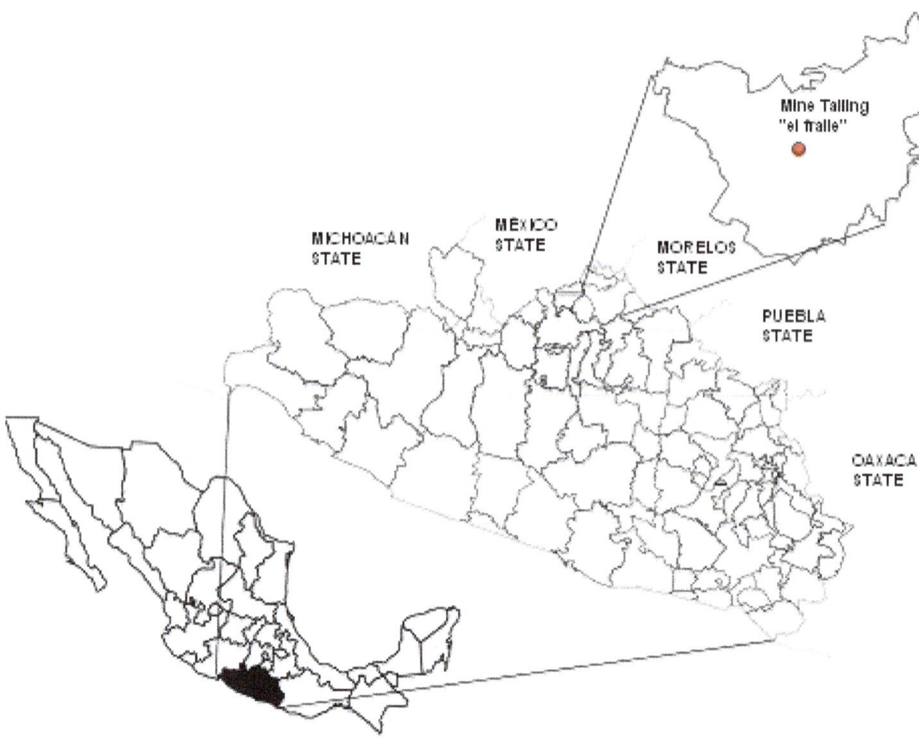

Figure 1. Map of Taxco mining district showing the localization of the El Fraile tailings deposits. Elaborated based on the Geostatistical Framework National Institute of Statistics and Geography (INEGI).

Biochemical characterization

Morphological and biochemical properties of the strains isolated were determined and compared by The Analytical Profile Index (API 20E) and VITEK2 Compact (bioMérieux). These systems use 8 conventional biochemical tests and 12 carbohydrate assimilation tests and were performed according to the manufacturer´s instructions. Gram staining, catalase and oxidase test were done with standard protocols as described in MacFaddin's Manual of Identification of Bacteria (MacFaddin, 2000; Kannan, 2002).

Minimum inhibitory concentration (MIC) determination

The MIC was used to define the lowest concentration of heavy metals and metalloids that inhibit the visible growth of bacteria (Andrews, 2001). The MIC values were determined using the agar dilution method (Kamala-Kannan and Krishnamoorthy, 2006). A log-phase culture of the isolates was inoculated onto LB agar plates supplemented with increasing concentrations (1-900 mM) of heavy metal salts and metalloids: $Pb(NO_3)_2$, $Cd(NO_3)_2$, $Cu(NO_3)_2$, $Fe(NO_3)_2$ $Zn(NO_3)_2$ and As IV. The plates were incubated at 30°C for a week, looking for bacterial growth.

DNA extraction, sequencing and phylogenetic analyses

Genomic DNA was extracted by using a GeneJeT™ Genomic DNA purification kit (Thermo Scientific) according the manufacturer´s instructions. Bacterial 16S rDNA was amplified by PCR using the universal bacterial 16S rDNA primers fd1 5´-AGAGTTTGATCCTGGCTCAG-3´ and rd1 5´-AAGGAGGTGATCCAGCC-3´ (Ndip et al., 2007; Rodicio et al., 2004). PCR was performed with a 25 µl reaction mixture containing 1 µl (10 ng) of DNA as a template, each primer at a concentration of 5 µM, 2.5 mM $MgCl_2$ and dNTP´s at a concentration of 2.5 µM, as well as, taq polymerase and buffer used as recommended by the manufacturer (Thermo Scientific). The PCR conditions consisted of an initial denaturing step at 95°C for 7 min, 30 cycles of 95°C for 40 s, 55°C for 40 s, 72°C for 2 min and a final extension at 72°C for 15 min. The PCR products were electrophoresed and purified from 1% agarose gels (0.56 TAE buffer; 2.42 g Tris-HCl, 0.57 ml glacial acetic acid, 1 ml 0.5 M EDTA (pH 8.0) and 1 L distilled and deionized water) using the GeneJET™ PCR Purification Kit (Thermo Scientific). DNA sequences were determined by the dideoxy chain termination method (Sanger et al., 1977) with a Perkin Elmer/Applied Biosystems DNA Sequencer. The 16S rDNA sequences were aligned and compared with other 16S rDNA genes in Ribosomal Database Project II (RDP) by using SEQMATCH (Wang et al., 2007) and GeneBank by using NCBI basic local alignment search tools BLASTn program (Benson and Karsch-Mizrachi, 2000). The evolutionary history was inferred using the Neighbor-Joining method (Saitou and Nei, 1987). The percentage of replicate trees in which the associated taxa clustered together in the bootstrap test, has 1000 replicates (Felsenstein, 1985). The evolutionary distances were computed using the Jukes-Cantor method (Jukes and Cantor, 1969). The rate variation among sites was modeled with a gamma distribution. Evolutionary analyses were conducted in MEGA6 (Tamura et al., 2013). GenBank accession numbers for *Enterobacter cancerogenus* 16S rDNA sequences are KF811178, KF811179, KF811181, KF811185, KF811186, KF811187 and KF811188, for *Klebsiella pneumonia* 16S rDNA gene sequence is KF811183. That for *Solibacillus silvestris* 16S rDNA gene sequence is KF811182; for *Arthrobacter luteolus* 16S rDNA gene sequence is: KF811177 and for *Pantoea vagans* 16S rDNA gene sequence is KF811192.

Table 1. Microbiological characterization.

Isolate	Samples pH	Gram staining	Oxidase assay	Catalase assay
Leachates				
M14D1-12	2.0	Bacillus Gram -	-	+
M14D2-13	2.0	Bacillus Gram -	-	+
M14-1	6.0	Bacillus Gram +	-	+
M14A-1	2.0	Bacillus Gram -	+	+
M14B	2.0	Bacillus Gram -	+	+
Mine tailings				
M2B-1	8.4	Bacillus Gram -	-	+
M2B-4	8.0	Bacillus Gram +	-	+
M15A-4	7.0	Bacillus Gram -	+	+
M4D-2	7.4	Bacillus Gram -	-	+
M10-7	8.4	Bacillus Gram -	-	+
PAO	8.4	Bacillus Gram -	-	+

All determinations were done in duplicate. (-), No activity, (+), activity.

Statistical analysis

The experiments were set up in duplicate. Data sets of resistance to metal among individual isolates were analysed by principal component analysis (PCA) using a statistical package STATA v 11.

RESULTS

Isolation, identification and characteristics of biosurfactant-producing bacteria

This study represents an attempt to identify the biosurfactant producing bacteria within mine tailings and leachates from El Fraile. Serial dilution plating of mine tailing suspensions and leachates on LB plates resulted in the isolation of a total of 103 morphologically distinct bacteria (23 isolates of leachates and 80 of mine tailings). Screening of the bacterial isolates for biosurfactant production resulted in the isolation of 11 strains (5 of leachates and 6 of mine tailings) which gave positive results for most of the screening assays, indicating that they are efficient biosurfactant producers. Strains were subjected to Gram staining to determine their structure and morphology. We found that nine of these bacteria are Gram negative bacillus and two are Gram positive bacillus. We also carried out the oxidase and catalase assays in order to determine the metabolism of the 11 strains isolated. Table 1 indicates the results obtained. All bacteria were catalase positive and most were oxidase negative.

Study of the emulsifying capacity of biosurfactants produced by the bacterial isolates

The 11 isolates were tested for biosurfactant production and activity (Table 2). In all assays and tests *Pseudomonas aeruginosa* was used as reference biosurfactant producing bacteria. We used the hemolytic activity assay as the first method for screening biosurfactant producers. In this assay, we looked for a clear zone around the wells produced by hemolysis. Although we did not observe hemolytic activity of the isolates in this assay, when we used PSVW plates to detect extracellular glycolipid production, the formation of dark blue halos around all the isolates was observed, and the halos produced by the isolates M14D1-12, M14D2-13 and M14-1 were similar to those produced by *P. aeruginosa* (Table 2). Biosurfactant activity was measured by the drop collapse test, by comparing an oil droplet treated with sterile medium as the control to an oil droplet treated with the culture (Table 2). The oil droplet collapsed when treated with cultures of all isolates, especially with cultures of the isolates M14D2-13 and M14-1 (Table 2). In addition, biosurfactant activity was also confirmed by the emulsifying activity of biosurfactant in the supernatant. The emulsifying capacity was evaluated by the E_{24} emulsification index using *P. aeruginosa* PAO1 as a control. Amongst the isolates from the mine tailings, the emulsifying capacity was higher with the isolates M10-7 and M4D-2, with a 52.27 and 26.09% respectively, whereas with the isolates from the leachates it was M14D1-12, with 44% (Table 2). No substantial emulsification was achieved with M14D2-13 (isolate from leachates) and M2B-4, M15A-4 and PAO (isolates from mine tailings). Foam formation was evaluated in the isolates on PPGAS medium, for a period of 7 to 15 days. We observed that the isolates from leachates that showed the higher production were M14D2-13, M14D1-12 and M14A-1, whereas the isolates M2B-1, M2B-4, M15A-4 and M4D-2 from the mine tailings showed the same production, except for strain M10-7 that produced less and the strain PAO who´s production was similar to that of the control strain (Table 2).

Table 2. Detection of biosurfactant production assays.

Strain	Hemolytic activity	PSVWagar test	Drop collapse test	E_{24} (%)	Foam-forming
Control					
Pseudomonas aeruginosa PA01	++++	++++	++++	60	++++
Leachates					
M14D1-12	-	+++	+	44	+++
M14D2-13	-	+++	+++	10	+++
M14-1	-	+++	++++	17	++
M14A-1	-	+	+	12	+++
M14B	-	+	+	12	+++
Mine tailings					
M2B-1	-	+	+	11.9	++
M2B-4	-	+	+	6.98	++
M15A-4	-	+	+	8.70	++
M4D-2	-	+	+	26.09	++
M10-7	-	+	+	52.27	+
PAO	-	+	+	0	++++

All determinations were done in duplicate using *P. aeruginosa* PAO1 as a control. (-), No produced hemolysis, (+), Foam formation and stability.

Effect of growth of bacteria on heavy metal

Diverse studies have shown that bacteria isolated from mine tailings show metal tolerance (Pepper et al., 2012; Koc et al., 2012). To determine the extent of tolerance through MIC values, the isolated bacteria were grown on LB agar plates containing different concentrations of heavy metals and metalloids (Pb, Cd and As). The results of the metal sensitivity tests (inhibition zone diameters) for the isolates and the control strain (*P. aeruginosa* PAO1), on $Pb(NO_3)_2$, $Cd(NO_3)_2$, $Cu(NO_3)_2$, $Fe(NO_3)_2$ $Zn(NO_3)_2$ and As IV, are provided in Table 3. The cut-off value of MIC was taken to be 17 mM for $Pb(NO_3)_2$, $Cd(NO_3)_2$, $Cu(NO_3)_2$, $Fe(NO_3)_2$ $Zn(NO_3)_2$ and 853 for As. All the isolates showed high tolerance to the various heavy metals and metalloids tested. The MIC levels were significantly higher as compared to the control (Table 3). Isolates from leachates exhibited MIC values of 11 mM on $Fe(NO_3)_2$, up to17 mM on $Cd(NO_3)_2$, 12 mM on $Pb(NO_3)_2$, 10.5 and 10.6 mM on $Zn(NO_3)_2$ and $Cu(NO_3)_2$ respectively; whereas on As IV, the MIC value was 853 mM. For the isolates from the mine tailings, we observed that the MIC values on $Pb(NO_3)_2$, $Cu(NO_3)_2$ and $Zn(NO_3)_2$ were similar to the values of isolates from the leachates, whereas the maximum MIC values on $Fe(NO_3)_2$ and $Cd(NO_3)_2$ were 22 and 8.5 mM, respectively, instead of 11 and 17 mM. Of these isolates from the mine tailings, only M15A-4 and M10-7 displayed a high MIC value on As IV (853 mM (Table 3). PCA was performed to compare resistance to metal among individual isolates. We observed the higher tolerance to Pb ascribed to the variability between the strains isolated, variability was not observed in the resistance to Cd, As, Fe and Zn, even though the As showed the highest resistance concentrations, but has no variability as the Pb resistance. So, according to data from the principal component analysis (PCA), the strains isolates showed higher tolerance to Pb followed by Cd and As and less tolerance to Fe and Zn, the analysis results were consistent with the values of the MICs.

16S rRNA sequence analysis

To further characterize the isolated bacteria, their 16S rDNA sequences were obtained by amplifying and sequencing their DNA fragments. The results from the analysis using the Ribosomal Database project II (RPD II), showed that the 16S rDNA sequences of all isolates group in five distant clusters (Figure 2). The first one belongs to the *Enterobacter* group and contains seven isolates showing a unique 16S rDNA sequence (M14D2-13, M14-A1, M2B-1, M14B, M15A-4, M4D-2 and PAO). The second one (M14D1-12) belongs to the *Pantoea* group and displays the same 16S rDNA sequence as *P. vagans*. The third (M10-7) belongs to the *Klebsiella* group, displaying the same 16S rDNA sequence as *K. pneumoniae.* The fourth one (M14-1) belongs to the *Solibacillus* group; displaying the same 16S rDNA sequence as *S. silvestris*, and the fifth one (M2B-4) belongs to the *Arthrobacter* group, displaying the same 16S rDNA sequence as *A. luteolus.* The seven *Enterobacter* strains, sharing a 100% identity of the 16S rDNA sequence between them, were closer to

Table 3. MICs of the isolates.

Isolate	$Pb(NO_3)_2$	$Cd(NO_3)_2$	$Cu(NO_3)_2$	$Fe(NO_3)_2$	$Zn(NO_3)_2$	As IV
Control						
Pseudomonas aeruginosa PA01	3	4	5	5	5	-
Leachates						
M14D1-12	12	17	10.6	11	10.5	853
M14D2-13	12	17	10.6	11	5.3	853
M14-1	12	8.5	10.6	11	5.3	853
M14A-1	12	17	10.6	11	5.3	853
M14B	12	17	10.6	11	5.3	853
Mine tailings						
M2B-1	6	8.5	10.6	22	10.5	426
M2B-4	6	4.2	10.6	22	10.5	426
M15A-4	6	8.5	10.6	22	10.5	853
M4D-2	6	4.2	10.6	22	5.3	426
M10-7	12	8.5	10.6	22	10.5	853
PAO	6	4.2	10.6	22	5.3	426

All determinations were done in duplicate in cells grown for a week on LB agar plates supplemented with increasing concentrations (1–900 mM) of heavy metal salts and metalloids: $Pb(NO_3)_2$, $Cd(NO_3)_2$, $Cu(NO_3)_2$, $Fe(NO_3)_2$ $Zn(NO_3)_2$ and As IV. (–), No MIC hence no growth. *P. aeruginosa* PAO1 Was used as a control, the results are expressed as MIC mM.

E. cancerogenus, than to *E. asburlae*.

DISCUSSION

The mining activity in Taxco, Guerrero, Mexico has resulted in the discharge of large amounts of mine wastes (Talavera et al., 2005, 2006). These deposits are major sources of heavy metal contamination affecting the surrounding area. We have interest in finding bacteria capable of absorbing or assimilating heavy metals as an alternative to bioremediation. We isolated eleven novel strains capable of growing in mine tailings and leachates, in addition to producing biosurfactants. Biosurfactant applications in environmental applications are promising due to their biodegradability, low toxicity and effectiveness in enhancing the biode-gradation and solubilization of low solubility compounds (Soberón-Chávez G and Maier, 2011; Mulligan et al., 1999). A number of studies have demonstrated that biosurfactants produced by bacteria, accelerate the bioremediation of hydrocarbon contaminated soils and organic contaminants (Moldes et al., 2001; Mahjoubi et al., 2013; Lawniczak et al., 2013). It has also been demonstrated that the production of biosurfactants greatly contributes to the metal removal from contaminated soils (Wang and Mulligan, 2009). It has been proposed that the rhamnolipids can remove arsenic and heavy metals through the formation of complex micelles with the metals, reducing the interfacial tension (Wang and Mulligan, 2009). However, the presence of biosurfactant producing bacteria isolated from metal contaminated soils has not been documented; therefore, these bacteria are the focus of attention in this work because the production of biosurfactants could possibly be a mechanism that the bacteria isolates of leachates of mine tailings use to resist high levels of heavy metals. Therefore, in this study, we characterized biosurfactant producing bacteria capable of growth on high levels heavy metals with the objective of creating a collection of strains for future studies on the role of biosurfactants on removal of heavy metals from the mine tailings and the possible mechanisms involved.

The high value of MIC that the bacterial isolates present, indicate their capacity to resist high concentrations of heavy metals. This should be taken into consideration if these bacteria are to be used on the remediation of arsenic and heavy metal contaminated mine tailings or soils. We observed that the best BS producing strain isolated from leachates was the M14D1-12, exhibiting an index of emulsification of 44%, whereas that the best BS producing strain isolated from the mine tailing was the M10-7, with an index of emulsification of52.27%, with respect to the control strain *P. aeruginosa* PA01 (Table 2). This strains were capable of tolerating more value of $Cd(NO_3)_2$ and $Fe(NO_3)_2$, therefore we observed the correlation between the production of biosurfactant, with the ability for growth in high concentrations of heavy metals. The RDP II 16S rRNA gene sequence analysis was carried out to identify the dominant bacterial population. The data showed that the most

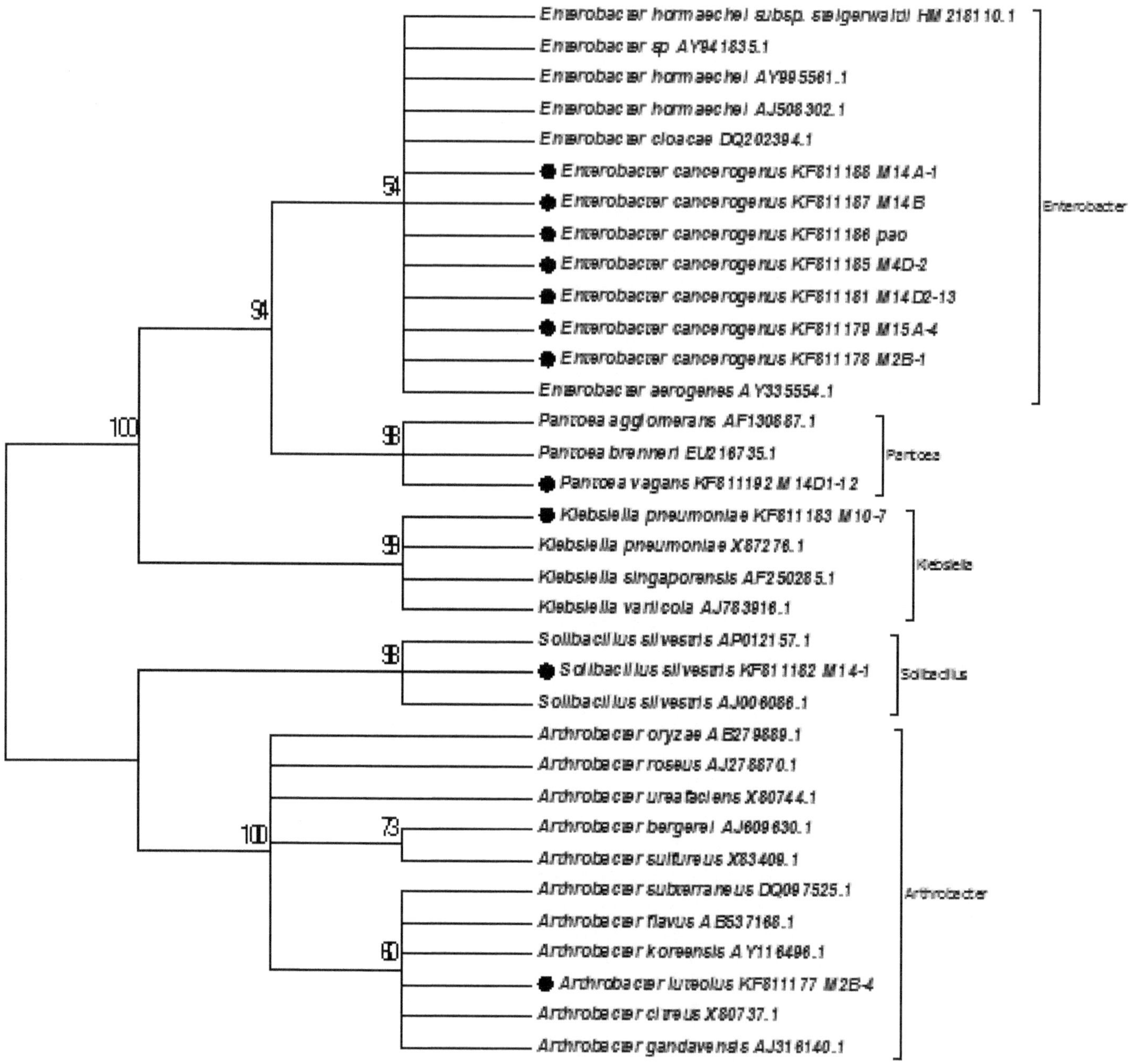

Figure 2. Molecular phylogenetic analysis by neighbor-joining method on 16s gene sequences. The evolutionary history was inferred using the neighbor-joining method (Saitou and Nei, 1987). The percentage of replicate trees in which the associated taxa clustered together in the bootstrap test (1000 replicates) are shown next to the branches (Felsenstein, 1985). The evolutionary distances were computed using the Jukes-Cantor method (Jukes and Cantor 1969) and are in the units of the number of base substitutions per site. The rate variation among sites was modeled with a gamma distribution (shape parameter = 1). All positions containing gaps and missing data were eliminated. Evolutionary analyses were conducted in MEGA6 (Tamura et al., 2013).

persistent bacterial populations were members of dominant soil bacterial group in distant clusters: *Enterobacter*, *Klebsiella*, *Arthrobacter*, *Silobacillus* and *Pantoea*, the strain isolated from leachates M14D1-12 was identified as *P. vagans*, whereas the strain isolated from the mine tailing M10-7 was identified as *E. cancerogenus* (Figure 1). Using denaturing gradient gel electrophoresis (DGGE) bacteria isolates were found in heavy metal contaminated soils belonging to the genus *Arthrobacter, Bacillus, Brevibacterium, Brochothrix, Comamonas, Cytophaga, Deinococcus, Enterobacter, Hafnia, Micrococcus, Mycobacterium, Nocardia, Pseudomonas, Rathayibacter, Rhodococcus, Salmonella, Serratia, Staphylococcus, Variovorax* and *Xanthomonas* (Ellis et al., 2003).

However, the existence of these microorganisms can be influenced by several factors such as the presence of nutrients, pH and specially the humidity and temperature. In this study, we found the genus: *Enterobacter*, *Silobacillus*, *Klebsiella*, *Arthrobacter* and *Pantoea* and we observed that they were capable of growth in a mineral-rich environment therefore these bacteria are remarkably tolerant to a wide range of soluble metal ions (Table 3). The capacity of these bacteria to tolerate high concentrations of heavy metals and the production of BS can be exploited for the remediation of heavy metal pollution.

Experiments are in progress to characterize the optimal conditions for biosurfactant production on heavy metal and to determine the type and the potential role biosurfactants may play in heavy metal metabolism by the producing organism. The current study showed that bacteria of genus: *Enterobacter*, *Silobacillus*, *Klebsiella*, *Arthrobacter* and *Pantoea* isolated from mine tailings and leachates of the "El fraile" mine tailing are capable of resisting high concentrations of heavy metal salts and metalloids: $Pb(NO_3)_2$, $Cd(NO_3)_2$, $Cu(NO_3)_2$, $Fe(NO_3)_2$ $Zn(NO_3)_2$ and As IV and produce biosurfactants (BS). This is the first study on bacteria isolated from mine tailing and leachates that showed the production of BS which can be a key factor for living in these ecosystems.

Conflict of Interests

The author(s) have not declared any conflict of interests.

ACKNOWLEDGEMENTS

This work was supported by convening 2012 of the University of Guerrero. We are grateful to Travis Ashworth for critical discussions of this work and Dr. Daniel Segura Gonzalez for sequencing the 16S rDNA sequences. We also thank Fernando Reyna, Ana Patricia Torres, Daysi Roman and Analy Lara for their technical support, Dr. Germán Bonilla Rosso for his constructive criticism on phylogenetic analysis.

REFERENCES

Andrews JM (2001). Determination of minimum inhibitory concentrations. J. Antimicrob. Chemother. 48:5-16.

Armienta MA, Talavera O, Morton O, Barrera M (2003). Geochemistry of metals from mine tailings in Taxco, Mexico. Bull Environ. Contam. Toxicol. 71:387-393.

Benson DA, Karsch-Mizrachi I (2000). GenBank. Nucleic. Acids. Res. 28:15-18.

Bodour AA, Drees KP, Maier RM (2003). Distribution of biosurfactant producing bacteria in undisturbed and contaminated arid southwestern soils. Appl. Environ. Microbiol. 69:3280-3287.

Das S, Patnaik S, Sahu H, Chakraborty A, Sudarshan M, Thatoi H (2013). Heavy metal contamination, physic-chemical and microbial evaluation of water samples collected from chromite mine environment of Sukinda, India. Trans. Nonferrous. Met. Soc. China. 23:484-493.

Dhakate R, Singh VS (2008). Heavy metal contamination in groundwater due to mining activities in Sukinda valley, Odisha -A case study. J. Geogr. Reg. Plann. 1(4):58-67.

Doni S, Macci C, Peruzzi E, Iannelli R, Ceccanti B, Masciandaro G (2013). Decontamination and functional reclamation of dredged brackish sediments. Biodegradation. 24:499-512.

Ellis RJ, Morgan P, Weightman AJ, Fry JC (2003). Cultivation-dependent and-independent approaches for determining bacterial diversity in heavy-metal-contaminated soil. Appl. Environ. Microbiol. 69:3223-3230.

Felsenstein J (1985). Confidence limits on phylogenies: An approach using the bootstrap. Evol. 39:783-791.

Gadd G (2004). Microbial influence on metal mobility and application for bioremediation. Geoderm. 122:109-119.

Hall G, Puhlmann T (2004). Spatial distribution of iron oxidation in the aerobic cells of the Wheal Jane Pilot passive treatment Plant. Sci. Total. Environ. 338:73-80.

Hall G, Swash P, Kotilainen S (2005). The importance of biological oxidation of iron in the aerobic cells of the Wheal Jane Pilot passive treatment system. Sci. Total. Environ. 338:41-51.

Johnson D, Hallberg K (2004). Acid mine drainage: remediation options: a review. Sci. Total. Environ. 338:3-14.

Jukes TH, Cantor CR (1969). Evolution of protein molecules. In Munro HN, editor, Mammalian Protein Metabolism. Academic Press, New York. pp. 21-132

Kamala-Kannan S, Krishnamoorthy R (2006). Isolation of mercury resistant bacteria and influence of abiotic factors on bioavailability of mercury – a case study in Pulicat Lake north of Chennai, south east India. Sci. Total. Environ. 367:341-353

Kannan N (2002). Laboratory Manual in General Microbiology. New Delhi, India: Panima Publishing Incorporation.

Kavamura V, Esposito E (2010). Biotechnological strategies applied to the decontamination of soils polluted with heavy metals. Biotech. Advances. 28:61-69.

Koc S, Kabatas B, Icgen B (2012). Multidrug and Heavy Metal-Resistant *Raoultella planticola* Isolated from Surface Water. Bull Environ. Contam. Toxicol. 91(2):177-183.

Lawniczak L, Marecik R, Chrzanowski L (2013). Contributions of biosurfactants to natural or induced bioremediation. Appl. Microbiol. Biotechnol. 97:2327-2339.

MacFaddin J (2000). Biochemical test for identification of medical bacteria. Lippincott Williams and Wilkins Ing. United States, 3a Editon.

Moldes AB, Paradelo R, Vecino X, Cruz JM, Gudiña E, Rodríguez L, Teixeira JA, Domínguez JM, Barral MT (2013). Partial Characterization of Biosurfactant from *Lactobacillus pentosusand* Comparison with Sodium Dodecyl Sulphate for the Bioremediation of Hydrocarbon Contaminated Soil. Biomed. Res. Int. 961842. 6p.

Moreno ME, Acosta-Saavedra LC, Meza-Figueroa D, Vera E, Cebrian ME, Ostrosky-Wegman P, Calderon-Aranda ES (2010). Biomonitoring of metal in children living in a mine tailings zone in Southern Mexico: A pilot study. Int. J. Hyg. Environ. Health. 213(4): 252-258.

Mahjoubi M, Jaouani A, Guesmi A, Ben Amor S, Jouini A, Cherif H, Najjari A, Boudabous A, Koubaa A, Cherif A (2013). Hydrocarbonoclastic bacteria isolated from petroleum contaminated sites in Tunisia: isolation, identification and characterization of the biotechnological potential. New Biotechnol. 30(6):723-33.

Mulligan CN, Yong RN, Gibbs BF, James S, Bennett HPJ (1999). Metal removal from contaminated soils and sediments by biosurfactants surfactin. Environ. Sci. Technol. 33:3812-3820.

Ndip RN, Beeching EC, Ndip LM, Mbacham WF, Titanji VP (2007). Molecular characterization of *pseudomonas aeruginosa* recovered in the Buea health district of Cameroon: implication for nosocomial spread. West. Afr. J. Med. 26(3):191-195.

Pepper IL, Zerzghi HG, Bengson SA, Iker BC, Banerjee MJ, Brooks JP (2012). Bacterial populations within copper mine tailings: long-term effects of amendment with Class A biosolids. J. Appl. Microbiol. 113:569-577.

Rodicio MR, Mendoza MC (2004). Identificación bacteriana mediante secuenciación del ARNr 16S: fundamento, metodología y aplicaciones en microbiología clínica. Enferm. Infecc. Microbiol. Clin. 22 (4):238-245.

Sanger F, Nicklen S, Coulson AR (1977). DNA sequencing with chain-terminating inhibitors. Proc. Natl. Acad. Sci. U. S. A. 74:5463-5467.

Seaker EM, Sopper WE (1988). Municipal sludge for minespoil reclamation: I. Effects on microbial populations and activity. J. Environ. Qual. 17:591-597.

Sen R (2010). Biosurfactants. Springer Science Business Media; Landes Bioscience, New York, N.Y.; Austin, Tex.

Saitou N, Nei M (1987). The neighbor-joining method: A new method for reconstructing phylogenetic trees. Mol. Biol. Evol. 4:406-425.

Sevgi E, Coral G, Gizir A, Sagun M (2009). Investigation of heavy metal resistance in some bacterial strains isolated from industrial soil (J). Turk. J. Biol. 34:423-431.

Soberón-Chávez G, Maier RM (2011) Biosurfactants: a General Overview. *Biosurfactants*, Microbiol. Monogr. 20:1-11.

Talavera O, Armienta MA, García J, Flores N (2006). Geochemistry of leachates from the El Fraile sulfide tailings piles in Taxco Guerrero, southern Mexico. Environ. Geochemis. Health. 28:243-255.

Talavera O, Yta M, Moreno-Tovar R, Almazán A, Flores N, Duarte C (2005). Mineralogy and geochemistry of sulfide-bearing tailings from silver mines in the Taxco, Mexico area to evaluate their potential environmental impact. Geofísica internacional. 44:49-64.

Tamura K, Stecher G, Peterson D, Filipski A, Kumar S (2013). MEGA6: Molecular Evolutionary Genetics Analysis version 6. Mol. Biol. Evol. 30:2725-2729.

Toribio J, Escalante AE, Caballero-Mellado J, González-González A, Zavala S, Souza V, Soberón-Chávez G (2011). Characterization of a novel biosurfactant producing *Pseudomonas koreensis* lineage that is endemic to Cuatro Ciénegas Basin. Syst. Appl. Microbiol. 34(7):531-535.

Umrania V (2006). Bioremediation of toxic heavy metal using acidothermophilic autotrophes. Bioresour. Technol. 97(10):1237-1242.

Wang Q, Garrity GM, Tiedje JM, Cole JR (2007). Naïve Bayesian Classifier for Rapid Assignment of rRNA Sequences into the New Bacterial Taxonomy. Appl. Environ. Microbiol. 73(16):5261-5267.

Wang S, Mulligan C (2009). Rhamnolipids biosurfactant-enhanced soil flushing for the removal of arsenic and heavy metals from mine tailings. Proc. Biochem. 44:296-301.

Whitehead PG, Prior H (2004). Bioremediation of acid mine drainage: an introductory overview of the Wheal Jane Wetlands project. Sci. Total. Environ. 338:15-21.

Will M, Caro AD, Miller RM, Soberón-Chávez G (1997). Chemical behavior of the Wheal Jane bioremediation system Selection and partial characterization of a *Pseudomonas aeruginosa* mono-rhamnolipid deficient mutant. FEMS. Microbiol. Lett. 153:279-285.

Zhang Y, Miller RM (2002). Enhancement of octadecane dispersion and biodegradation by a *Pseudomonas* rhamnolipid surfactant (biosurfactant). Appl. Environ. Microbiol. 58:3276-3282.

Extended-spectrum-beta-lactamase producing uropathogenic *Escherichia coli* infection in Dhaka, Bangladesh

M. Saiful Islam[1], M. Abdullah Yusuf[2], Shaheen Ara Begum[3], AFM Arshedi Sattar[2], Afzal Hossain[4] and Sushmita Roy[5]

[1]Department of Microbiology, Shaheed Suhrawardy Medical College, Dhaka, Bangladesh.
[2]Department of Microbiology, National Institute of Neurosciences & Hospital, Dhaka, Bangladesh.
[3]Department of Microbiology, Comilla Medical College, Comilla, Bangladesh.
[4]Department of Microbiology, ZH Shikder Women's Medical College, Dhaka, Bangladesh.
[5]Department of Microbiology, Enam Medical College, Dhaka, Bangladesh.

Extended-spectrum-beta-lactamase (ESBL) producing *Escherichia coli* that cause urinary tract infection (UTI) is a burning issue. This study was carried out to detect extended spectrum beta lactamase producing *E. coli* isolated from patients presented with UTI. This cross sectional study was conducted in the Department of Microbiology at Dhaka Medical College, Dhaka from January to December 2005, a period of one year. Clinically diagnosed cases of infected (UTI) patients were included in this study. The clean catch mid-stream (CCMU) technique was employed to collect urine sample. Microscopical examination of urine was done and pus cell ≥5/HPF was included in the study. Urine samples were inoculated into blood agar and MacConkey agar media. All the organisms were identified by their colony morphology, staining character, pigments production, haemolysis, motility and other relevant biochemical tests as per standard methods. Antibiogram for all bacterial isolates were done by disc diffusion method of modified Kirby-Bauer technique using Mueller Hinton agar plates. Detection of ESBL producers was performed by double disc diffusion test. Phenotypic confirmatory test was done by E test. A total of 250 samples of urine were collected and within this, 103 (41.2%) samples were shown in positive culture. Out of 103 positive urine samples, majority were *E. coli* (67.0%) followed by *Klebsiella* species (19.4%), *Pseudomonas* species (7.8%) and *Proteus* species (5.8%). Out of 69 *E. coli* isolates, ESBL producers were found in 22 (31.9%) urine samples. The difference between the rate of isolation of *E. coli* with ESBL and other than *E. coli* with ESBL is statistically significant (p=0.0001). *E. coli* strains showed 100.0% resistance to amoxicillin, aztreonam, cefotaxim, ceftazidime, ceftriaxone and cephradine. However, more than 80.0% resistant was observed in cotrimoxazole, amikacin and nalidixic acid. Nitrofurasntoin and mecillinam were more than 50.0% resistant. All strains were sensitive to imipenem. A considerable number of ESBL producing *E. coli* was detected from UTI cases, indicating it as the major challenge for future antibiotic therapy.

Key words: Extended-spectrum-beta-lactamase (ESBL), bacterial agents, urinary tract infection, antimicrobial susceptibility.

INTRODUCTION

Drug resistance is a burning problem in the field of medicine (Altoparlak et al., 2004). There are many ways to mediate resistance, among them beta-lactamases play a major role in developing resistance against Gram negative organisms (Sirot, 1995). During the last three decades many betalactam drugs have been used against the hydrolytic action of betalactamases for the treatment of Gram negative bacterial infections. Microorganisms are gradually developing resistance to these beta-lactam antibiotics by producing betalactamases (Altoparlak et al., 2004). Due to increased spectrum of activity of betalactamases especially against the oxyimino cephalosporins, they are called extended spectrum betalactamases (ESBLs) (Sirot, 1995).

ESBL was first reported in 1983 from Germany in isolates of *Klebsiella pneumoniae* (Shukla et al., 2004). Gradually, many more ESBLs have been described (Bradford, 2001). ESBL producing bacteria are *Escherichia coli, Klebsiella* spp., *Salmonella* spp., *Morganella morganii, Proteus mirabilis, Serrati marcescens, Pseudomonas aeruginosa* and many more. Frequency of ESBL production is high among *E. coli and Klebsiella* species (Nathisuwan et al., 2001). This has created a worldwide problem resulting in treatment failure. ESBLs are enzymes that mediate resistance to extended spectrum (3rd generation) cephalosporins like ceftazidime, ceftriaxone, cefotaxime and monobactams like aztreonam but do not affect 2nd generation cephalosporins such as cephamycin (CDC, 1999).

In United States, occurrence of ESBL production in Enterobacteriaceae ranges from 0 to 25.0% depending on institutions (Bradford, 2001). In Europe, the prevalence is 23.0-25.0% for *Klebsiella* spp. and 5.4% for *E. coli* (Foury and Araj, 2003). In Asia, the proportion of ESBL production in *E. coli* varies from 4.8% in Korea to 8.5% in Taiwan and up to 12.0% in Hongkong (Bradford, 2001). In Singapore, 16.1% *E. coli* are ESBL producers (Chlebicki and Oh, 2004). In India, 58.1% *E. coli* are ESBL producers (Anathakrishnan et al., 2000). In Bangladesh, 26.9-43.2% *E. coli* are ESBL producers (Rahman et al., 2004; Alim, 2005).

This study has been designed to isolate ESBL producing *E coli* from urine samples by double disc diffusion method and to confirm them by E-test ESBL method. Susceptibility pattern of ESBL producing *E coli* were also observed.

METHODOLOGY

This cross sectional study was conducted in the Department of Microbiology at Dhaka Medical College, Dhaka from January to December, 2005, a period of one year. Clinically diagnosed cases of infected (UTI) patients were included in this study.

Urine collection

The clean catch mid-stream (CCMU) technique was employed to collect urine sample. Urine was collected into two sterile containers of which one for microscopy and another for culture (Cheesbrough, 2000). In the case of catheterized patients, urine was collected from the catheter by sterile disposable syringe after proper cleaning of the catheter. Approximately 20 ml of urine was collected aseptically (Cheesbrough, 2000). The containers were properly labeled with patient's name and ID number. The specimens were then transported to the laboratory as quickly as possible.

Microscopical examination

Urine was transferred into a clean and dry 15 ml centrifuge tube and was centrifuged at 1000 rpm for 5 min. The supernatant was discarded and one drop of sediment was taken into a clean glass slide, a cover slip was placed over it and then examined for pus cells under light microscope using 10 and 40x objectives. Samples with pus cell ≥5/HPF were included in the study (Colle et al., 1996).

Culture of urine

Urine samples were inoculated into Blood agar and Mac Conkey agar media by calibrated loop technique (Colle et al., 1996; Hoeprich, 1960).

Isolation and identification of organisms

All the organisms were identified by their colony morphology, staining character, pigments production, haemolysis, motility and other relevant biochemical tests as per standard methods (Cheesbrough, 2000; Colle et al., 1996).

Antimicrobial susceptibility test (Bauer et al., 1966)

Antibiogram for all bacterial isolates were done by disc diffusion method of modified Kirby-Bauer technique using Mueller Hinton agar plates and commercially available antimicrobial disc (Oxoid Ltd. UK). For *E. coli,* the following discs were used amoxicillin (Amx), co-trimoxazole (SXT), gentamicin (CN), amikacin(AK), nalidexic acid (Na), nitrofurantoin (Nf), netilmycin (NET), ciprofloxacin (CIP), cephradine (CL), ceftriaxone (CRO), ceftazidime (CAZ), imipenem (I) and aztreonam (ATM).

For ESBL producers

To see the susceptivility against quinolenoe, fluoroquinolones and cephamycin, the following discs were used: ciprofloxacin (CIP), gatifloxacin (GTX), levofloxacin (LEV), ofloxacin (OFX), sparfloxacin (Sp), lornefloxacin (LOM), pefloxacin (PEF) and cephamycin (CF).

Quality control

Discs from each batch were first standardized by testing against reference strain of *E. coli* ATCC-25922 and zones of inhibition were compared with standard value (Cheesbrough, 2000; Jeong et al., 2004).

Inoculum standardization

With a sterile wire loop, three isolated colonies were transferred to screw-capped tube containing 4 ml of sterile normal saline and standardized with 0.5 Mac Farland standards by adding more organisms or more saline, the solution approximately corresponds to 1.5×10^8 organisms/ml (Cheesbrough, 2000).

Inoculation of test plate and disc placement

Before use, the Mueller-Hinton agar plates were dried in an incubator at 37°C for 30 min. Within 15 min after standardization of inoculums, a sterile cotton swab was immersed into the bacterial suspension. The excess broth was removed by retaining the swab with firm pressure against the inner side of the tube above fluid level. The swab was then streaked evenly on the surface of the plate in 3 different planes by rotating the plate approximately 60° each time to get a uniform distribution of inoculum. The inoculum was allowed to dry for 10-15 min at room temperature keeping the lid closed. The discs were then placed on the inoculum surface by a disposable niddle 15 mm away from the edge of the Petridish and having 20-25 mm gap between the discs. The plates were incubated at 37°C for 24 h (Cheesbrough, 2000).

Reading of the sensitivity tests

After over right incubation, each plate was examined and diameter of the complete zones of inhibition was measured in mm with the help of scale placed under surface of the Petridish. Zones of inhibition were measured in two directions of right angles to each other through the center of each disc and the average of the two readings was taken and compared with the standard (Washington, 1985).

Interpretation of zone

The zone of inhibition in growth produced by each antimicrobial agent on the test organisms was compared with that produced on the control organisms (CLSI, 2007).

Detection of ESBL producers by double disc diffusion test (Jarlier et al., 1988)

Synergy between a disc of augmentin (amoxycillin and clavulamic acid) and 3rd generation cephalosporins was detected as clavulanic acid in augmentin inhibits β-lactamases around 3rd generation cephalosporin disc in agar plate.

Procedure

Mueller-Hinton agar plates were prepared and inoculated with standardized inoculum (compared with 0.5 McFarland standards) by sterile cotton swab. Augmentin disc (20 µg amoxycillin+10 µg clavulanic acid) was placed in the center of the plate. 3rd generation cephalosprins (ceftazidime, ceftriaxone, cefotaxime) and aztreonam discs were placed 20-30 mm apart (center to center) from the augmentin disc. The plate was observed after over night incubation at 37°C (Linscott and Brown, 2005).

Interpretation

Inhibition around the 3rd generation cephalosporins or aztreonam disc were increased towards the augmentin disc or neither disc were inhibitory alone but bacterial growth was inhibited where two antibiotics diffuse together, interpreted as ESBL positive (Shukla et al., 2004).

Quality control

E. coli ATCC-25922 were used as negative control and *K. pneumoniae* ATCC 700603 were used as positive control collected from BSMMU (Jeong et al., 2004).

Phenotypic confirmatory test (Chaudhary and Aggarwal, 2004)

E test ESBL

Principle: E-test ESBL strip consists of a thin, inert and non-porous plastic carrier (5x60 mm). One side of the strip is calibrated with MIC reading scales in µg/ml while the reverse surface carries two predefined exponential gradients. TZ code for ceftazidime (0.5-32 µg/ml) gradient and TZL for ceftazidime (0.064-4 µg/ml) plus 4 µg/ml clavulanic acid. The test is set up like a standard E test MIC procedure, two inhibition ellipses in opposing alignment appears at the end of the strip. The presence of ESBL was confirmed by the appearance of a phantom zone or deformation of the TZ ellipse or when TZ MIC was reduced by >3 log (8 times) dilutions in the presence of clavulanic acid.

Reagents: E-test ESBL strips containing (0.5-32 µg/ml) ceftazidim alone and (0.064-4 µg/ml) ceftazidim plus (4 µg/ml) clavulanic acid were collected from AB BIODISK, Solna, Sweden.

Procedure: Inoculum was prepared and standardized with 0.5 McFarland standard and streaked by sterile swab over the entire agar surface of Mueller Hinton agar plates three times, rotating the plate approximately 60° each time to ensure even distribution of inoculum and kept for 15 min to absorb excess moisture.

Application of E test ESBL strip: With a forcep the ESBL strip was gripped at areas labeled TZ or TZL and placed at the middle of the inoculated plate in such a manner that the whole length of the strip was in complete contact with the agar surface, so that no bubble appeared under the strip and incubation was done at 35°C for 16-18 h.

Reading and interpretation (AB BIODISK, Solna, Sweden): After over might incubation, the plates were examined to see the values. TZ and TZL MIC values were recorded where the respective inhibition ellipses intersect the strip. Growth along the entire gradient, that is no inhibition ellipse indicates that the MIC was greater than the highest value on the reading scale. An inhibition ellipse below the gradient indicates a MIC less than the lowest value on the scale. When mutant colonies are present in the inhibition ellipse, MIC value was read where the colonies were completely inhibited. ESBL production was determined by a ≥3 two

Table 1. Rate of isolation of bacteria from urine samples (n=250)

Isolated bacteria	Frequency	Percentage
Growth of bacteria	103	41.2
No Growth	147	58.8
Total	250	100.0

Table 2. Isolation rate of different bacterial species from urine samples (n=250).

Isolated bacteria	Frequency	Percentage
Esch. coli	69	67.0
Klebsiella species	20	19.4
Pseudomonas aeregenosa	8	7.8
Proteus species	6	5.8
Total	103	100.0

fold concentration decrease in the MIC of TZ in presence of clavulanic acid (Linscott and Brown, 2005). Occasionally a 'rounded' zone (phantom zone) was seen below the TZL gradients while no ellipse was seen around the TZ end. The TZ inhibition ellipse may also be deformed at the tapering end. Presence of a phantom zone or ellipse deformation also indicates ESBL production due to synergy between TZ and clavulanic acid diffusing across from the TZL sections.

Interpretation: ESBL was positive if TZ ≥ 1 and TZ / TZL ≥8 or phantom zone or deformation of TZ ellipse and ESBL producer and resistant to all penicillins, cephalosporins and aztreonam (NCCLS M 100-S series). ESBL was negative if TZ < 1 or TZ /TZL < 8. ESBL non-producer and report actual MIC of all relevant drugs as determined by an MIC method. ESBL was non-determinable (ND) if TZ >32 or TZL >4. ESBL non-determinable and report actual MIC of all relevant drugs as determined by an MIC method. If ESBL is suspected, this was confirmed by CLSI methods and genotyping.

Quality control: *E. coli* ATCC 25922 as negative control and *K. pneumonia* ATCC 700603 as positive control collected from BSMMU were used.

RESULTS

A total of 250 samples of urine were collected and within this, 103 (41.2%) samples were shown in positive culture (Table 1).

Out of 103 positive urine samples majority were *E. coli* (67.0%) followed by *Klebsiella* species (19.4%), *Pseudomonas* species (7.8%) and *Proteus* species (5.8%) (Table 2). Out of 69 *E. coli* isolates, ESBL producers were found in 22 (31.9%) urine samples. The difference between the rate of isolation of *E. coli* with ESBL and other than *E. coli* with ESBL is statistically significant (p=0.0001) (Table 2).

E. coli strains showed 100.0% resistance to amoxicillin, aztreonam, cefotaxim, ceftazidime, ceftriaxone and cephradine. However, more than 80.0% resistant was observed in cotrimoxazole, amikacin and nalidixic acid. Nitrofurasntoin and mecillinam were more than 50.0% resistant. All strains were sensitive to imipenem (Figure 1).

DISCUSSION

ESBLs are the enzymes produced by a variety of organisms like enterobacteriacace as well as *P. aeruginosa* (Bradford, 2001). Failure to detect these enzymes has contributed to their uncontrolled spread and therapeutic failure (Thomson, 2001). In this study, ESBL producing *E. coli* was isolated and the sensitivity pattern was recorded. Regarding that issue, a total number of 250 urine samples were analyzed and among thesem 103 (41.2%) samples was yielded bacterial growth. A study was conducted at Bangabandhu Sheik Mujib Medical University by Mostaquim (2007) who found 69.4% bacteria from various samples which is not consistent with the present study. The reason of low bacterial growth in this study may be use of urine sample only; on the other hand, culture of different samples was performed by the the previous author. In addition to that, majority of the patients with UTI usually take the antibiotic from the pharmacy before coming to hospital and this irrational use of antibiotic may hinder the growth of bacteria from urine (Table 3).

In this study, majority of isolated bacteria were *E. coli* (67.0%) followed by *Klebsiella* species (19.4%), *Pseudomonas* species (7.8%) and *Proteus* species (5.8%). In the present study, *E. coli* were the predominant

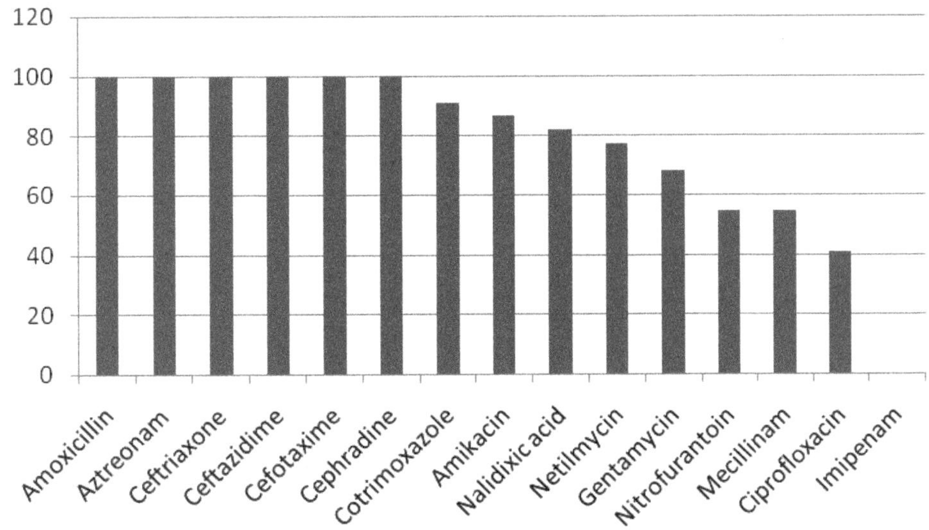

Figure 1. Antimicrobial drug resistance among the ESBL producing *E. coli.*

Table 3. Rate of isolation of ESBL among *Escherechia coli* isolates (n=69).

Isolated bacteria	ESBL producer		Total	p value
	Positive	Negative		
Escherechia coli	22(31.9)	47(68.1)	69(100.0)	
Other bacteria	9(25.0)	25(75.0)	34(100.0)	0.0001
Total	31(30.1)	72(69.9)	103(100.0)	

*Chi-square test has been performed. *Figure within parenthesis indicates percentage.

bacteria isolated from urine samples. From this result, it has been cleared that *E. coli* are the most prevalent bacteria isolated from urine. Similar result was reported by Mansour et al. (2009) and stated that *E. coli* is still the most common cause of UTI. Ferry et al (1988) published that *E. coli* were the most common isolated bacteria from the urine samples of the patients presented with UTI. This result also corresponds with the result obtained by other investigator (Jakobsen et al., 2012). Some have shown, however, that the percentage of *E. coli* is slowly declining, being replaced by other members of the Enterobacteriaceae and enterococci (Weber et al., 1997).

Out of all *E. coli* isolates, ESBL producers were found in 31.9% urine samples. The difference between the rate of isolation of *E. coli* with ESBL and other than *E. coli* with ESBL is statistically significant (p=0.0001). Mostaqim (2007) found 56.7% *E. coli* and among them, 34.1% were ESBL producers which corelate with the findings of the present study. Again, Rahaman et al. (2004) reported that *E. coli* was the predominant organism in urine and among them majority were ESBL producers which are similar to this present study. In another study, Alim (2005) found that 26.9% *E. coli* isolates from urine were ESBL producers which are more or less similar to the present result. In Bangladesh Rahaman et al (2004b) found that 43.20% *E. coli* were ESBL producer which is higher than the present study. The explanation regarding high rate is that they are isolated from pus samples.

In this study, sensitivity pattern of ESBL producing *E. coli* was done against different antibiotics and has shown that *E. coli* strains have 100.0% resistance to amoxicillin, aztreonam, cefotaxim, ceftazidime, ceftriaxone and cephradine (Table 4). However, more than 80.0% resistant was observed in cotrimoxazole, amikacin and nalidixic acid. Nitrofurantoin and mecillinam were more than 50.0% resistant. All strains were sensitive to imipenem. Higher resistance to other antibiotics like cephradine, cotrimoxazole, gentamycin, amikacin against ESBL producing *E. coli* was observed in this study which indicates that ESBL produing organisms are multidrug resistant and genes that code for ESBL are linked to other resistance genes (Ahmed and Salam, 2002). ESBL strains were 100% sensitive to imipenem. According to CDC, ESBL are defined as enzymes which hydrolyze 3rd generation cepholosporins and aztreonam but sensitive to cephamycin and imipenem (CDC, 1999). Nakamura

Table 4. Antimicrobial susceptibility profiles of ESBL *E coli* and non ESBL *E coli* isolates.

Antibiotic name	ESBL *E. coli* (n=22)	Non ESBL *E. coli* (n=47)	OR (95% CI)	P value
Amoxicillin	22 (100.0%)	38 (80.8%)	1.24 (1.08-1.40)	0.049
Aztreaonam	22 (100.0%)	26 (55.3%)	1.81 (1.40-2.34)	0.0001
Ceftriaxone	22 (100.0%)	28 (59.6%)	1.68 (1.33-2.12)	0.0001
Ceftazidime	22 (100.0%)	29 (61.7%)	1.62 (1.29-2.03)	0.0001
Cefotaxime	22 (100.0%)	30 (63.8%)	1.57 (1.26-1.94)	0.001
Cephradine	22 (100.0%)	42 (89.4%)	1.12 (1.12-1.23)	0.169
Cotrimoxazole	19 (86.4%)	33 (70.2%)	0.37 (0.095-1.46)	0.231
Amikacin	11 (50.0%)	41 (87.2%)	6.83 (2.06-22.61)	0.001
Nalidixic Acid	18 (81.8%)	28 (59.6%)	0.327 (0.096-1.12)	0.058
Netilmycin	16 (72.7%)	33 (70.2%)	0.884 (0.286-2.73)	0.830
Gentamycin	15 (68.2%)	24 (51.1%)	0.487 (0.168-1.41)	0.181
Nitrofurantoin	12 (54.5%)	17 (36.2%)	0.472 (0.169-1.32)	0.150
Mecilinum	12 (54.5%)	18 (38.3%)	0.517 (0.186-1.44)	0.205
Ciprofloxacin	8 (36.4%)	18 (38.3%)	1.09 (0.380-3.10)	0.887
Imipenem	0 (0.0%)	0 (0.0%)	NA	NA

and Komatsu (2005) found that 61.1% *E. coli* were ciprofloxacin resistant among the ESBL producers which is similar to the present study.

When ESBL producing organisms are confirmed by NCCLS guidelines, results should be reported as resistance to all penicillins, aztreonam and cephalosporins excluding cephamycin (CDC, 1999). Treatment of ESBL producing *E. coli* can be done by imipenem or cephamycin. Imipenem is costly and not within the reach of the people of developing country like Bangladesh. But quinolone (ciprofloxacin) and fluoroquinolones (levofloxacin, gatifloxacin etc) are cheap, available and single dose drug which be may be used in the treatment against ESBL producing organisms. So early correct detection of ESBL producing *E. coli* by E test ESBL method and rational use of antibiotics can limit the spread of multidrug resistant pathogens (Medeiros, 1993).

Conclusion

A considerable number of ESBL producing *E. coli* from urine samples were isolated. Sensitivity was higher in the case of imipenem as these are expensive and dose inconvenient so may be used for reserve antibiotics. Awareness regarding proper identification of ESBL producing *E. coli* should be carried out for appropriate antibacterial agent against them can reduce the hospital stay and sufferings of the patients.

Conflict of Interests

The author(s) have not declared any conflict of interests.

REFERENCES

Ahmed I, Salam A (2002). Extended Spectrum β-lactamases and Bacterial Resistance. Pak. J. Med. Sci. 18(2):151-155.

Alim R (2005). Detection of Extended Spectrum β-Lactamases (ESBL) Producing Bacteria. [M.Phil Thesis, Microbiology] BSMMU, Dhaka.

Altoparlak U, Erol S, Aklay MN, Celebi F, Kadanali A (2004). Time related changes of antimicrobial resistance patterns and predominant bacterial profiles of burn wounds and body flora of burned patients. Burns 30:660-664.

Anathakrishnan AN, Kanungo R, Kumar A, Badrinath S (2000). Detection of Extended Spectrum β-lactamase producers among Surgical wound infections and Burn Patients in Jipmer. Indian J. Med. Microbiol.18(4):160-165.

Bauer AW, Kirby WMM, Sherris JC, Turck S (1966). Antibiotic susceptibility testing by a standardized single disc method. Am. J. Clin. Path 36:493-496

Bradford PA (2001). Extended spectrum β-lactamases in the 21st century: Characterization, epidemiology and detection of this important resistance threat. Clin. Microbiol. Rev. 14(4):933-951.

CDC (1999). Laboratory Detection of Extended Spectrum β-lactamases (ESBL). http://www.cdc.gov/ncidod/hip

Chaudhary U, Aggarwal R (2004) Extended spectrum β-lactamases (ESBL) – An emerging threat to clinical therapeutics. Indian J. Microbiol. 22(2):75-80.

Cheesbrough M (2000). District laboratory practice in tropical countries, ELBS, Cambridgeshire, England, 2:175-180.

Chlebicki MP, Oh HML (2004). Extended Spectrum Beta-Lactamases in Clinical Isolates of *Escherichia coli* and *Klebsiella* spp. in a Singapore Hospital: clinical spectrum. Ann. Acad. Med. 33(3):302-306.

CLSI (2007). Performance standards for antimicrobial susceptibility testing. NCCLS approved standard. M100-S9,; Wayne, P A.

Colle JG, Miles RS, Watt B (1996). Tests for the identification of bacteria. In: Collee JG, Fraser AG, Marmion BP, Simmons A. editors, Mackie and McCartney Practical Medical Microbiology, 14th ed. 1996; Churchhill Livingstone, New York. pp. 553-559.

Ferry S, Burman LG, Holm SE (1988). Clinical and bacteriological effects of therapy of urinary tract infection in primary health care: relation to *in vitro* sensitivity testing. Scandinavian J. Infect. Dis. 20(5):535-44.

Foury K, Araj GF (2003). Recent developments in β-lactamase and Extended spectrum β-lactamases. Bangladesh Med. J. 327:1209-1213.

Hoeprich PD (1960). Culture of the urine. J. Lab. Clin. Med. 56(6):899-906.

Jakobsen L, Cattoir V, Jensen KS, Hammerum AM, Nordmann P, Frimodt-MÃ¸ller N. (2012). Impact of low-level fluoroquinolone resistance genes qnrA1, qnrB19 and qnrS1 on ciprofloxacin treatment of isogenic Escherichia coli strains in a murine urinary tract infection model. J. Antimicrob. Chemother. 67(10):2438-2444.

Jarlier V, Nicolas MH, Fournier G, Philippon A (1988). ESBLs conferring Transferable resistance to newer β-lactam agents in *Enterobacteriaceae*: hospital prevalence and susceptibility patterns. Rev. Infect. Dis. 10:867-878.

Jeong SH, Bae KII, Lee JH, Sohn SG, Kang GH, Jeon GJ, Kim YH, Jeong BC, Lee SH (2004). Molecular Charecterization of Extended Spectrum β-Lactamase Produced by Clinical Isolates of *Klebsiella pneumoniae* and *E. coli* from Korean Nationwide Survey. J. Clin. Microbiol. 2902-2906.

Linscott AJ, Brown WJ (2005). Evaluation of Four Commercially Available Extended Spectrum Beta-Lactamase Phenotypic Confirmation Tests. J. Clin. Microbiol. 43(3):1081-1084.

Mansour A, Manijeh M, Zohreh P (2009). Study of bacteria isolated from urinary tract infections and determination of their susceptibility to antibiotics. Jundishapur J. Microbiol. 2(3):118-23.

Medeiros AA (1993). Nosocomial outbreaks of multiresistant bacteria: extended-specreum beta-lactamase have arrived in North America. Ann. Intern. Med. 119(5):428-429.

Mostaquim R (2007). Rapid Detection of Extended Spectrum Beta-Lactamases (ESBL) Production Directly From Primary Culture. M.Phil.Thesis, BSMMU

Nakamura T, Komatsu M (2005). Susceptivility of ESBL producing *E. coli* and *Klebsiella pneumoniae* to various antibacterial agents. Japanese J. Antibiot. 58(1):1-10.

Nathisuwan S, Burgess DS, Lewis IIJS (2001). ESBLs: Epidemiology, Detection and Treatment. Pharmacother. 21(8):920-928.

Rahman M, Haq JA, Hossain MA, Sultana R, Islam F, Islam S (2004). Prevalence of extended-spectrum β-lactamase- producing *Escherichia coli and Klebsiella pneumoniae* in an urban hospital in Dhaka, Bangladesh. Int. J. Antimicrob. Agents 24:508-510.

Rahman M, Shoma S, Rashid H, Siddique AK, Nair GB, Sack DA (2004b). Extended-spectrum Î²-lactamase-mediated third-generation cephalosporin resistance in Shigella isolates in Bangladesh. J. Antimicrob. Chemother. 54(4):846-847.

Shukla I, Tiwari R, Agrawal M (2004). Prevalence of Extended Spectrum β-Lactamase-Producing *Klebsiella pneumoniae* In Tertiary Care Hospital. Indian J. Med. Microbiol. 22(2):87-91.

Sirot D (1995). Extended-spectrum plasmid-mediated β-lactmases. J. Antimicrob. Chemother. 36:Suppl A S19-34.

Thomson KS (2001). Controversies about Extended-spectrum and AmpC β-lactamases, Emerg. Infect. Dis. 7(2):333-336.

Washington JAII (1985). Susceptibility tests. In: Lennette E H, Balows A, Hauser WJ. Eds. *Manual of clinical microbiology*. 4th ed. American Society of Microbiology. Washington D C. pp. 967-971.

Weber G, Riesenberg K, Schlaeffer F, Peled N, Borer A, Yagupsky P (1997). Changing trends in frequency and antimicrobial resistance of urinary pathogens in outpatient clinics and a hospital in Southern Israel, 1991-1995. European J. Clin. Microbiol. Infect. Dis.16(11):834-838.

Antibacterial activities of two medicinal herbs on *Salmonella typhi* isolates in Abakaliki, Ebonyi State, Nigeria: Improvement to herbal medicine

Odikamnoro O. O.[1], Uhuo C. A.[1], Ikeh I. M.[1], Ogiji E. D.[2], Ibiam, G.A.[3], Azi, S.o.[3], Akpam L. J.[2] and Okoh N. F.[4]

[1]Department of Applied Biology, Ebonyi State University, Abakaliki, Ebonyi State, Nigeria.
[2]Department of Pharmacology/Therapeutics, Faculty of Basic Medicine, Ebonyi State, University.
[3]Department of Medical Laboratory, Ebonyi State University, Abakaliki, Ebonyi State.
[4]Department of Biological Sciences, Parasitology Unit, Evangel University, Abakaliki, Ebonyi State, Nigeria.

The study on the antibacterial activities of some medicinal plants on *Salmonella typhi* isolates in Abakaliki Ebonyi State was conducted using agar diffusion method. The study shows that there is high level yield in the dried leaves than in the wet leaves of the experimental plants (W/D). The investigation reveals that phenol yield is 296% W and 67% D in bitter leaf while oil yield in Utazi is 1.900 W and 1.034 D; phenol is 276.00 W and 38.480 D. The phytochemical analysis showed the presence of alkaloids, tannins, glycosides, flavonoids, steroids, phenolic compounds, oil, proteins, terpenoids, fructose, glucose and essential oils. Antimicrobial test screening showed that the extract had a strong inhibitory effect against; *Staphylococcus aureus, Escherichia coli,* and *Klebsiella pneumonia.* The observation tends to confirm the medicinal value of these plant leaves and justify the claim of its local use in health remedies for diarrhea and kidney failure caused by *E. coli,* skin infection caused by *S. aureus* and pneumonia which is caused by *K. pneumonia*. However, this had proved inhibitory on *S. typhi* which could be synergistic or antagnostic in action and therefore suggest to be effective herbally when orally administered.

Key words: Antibacterial, medicinal herbs, *Salmonella species*, Ebonyi State.

INTRODUCTION

Medical herbs are plants which contain substance that can be used for therapeutic purpose or which are the precursor for the synthesis of useful drugs (Sato et al., 1997). Also herbs can be described as those plants which are valuable in medicinal production and could be used as food seasoned nutrition and other useful purpose to man (Byarygaba, 2004). Although, the practice of traditional medicine differs according to the locality, roots from which the certain geographical areas or climate makes traditional medicine a more specific indigenous cultural practices (Walker, 2001).

In Nigeria, the use of local herbs in the form of concoction and decortion treatment of various infections has been a common practice and about 75% of the people living in the rural areas depend on this practice (Yoshida et al., 1998). For instance, *Uvaria chamas* has been used in the South-East of Nigeria to treat feverish condition, malaria and body pains (Amadi et al., 2003). Other herbs such as the *Azadirachta indica, Aleovera* spp*., Psiduim guava, Carica papaya* and *Cymbopogen citrates* are used for various ailments such as malaria fever, headache, body pains, skin infections, stomach, diarrhea, etc simply by use of different methods of applications.

Moreover, it will serve as a cheaper means of treatment to the low income earner patients that cannot afford the cost of modern therapeutic used for treatment of typhoid fever. The germs *Salmonella* is among the most common cause of foods and water borne infectious disease in the world. A number of studies in Nigeria have shown that *Salmonella* infection is endemic in many parts of the country and its endemicity increases especially in areas with low environmental hygiene, and rural communities with without drinking water.

Garlic and onion is a common food for flavor and spice. Garlic contain a lot of health chemical compounds that are antioxidants, they reduce blood pressure, lower cholesterol, fight bacterial, fungal and viral infections, even detoxify cancer and protect liver from toxins including alcohol and synthetic drugs (Chen et al., 2002), (Aureli et al., 2006). *Allicin,* a penicillin like antibiotic hinders the growth of *Staphylococcus aureus and Salmonella typhi.* Another component of garlic called ajoene, has been found to thin the blood and contribute to lower cholesterol level (Mbata et al., 2006).

The onions bulb has been chemically analyzed and it is shown to contain an acrid, volatile oil, uncrystallizable sugar, gum, albumen, acetic and phosphoric acids, phosphate and citrate of calcium and water (Byargada et al., 1996). Onion possess properties allied to those of garlic, but in a milder degree, medically, it has been noted to be antihelminths, anti-inflammatory, antiseptic, antispasmodic, hypoglycaemic and tonic; sugar and onion juice form a syrup, much used in domestic practice for cough and other infections of the air-tube among children (Nweze et al., 2002). It is also useful in preventing oral infection and tooth decay. Fresh onion juice is a very useful first aid treatment for bee and wasp stings and bite of scorpion. Apart from all these usefulness of onion, it is commonly used for cooking of dishes, eating raw in making salad.

The bitter leaf, *Vernonia amygdalina* which is the type used in this research; contains the following chemical compound which include vernonioside which is said to have antimalaria effect, sesquiterpene and lactone that have anti tumor activities.

Traditionally, bitter leaf are commonly used for cooking of soup and porridge, but medicinally, it has been extensively used for the treatment of stomach ache, parasitic infection and has proven to be effective against drug resistant malaria parasites (Ohigashi et al., 1994).

Laboratory identification of salmonella can be typically isolated from stool of a diarrhea patient on MacConkey agar or moderately selective media for patient with enteric fever, this organism can be isolated from their blood, urine and bone narrow specimen. This study is aimed to determine the viability of some plants which have the ability to inhibit the growth of *S. typhi* and to determine the level at which these herbs can be used locally in prophylactic treatment.

MATERIALS AND METHODS

The herbs used include *V. amygdalina, Allium sativum and Allium cepa.* All these herbs except *A. sativum* were purchased from Abakpa Main Market Abakaliki, and *A. cepa* were also purchased at local market in Igbeagu Development Centre whereas *V. amygdalina,* the bitter species was collected from various home and farmland at Nkaliki Community in Abakaliki Local Government Area all in Ebonyi State. It was confirmed that the cultivation of *A. cepa* and *A. sativum* was common in the Northern part of Nigeria, from where the greater quantities were purchased from the sellers. The identification of these herbs was done by Professor S. S. C. Onyekwelu (taxonomist) of Applied Biology Department of Ebonyi State University, Abakaliki, Nigeria.

Preparation of materials

The herbs were washed with clean sterile distilled water and allowed to air dry for one hour before grinding them into powder using pestle and mortar. After this, blended herbs was collected and 40 g of each was measured and introduced into 200 ml of 95% ethanol, cold and hot water (Aureli et al., 2006), (Walker, 2000).

Ethanol extraction

40 g each of these herbs was weighed and introduced into 200 ml of 25% ethanol. This was allowed to stay for 24 h with intermittent shaking. After this period, they were filtered and the liquid extract was allowed to evaporate so that a concentrated extract will be obtained.

Cold water extraction

The same procedure described above was adopted for both cold and hot water extraction, but the extraction of *A. cepa* and *A. sativum* took the following processes.

Extraction of *A. cepa* (onion)

The onion were washed with distilled water and allowed to air dry for one hour. The outer covering of the onion were manually peeled off. The onion bulb being separated were washed and extracted in the following ways:

Table 1. The Antibacterial effect of the extracts against *Salmonella typhi.*

Botanical Names	Common names	Parts used	Results of the extracts		
			Cold H_2O	Hot H_2O	Ethanol
Allium sativum	Garlic	Bulbs	Sensitive	N.I	Sensitive
Allium cepa	Onion	Bulbs	Sensitive	N.I	Sensitive
Vernonia amygdalina	Bitter leaf	Leaves	N.I.	N.I.	N.I.

N.I. = No inhibition to the growth of the organism; sensitive means that extract inhibited growth of the organism.

1. Exactly 200 g of fresh onion bulbs were blended into fine powder and soaked in 100 ml of distilled water for 24 h. The pulp obtained was left in a clean sterile container and shaken vigorously to allow for proper extraction and it was filtered using a sterile muslin cloth after which the extract was obtained, air dried and stored below ambient temperature until required.
2. Exactly 200 g of fresh onion bulb were blended and soaked in100 ml of hot water for 24 hr; the resultant juice was extracted, air dried and stored.
3. Exactly 200 g of fresh onion bulb were blended and soaked in100 ml of 95% ethanol for 24 h and the extract was obtained, air dried and stored.
4. Exactly 200 g of fresh onion bulb were blended and the raw juice was extracted after standing in a clean glass for 25 h, it was extracted using a sterile muslin cloth and extract was air dried and stored.

Extraction of *A. staivum* (garlic)

The extraction of *A. sativum* is the same as the extraction of onion as stated above, whereas, the *V. amydalina* extract was obtained by squeezing the bitter leaf which released the pigment into a sterile bottles, respectively.

Media preparation

The media used includes Muller Hinton agar and deoxychocolate citrate agar (DCA). The media was prepared as instructed by the manufacturer. This was mixed very well before sterilizing it, using oven for 121°C for 15 min. It was allowed to cool for about 140°C before pouring about 20 ml into each of sterile petri dish and then allowed to gel before incubating them for 24 h to test for sterility.

Organism inoculum preparation

In the first eighteen hours, fresh culture of the organism *S. typhi* was used. The colony was collected from the media with sterile wire loop and introduced into the test tubes containing 2 ml of distilled water.

This was dissolved and the same turbidity with that of Macfarland solution standard (0.5) was obtained which made organism solution to yield 1 x 10^8 cells colony units.

Then about 0.2 ml of the diluted organism solution which is equivalent to two drops of the solution was added into the molten Muller Hinton. This was spread all over the media with the help of the glass spreader. After this, the sterile cork borer was used to make about five cups well (5 mm in diameter) in the prepared media, for the inoculation of the extract, which its procedure of preparation is described.

RESULTS

The sensitivity of the tested organism, *S. typhi* to the cold H_2O, hot H_2O and ethanol extract of these herbs, were evaluated by the cup plate agar diffusion method. The following plants and its parts were used for the sensitivity test against the organism. Table 1 shows result of the three herbs used, it was only two namely, garlic and onions that showed inhibitory effects on *S. typhi.*

The zone of inhibition of the control drug chloramphenicol differs depending on the herbs. In *A. sativum*: 29 mm with 0.4 g/ml while *A. cepa* is 23 mm. The result of the effect shows that *A. sativum* has the highest inhibitory effects on the tested organism followed by *A. cepa*. Also, the MIC of *A. sativum* is 0.4 g/ml, *A. cepa* also has 0.4 ml/ml (Table 2).

The combination of the extracts showed that when garlic was mixed with onion using the same concentration, 0.2 ml, its zone of inhibition was 28 mm which is higher in different zone of the concentration of the mixture. But when the three herbs, onion, garlic and bitter leaf were mixed, there was no reaction, the factor responsible for this susceptibility of *S. typhi* to the extracts are not exactly known but may be attributed to the presence of secondary plants metabolites. The diameter of the concentration of drugs CAC was 0.4 g/ml and it zones of inhibition was 34 mm (Table 3).

DISCUSSION

The sensitivity of the test organism to the cold water, hot water and ethanol extract of the three different herbs were evaluated using the cup plate agar diffusion method (Mbata et al., 2006). From the result obtained, it is shown that among the herbs used, it was only two, namely, *A. cepa* and *A. sativum* juice that shows inhibitory effect on the test organism *S. typhi,* whereas the remaining one herb were inactive to the organism. This has not only confirmed the facts that *S. typhi* possesses spores, unlike other species, and this peculiar feature has contributed to its stability to be highly resistant to many antibiotic such as carbenicillin and ciprofloxacin (De Pasquale et al., 2001; WHO, 2006).

Table 2. The MIC and inhibition zone diameter of the extracts on *S. typhi.*

Plant extract concentration (g/ml)	Inhibition zone/diameter (mm)	
	Allium sativum	*Allium cepa*
0.2	20 mm	11 mm
0.4	28 mm	13 mm
0.6	20 mm	10 mm
0.8	19 mm	9 mm
0.4. Drug control (CAC)	29 mm	23 mm

Table 3. Combination of extracts and its zone of inhibition.

Allium sativum + Allium cepa	
Zone of inhibition/diameter (mm)	Different concentration (g/ml)
28 mm	0.2 ml
11 mm	0.4
12 mm	0.6
34 mm	0.4 ml Drug control (CAC)

It was clear from this study that solvent extraction affected the degree of antibacterial activity of the extracts. It was also observed that the ethanolic extract of garlic gave the highest zone of inhibition (28 mm) using the concentration of 0.4 g/ml while ethanolic extract of onion gave 13 mm with 0.4 g/ml each against *S. typhi.* This credit to ethanol extraction was because it is an organic solvent and will dissolve organic compounds better, hence the active component required for antimicrobial activity.

It was also observed that raw onion and garlic had activity on *S. tyhpi.* The reason for this is not clear because the raw juice is thought to be more concentrated than the other extract since in some, like when *Escheichia coli* was used, there was no inhibition at all.

The hot water extracts of onions did not inhibit the growth of *S. typhi.* This may be explained by the fact that the antimicrobial substance in the onions extracts, which are mainly phenolic compound are destroyed by heat from the hot water which might have raised the temperature of the extracts inactivating them. In general, antibacterial components in the species plant are heat-labile; hence all spices lost their antibacterial activities by 20 min at 100°C.

The cold water extract inhibited the growth of *S. typhi* at 0.4 g/ml. It could be said that cold water as an extract could liberate the active constitution of garlic better as compared to onions. It is worthy to note that the antimicrobial activities of these plants extracts were dependent on the concentration of the extracts as reported early. Also, if the extract has high molecular weight, the rate of diffusion is always slow, reduced and also takes longer time, whereas of low molecular weight diffuses faster and at a quicker rate. The result obtained is evidence that garlic produced marked inhibitory effect on the tested organism.

Further, the allicin compound is an oily yellow liquid which was assumed to give garlic its characteristic odour. It also has a range of medical properties like synthesis when garlic is crushed this reaction is catalyzed by an enzymes alliinase. This allicin is responsible for the growth inhibition of *S. typhi, S. aureus* and other bacteria growth as reported by De Pasquale et al. (2001). It was observed that *A. sativum* in this study showed activity against the tested organism with minimum inhibition concentration (MIC) of 0.4 g/ml. This is an indication of possible high activity, since this tested organism is reported to be resistance to most antibiotic and herbs. Lastly, the *V. amydalina* recorded the least activity against this organism among the three herbs, since it show no zone of inhibition at the same concentration of 0.4 g/ml. Their individual zone of inhibition which has been recorded: 28 and 13 mm respectively to garlic and onion suggest an antagonistic effect.

Conclusion and recommendation

The effect of *A. cepa and A. sativum* in this study has shown that they really possess the same properties that are highly of antimicrobial importance. This study tends to collaborates the local use of these herbs in the treatment of diseases for instance, lime juice used to treat stomach upset, dairrhoea and skin related diseases, whereas garlic has been used to treat cancer, malaria, candidiasis and was able to inhibit the growth of *S. aureus,* and the result of this research showed that these herbs have high activity against *S. typhi.*

Therefore, it is recommended that efforts should be geared towards identifying the active ingredient of these herbs and other unknown medicinal herbs, since they are common to obtain, cheap and available in our environment and has a lesser side effect when used for the treatment of various diseases.

However, people should increase the use of spices such as the garlic and onion in preparation of their dishes and eat them raw, likewise the use of *V. amydalina* should be encouraged in the family because of the role it plays in the liver. All these will help to prevent and equally inhibit the manifestation of *Salmonella* infection in the body system of an individual.

Conflict of interests

The authors have not declared any conflict of interests.

REFERENCES

Amadi ES, Nwakpu KO, Nworie O (2003). Effect and Interaction of Local Medicinal Herbs on Microorganism. J. Biomed. Sci. Afr. 5:42-67.

Aureli P, Costantini A, Zolea S (2006). Antimicrobial activity of some plant essential oils against listeria monocytogenes. J. Food Prot. 55:344-348.

Byarygaba DK (2004). A review on Antimicrobial resistance in Developing Countries and responsible risk factors. Int. J. Antimicrob. Agent 24:367-375.

Chen SJ, Lee TE, Wang EM, Cho TJ Wang CH (2002). Monitoring hygiene of chicken hatcheries in Taiwan during 1999-2001. J. Microbial Immunol. Infect. 4:236-242.

De Pasquale R, Germano MP, Keita A, Sanogo R, Lauk U (2001). Anti ulcer activity of pteleopsis suberosa. J. Enthnopharmcol. 47:55-558. http://dx.doi.org/10.1016/0378-8741(95)01256-D

Mbata TI, Orji MU, Anyaegbunam BC, Ike NP (2006). Presence and resistance of Salmonella in commercial broiler hatchery in Awka, Nigeria, Nigeria J. Microbiol. 20:911-916.

Nweze EI, Okafor JI, Njoku O (2004). Antimicrobial activities of methanolic extracts of Tremaguinesis (Schumm and Thorn) and Moringa Lucida Benth used in Nigeria Herbal medicinal practice. J. Biol. Res. Biotechnol. 2:39-49.

Ohigashi H, Yeh YY (1994). Garlic extract reduce plasma concentration of homolysteine in rat rendened folic acid deficiency. FASEB J.13:209-306.

Sato Y, Odetani H, Singyouchi K, Ohtsubo T, Kihara M, Shibata H, Higuti T (1997). Extraction and purification of effective antimicrobial constituents of Terminalia Chebula RETS against methicillin-resistant Staphylococcus aureus. J. Biol. Pharm. Bull. 20:401-404. http://dx.doi.org/10.1248/bpb.20.401

Walker CB (2000). Microbiological effects of mouth rinses containing antimicrobial. J. Clin. Periodontol. 15:199-505.

World Health Organization (2006). Global strategy for containment of antimicrobial resistance. Geneva, Yoshida S, Kasuga S, Hayashi N, Ushiroguchi T, Matsuura H. Nakagawa S (1998). Antifungal activity of ajoene derived from garlic J. Appl. Environ. Microbial. 53:615-617.

Molecular and biochemical study of *Anaplasma marginale* in cattle in Wassit Province of Iraq

Ghaidaa Abbas Jassem[1] and Olaa A. Agaar[2]

[1]Department of Microbiology, Veterinary Medicine College, AL-Qadissiya University, Iraq.
[2]Department of Microbiology, Veterinary Medicine College, Wasit University, Iraq.

This study was designed to investigate the prevalence of bovine anaplasmosis among cattle from various areas in Wassit governorate; the investigation was performed on 184 blood samples collected from suspected cattle suffering from fever (41°C), severe anemia, pale mucus membrane, progressive emaciation and drop in milk yield, including 85 male and 99 female cattle, aged from < 1 year to > 2 years, the samples were collected during the period of October 2012 - April 2013 from AL-Kut, AL-hayy, AL-Bashair, AL-Moufaqia and AL-Noamania areas to investigate antibodies against Anaplasma parasite by using indirect enzyme-linked immunosorbent assay (ELISA) test and to determine the species of genus *Anaplasma* by using RFLP-PCR technique and by also measuring some biochemical parameters to indicate the effect of the disease on liver function. The results of ELISA test showed that the rate of infection was 13.04%, the rate of infection was different between age groups and were 8, 11.25 and 16.45% in ages < 1, 1-2 and 2 - 3 years, respectively. The study revealed that females were given higher percent of infection 14.14% than males 11.7%, there is no significant differences under p > 0.05 according to age groups and sex. The highest rate of infection was recorded in AL-Kut, 17.14% followed by AL-hayy, 14% and AL-Bashair, 10% and the lowest rate was recorded in AL-Noamania, 8.33% and AL-Moufaqia, 5%; the study showed significant differences in incidence of disease between study districts and area in Wassit governorate at p > 0.05. The most sensitive method for the diagnosis of anaplasmosis is the method of polymerase chain reaction, DNA extraction was performed only on 24 blood samples which were positive for *Anaplasma* spp. by ELISA test, the extracted DNA from blood cells were analyzed by PCR and PCR-RFLP technique using primers derived from 16S rRNA gene and restriction endonuclease *Bst*1107I enzyme which can recognize the sequence (GTATAC) in corresponding PCR product of *Anaplasma marginale* and cut it in the position 68 and 509, whereas the used restriction enzyme cannot cut the corresponding PCR product of other *Aanaplasma* spp. and the result was 20 from 24 which were positive for *Aanaplasma* spp. by PCR and 18 from 20 was positive for *A. marginale*. The results of liver enzymes activity showed significant increase in serum AST, ALT, CK and TBIL level in infected cattle (96.8±0.97, 42±0.52, 406±2.06 and 0.95±0.24, respectively) as compared to the control (65.5±1.26, 21.4±0.45, 142±14.17 and 0.27±0.05, respectively).

Key words: *Anaplasma marginale*, indirect enzyme-linked immunosorbent assay (ELISA), RFLP-PCR, biochemical parameters.

INTRODUCTION

Bovine anaplasmosis is one of the important hemoparasitic tick-borne disease of cattle and other ruminants with tropical and sub-tropical regions of the world, caused by intraerythrocytic parasite of the genus Anaplasma (order: Rickettisale; family: Anaplasmataceae) (Dumler et al., 2001). Based on location within the infected erythrocyte, two species of *Anaplasma* that infect cattle and cause bovine

Table 1. Oligonucleotide sequences that were used for PFLP-PCR.

Primer	sequence		PCR product size
16S rRNA	F	AGAGTTTGATCCTGGCTCAG	577 bp
	R	GTTAAGCCCTGGTATTTCAC	

anaplasmosis have been described, *Anaplasma marginale* (Theiler, 1910) and *Anaplasma centrale* (Theiler, 1911). The most common etiological agent is *A. marginal*e which cause acute anaplasmosis and responsible for severe morbidity and mortality in temperate tropical and subtropical regions worldwide (Palmer et al., 2000). According to Theiler (1911), *A. centrale* is less pathogenic to cattle than *A. marginale* but, most importantly, gives resistance against the latter; hence it is used for the preparation of live vaccine strains, assuring immunological protection against bovine anaplasmosis, such vaccines are produced in Africa, Australia, Latin America (Kocan et al., 2003). Bovine anaplasmosis occurs in tropical and sub-tropical regions of the world including 40 states of the USA, south and Central America, Asia, Africa, southern Europe, Middle East and Australia (Cynthia and Susan, 2008). The infectious agent is transmitted either biologically by ticks or mechanically via contaminated mouth part of biting insects or by contaminated fomites such as needles, castrating knives, ear taggers and other surgical instruments (Ewing et al., 1997).

Bovine anaplasmosis can be seen at any age group in cattle, however, the severity of disease and death rate increase with the advance in age, with clinical anaplasmosis being more commonly in cattle older than 1 year of age (Kocan et al., 2003). Acute anaplasmosis is characterized by a progressive hemolytic anemia associated with fever, weight loss, abortion, decreased milk production and in some cases death of infected cattle (Kocan et al., 2000). Animals that recover from acute anaplasmosis become persistently infected with *A. marginale* and serve as reservoir and source of infection within the herd (Aubry and Geale, 2011). The daignosis of bovine anaplasmosis is performed routinely by Giemsa stained blood smears which can indeed be used as a suitable method to detect *Anaplasma* in the animals clinically suspected for acute diseases, but it is not applicable to the determination of pre-symptomatic and carrier animals (Carelli et al., 2007). Therefore, several serological tests have been used to measure *Anaplasma*-specific antibodies, including the complement fixation test, card agglutination test, indirect fluorescent antibody test and enzyme-linked immunosorbent assay (ELISA) (Goff et al., 1990). Unfortunately, because of antigen cross reactivity, these tests do not discriminate between different *Anaplasma* species (Dreher et al., 2005). PCR is an alternative diagnostic technique, which offers a high degree of sensitivity and specificity, and has been developed to identify *A. marginale* DNA in the blood of infected animal even if present in very low numbers (Carelli et al., 2007).

MATERIALS AND METHODS

Blood sample collection

From October 2012-April 2013, 184 venous blood samples were collected from cattle suspected to have anaplasmosis aged from less than 1 year to 2 - 3 years and based on the clinical manifestation of severe anemia and jaundice from districts and the areas of Wasiit governorate (AL-kut, AL-hayy, AL-Bashair, Al-Moufaqia and Al-Noamania). 5 ml of blood samples were collected from the jugular vein into 2 vaccutainer tubes, one with anticoagulant for PCR, and the other without for the ELISA.

Serum preparation

The serum was separated by centrifugation at 3000 rounds per minute for 10 min, then aspirated carefully by pipette into dry, sterile and labeled test tubes, which were stored frozen then transported under cold conditions to the laboratory of Al-Karama hospital, where indirect ELISA test was conducted for detection of *A. marginale* antibodies using an *A. marginale*-Ab ELISA Kit (Svanova Biotech AB, Sweden).

Primers

The 16S rRNA primers used in REFLP-PCR for genotyping of *Anaplasma* spp. were designed by Noaman et al. (2009) and provided by Bioneer company, Korea and are listed in Table 1.

Serological analysis

A. marginale-Ab ELISA Kit (Svanova Biotech AB, Sweden), was used to detect specific antibodies against *A. marginale* in bovine serum samples. This technique was done according to the manual.

Molecular analysis

PCR followed by RFLP was performed for detection and genotyping of *Anaplasma* spp. in blood samples that were give positive results

Table 2. Seropositivity rate of bovine anaplasmosis based on ELISA.

No. of examined cattle	No. of ELISA positive cases	Infection rate (%)
184	24	13.04

Table 3. Infection rates according to age based on ELISA.

Age group	No. of animals examined	No. of ELISA positive cases	Infection rate (%)
Less than 1 year	25	2	8a
1-2 years	80	9	11.25a
Higher than 2 years	79	13	16.45a
Total	184	24	13.04

Similar letters refers to the non-significant differences between ages (p <0.05).

Table 4. Infection rates according to sex of animals based on ELISA.

Cattle	No. of animals examined	No. of ELISA positive cases	Infection rate (%)
male	85	10	11.76a
Female	99	14	14.14a
Total	184	24	13.04

The similar letters refers to the non-significant differences between sexes (p <0.05).

Table 5. Infection rates according to location based on ELISA.

Area	No. of animals examined	No. of ELISA positive cases	Infection rate (%)
Al-Kut	70	12	17.14a
Al-hayy	50	7	14a
Al-Bashair	20	2	10ab
Al-Noamania	24	2	8.33ab
Al-Moufaqia	20	1	5b
Total	184	24	13.04

Similar letters refers to the non-significant differences while different letters refers to the statistically significant differences among areas (p <0.05).

in ELISA. This technique was based on the digestion of 16S rRNA gene by restriction endonuclease (*Bst*1107I) and was carried out according to methods previously described by Noaman et al. (2009).

Genomic DNA extraction

Genomic DNA from blood samples were extracted by using genomic DNA mini kit extraction kit (frozen blood) Geneaid, USA, and done according to the company's instructions.

Biochemical tests

AST, ALT, CK and TBIL were measured using biochemical test kits and according to the manufacturer's instructions.

Statistical analysis

Chi-square (X^2) and t-test were used to detect statistically significant differences between age groups, sex and study area of data prevalence of disease and the effect of other factors. The differences were considered statistically significant at P > 0.05 (Al-Rawi, 2000).

RESULTS

Prevalence of bovine anaplasmosis in cattle using ELISA test

Out of 184 serum samples collected from suspected cattle examined by ELISA, there were 24 (13.04%) positive cases (Table 2). The study showed that higher infection rate was recorded in animals aged 2-3 years, 16.45% (13/79), while the lowest infection rate was in animals less than 1 year of age, 8% (2/25) (Table 3). Also, the study revealed that females were given higher percent of infection, 14.14% (14/99) than males, 11.7% (10/85) (Table 4). Higher percent of infection was record in AL-Kut, 17.14% (12/70), while the lower percent was record in Al-Moufaqia, 5% (1/20) (Table 5).

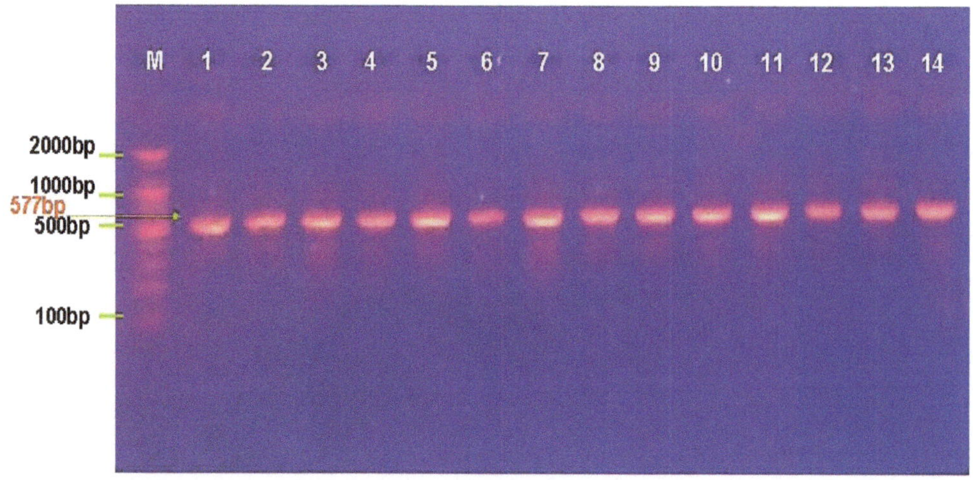

Figure 1. Agarose gel electrophoresis image that shows the PCR product of 16S the rRNA gene of *Anaplasma* spp. isolated from blood samples. Where M: marker 100 bp, lane (1-14) positive blood samples at expected size of 577 bp PCR product.

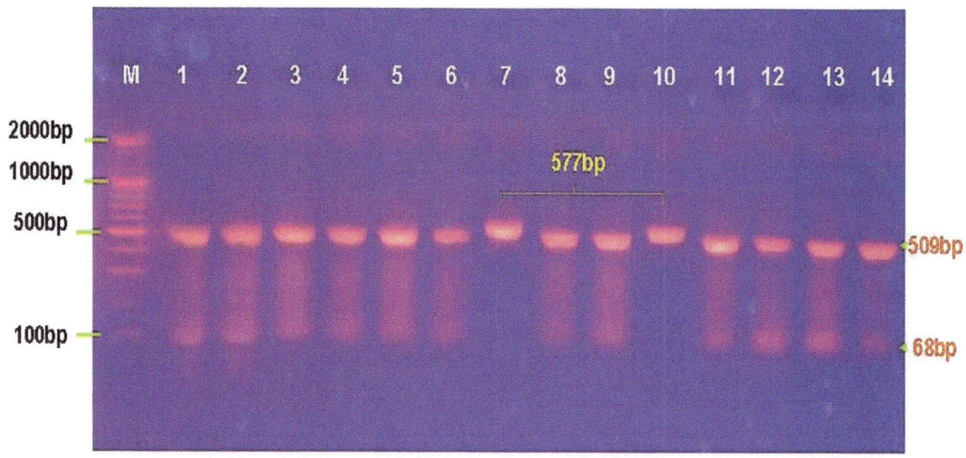

Figure 2. Agarose gel electrophoresis following restriction digestion of the 557 bp amplicon of the 16S rRNA gene of *Anaplasma* spp. using the restriction endonuclease (*Bst*1107I). Where M: marker 100 bp, (lanes 1, 2, 3, 4, 5, 6, 8, 9, 11, 12, 13 and 14) positive *Anaplasma marginale* that gave cut at 509 and 68 bp, while lanes 7 and 10 was negative for RFLP PCR that was still kept at 577 bp PCR product.

Bovine anaplasmosis according to RFLP-PCR technique

This technique was used to determine *Anaplasma* species present in the study area. PCR of the DNA isolated from blood samples showed that 20 out of the 24 blood samples were *Anaplasma* spp. positive and resulted in an expected product of 577 bp (Figure 1).

The restriction endonuclease *Bst*1107I when added to *A. marginale* 16S PCR products resulted in DNA fragments with the expected sizes of 509 and 68 bp. Thus, 18 out of 20 blood samples were confirmed to have *A. marginale* (Figure 2).

DISCUSSION

Prevalence of bovine anaplasmosis according to the ELISA test

In the current study, the prevalence of antibodies against *A. marginale* in cattle from Wassit province was investigated using an ELISA. The results demonstrated that these ruminants are exposed to this hemoparasite in the region studied and this result is agreement with worldwide distribution of anaplasmosis in cattle (Kocan et al., 2010). The results of this study revealed that the rate of infection in cattle was 13.04% and was comparative with other studies done in Iraq, this result was higher than the result recorded by Ameen et al. (2012) who showed that 9.09% of cattle in Erbil were infected with anaplasmosis but lower than the result (30.4%) which was reported by Al-Mossawy (2012) in Al-Diwanyia governorate, this difference in results is due to the differences in weather conditions and geographic distribution of tick vectors between Erbil and AL-Diwanyia with Wasit governorates. Specifically, the weather is cooler in the north which results in low distribution of tick vectors available to transmit the disease, while AL-Diwanyia is a big agricultural area with concentrated animal husbandry especially cows. The warm climate in this area allows for the presence of the tick vectors in huge number, thus a higher prevalence of infection. Regarding age wise prevalence, the results of this study revealed that the highest rate of infection was in cattle of over 2 years old (16.45%), while the lowest rate was in calves less than 1 year (8%). However, the differences between groups were not statically significant. This result agrees with many similar studies (Al-Mossawy, 2012; Chowdhury et al., 2006; Al-Khaledi, 2008) that reported that the highest prevalence of anaplasmosis is in cattle of 2 years of age and over. Calves from immune dams receive temporary protection from the colostrum which prevents anaplasmosis, this protection lasts about 3 months and in most cases, is followed by age resistance which lasts until the animals are approximately 9-12 months of age. The age resistance in calves gradually wanes after 1 year of age and these animals become increasingly susceptible to the disease (Kocan et al., 2000; Tassi et al., 2002).

In relation to the sex wise prevalence, the results of the study indicated that females are more susceptible than male to anaplasmosis infection with the rate 14.14 and 11.7%, respectively but the statistical analysis show no significant differences. The higher prevalence of anaplasmosis in female animals may be due to the fact that contaminated needles are commonly used for injecting drugs for milk let down (Kocan et al., 2010).

The immunosuppression in advanced pregnancy and or lactation in high producing animals are the possible reasons for the higher prevalence of *A. marginale* in female cattle. Also, differences in management practices in which male receive high quality feed, treatment with ectoparasitic drugs as well as housing few cohorts in relatively clean conditions yard in order to reach high body weight for sale, all can lead to decrease exposure of males to *A. maginale* (Al-Mossawy, 2012).

The present study showed significant differences in incidence of disease between study districts and area in Wasit governorate, this difference may be attributed to the differences in climatic conditions and intensity of ticks' infestation in the areas, the highest rate of infection was record in AL-Kut, 17.14% followed by AL-hayy, 14% and AL-Bashair, 10%. The areas with high percentage of infection are agricultural areas where available animal husbandry and a wet and warm climate are conducive for tick transmission, which are the key factors in *A. marginale* transmission. These areas are situated near Iran in which many casualties among the herds of cows was recorded and from where there is possible entry of infected animals into the country through trade or animal movement. The lowest prevalence was 8.33 and 5% in Noamania and Mouafaqia, respectively. This may be attributed to the desert climate of these two regions in which animals and tick vectors are fewer.

Prevalence of bovine anaplasmosis according to RFLP-PCR method

In the present study, DNA was extracted from 24 blood samples which were positive for *Anaplasma* by ELISA. The extracted DNA from blood cells was analyzed by PCR and RFLP using primers derived from 16S rRNA gene and restriction endonuclease Bst1107I. The restriction endonuclease Bst1107I only recognizes the sequence (GTATAC), which is specific for *A. marginale* (Noaman and Shayan, 2010). Based on the PCR, 20 of 24 samples were positive and this suggested that the rest 4 negative sample was collected from cattle in the late stage of infection when all parasite's DNA was faded. Analysis of all 20 Anaplasma positive PCR products with the restriction endonuclease Bst1107I showed that 18 PCR products could be cut in two expected DNA fragments with 509 and 68 bp in length, respectively, and this confirm diagnosis that cattle in this study were infected with *A. marginale*. The failure of restriction endonuclease cutting in the remaining two samples suggests these animals may have been infected with *A. centrale* or *A. bovis*. This simple PCR method based on 16S rRNA gene flowed by RFLP-PCR is able to rapidly discriminate between *A. marginale* and another species within the genus Anaplasma that infect cattle. The results of this study are in agreement with the results reported by Noaman and Shayan (2010) in Iran, which revealed the possibility of developing a new PCR-RFLP method based on 16S rRNA gene able to differentiate between *A. marginale*, *A. centrale* and *A. bovis* that infect cattle.

Biochemical parameters

Concerning the effect of *A. marginale* infection on activity of liver enzymes, the obtained results of the study revealed significant increase in serum AST and ALT in infected cattle when compared with healthy cattle. These results were in agreement with other studies reported by Allen et al. (1981) and (Turgut 2000). Allen et al. (1981) stated that the increase in enzyme activity may be attributed to severe anemia that leads to hypoxic and liver damage. Also, massive hemolysis may occur which in conjunction with hypoxia may lead to hepatic cell degeneration and glomerular dysfunction leading to increase in AST and ALT enzymes. In the present study,

Table 6. Levels of biochemical parameters in infected and control.

Parameter	Control (n = 10)	Infected cows (n = 10)
AST	65.5±1.26a	96.8±0.97b
ALT	21.4±0.45a	42±0.52b
CK	142±14.17a	406±2.06b
TBIL	0.27±0.05a	0.95±0.24b

Values were shown as Mean ± SE. The different letters denote the statistically significant differences at $p > 0.05$.

the results revealed an increase in creatine kinase (CK) and total serum bilirubin concentration in infected cattle as compared to healthy cattle. The result agrees with Coskun et al. (2012) which explain that this increase may have been caused by muscular trauma as a result of recumbency due to anaplasmosis (Sandhu et al., 1998) and Hofmann-Lehmann et al. (2004) mentioned that anaemia results from extra-vascular phagocytosis and hemolysis of parasitized erythrocytes by the reticulo-endothelial system may cause increase in serum bilirubin concentration and may be recognized by yellow discolouration of the plasma. Furthermore, Manna (1990), Omer et al. (2003) and Khan et al. (2011) reported that hyperbilirubinemia in anaplasmosis is due to excessive destruction of RBCs and hepatic dysfunction (Table 6).

Conflict of interests

The authors did not declare any conflict of interest.

REFERENCES

Al-Khaledi MJ (2008). Epidemiological study of (Theileriosis, Babesiosis and Anaplasmosis) in cattle of Al-Qadisiya province.M.S. thesis. Vet. Med. College Baghdad Univ. In Arabic.

Allen PC, Kuttler KL, Amerault BS (1981). Clinical chemistry of anaplasmosis: Blood chemical changes in infected mature cows. Am. J. Vet. Res. 42: 322-325.

Al-Mossawy HS (2012). A Serological Investigation of Bovine Anaplasmosis in Al- Diwanyia City. Diploma thesis of Veterinary Medicine/Internal and Preventive Veterinary Medicine.

Al-Rawi K (2000). Introduction for biostatic, Al-Mousul University.

Ameen KAH, Abdullah BA and Abdul-Razaq RA (2012). Seroprevalence of *Babesia bigemina* and *Anaplasma marginale* in domestic animals in Erbil, Iraq. Iraqi J. Vet. Sci. 26 (3): 109-114.

Aubry P, Geale DW (2011). A Review of Bovine Anaplasmosis Transboundary and Emerging Diseases. Canadian Food Inspection Agency, Ottawa, ON, Canada. 58:1–30.

Carelli G, Decaro N, Lorusso A, Elia G, Lorusso E, Mari V *et al.* (2007). Detection and quantification of *Anaplasma marginale* DNA in blood samples of cattle by real-time PCR. Vet. Microbiol. 124: 107-114.

Chowdhury S, Hossain MA, Barua SR, Islam S (2006). Occurrence of common blood parasites of cattle in Sirajgoni Sadar of Bangladesh. Bangl. J. Vet. Med. 4 (2): 143–145.

Coskun A, Ozlem DE,Hassan G, Uğur A, İsmail ŞK (2012). Acute Phase Proteins, Clinical, Hematological and Biochemical Parameters in Dairy Cows Naturally Infected with *Anaplasma marginale*. Univ. Vet. Fak. Derg.18 (3): 497-502.

Cynthia MK, Susan EA (2008). The Merck Veterinary Manual, 9th edn. Whitehouse station NJ, USA.

Dreher UM, Hofmann-Lehmann R, Meli ML, Regula G, Cagienard AY, Stark KDC, Doherr MG, Filli F, Hassig M, Braun U, Kocan KM, Lutz H (2005). Seroprevalence of anaplasmosis among cattle in Switzerland in 1998 and 2003. No evidence of an emerging disease. Vet. Microbiol. 107: 71-79.

Dumler JS, Barbet A F, Bekker C P J, Dasch G A, Palmer G H, Ray S C, Rikihisa Y, Rurangirwa FR (2001). Reorganization of the genera in the families Rickettsiaceae and Anaplasmataceae in the order Rickettsiales: unification of some species of Ehrlichia with Anaplasma, Cowdria with Ehrlichia and Ehrlichia with Neorickettsia, descriptions of six new species combinations and designation of *Ehrlichia equi* and "HGE agent" as subjective synonyms of *Ehrlichia phagocytophila*. Int. J. Syst. Evol. Microbiol. 51:2145-2165.

Ewing SA, Panciera RJ, Kocan KM, Ge NL, Welsh RD, Olson RW, Barker RW, Rice LE (1997). A winter outbreak of anaplasmosis in a nonendemic area of Oklahoma: a possible role for *Dermacentor albipictus*. J. Vet. Invest. 9:206-208.

Goff WL, Stiller D, Roeder RA, Johnson LW, Falk D, Gorham JR, McGuire TC (1990). Comparison of a DNA probe, complement fixation and indirect immunofluorescence tests for diagnosing *Anaplasma marginale* in suspected carrier cattle. Vet. Microbiol. 24:381-390.

Hofmann-Lehmann R, Meli ML, Dreher UM, Gnczi E, Deplazes P, Braun U, Engels M, Schüpbach J, Jrger K, Thoma R, Christian G, Katharina D C S, Willi B, Schmidt J, Kocan KM, Hans L. (2004). Concurrent Infections with Vector-Borne Pathogens Associated with Fatal Hemolytic Anemia in a Cattle Herd in Switzerland. J. Clin. Microbiol. 42(8): 3775-3780.

Khan IA, Khan A, Hussain A, Riaz A, Aziz A. (2011). Hemato-biochemical alterations in cross bred cattle affected with bovine theileriosis in semi-arid zone. Pak. Vet. J. 31: 137-140.

Kocan KM, De la Fuente J, Guglielmone AA, Melendez RD (2003). Antigens and Alternatives for Control of *Anaplasma marginale* infection in Cattle. Clin. Microbiol. Rev. 16(4):698-712.

Kocan KM, Blouin EF, Barbet AF (2000). Anaplasmosis control: Past, present, and future. Ann. New York Acad. Sci. 916: 501-509.

Kocan KM, de la Fuente J, Bouin EF, Coetzee JF, Ewing SA (2010). The natural history of *Anaplasma marginale*. Vet. Parasitol. 167: 95-10.

Kocan KM, de la Fuente J, Step DL, Blouin EF, Coetzee JF, Simpson KM, Genova SG, Boileau MJ (2010). Current Challenges of the Management and Epidemiology of Bovine Anaplasmosis. Bovine Practitioner 44 (2): 94-96.

Manna AMM (1990). Clinical and some bio-chemical changes in healthy and diseased camels. Ph.D Thesis, Fac Vet Med, Assui University.

Noaman V, Shayan P (2010). A new PCR-RFLP method for detection of *Anaplasma marginale* based on 16S rRNA. Vet. Res. Commun. 34: 43-50.

Noaman V, Shayan P, Amininia N (2009). Molecular Diagnostic of *Anaplasma marginale* in Carrier Cattle. Iran. J. Parasitol. 4(1): 26-33.

Omer OH, El-Malik KH, Magzoub M, Mahmoud OM, Haroun EM, Hawas A, Omar HM (2003). Biochemical profiles in Friesian cattle naturally infected with *Theileria annulata* in Saudi Arabia. Vet. Res. Commun. 27:15-25.

Palmer GH, Brown WC, Rurangirwa FR (2000). Antigenic variation in the persistence and transmission of *Anaplasma marginale*. Microb. Infect. 2: 167-176.

Sandhu GS, Grewal AS, Singh A, Kondal JK, Singh J, Brar RS (1998). Haematological and biochemical studies on experimental *Theileria annulata* infection in crossbred calves. Vet. Res. Commun. 22: 347-354.

Tassi P, Carelli G, Ceci L (2002). Tick-borne diseases (TBDs) of dairy cows in a Mediterranean environment: a clinical serological and hematological study. Ann. N. Y. Acad. Sci. 969: 314-316.

Theiler A (1910). *Anaplasma marginale,* the marginale points in the blood of cattle suffering from a specific disease, In: A. Theiler (ed.), Report of the government veterinary bacteriologist, 1908-9. Transvaal, South Africa. pp. 7-64.

Theiler A (1911). Further investigations into anaplasmosis of South African cattle, In: 1st Report of the Director of Veterinary Research, Department of Agriculture of the Union of South Africa. pp. 7-46.

Turgut K (2000). Veteriner Klinik Laboratuvar Teşhis. s. 157-158, BahçıvanlarBasımsanayi, Konya.

Purification and characterization of tryptophan and indole-3-acetic acid produced by *Serratia marcescens* strain *MCB* associated with *Oscheius* sp. MCB (Nematoda: Rhabditidae) obtained from South African soil

Mahloro H. Serepa[1], Nikita T. Tavengwa[2] and Vincent M. Gray[1]

[1]School of Molecular and Cell Biology, University of the Witwatersrand, P Bag 3, Wits 2050, Johannesburg, South Africa.

[2]School of Chemistry, University of the Witwatersrand, P Bag 3, Wits 2050, Johannesburg, South Africa.

Entomopathogenic nematodes (EPNs) of the genera *Steinernema* and *Heterorhabditis* are lethal to insect pests that attack plants. These EPNs are associated symbiotically with the two enterobacteria *Xenorhabdus* and *Photorhabdus* sp., respectively. The bacteria synthesize a range of insecticidal and antimicrobial metabolites which may be useful as agricultural pest control and medical disease control. Recently, EPNs characterization and explanation have been expanded to include genera, *Oscheius*. We isolated, characterized and identified a symbiotic bacterium and the metabolites it produces. The symbiotic bacterium was isolated from a South African nematode, *Oscheius* sp. *MCB* (GenBank accession number: KF684370). The symbiotic bacterium was identified to be a *Serratia marcescens* strain *MCB* (GenBank accession number: KF793930). Two metabolites it produces are indole-3-acetic acid and tryptophan, with tryptophan as an aromatic amino acid assumed to be a precursor for indole-3-acetic acid production. From this study it is evident that *S. marcescens* strain *MCB* isolated from *Oscheius* sp. *MCB* shares similar attributes with *Xenorhabdus* and *Photorhabdus* sp., thus its host can be accepted as an EPN.

Key words: Entomopathogenic nematodes, *Serratia*, *Oscheius*, *Xenorhabdus, Photorhabdus*.

INTRODUCTION

Entomopathogenic nematodes (EPNs) have been found to be advantageous as biocontrol agents (Kaya and Stock, 1997; Poinar, 1975). The biocontrol advantages of EPNs have been well established and encouraged further surveys of EPNs on a global basis as it has been shown that EPNs are widely distributed in various soils throughout the world (Adams et al., 2006; Hominick et al., 1996; Hominick, 2002). With the biocontrol advantages and useful antimicrobial products produced by the symbiotic bacteria associated with EPNs (Wang et al., 2011), there exists a strong incentive to discover new EPN species associated with new species of symbiotic bacteria. Recent isolated species of *Oscheius* genus EPNs have been described (Liu et al., 2012; Ye et al., 2010). The genus *Oscheius*, shares similar attributes with two other well-known EPN genera (*Steinernema* and *Heterorhabditis*) and has all the attributes associated with these EPNs (Dillman et al., 2012; Torres-Barragan et al., 2011) and was first erected by Andrassy in 1976. Liu et al. (2012) and Ye et al. (2010) used *Galleria mellonella* larvae to test the entomopathogenicity of *Oscheius chongmingensis* and *Oscheius carolinensis* which occurred in 56 h with 35% larvae mortality and 48 h with 100% larvae mortality, respectively.

The *Oscheius* genus comprises of the following recently identified novel species of EPNs, *O. chongmingensis*, *O. rugaoensis* and *Oscheius carolinensis* (Liu et al., 2012; Torres-Barragan et al., 2011; Ye et al., 2010). The EPNs in all three genera have evolved mutualistic associations with three different genera of Gram negative, motile, rod-shaped bacteria belonging to the family Enterobactericeae, all of which are lethal pathogens to a wide range of insects. *Xenorhabdus* bacteria are located in the intestinal vesicle of the infective juveniles (IJs) of *Steinernema* (Bird and Akhurst, 1983); in contrast *Photorhabdus* bacteria are located throughout the whole intestine of *Heterorhabditis* IJs (Endo and Nickle, 1991) whereas bacteria belonging to the genus *Serratia* have been shown to be symbiotically associated with *O. chongmingensis* and are located within a pouch next to the pharyngeal bulb (Liu et al., 2012).

Infective juveniles (IJs) penetrate the insect host through natural openings such as mouth, anus and respiratory spiracles and release the bacteria in the haemocel of the host (Poinar, 1975). In this relationship, EPNs are vectors for the bacteria, which are transported into the haemolymph of insect hosts (Boemare et al., 1993). After releasing the symbiotic bacteria into the insect haemolymph, the bacteria replicates and suppress the insect host's immune response system by producing various metabolites which hamper and kill the insect host and further inhibit the growth of competing microorganism. In doing so, a monoxenic environment is maintained for the EPNs to reproduce, grow and develop (Hussien and Hanan, 2008; Wang et al., 2011). The metabolites produced by the symbiotic bacteria have shown to have antibiotic, antimycotic, insecticidal, nematicidal, antiulcer, antineoplastic and antiviral properties and may have useful medical and agricultural applications (Wang et al., 2011). Various metabolites have been reported from entomopathogenic bacteria cell cultures, and these include: indole, dithiolopyrrolones and xenocoumacins, however, very little is known about the ones secreted by *Serratia* species. In addition to the work done thus far, we strongly believe that in further accepting the *Oscheius* as an EPN genus, the metabolites produced by its symbiotic bacteria can be studied and compared with the ones produced by *Xenorhabdus* and *Photorhabdus* sp., thereby examining the similarities and differences in metabolites produced by the three bacteria. In this paper, the authors reported on metabolites produced by a new strain of *Serratia marcescens; S. marcescens* strain *MCB* isolated from *Oscheius* sp. *MCB* which was obtained from South African soil.

MATERIALS AND METHODS

Isolation and morphological identification of the symbiotic bacterium

The symbiotic bacteria was isolated from *Oscheius* sp. *MCB* (GenBank accession number: KF684370) obtained from soil samples collected in North West province, South Africa. The methods used for isolating the bacteria were those described by Kaya and Stock (1997). The isolated bacteria were then grown and maintained on solid media namely nutrient agar supplemented with 0.004% (w/v) triphenyltetrazolium chloride (TTC) and 0.0025% (w/v) bromothymol blue (BTB) (NBTA) and MacConkey agar plates and sub-cultured on the same media on a monthly basis at 25°C. Long term maintenance of the bacterial cultures was done by storing the cultures in 1 mL of 30% glycerol and stored in -70°C refrigerator. Red colonies were isolated, characterized morphologically and Gram staining performed (Zhang et al., 2008; Torres-Barragan et al., 2011; Bergey et al., 1994). To deduce if the symbiotic bacteria from *Oscheius* sp. *MCB* occurs in growth phases like *Xenorhabdus* and *Photorhabdus* sp., NBTA and MacConkey agar plates were observed for the first 24 h of incubation and compared with colonies at 72-120 h. The fluorescence ability of the symbiotic bacteria was determined with Fluorescent Microscope equipped with an Olympus DP 80 video camera using the CellSens Dimension 1.9.1 software.

PCR amplification of the 16S rDNA from symbiotic bacteria

The DNA of the 16S rRNA gene of the symbiotic bacteria was amplified. Genomic DNA was isolated from the NBTA colonies using the ZR fungal/bacterial DNA MiniPrep kit (Zymo Research, catalog # D3050). The DNA extracted from the bacterial colonies was quantified using the NanoDrop ND-1000® spectrophotometer (Bio-Rad). The primers used for 16S rDNA amplification and sequencing were the 5`-AGAGTTTGATCCTGGCTCAG-3`f and 5`-AAGGAGGTGATCCAAGCCGCA-3`r (Brosius et al., 1978), corresponding to positions 8-27 and 1521- 1540 in the 16S rDNA sequence of *Escherichia coli*. The PCR products were cleaned using ExoSAP-it® and sequenced by Inqaba Biotechnical Industries (Pty) Ltd, South Africa.

Sequence and phylogenetic relationship analysis of the symbiotic bacteria

The sequence was screened for chimeras by using DECIPHER tool (Wright et al., 2012). The NCBI-BLAST search tool was used to establish the degree of similarity of the sequence with other sequences of known bacteria taxa in the GenBank database. Bacterial species from GenBank with the highest matches after a BLAST search were selected for phylogenetic analysis. We used *Serratia nematodiphila* strain *DR186* (JQ002568), *Serratia* sp. *R26*

Table 1. HPLC determination of tryptophan and indole-3-acetic acid including the concentrations.

Parameter	Metabolites	
	Tryptophan	Indole-3-acetic acid
Retention time (min)	2.00	3.13
Equation	y = 1.9714x + 976.16	y = 25.054x - 338.8
R^2	0.995	1.000
$^{\Psi}$LOD (μg L^{-1})	46.7	7.99
$^{\chi}$LOQ (μg L^{-1})	155.7	26.6

ψLOD is limit of detectation and χLOQ limit of quantification.

(2012) (JX082294), *S. marcescens WW4* (NR_102509), *S. marcescens* strain *AU1209* (AY043387), *S. marcescens* clone *HUWU3B* (KC583444), *S. marcescens* strain *DAP33* (EU302858), *Serratia* sp. *C8* (JX258135), *Serratia urelytica* (AJ854062), *Serratia rubidaea* strain *DSM 4480* (AJ233436), *S. entomophila* (NR025338), *Serratia ficaria* strain *DSM 4569* (NR_041979), *P. luminescens* subp. *luminescens* strain *Hb* (NR_037074) and *X. bovienii DSM 4766* (X82252) species to observe the relationship that the current strain (*S. marcescens* strain *MCB* KF793930) has with the other species. *Bacillus cereus* (AB334763) was used to root the tree. The sequences were aligned using MUSCLE (Edgar, 2004), with default options and the distance analysis with closest taxa were conducted in MEGA6 (Tamura et al., 2013). The evolutionary distances were computed using Jukes-Cantor model (Jukes and Cantor, 1969). Neighbour-joining tree of the homologous sequences in this study obtained from NCBI GenBank was inferred (Saitou and Nei, 1987).

Selection of the optimal nutrient medium for metabolite production

Five times loopful of log phase of the symbiotic bacteria obtained from NBTA/MacConkey agar plates were inoculated into each 100 mL flasks containing 20 mL of Luria-Bertani broth (LB), nutrient broth (NB), nutrient broth supplemented with canola oil and glucose and nutrient broth supplemented with canola oil only (Torre, 2003). The media were adjusted to final pH of 7.0 (Wang et al., 2011). The flasks were incubated at 28°C for 48 h in the dark on an orbital shaker at 160 rpm. At an hour interval, cell culture was centrifuged and supernatant removed, then bacterial concentration measured at 600 nm with an S-20 Boeco Spectrophotometer for 24 h. The media with high bacterial yield was deduced to be the one suitable for the bacterial growth. This experiment was carried out in triplicates.

Metabolite identification and characterization

For purification and characterization of the metabolites, all media were analyzed with an HPLC system (Model 510, waters, USA) and the metabolites were separated using a C18 reversed phase column (25 cm x 4 mm x 0.5 μm). A flow rate of 1.0 mL min^{-1} with detection of separated metabolites was achieved using a fluorescence detector at excitation and emission 280 and 435 nm, respectively. The solvent used was a mixture of acetonitrile: water with 0.3% acetic acid (50:50 v/v). After the HPLC development, the method was validated in terms of limit of detectation (LOD), limit of quantification (LOQ) and other parameters (Table 1).

Role of metabolite on the bacterial growth

Different concentrations (mg L^{-1}) of the identified metabolites were used to determine effects on bacterial growth. Luria-Bertani broths with the metabolites were inoculated with a loopful of *S. marcescens* strain *MCB*. The produced metabolites were then detected with HPLC and the concentrations determined. LB without the metabolites were used as a negative control.

RESULTS

Morphological characteristics

Colony morphology and cell morphology based on Gram staining of these bacteria were determined (Table 2). *Serratia marcescens* strain MCB was rod shaped with fluorescing ability and was located next to the pharyngeal bulb as well as throughout the intestines of adult females (Figure 1a-c).

Molecular identification

The 16S rDNA sequence product was 1152 base pairs (bp) and was deposited in GenBank and assigned the accession number KF793930. The BLAST search of this sequence showed that it belongs to the genus *Serratia,* with high sequence similarity identity with *S. marcescens* strain DAP33, *S. marcescens* strain AU1209, *S. marcescens* sp. clone HUWU3B, *S.* marcescens sp. C816, *S. marcescens* sp. R26 (2012) and *S. marcescens* WW4. Phylogenetic tree revealed the close relationship of *S. marcescens* strain *MCB* with *S. marcescens* strain DAP33 (bootstrap of 70%) as shown in Figure 2. Neighbour-joining tree indicated that *S. marcescens* strain *MCB* and *S. marcescens* strain *DAP33* form a polytomy clade with four strains of *S. marcescens* and one *S. nematodiphila* (bootstrap of 52%).

Selection of the optimal nutrient medium for antibiotic production

Distinctive colour changes were observed in all media types, that are, NB, NB with canola oil, NB with canola oil and glucose and LB inoculated with *S. marcescens* strain *MCB* (results not shown). The colour changes were a

Table 2. Morphological characteristics of symbiotic bacteria isolated from *Oscheius* sp. *MCB.*

Characteristic	Symbiotic bacteria
Host EPN	*Oscheius sp. MCB*
Symbiotic bacteria	*Serratia marcescens* strain *MCB*
Colony diameter on NBTA agar plate (mm)	1-4 mm
Colony morphology and pigmentation on NBTA agar (Phase I)	Deep red circular, with regular margins and smaller diameter
Colony morphology and pigmentation on NBTA agar (Phase II)	Brown, flat with irregular margins and larger in diameter
Pigmentation on MacConkey agar	Light pink
Pigmentation in Nutrient broth	Light Maroon
Pigmentation in Nutrient broth supplemented with canola oil	Light pink
Pigmentation in Nutrient broth supplemented with canola oil and glucose	Light pink
Pigmentation in Luria-Bertani broth	Maroon
Optimum temperature	28°C
Gram stain	Gram negative
Fluorescence ability	Present
Morphology	Rod

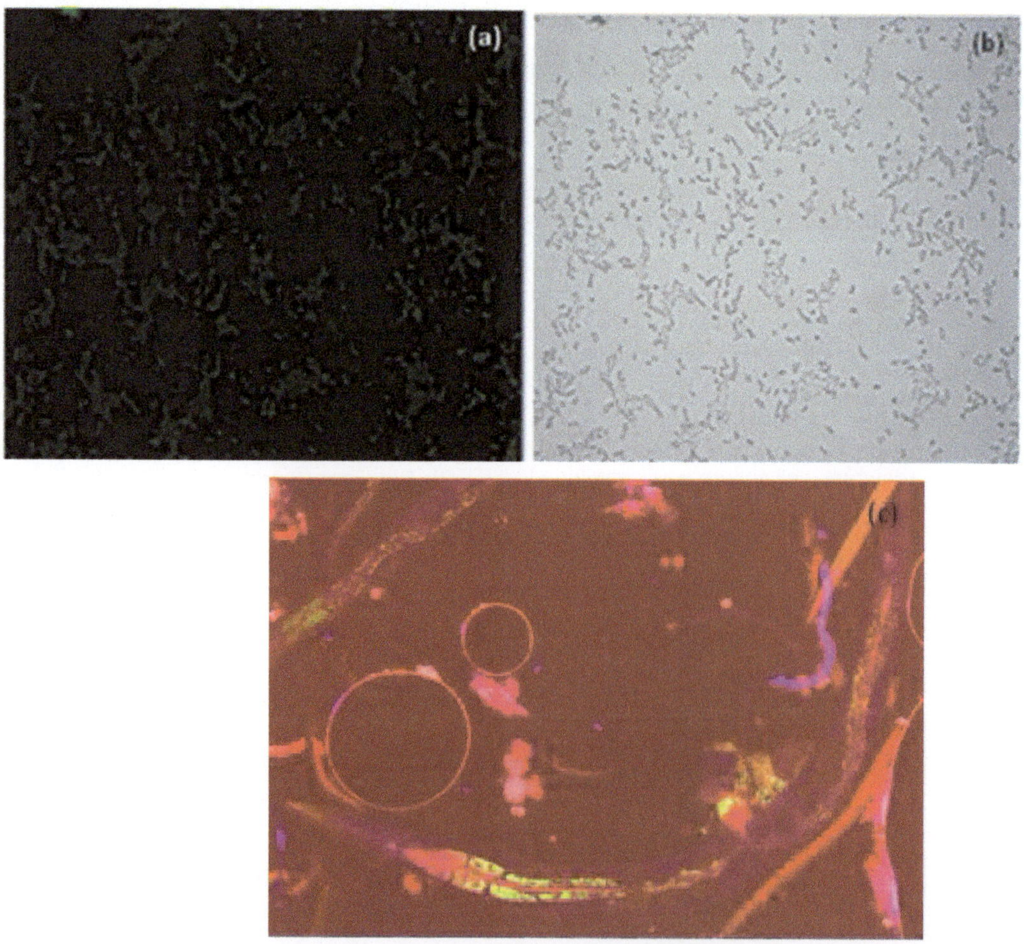

Figure 1. *S. marcescens* strain *MCB* under the light microscope (a) fluorescing rod shaped bacteria, (b) bacteria without fluorescence and (c) fluorescing bacteria within the *Oscheius* sp. *MCB* adult nematode.

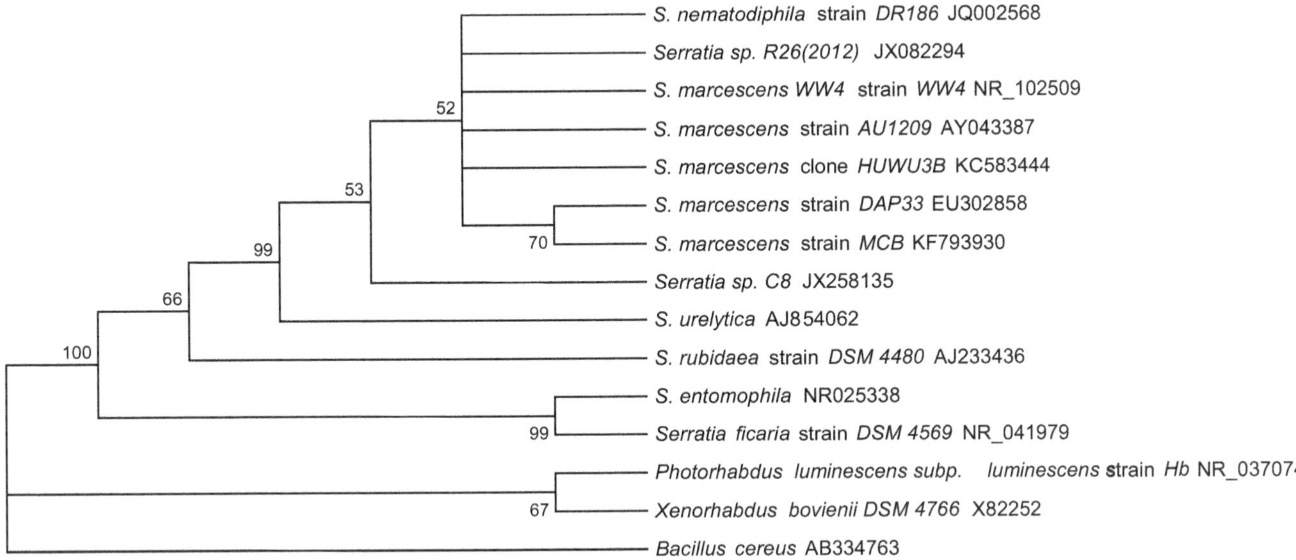

Figure 2. Neighbour-joining tree of *S. marcescens* strain *MCB* with closely related bacteria inferred by homologous sequences based on 16S rRNA gene sequence data. The nodal supports are bootstrap values from 1000 replications.

result of metabolites secreted by the bacteria as it proliferates and some metabolites have antibiotic activity against opportunistic bacterial and fungi that may be in the media. The colour changes in the media thus serve as a method to visually verify purity of the bacterial culture.

The different growth media had significant effects on bacterial growth and metabolite production. Luria-Bertani broth (LB) showed a maximum bacterial yield, thus it increases bacterial growth and metabolism, increased bacterial yield results in an even increased antibiotic production due to nutrients available and a highly active metabolism and most antibiotics or metabolites get secreted in the stationery phase of bacterial growth (Thomashow, 1996). LB had a higher bacterial yield with an optical density of 1.5, followed by NB with an optical density of 0.9 at the 24th h and for NB supplemented with canola oil had the highest optical density of 0.6 at 19th h and NB supplemented with canola oil and glucose had optical density of 0.4 at the 16th h (Figure 3a and b).

Metabolite identification and characterization

Metabolites were analyzed from all the four different media used. Three metabolites; tryptophan, indole-3-acetic acid (IAA) and indole were expected to be produced; in this study only two were identified, tryptophan and IAA produced by *S. marcescens* strain *MCB* (Figure 4a to e). Tryptophan was detected first with high quantity concentration as compared to IAA (Table 3) and it was only produced in all the media whilst IAA was produced in LB and NB + canola oil.

Role of metabolite on the bacterial growth

The bacterial growth increased in with an increment in tryptophan concentration (results not shown). It was assumed that the aromatic amino acid tryptophan was the precursor for IAA production (Dewick, 2002), because from the metabolite analysis, the IAA production increased with an increase in tryptophan, however there was no relationship between IAA in the broth and tryptophan production (Tables 4 and 5).

DISCUSSION

A new species of *Oscheius* genus of entomopathogenic nematodes (EPNs) (Ye et al., 2010) had been accepted as a genus displaying characteristic EPN adaptation which have been acquired through convergent evolution (Dillman et al., 2012); however more research needs to be done to support its acceptance as an EPN genus. One of the criteria for EPNs, is for their symbiotic bacteria to produce metabolites which not only kill the insect host but also other competitive microorganisms, thus, we strongly believe that analyzing metabolites produced by symbiotic bacteria from *Oscheius* species, will shed light as to whether *Oscheius* symbiotic bacteria is similar to *Xenorhabdus* and *Photorhabdus* sp. In this paper, we managed to identify two compounds produced by *S. marcescens* strain *MCB*, namely tryptophan and IAA. Indole derivative compounds produced by *Xenorhabdus* and *Photorhabdus* sp. have been identified and shown to inhibit growth of insect larvae and other microorganisms (Sundar and Chang, 1993; Hu and Webster, 2000; Paul et al., 1981). Tryptophan was recently added to growth

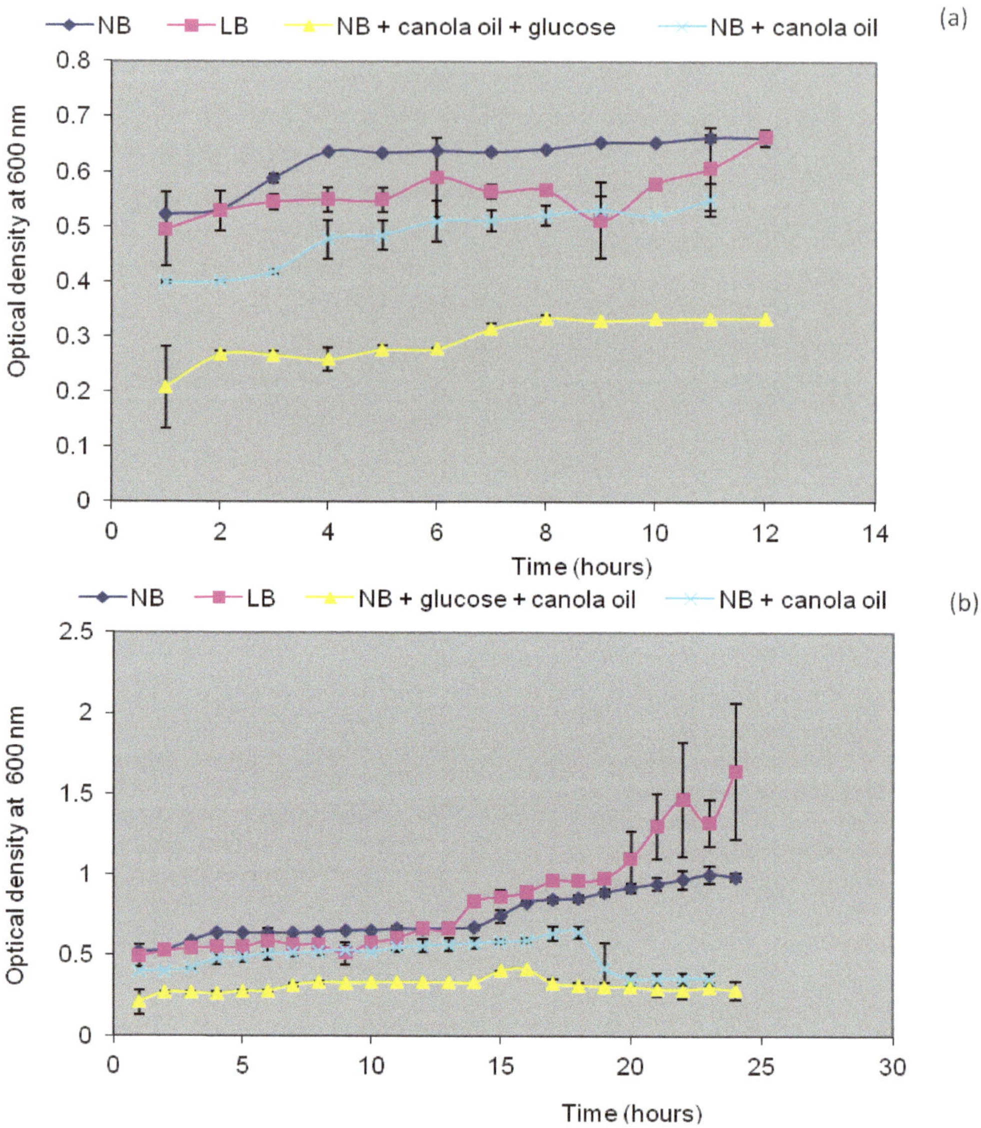

Figure 3. Absorbance over time at 600 nm of *S. marcescens* strain *MCB* in Luria-Bertani broth (LB), nutrient broth (NB), nutrient broth supplemented with canola oil and nutrient broth supplemented with glucose and canola oil at approximately 25°C on a shaker at 160 rpm for (a) 12 h and (b) 24 h. Each point represents the average of three replicates. Bars indicate size exclusion.

medium for bacteria to induce production of other antimicrobial metabolites (Sundar and Chang, 1993), in this paper we identified tryptophan as one of the metabolites produced and we strongly believe that tryptophan is a precursor for IAA production and other compounds (Sundar and Chang, 1993; Manulis et al., 1991; Paul et al., 1981; Martino et al., 2003). Previous studies indicate that both *Xenorhabdus* and *Photorhabdus* sp. are biochemically and physiologically suitable as symbionts of entomopathogenic nematodes (Thaler et al., 1998; Thomas and Poinar, 1979; Akhurst, 1980; 1982), due to production of these metabolites. The metabolites produced by these bacteria also contribute towards the symbiotic relationship with the entomopathogenic nematodes in providing a monoxenic growth environment and nutrients for the nematodes which in turn act as vectors.

In this study, various liquid media for *S. marcescens* strain *MCB* were used and it was found that LB resulted in higher bacterial yield which in turn lead to an increased metabolite synthesis and production (Figures 3a and b). Increased bacterial yield results in an increased level of metabolite production due to nutrient availability and a highly active metabolism (Thomashow, 1996). Luria-Bertani broth contains sodium chloride, yeast extract and tryptone, whereas nutrient broth contains meat extract,

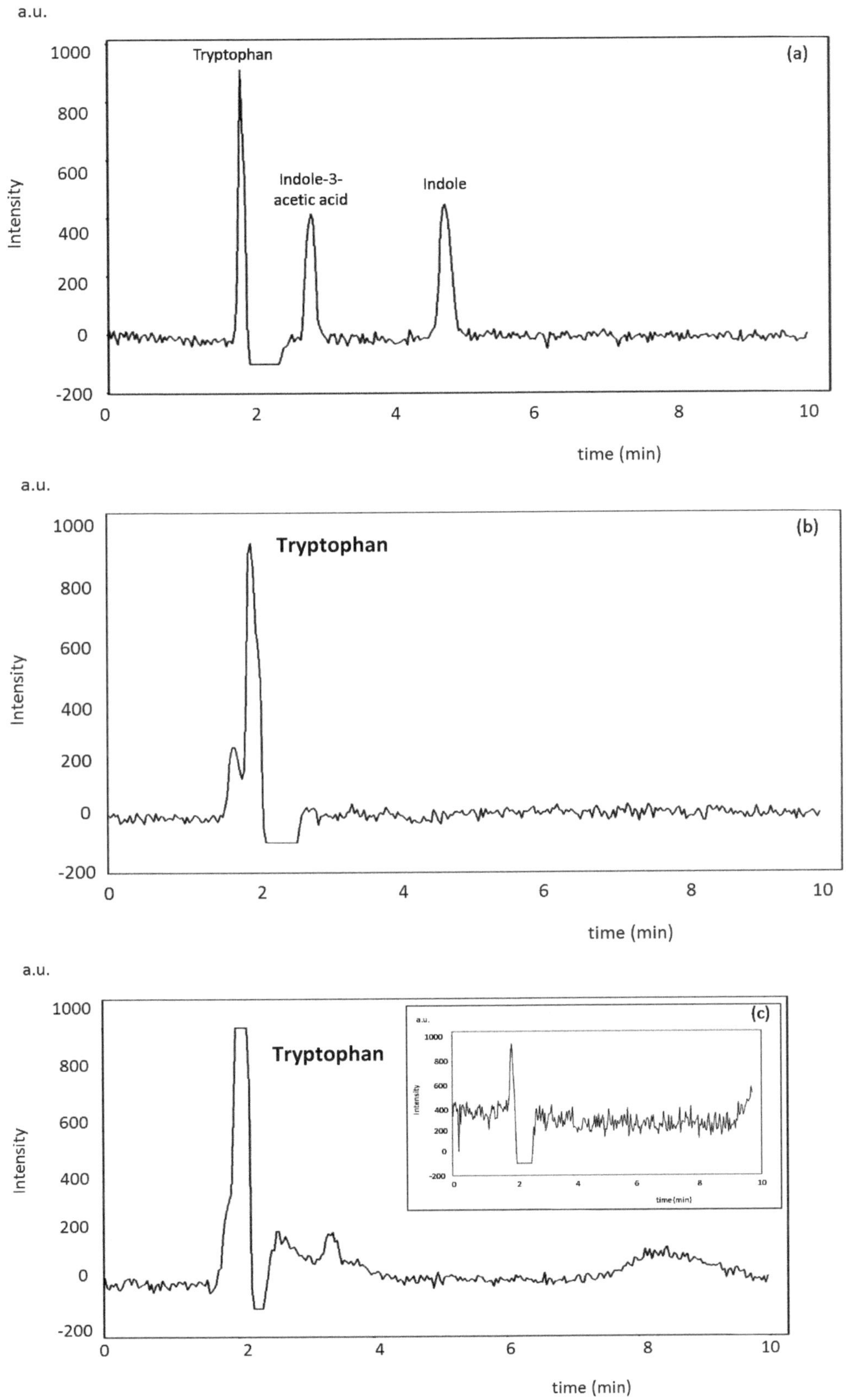

Figure 4. Chromatograms showing: (a) retention times of tryptophan, indole-3-acetic acid and indole standards, (b) 250X diluted NB sample, (c) 100X diluted NB + canola oil + glucose sample with 200X diluted, (d) 100X diluted LB sample and (e) 150X diluted NB + canola oil sample with 250X diluted.

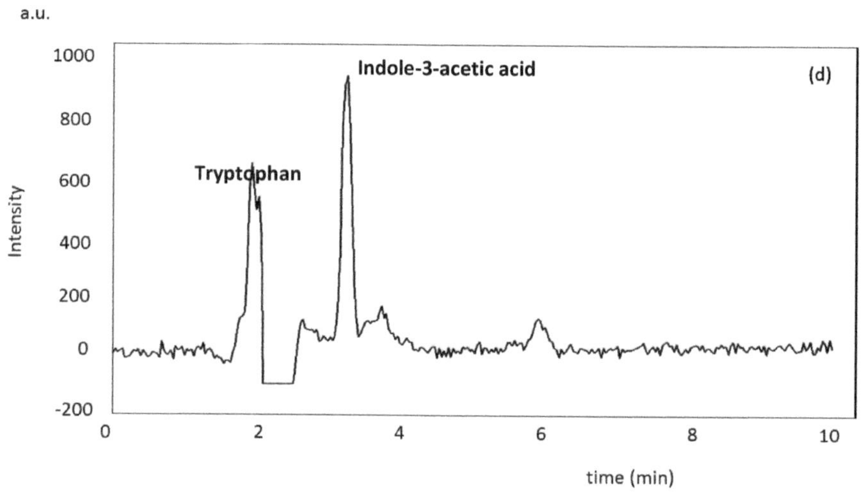

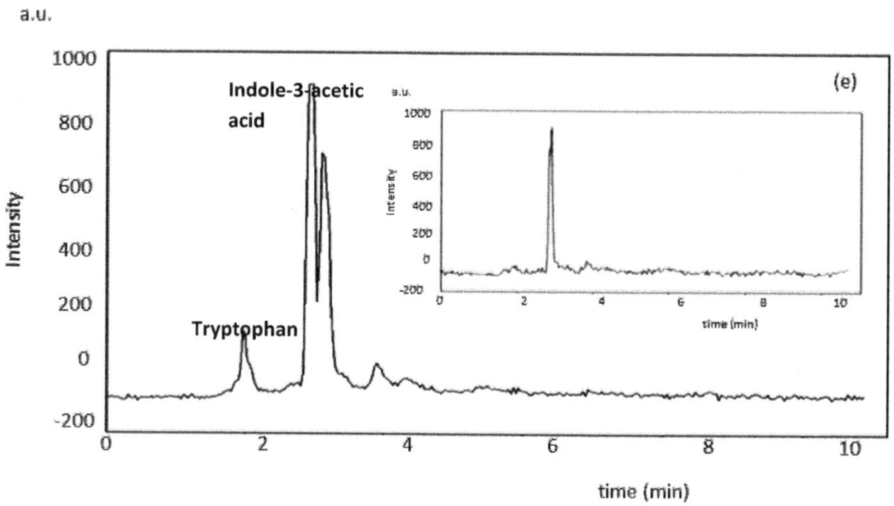

Figure 4. Contd.

Table 3. Concentrations of the metabolite from each media.

Media name	Concentration (mg L^{-1})		
	Tryptophan	Indole-3-acetic acid	Indole
LB	380	32	-
NB+ canola oil + glucose	42	-	-
NB	1194	-	-
NB+ canola oil	148	54	-

Table 4. Effects of tryptophan concentrations on tryptophan and indole-3-acetic acid production.

Concentration of tryptophan in LB (mg L^{-1})	Metabolites (mg L^{-1})	
	Tryptophan	Indole-3-acetic acid
0	1123	45
10	1342	78
20	1402	83
30	1409	94

Table 5. Effects of indole-3-acetic acid concentrations on tryptophan and indole-3-acetic acid production.

Concentration of indole-3-acetic acid in LB (mg L^{-1})	Metabolites (mg L^{-1})	
	Tryptophan	Indole-3-acetic acid
0	1124	45
10	1124	51
20	1134	55
30	1123	69

yeast extract, peptone and sodium chloride. Sodium chloride increases metabolite production due to its osmolarity and peptone, tryptone as well as yeast extract are good nitrogen sources (Wang et al., 2011), which explains the high bacterial growth which resulted in high metabolite quantity and concentrations in both media (Table 3). Peptone has been found to be a better nitrogen source as compared to tryptone (Wang et al., 2011), this is validated by the high concentration of IAA produced in NB as compared to LB. Results reported here are similar to those reported in previous studies which indicate that both NB and LB are suitable for metabolite production of *S. marcescens* strain *MCB* due to presence of NaCl, peptone, tryptone and yeast extract which are good sources of nitrogen and also increase metabolite production (Wang et al., 2011).

Conclusions

In this study, we showed that media composition affects bacterial yield and quality and quantity of metabolite production. Tryptophan is a precursor of IAA which is produced by *S. marcescens* strain *MCB*. *S. marcescens* strain *MCB* isolated from *Oscheius* sp. *MCB* has similar attributes with *Xenorhabdus* and *Photorhabdus* sp., thus its host can be accepted as an EPN. Future studies may be focused on determining the genetic characteristics, mechanism and the pathway that are responsible for the production of metabolites in the symbiotic bacteria.

ACKNOWLEDGEMENTS

We thank the National Research Foundation (NRF) and Gauteng Department of Agriculture and Rural development (GDARD), SAB-FoodBev SETA for providing us with financial support. Mahloro Hope Serepa and Nikita Tavengwa received Innovation Doctoral Scholarships from the NRF (grant ID: SFH1208157967 and NFS13091843717, respectively, National Research Foundation of South Africa). The opinions expressed and conclusions arrived at are those of the authors and are not necessarily to be attributed to the NRF, GDARD and SAB-FoodBev SETA.

REFERENCES

Adams BJ, Fodor A, Koppenhöfer HS, Stackebrandt E, Stock, SP, Klein MG (2006). Reprint of biodiversity and systematics of nematode-bacterium entomopathogens. Biol. Control 38: 4-21.
Akhurst RJ (1980). Morphological and functional dimorphism in *Xenorhabdus* spp., bacteria symbiotically associated with the insect pathogenic nematodes, *Neoaplectana* and *Heterorhabditis*. J. Gen. Microbiol. 121: 303-309.
Akhurst RJ (1982). Antibiotic activity of *Xenorhabdus* spp., bacteria symbiotically associated with insect pathogenic nematodes of the families Heterorhabditidae and Steinernematidae. J. Gen. Microbiol. 128(12): 3061-3065.
Andrassy I (1976). Evoluation as a basis for the systematization of nematodes. London: Pitman.
Bergey DH, Holt JG, Krieg NR, Sneath PHA (1994). Bergey's Manual of Determinative Bacteriology. (9th ed.). Lippincott Williams & Wilkins. ISBN 0-683-006037.
Bird AF, Akhurst RJ (1983). The nature of the intestinal vesicle in nematodes of the family Steinernematidae. Int. J. Parasitol. 13:599-606.
Boemare NE, Akhurst RJ, Mourant RG (1993). DNA relatedness between *Xenorhabdus sp.* (Enterobacteriaceae), symbiotic bacteria of entomopathogenic nematodes, and a proposal to transfer *Xenorhabdus luminescens* to a new genus, *Photorhabdus gen.* nov. Int. J. Syst. Bacteriol. 4: 249-255.
Brosius J, Palmer JL, Kennedy JP, Noller HF (1978). Complete nucleotide sequence of 16S ribosomal RNA gene from *Escherichia coli*. Proc. Natl. Acad. Sci. USA. 75:4810–4805.
Dewick PM (2002). Medicinal Natural Products: The Shikimate Pathway: Aromatic amino acids and phenylpropanoids. ISBNs: 0471496405, pp. 121-166.
Dillman AR, Chaston JM, Adams BJ, Ciche TA, Goodrich-Blair H, Stock SP, Sternberg PW (2012). An entomopathogenic nematode by any other name. Public Library of Science Pathogens 8: e1002527.
Edgar RC (2004). Muscle: Multiple sequence alignment with high accuracy and high throughput. Nucleic Acids Res. 32:1792-1997.
Endo BY, Nickle WR (1991). Ultrastructure of the intestinal epithelium, lumen and associated bacteria in *Heterorhabditis bacteriophora*. J. Helminthol. Soc. Washington 58:202-212.
Hominick WM (2002). Entomopathogenic Nematology: Biogeography. CABI: Wallingford, pp. 115-143.
Hominick WM, Reid AP, Bohan DA, Briscoe BR (1996). Entomopathogenic nematodes: biodiversity, geographical distribution and the convention on biological diversity. Biocontrol Sci. Technol. 6:317-331.
Hu K, Webster JM (2000). Antibiotic production in relation to bacterial growth and nematode development in *Photorhabdus-Heterorhabditis* infected *Galleria mellonella* larvae. FEMS Microbiol. Lett. 189:219-223.
Hussien AAE, Hanan AE (2008). Xenoprotec: Antimicrobial Agent Derived from *Xenorhabdus indica*. J. Agric. Environ. Sci. 2:568-576.
Jukes TH, Cantor CR (1969). Evolution of protein molecules. New York: Academic Press, pp. 21-132.
Kaya HK, Stock SP (1997). Manual of Techniques in Insect Pathology: Entomopathogenic nematodes. New York: Academic Press, pp. 259-262.

Liu Q, Mracek Z, Zhang L, Puza V, Dong L (2012). Re-description *of Oscheius chongmingensis* (Zhang et al., 2008) (Nematoda: Rhabditidae) and its entomopathogenicity. Nematology 14:139-149.

Manulis S, Valinski L, Gafni Y, Hershenhorn J (1991). Indole-3-acetic acid biosynthesis pathways in *Erwinia herbicola* in relation to pathogenicity on *Gypsophila paniculata*. Physiol. Mol. Plant Pathol. 39:161-171.

Martino PD, Fursy R, Breta L, Sundararaju B, Phillips RS (2003). Indole can act as an extracellular signal to regulate biofilm formation of *Escherichia coli* and other indole-producing bacteria. Can. J. Microbiol. 49: 443-449.

Paul V, Frautschy S, Fenical W, Nealson KH (1981). Antibiotics in Microbial Ecology Isolation and Structure Assignment of Several New Antibacterial Compounds from the Insect-Symbiotic Bacteria *Xenorhabdus* spp. J. Chem. Ecol. 7:549-597.

Poinar Jr GO (1975). Description and biology of a new insect parasitic rhabditoid, *Heterorhabditis bacteriophora n.gen., n. sp. (Rhabditida; Heterorhabditidae n.fam.)*. Nematologica 21:463-470.

Saitou N, Nei M (1987). The neighbour-joining method: a new method for reconstructing phylogenetic trees. Mol. Biol. Evol. 4:406-425.

Sundar L, Chang FN (1993). Antimicrobial activity and biosynthesis of indole antibiotics produced *Xenorhabdus nematophilus*. J. Gen. Microbiol. 139:3139-3138.

Tamura K, Stecher G, Peterson D, Filipski A, Kumar S (2013). MEGA6: Molecular Evolutionary Genetics Analysis version 6.0. Mol. Biol. Evol. 30:2725-2729.

Thaler J, Duvic B, Givaudan A, Boemare N (1998). Isolation and Entomotoxic Properties of the *Xenorhabdus nematophilus* F1 Lecithinase. Appl. Environ. Microbiol. 64(7):2367-2373.

Thomas GM, Poinar GO (1979). *Xenorhabdus gen. nov*., a Genus of Entomopathogenic, Nematophilic Bacteria of the Family Enterobacteriaceae. Int. J. Syst. Bacteriol. 29: 352-360.

Thomashow LS (1996). Biocontrol of plant root pathogens. Curr. Opin. Biotechnol. 7:343-347.

Torre M (2003). Challenges for mass production of nematodes in submerged culture. Biotechnol. Adv. 21:407-416.

Torres-Barragan A, Suazo A, Buhler WG, Cardoza YJ (2011). Studies on the entomophathogenicity and bacterial associates of the nematode *Oscheius carolinensis*. Biol. Control 59:123-129.

Wang H, Liu Y, Dong H, Qin L, Cong B, Li T (2011). Antibiotic activity of bacterial isolates associated with entomopathogenic nematodes. Afr. J. Microbiol. Res. 5:5039-5045.

Wang Y, Fang X, An F, Wang G, Zhang X (2011). Improvement of antibiotic activity of *Xenorhabdus bovienii* by medium optimization using response surface methodology. Microb. Cell Fact. 10:1-15.

Wright ES, Yilmaz LS, Noguera DR (2012). Decipher, a search-based approach to chimera identification for 16S rRNA sequences. Appl. Environ. Microbiol. 78: 717-725.

Ye W, Torres-Barragan A, Cardoza Y (2010). *Oscheius carolinensis n. sp.* (Nematoda: Rhabditidae), a potential entomopathogenic nematode from vermicompost. Nematology 12:121-135.

Zhang C, Liu J, Xu M, Sun J, Yang S, An X, Gao G, Lin M, Lai R, He Z, Wu Y, Zhang K (2008). *Heterorhabditidoides chongmingensis gen. nov., sp. nov.* (Rhabditida: Rhabditidae), a novel member of the entomopathogenic nematodes. J. Invertebrate Pathol. 98:153-168.

Permissions

All chapters in this book were first published in JBR, by Academic Journals; hereby published with permission under the Creative Commons Attribution License or equivalent. Every chapter published in this book has been scrutinized by our experts. Their significance has been extensively debated. The topics covered herein carry significant findings which will fuel the growth of the discipline. They may even be implemented as practical applications or may be referred to as a beginning point for another development.

The contributors of this book come from diverse backgrounds, making this book a truly international effort. This book will bring forth new frontiers with its revolutionizing research information and detailed analysis of the nascent developments around the world.

We would like to thank all the contributing authors for lending their expertise to make the book truly unique. They have played a crucial role in the development of this book. Without their invaluable contributions this book wouldn't have been possible. They have made vital efforts to compile up to date information on the varied aspects of this subject to make this book a valuable addition to the collection of many professionals and students.

This book was conceptualized with the vision of imparting up-to-date information and advanced data in this field. To ensure the same, a matchless editorial board was set up. Every individual on the board went through rigorous rounds of assessment to prove their worth. After which they invested a large part of their time researching and compiling the most relevant data for our readers.

The editorial board has been involved in producing this book since its inception. They have spent rigorous hours researching and exploring the diverse topics which have resulted in the successful publishing of this book. They have passed on their knowledge of decades through this book. To expedite this challenging task, the publisher supported the team at every step. A small team of assistant editors was also appointed to further simplify the editing procedure and attain best results for the readers.

Apart from the editorial board, the designing team has also invested a significant amount of their time in understanding the subject and creating the most relevant covers. They scrutinized every image to scout for the most suitable representation of the subject and create an appropriate cover for the book.

The publishing team has been an ardent support to the editorial, designing and production team. Their endless efforts to recruit the best for this project, has resulted in the accomplishment of this book. They are a veteran in the field of academics and their pool of knowledge is as vast as their experience in printing. Their expertise and guidance has proved useful at every step. Their uncompromising quality standards have made this book an exceptional effort. Their encouragement from time to time has been an inspiration for everyone.

The publisher and the editorial board hope that this book will prove to be a valuable piece of knowledge for researchers, students, practitioners and scholars across the globe.

List of Contributors

Hongbao Ma
Bioengineering Department, Zhengzhou University, Zhengzhou, Henan 450001, China
Brookdale University and Medical Center, Brooklyn, New York 11212, USA

Jenny Young
Brookdale University and Medical Center, Brooklyn, New York 11212, USA

Shen Cherng
Department of Electrical Engineering, Chengshiu University, Niaosong, Taiwan 833, China

Sonia Chacko
Department of Biotechnology, Allahabad Agricultural Institute - Deemed University, Allahabad - 211007, India

Pramod W. Ramteke
Department of Biological Sciences, Allahabad Agricultural Institute - Deemed University, Allahabad - 211007, India

Suchit A. John
Department of Biological Sciences, Allahabad Agricultural Institute - Deemed University, Allahabad - 211007, India

Arifa Tahir
Microbiology Research Centre, Lahore College for Women University, Lahore Pakistan

Bushra Mateen
Microbiology Research Centre, Lahore College for Women University, Lahore Pakistan

Salih Univerdi
Science department, PakTurk International Colleges, Islamabad, Lahore Pakistan

Ozer KaraGoban
Science department, PakTurk International Colleges, Islamabad, Lahore Pakistan

Mehmet Zengin
Science department, PakTurk International Colleges, Islamabad, Lahore Pakistan

Pilanee Vaithanomsat
Enzyme Technology and Waste Management Research Unit, Kasetsart Agricultural and Agro-Industrial Product Improvement Institute (KAPI), Kasetsart University, 50, Chatuchak, Bangkok 10900, Thailand

Poom Phusanakom
Toyota Motor Asia Pacific Engineering and Manufacturing Company Limited, 99 Moo 5, Ban Ragad, Amphur Bang Bo, Samutprakan 10560, Thailand

Waraporn Apiwatanapiwat
Enzyme Technology and Waste Management Research Unit, Kasetsart Agricultural and Agro-Industrial Product Improvement Institute (KAPI), Kasetsart University, 50, Chatuchak, Bangkok 10900, Thailand

Molnapat Songpim
Enzyme Technology and Waste Management Research Unit, Kasetsart Agricultural and Agro-Industrial Product Improvement Institute (KAPI), Kasetsart University, 50, Chatuchak, Bangkok 10900, Thailand

Amirmorteza Ebrahimzadeh Namvar
Antimicrobial Research Center, Faculty of Medicine, Iran University of Medical Science, Iran

Babak Asghari
Antimicrobial Research Center, Faculty of Medicine, Iran University of Medical Science, Iran

Shoshtari Zadeh
Hashemi Nezhad Hospital of Tehran, Iran University of Medical Science, Iran

Abdolaziz Rastegar Lari
Antimicrobial Research Center, Faculty of Medicine, Iran University of Medical Science, Iran
Medical Microbiology Department, Faculty of Medicine, Iran University of Medical Science, Iran

G. P. S. Jadaun
Department of Microbiology and Molecular Biology, National JALMA Institute for Leprosy and Other Mycobacterial Diseases(Indian Council of Medical Research), P. O. Box 1101, Dr. M. Miyazaki Marg, Tajganj, Agra 282001, India

Prashant Upadhyay
Department of Microbiology and Molecular Biology, National JALMA Institute for Leprosy and Other Mycobacterial Diseases(Indian Council of Medical Research), P. O. Box 1101, Dr. M. Miyazaki Marg, Tajganj, Agra 282001, India

Zafar Ahmed
Department of Microbiology and Molecular Biology, National JALMA Institute for Leprosy and Other Mycobacterial Diseases(Indian Council of Medical Research), P. O. Box 1101, Dr. M. Miyazaki Marg, Tajganj, Agra 282001, India

Ram Das
Department of Microbiology and Molecular Biology, National JALMA Institute for Leprosy and Other Mycobacterial Diseases(Indian Council of Medical Research), P. O. Box 1101, Dr. M. Miyazaki Marg, Tajganj, Agra 282001, India

Deepti Parashar
National Institute of Virology (ICMR) Pune, India

D. S.Chauhan
Department of Microbiology and Molecular Biology, National JALMA Institute for Leprosy and Other Mycobacterial Diseases(Indian Council of Medical Research), P. O. Box 1101, Dr. M. Miyazaki Marg, Tajganj, Agra 282001, India

V. D. Sharma
Department of Microbiology and Molecular Biology, National JALMA Institute for Leprosy and Other Mycobacterial Diseases(Indian Council of Medical Research), P. O. Box 1101, Dr. M. Miyazaki Marg, Tajganj, Agra 282001, India

V. M. Katoch
Department of Health Research, Ministry of Health and Family Welfare, Government of India and Director General, Indian Council of Medical Research Post Box No. 4911 Ansari Nagar, New Delhi-110029

Sanaa O. Yagoub
Department of Microbiology and Molecular Biology, Faculty of Science and Technology, El Neelain University, P. O. Box 12702, Khartoum, Sudan

Mohammad Reza Shakibaie
Department of Microbiology and Immunology and Infectious Diseases and Tropical Medicine Research Comittee, Kerman University of Medical Sciences, Kerman, Iran

Pershant Dhakephalker
Department of Microbiology, Agarkar Agriculture Institute, Pune, India

Balu Parnab Kapadnis
Department of Microbiology, University of Pune, Ganeshkhind Road, Pune - 411007, India

Balu Ananda Chopade
Institute of Bioinformatics and Biotechnology, University of Pune, Ganeshkhind Road, Pune - 411007 Maharashtra State, India

Mohammad Shahriar
Department of Pharmacy, University of Asia Pacific, House 73, Road 5A, Dhanmondi R/A, Dhaka, Bangladesh

Nishat Zareen Khair
Department of Pharmacy, University of Asia Pacific, House 73, Road 5A, Dhanmondi R/A, Dhaka, Bangladesh

Mahmoud M. M. Zaky
Botany Department, Faculty of Science, Port-Said University, Egypt

Fathy A. Mansour
Botany Department, Faculty of Science, Mansoura University, Egypt

Kenneth M.Persson
Water Resources Engineering, Lund University, Sweden

Olav Rosef
Telemark University College, Hallvard Eikas Plass 1, 3800 Bø i Telemark, Norway

Algimantas Paulauskas
Vytautas Magnus University, Kaunas, Lithuania

Nils Grude
Unilabs Telelab AS, Skien, Norway

Camilla Haslekås
Telemark University College, Hallvard Eikas Plass 1, 3800 Bø i Telemark, Norway

Andrew Jenkins
Unilabs Telelab AS, Skien, Norway

Mohammad A. Alzohairy
Department of Medical Laboratories, College of Applied Medical Sciences, Qassim University, Qassim, Saudi Arabia

C. P. Ukpaka
Department of Chemical/Petrochemical Engineering, Rivers State University of Science and Technology, Nkpolu, P.M.B. 5080, Port Harcourt, Nigeria

T. O. Elufisan
National Centre for Technology Management, P.M.B. 012, Obafemi Awolowo University, Ile-Ife, Osun State, Nigeria

Oyelade Busayo
National Centre for Technology Management, P.M.B. 012, Obafemi Awolowo University, Ile-Ife, Osun State, Nigeria

J. K. Oloke
National Centre for Technology Management, P.M.B. 012, Obafemi Awolowo University, Ile-Ife, Osun State, Nigeria

K. Thirumalai Raj
Department of Fish Processing Technology, Fisheries College and Research Institute, Tamil Nadu Veterinary and Animal Sciences University, Tuticorin 628 008, India

G. Jeyasekaran
Department of Fish Processing Technology, Fisheries College and Research Institute, Tamil Nadu Veterinary and Animal Sciences University, Tuticorin 628 008, India

R. Jeya Shakila
Department of Fish Processing Technology, Fisheries College and Research Institute, Tamil Nadu Veterinary and Animal Sciences University, Tuticorin 628 008, India

A. Jemila Thangarani
Department of Fish Processing Technology, Fisheries College and Research Institute, Tamil Nadu Veterinary and Animal Sciences University, Tuticorin 628 008, India

D.Sukumar
Department of Fish Processing Technology, Fisheries College and Research Institute, Tamil Nadu Veterinary and Animal Sciences University, Tuticorin 628 008, India

Deepti Chachra
Department of Veterinary Microbiology, College of Veterinary Science, Guru Angad Dev Veterinary and Animal Sciences University (GADVASU), Ludhiana, 141004 India

Hari Mohan Saxena
Department of Veterinary Microbiology, College of Veterinary Science, Guru Angad Dev Veterinary and Animal Sciences University (GADVASU), Ludhiana, 141004 India

Gurpreet Kaur
Department of Veterinary Microbiology, College of Veterinary Science, Guru Angad Dev Veterinary and Animal Sciences University (GADVASU), Ludhiana, 141004 India

Mudit Chandra
Department of Veterinary Microbiology, College of Veterinary Science, Guru Angad Dev Veterinary and Animal Sciences University (GADVASU), Ludhiana, 141004 India

Qurban Ali
Department of Plant Breeding and Genetics, University of Agriculture, Faisalabad, Pakistan

Muhammad Hammad
Department of Plant Breeding and Genetics, University of Agriculture, Faisalabad, Pakistan

Nadeem Tahir
Department of Plant Breeding and Genetics, University of Agriculture, Faisalabad, Pakistan

Muhammad Ahsan
Department of Plant Breeding and Genetics, University of Agriculture, Faisalabad, Pakistan

Shahzad Maqsood Ahmed Basra
Department of Crop Physiology, University of Agriculture, Faisalabad, Pakistan

Jehanzeb Farooq
Department of Plant Breeding and Genetics, University of Agriculture, Faisalabad, Pakistan

Muhammad waseem
Department of Agronomy, University of Agriculture, Faisalabad, Pakistan

Mehboob Elahi
Department of Agronomy, University of Agriculture, Faisalabad, Pakistan

R. Pramila
Department of Plant Biology and Plant Biotechnology, Quaid-e-Millath Government College for Women, Chennai 600 002, Tamil Nadu South India

Kesavaram Padmavathy
Department of Microbiology, Sree Balaji Dental College and Hospital, Chennai 600 100, Tamil Nadu, South India

K. Vijaya Ramesh
Department of Plant Biology and Plant Biotechnology, Quaid-e-Millath Government College for Women, Chennai 600 002, Tamil Nadu South India

Krishnan Mahalakshmi
Department of Microbiology, Sree Balaji Dental College and Hospital, Chennai 600 100, Tamil Nadu, South India

M. F. Abdel-Monaim
Legume Diseases Research Department, Plant Pathology Research Institute, Agricultural Research Center, Giza, Egypt

M. R. Gabr
Plant Pathology Department, Faculty of Agriculture, El-Minia University, Minia, Egypt

S. M. El-Gantiry
Legume Diseases Research Department, Plant Pathology Research Institute, Agricultural Research Center, Giza, Egypt

M. N. Shaat
Plant Pathology Department, Faculty of Agriculture, El-Minia University, Minia, Egypt

A. A. El-Bana
Plant Pathology Department, Faculty of Agriculture, El-Minia University, Minia, Egypt

O. Olajuyigbe Olufunmiso
Phytomedicine Research Centre, University of Fort Hare, Alice 5700, South Africa
Biosciences and Biotechnology Department, Babcock University, P. M. B. 21244, Ikeja, Lagos, Nigeria

Adeoye Oluwaseun
Biosciences and Biotechnology Department, Babcock University, P. M. B. 21244, Ikeja, Lagos, Nigeria

Lawrence A. Klobutcher
Department of Molecular, Microbial and Structural Biology, University of Connecticut Health Center, Farmington, CT 06030-3305, USA

Elena Gaidamakova
Department of Pathology, School of Medicine, Uniformed Services University of the Health Sciences, Bethesda, MD 20814, USA

Michael J. Daly
Department of Pathology, School of Medicine, Uniformed Services University of the Health Sciences, Bethesda, MD 20814, USA

Barbara Setlow
Department of Molecular, Microbial and Structural Biology, University of Connecticut Health Center, Farmington, CT 06030-3305, USA

Peter Setlow
Department of Molecular, Microbial and Structural Biology, University of Connecticut Health Center, Farmington, CT 06030-3305, USA

Muhonja Christabel
Jomo Kenyatta University of Agriculture and Technology, P. O. Box 62000-00200, Nairobi, Kenya

Nancy Budambula
Jomo Kenyatta University of Agriculture and Technology, P. O. Box 62000-00200, Nairobi, Kenya

John Kiiru
Centre for Microbiology Research, Kenya Medical Research Institute. P. O. BOX 54840-00200, Nairobi, Kenya

Samuel Kariuki
Centre for Microbiology Research, Kenya Medical Research Institute. P. O. BOX 54840-00200, Nairobi, Kenya

Mohammed A. Mohammed
Department of Medical Microbilogy, Faculty of Medicine, Medical Institute Center, Al Neelain University, Khartoum, Sudan

Al Fadhil A. Omer
Department of Medical Microbilogy, Faculty of Medical Laboratory Science, Al Neelain University Khartoum, Sudan

Kalantar Enayat
School of Health, Environmental Health Research Center, Kurdistan University of Medical Sciences, Sanandaj, Iran
Department of Pathobiology, Alborz University of Medical Sciences, Karaj, Iran

Amin Mansour
Department of Microbiology, School of Medicine, Infectious and Tropical Diseases Research Center, Ahvaz Jundishapur University of Medical Sciences, Ahvaz, Iran

Bahmani Nasrin
Department of Microbiology, School of Medicine, Kurdistan University of Medical Sciences, Sanandaj, Iran

Tamimi Mohammad
Food and Drug Deputy, Kurdistan University of Medical Sciences, Sanandaj, Iran

Hashemzadeh Mohammad
Department of Microbiology, School of Medicine, Infectious and Tropical Diseases Research Center, Ahvaz Jundishapur University of Medical Sciences, Ahvaz, Iran

Narenji Hanar
Department of Microbiology, School of Medicine, Kurdistan University of Medical Sciences, Sanandaj, Iran

Jiao Zhang
Jiangsu University of Science and Technology, Zhenjiang 212018, Jiangsu Province, China

Zhong-yuan Shen
Jiangsu University of Science and Technology, Zhenjiang 212018, Jiangsu Province, China
The Sericultural Research Institute, Chinese Academy of Agricultural Sciences, Zhenjiang 212018, Jiangsu Province, China
The Key Laboratory of Genetic Improvement of Silkworm and Mulberry of Agricultural Ministry, Zhenjiang 212018,Jiangsu Province, China

Xu-dong Tang
Jiangsu University of Science and Technology, Zhenjiang 212018, Jiangsu Province, China
The Sericultural Research Institute, Chinese Academy of Agricultural Sciences, Zhenjiang 212018, Jiangsu Province, China

Li Xu
Jiangsu University of Science and Technology, Zhenjiang 212018, Jiangsu Province, China
The Sericultural Research Institute, Chinese Academy of Agricultural Sciences, Zhenjiang 212018, Jiangsu Province, China

Feng Zhu
Jiangsu University of Science and Technology, Zhenjiang 212018, Jiangsu Province, China

Md. Wahidul Alam
Institute of Marine Sciences and Fisheries, University of Chittagong, Chittagong-4331, Bangladesh

Mohammad Zafar
Institute of Marine Sciences and Fisheries, University of Chittagong, Chittagong-4331, Bangladesh

Ali M. Al-Mohana
Department of Microbiology, College of Medicine, Kufa University, Iraq

Alaa H. Al-Charrakh
Department of Microbiology, College of Medicine, Babylon University Babylon province, Iraq

Fadhil H. Nasir
Department of Microbiology, College of Medicine, Kufa University, Iraq

Miaad K. Al-Kudhairy
Department of Community Health, Kufa Technical Institute, Iraq

Chijioke A. Nsofor
Department of Microbiology, University of Nigeria Nsukka, Enugu State, Nigeria
Department of Veterinary Microbiology and Pathology, Washington State University, Pullman, USA

Christian U. Iroegbu
Department of Microbiology, University of Nigeria Nsukka, Enugu State, Nigeria

Douglas R. Call
Department of Veterinary Microbiology and Pathology, Washington State University, Pullman, USA

Margaret A. Davies
Department of Veterinary Microbiology and Pathology, Washington State University, Pullman, USA

Sixtus Bieranye Bayaa Martin Saana
Department of Pharmaceutics, Faculty of Pharmacy and Pharmaceutical Sciences, Kwame Nkrumah University of Science and Technology, Kumasi, Ghana

Francis Adu, Christian Agyare
Department of Pharmaceutics, Faculty of Pharmacy and Pharmaceutical Sciences, Kwame Nkrumah University of Science and Technology, Kumasi, Ghana

Stephen Yao Gbedema
Department of Pharmaceutics, Faculty of Pharmacy and Pharmaceutical Sciences, Kwame Nkrumah University of Science and Technology, Kumasi, Ghana

Vivian Etsiapa Boamah
Department of Pharmaceutics, Faculty of Pharmacy and Pharmaceutical Sciences, Kwame Nkrumah University of Science and Technology, Kumasi, Ghana

Duredoh Freeman George
Department of Pharmaceutics, Faculty of Pharmacy and Pharmaceutical Sciences, Kwame Nkrumah University of Science and Technology, Kumasi, Ghana

Chijioke A. Nsofor
Department of Microbiology, University of Nigeria, Nsukka, Enugu State, Nigeria

Christian U. Iroegbu
Department of Microbiology, University of Nigeria, Nsukka, Enugu State, Nigeria

Hatami saeideh
Science and Research Branch, Islamic Azad University, Department of Biology Tehran, Iran

Esmaeili Davoud
Applied Microbiology Research center and Department of Microbiology, Faculty of Medical science, Baqyiatallah University, Tehran, Iran

I. Fallah Jali
Applied Microbiology Research center and Department of Microbiology, Faculty of Medical science, Baqyiatallah University, Tehran, Iran

Rezaee Jafar
Applied Microbiology Research center and Department of Microbiology, Faculty of Medical science, Baqyiatallah University, Tehran, Iran

F. Abuldahab Faten
Department of Biology, Faculty of Science, King Abdul Aziz University, Jeddah, Saudi Arabia

Y. Abozinadah Najlaa
Department of Biology, Faculty of Science, King Abdul Aziz University, Jeddah, Saudi Arabia

S. Al- Haiqi Nawal
Department of Biology, Faculty of Science, King Abdul Aziz University, Jeddah, Saudi Arabia

M. Abebe
College of Veterinary Medicine, Mekelle University, Tigray, Ethiopia

A. Hailelule
College of Veterinary Medicine, Mekelle University, Tigray, Ethiopia

B. Abrha
College of Veterinary Medicine, Mekelle University, Tigray, Ethiopia

A. Nigus
College of Veterinary Medicine, Mekelle University, Tigray, Ethiopia

M. Birhanu
College of Veterinary Medicine, Mekelle University, Tigray, Ethiopia

H. Adane
Wachemo University, SNNP, Ethiopia

T. Genene
Institute of Biodiversity Conservation, Addis Ababa, Ethiopia

H. Daniel
Asella School of Health Science, Adama University of Science and Technology, Asella, Ethiopa

G. Getachew
College of Veterinary Medicine, Mekelle University, Tigray, Ethiopia

G. Merga
Asella School of Health Science, Adama University of Science and Technology, Asella, Ethiopa

A. Haftay
College of Veterinary Medicine, Mekelle University, Tigray, Ethiopia

Dereje Abera
School of Veterinary Medicine, College of Medical and Health Sciences, Wollega University, P.O. Box 395, Nekemte, Ethiopia

Tesfaye Sisay
Unit of Health Biotechnology, Institute of Biotechnology, College of Natural and Computational Sciences, Addis Ababa University, P.O.Box 34, Debre Zeit, Ethiopia

Tadesse Birhanu
School of Veterinary Medicine, College of Medical and Health Sciences, Wollega University, P.O. Box 395, Nekemte, Ethiopia

Jeiry Toribio-Jiménez
Laboratorio de Investigación en Biotecnología y Genética Microbiana, Unidad Académica de Ciencias Químico Biológicas, Universidad Autónoma de Guerrero, Chilpancingo, Guerrero, México

Miguel Ángel Rodríguez-Barrera
Laboratorio de Investigación en Biotecnología y Genética Microbiana, Unidad Académica de Ciencias Químico Biológicas, Universidad Autónoma de Guerrero, Chilpancingo, Guerrero, México

Monserrat Valdez Lucena
Laboratorio de Investigación en Biotecnología y Genética Microbiana, Unidad Académica de Ciencias Químico Biológicas, Universidad Autónoma de Guerrero, Chilpancingo, Guerrero, México

Ashanti Barrera Flores
Laboratorio de Investigación en Biotecnología y Genética Microbiana, Unidad Académica de Ciencias Químico Biológicas, Universidad Autónoma de Guerrero, Chilpancingo, Guerrero, México

Daniel Segura
Departamento de Microbiología Molecular, Instituto de Biotecnología, Universidad Nacional Autónoma de México,Cuernavaca, Morelos, México

Víctor Wilson-Corral
Unidad Académica de Ingeniería, Universidad Autónoma de Sinaloa, Culiacan, México

Eugenia Flores Alfaro
Laboratorio de Investigaciones de Epidemiología clínica y Molecular, Unidad Académica de Ciencias Químico Biológicas, Universidad Autónoma de Guerrero, Chilpancingo, Guerrero, México

Yanet Romero
Laboratorio de Investigación en Biotecnología y Genética Microbiana, Unidad Académica de Ciencias Químico Biológicas, Universidad Autónoma de Guerrero, Chilpancingo, Guerrero, México

M. Saiful Islam
Department of Microbiology, Shaheed Suhrawardy Medical College, Dhaka, Bangladesh

M. Abdullah Yusuf
Department of Microbiology, National Institute of Neurosciences & Hospital, Dhaka, Bangladesh

Shaheen Ara Begum
Department of Microbiology, Comilla Medical College, Comilla, Bangladesh

AFM Arshedi Sattar
Department of Microbiology, National Institute of Neurosciences & Hospital, Dhaka, Bangladesh

Afzal Hossain
Department of Microbiology, ZH Shikder Women's Medical College, Dhaka, Bangladesh

Sushmita Roy
Department of Microbiology, Enam Medical College, Dhaka, Bangladesh

O. O. Odikamnoro
Department of Applied Biology, Ebonyi State University, Abakaliki, Ebonyi State, Nigeria

C. A. Uhuo
Department of Applied Biology, Ebonyi State University, Abakaliki, Ebonyi State, Nigeria

I. M. Ikeh
Department of Applied Biology, Ebonyi State University, Abakaliki, Ebonyi State, Nigeria

E. D. Ogiji
Department of Pharmacology/Therapeutics, Faculty of Basic Medicine, Ebonyi State, University

G.A. Ibiam
Department of Medical Laboratory, Ebonyi State University, Abakaliki, Ebonyi State

S.o. Azi
Department of Medical Laboratory, Ebonyi State University, Abakaliki, Ebonyi State

L. J. Akpam
Department of Pharmacology/Therapeutics, Faculty of Basic Medicine, Ebonyi State, University

N. F. Okoh
Department of Biological Sciences, Parasitology Unit, Evangel University, Abakaliki, Ebonyi State, Nigeria

Ghaidaa Abbas Jassem
Department of Microbiology, Veterinary Medicine College, AL-Qadissiya University, Iraq

Olaa A. Agaar
Department of Microbiology, Veterinary Medicine College, Wasit University, Iraq

Mahloro H. Serepa
School of Molecular and Cell Biology, University of the Witwatersrand, P Bag 3, Wits 2050, Johannesburg, South Africa

Nikita T. Tavengwa
School of Chemistry, University of the Witwatersrand, P Bag 3, Wits 2050, Johannesburg, South Africa

Vincent M. Gray
School of Molecular and Cell Biology, University of the Witwatersrand, P Bag 3, Wits 2050, Johannesburg, South Africa